Biomechanics

Circulation

Second Edition

Springer

New York
Berlin
Heidelberg
Barcelona
Budapest
Hong Kong
London
Milan
Paris
Santa Clara
Singapore
Tokyo

Y. C. Fung

Biomechanics
Circulation

Second Edition

With 289 Illustrations

 Springer

Y.C. Fung
Department of Bioengineering
University of California, San Diego
La Jolla, CA 92093-0419
USA

Cover illustration by Frank DeLano for the First World Congress of Biomechanics.

Library of Congress Cataloging in Publication Data
Fung, Y.C. (Yuan-cheng), 1919–
    Biomechanics : circulation / Y.C. Fung. — 2nd ed.
        p.   cm.
    Includes bibliographical references and index.
    ISBN 0-387-94384-6 (alk. paper)
    1. Hemodynamics.   I. Title.
QP105.F85   1996
612.1—DC20   96-11887

Printed on acid-free paper.

The first edition of this book was published as *Biodynamics: Circulation.*

Acquiring editor: Robert Garber.
Production coordinated by Chernow Editorial Services, Inc. and managed by Terry Kornak;
    manufacturing supervised by Joe Quatela.
Typeset by Best-set Typesetter Ltd., Hong Kong.
Printed and bound by Maple-Vail, York, PA.
Printed in the United States of America.

9 8 7 6 5 4 3 2 1

ISBN 0-387-94384-6   Springer-Verlag New York Berlin Heidelberg   SPIN 10480464

Preface to the Second Edition

The theory of blood circulation is the oldest and most advanced branch of biomechanics, with roots extending back to Huangti and Aristotle, and with contributions from Galileo, Santori, Descartes, Borelli, Harvey, Euler, Hales, Poiseuille, Helmholtz, and many others. It represents a major part of humanity's concept of itself. This book presents selected topics of this great body of ideas from a historical perspective, binding important experiments together with mathematical threads.

The objectives and scope of this book remain the same as in the first edition: to present a treatment of circulatory biomechanics from the standpoints of engineering, physiology, and medical science, and to develop the subject through a sequence of problems and examples. The name is changed from *Biodynamics: Circulation* to *Biomechanics: Circulation* to unify the book with its sister volumes, *Biomechanics: Mechanical Properties of Living Tissues*, and *Biomechanics: Motion, Flow, Stress, and Growth*. The major changes made in the new edition are the following: When the first edition went to press in 1984, the question of residual stress in the heart was raised for the first time, and the lung was the only organ analyzed on the basis of solid morphologic data and constitutive equations. The detailed analysis of blood flow in the lung had been done, but the physiological validation experiments had not yet been completed. Now, the residual stress is well understood, the zero stress states of the heart and blood vessels are well documented, and the morphometry of the blood vessels of the heart and skeletal muscles has been advanced sufficiently to allow an analysis of the blood flow in these organs on the basis of realistic geometric descriptions. Thus, two new chapters were added to discuss coronary blood flow and skeletal muscle microcirculation.

Chapters 6, 7, and 8 together illustrate a biomechanical approach to circulatory physiology, with emphasis on formulating questions in the form of boundary-value problems, and predicting outcome by solving the problems. In these chapters, the mechanics of vascular smooth muscles stands out as particularly important. In 1984, the mechanical properties of the vascular smooth muscle were mysterious. The autoregulation phenomenon was known, as was the phenomenon of hyperemia. The importance of local control of blood flow was appreciated, but there was an unresolved debate about whether the mechanism was myogenic, neurogenic, or metabolic. Now, however, we have extracted a length-tension relationship of the vascular smooth muscle from the results of a series of outstanding experiments on coronary arterioles by Kuo et al., who measured the diameters of the vessels in response to pressure and flow. We now know how the length-tension curve is shifted to the right by the shear stress acting on the endothelial cells. The length-tension curves of the vascular smooth muscle are found to be arch-like and quite similar to those of the heart and skeletal muscles. However, whereas the working range of the lengths of the sarcomeres of the heart and skeletal muscles lie on the left leg of the length-tension arch, that of the vascular smooth muscle lies on the right leg. This difference is revealed in the contrasting behavior of the muscles: while the heart muscle obeys Starling's law, the vascular smooth muscle exhibits the Bayliss phenomenon. Thus the mechanical properties of the vascular smooth muscle are no longer so strange. I believe that these differences are due to the different patterns of actin-myosin relationship in these muscles. Chapter 2, on the heart, was revised extensively, with the addition of new results and methods on the analysis of strain distribution in the ventricles, especially in association with in vivo experiments. Chapter 3 now includes a long section on fluid mechanics and solid mechanics of atherogenesis. It explains why atherogenesis is a problem of biomechanics, and discusses contemporary thinking on the subject. The significance of the shear stress acting on the endothelial cells due to blood flow, and the tensile stress in the blood vessel wall due to blood pressure are discussed from the point of view of gene expression and tissue remodeling of the vessel.

Chapter 4, on veins, presents advances on the stability of flow in collapsible tubes, with an emphasis on the nonlinear effects of longitudinal tension. Forced oscillations in collapsible tubes due to flow separation or turbulence are discussed. In Chapter 5, on microcirculation, a classical solution of a sheet-flow model of capillaries has been added. Chapter 6, on pulmonary circulation, presents the details of an analysis of the waterfall phenomenon in the lung, along with experimental validations. These are given because the causes of waterfall in different organs are different. For example, the waterfall in the lung occurs at the junctions of the capillaries and venules. When the pressure at the end of the capillaries is equal to the alveolar gas pressure, the pressure in the venule is below alveolar gas pressure, and the sluicing gates are kept open by the tension in the interalveolar septa. On the other hand, in the heart, the waterfalls are believed to be located in the coronary capillaries that are squeezed by the heart muscle. The waterfalls in the skeletal muscle are similar to those in the heart, but the muscle squeeze is milder. The interaction between the muscle cells and the capillary blood vessels is a subject of major interest.

Many other advances have been made in the field of circulation in the last ten years. To survey all of the field is beyond my ability. I discussed only those topics familiar to me. Even in these, I may have missed some important references. To those authors, I apologize. To people who are looking for a handbook, a compendium of solved problems, a review of the current literature, or a record of computational methods, I also apologize, because this book is not designed to serve those functions. But if a reader finds the book interesting, lucid, and useful, then I shall be very grateful. I wish to thank friends and readers who have offered suggestions for improving this book. I want to thank Dr. Geert Schmid-Schönbein for providing most of the materials in the last chapter; Dr. Ghasson Kassab for providing the morphological data of the coronary vasculature; Dr. Michael Yen for new results on pulmonary circulation; and Drs. Andrew McCulloch and Lew Waldman for advances in the analysis and experimentation on the mechanics of the heart and for materials used in the last three sections of Chapter 2. I want to express my pleasure to Dr. Shu Qian Liu for our close working relationship during the past nine years, and to Eugene Mead for our cooperative work during the past 31 years. I also enjoyed and benefited from working with Drs. Paul Zupkas, Yasuyuki Seguchi, Yuji Matsuzaki, Mitsumasa Matsuda, Takaaki Nakagawa, Jun Tomioka, Maw Chang Lee, Jeffrey Omens, Jack Debes, Jainbo Zhou, Shanxi Deng, Qilian Yu, Zong Jie Li, Zong Lai Jiang, Hai Chao Han, Gong Rui Wong, Win Peng Wu, Rui Fang Yang, Yun Qin Gao, Kegan Dai, Hao Xue, Rong Zhu Gan, Wei Huang, and Hans Gregersen, whose work is mentioned in this book. Finally, to my friends and colleagues Drs. Geert Schmid-Schönbein, Sidney Sobin, Shu Chien, Richard Skalak, David Gough, John Frangos, Andrew McCulloch, Paul and Amy Sung, Benjamin Zweifach, and Marcos Intaglietta, I want to offer sincere thanks for creating a most remarkable, stimulating, and pleasant environment, in which I was surrounded by bright students. I am extremely lucky to have good friends and a warm family. Luna has always supported me and my work. Conrad and Brenda make me proud of their character, achievements, and strength. Chia Shun Yih delights me with frequent letters announcing his exact solutions of difficult problems in fluid mechanics.

I dedicate this book to Luna.

YUAN-CHENG FUNG

Preface to the First Edition

This book is a continuation of my *Biomechanics*. The first volume deals with the mechanical properties of living tissues. The present volume deals with the mechanics of circulation. A third volume will deal with respiration, fluid balance, locomotion, growth, and strength. This volume is called *Biodynamics* in order to distinguish it from the first volume. The same style is followed. My objective is to present the mechanical aspects of physiology in precise terms of mechanics so that the subject can become as lucid as physics.

The motivation of writing this series of books is, as I have said in the preface to the first volume, to bring biomechanics to students of bioengineering, physiology, medicine, and mechanics. I have long felt a need for a set of books that will inform the students of the physiological and medical applications of biomechanics, and at the same time develop their training in mechanics. In writing these books I have assumed that the reader already has some basic training in mechanics, to a level about equivalent to the first seven chapters of my *First Course in Continuum Mechanics* (Prentice Hall, 1977). The subject is then presented from the point of view of life science while mechanics is developed through a sequence of problems and examples. The main text reads like physiology, while the exercises are planned like a mechanics textbook. The instructor may fill a dual role: teaching an essential branch of life science, and gradually developing the student's knowledge in mechanics.

The style of one's scientific approach is decided by the way one looks at a problem. In this book I try to emphasize the mathematical threads in the study of each physical problem. Experimental exploration, data collection, model experiments, in vivo observations, and theoretical ideas can be wrapped together by mathematical threads. The way problems are formulated, the kind of questions that are asked, are molded by this basic thought. Much of the book can be read, however, with little mathematics. Those passages in which mathematics is essential are presented with sufficient details to make the reading easy.

This book begins with a discussion of the physics of blood flow. This is followed by the mechanics of the heart, arteries, veins, microcirculation, and pulmonary blood flow. The coupling of fluids and solids in these organs is the central feature. How morphology and rheology are brought to bear on the analysis of blood flow in organs is illustrated in every occasion. The basic equations of fluid and solid mechanics are presented in the Appendix. The subject of mass transfer, the exchange of water, oxygen, carbon dioxide, and other substances between plasma and red cells and between capillary blood vessels and extravascular space, is deferred to the third volume, *Biodynamics: Flow, Motion, and Stress*, in order to keep the three volumes at approximately the same size.

Circulation is a many-sided subject. What we offer here is an understanding of the mechanics of circulation. We present methods and basic equations very carefully. The strengths and weaknesses of various methods and unanswered questions are discussed fully. To apply these methods to a specific organ, we need a data base. We must have a complete set of morphometric data on the anatomy, and rheological data on the materials of the organ. Unfortunately, such a data base does not exist for any organ of any animal. A reasonably complete set has been obtained for the lungs of the cat. Hence the analysis of the blood flow in the lung is presented in detail in Chapter 6. We hope that a systematic collection of the anatomical and rheological data on all organs of man and animals will be done in the near future so that organ physiology can be elevated to a higher level.

Blood circulation has a vast literature. The material presented here is necessarily limited in scope. Furthermore, there are still more things unknown than known. Progress is very rapid. Aiming at greater permanency, I have limited my scope to a few fundamental aspects of biomechanics. For handbook information and literature survey, the reader must look elsewhere. Many exercises are proposed to encourage the students to formulate and solve new problems. The book is not offered as a collection of solved problems, but as a way of thinking about problems. I wish to illustrate the use of mechanics as a simple, reliable tool in life science, and no more. A reasonably extensive bibliography is given at the end of each chapter, some with annotations from which further references can be found. Perhaps the author can be accused of quoting frequently papers and people familiar to him; he apologizes for this personal limitation and hopes that he can be forgiven because it is only natural that an author should talk about his own views. I have tried, however, never to forget mentioning the existence of other points of view.

I wish to express my thanks to many authors and publishers who permitted me to quote their publications and reproduce their figures and data in this book. I wish to mention especially Drs. Michael Yen, Sidney Sobin, Jen-Shih Lee, Benjamin Zweifach, Paul Patitucci, Geert Schmid-Schoenbein, William Conrad, Lawrence Talbot, H. Werlé, John Maloney, Paul Stein, and John Hardy who supplied original photographs for reproduction. I wish also to thank many of my colleagues, friends, and former students who read parts of the manuscripts and offered valuable suggestions. To Virginia Stephens I am grateful for typing the manuscript. Finally, I wish to thank the editorial and production staff of Springer-Verlag for their care and cooperation in producing this book.

In spite of great care and effort on my part, I am sure that many mistakes and defects remain in the book. I hope you will bring these to my attention so that I can improve in the future.

YUAN-CHENG FUNG

La Jolla, California

Contents

Chapter 5
Microcirculation   266

Chapter 6
Blood Flow in the Lung   333

Chapter 7
Coronary Blood Flow   446

Chapter 8
Blood Flow in Skeletal Muscle   514

1
Physical Principles of Circulation

1.1 Conservation Laws

Blood flow must obey the principles of conservation of mass, momentum, and energy. Applied to any given region of space, the principle of conservation of mass means that whatever flows in must flow out. If flow is confined to blood vessels, then we obtain a rule similar to Kirchhoff's law of electric circuits: At any junction the summation of current flowing into a junction must be equal to the sum of the currents flowing out of that junction. In a single tube of variable cross section, a steady flow implies that the average local speed of flow is inversely proportional to the local cross-sectional area.

Conservation of momentum means that the momentum of matter cannot be changed without the action of force. If there is force acting on a body, then according to the Newton's law of motion, the rate of change of momentum of the body is equal to the force.

Everyone knows these principles. Don't let familiarity breed contempt. Think of this: We are able to fly to the moon precisely because we know Newton's law. The ancient Greeks, with their superior knowledge about anatomy of the heart and blood vessels, were robbed of the glory of discovering blood circulation because they did not think of the principle of conservation of mass! The Western world had to wait for William Harvey (1578–1657) to establish the concept of circulation. Harvey remembered the principle of conservation of mass, and all he needed was to show that in every heart beat there is a net flow of blood out of the heart!

There is nothing more cost effective than learning the general principles and remembering to apply them.

1.2 Forces That Drive or Resist Blood Flow

What are the forces that drive blood flow? They are the gravitational and the pressure gradient forces. The pressure in a blood vessel varies from point to point. The rate of change of pressure with distance in a specific

1

direction is the pressure gradient in that direction. Pressure itself does not cause the blood to move; the pressure gradient does.

What are the forces that oppose blood flow? They are the shear forces due to the viscosity of the blood and turbulences.

We use the term *stress* to mean the force acting on a surface divided by the area of the surface. In the International System of Units, the unit of stress is Newton per square meter, or Pascal. The stress acting on any surface can be resolved into two components: the shear stress, which is the component tangent to the surface, and the normal stress, which is the component perpendicular (i.e., normal) to a surface. Pressure is a normal stress. A positive pressure is regarded as a negative normal stress. A positive normal stress is a tensile stress.

Figure 1.2:1(a) shows a small rectangular element of blood subjected to equal pressure on all sides. The element remains in equilibrium. Figure 1.2:1(b) shows an element subjected to shear stresses that are equal on all sides. Because the shear stresses are equal in this case, the fluid element will

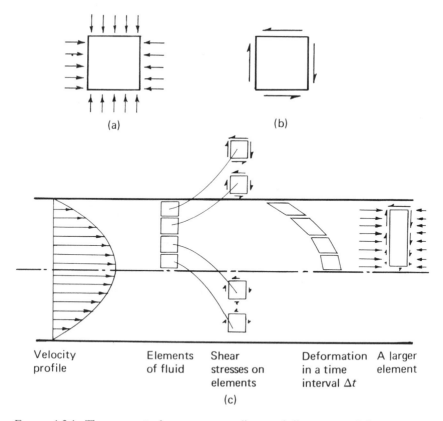

(a)   (b)

| Velocity profile | Elements of fluid | Shear stresses on elements | Deformation in a time interval Δt | A larger element |

(c)

FIGURE 1.2:1. The concept of a pressure gradient and divergence of shear stresses as motive forces for motion.

be distorted by the shear stresses, but there will be no tendency to accelerate.

Figure 1.2:1(c) shows a steady laminar flow of a fluid in a tube. The velocity distribution is indicated on the left. To the right, a series of fluid elements is shown. The shear stresses acting on these elements are indicated by arrows, with the magnitude proportional to the length. These shear stresses distort the elements, whose shapes after a certain interval of time are shown. If these elements are stacked together, as shown by the larger rectangular element on the right, we see that the horizontal shear force (= stress × area on which the shear stress acts) at the top is larger than the shear force on the bottom. The resultant of these shear forces is the force that resists the motion of this element. The resultant of the vertical shear forces acting on the two sides of the element is zero. On the other hand, the pressure on the left-hand side of the element is larger than that on the right-hand side. The difference of pressure per unit axial distance is the pressure gradient. The resultant pressure force (= pressure × area on which the pressure acts) is the force that drives the fluid to flow. In a blood vessel, the pressure and shear forces coexist, and they tend to balance each other. If, together with the gravitational force, an exact balance is achieved, then the flow is steady. If the forces are out of balance, then the fluid either accelerates or decelerates.

1.3 Newton's Law of Motion Applied to a Fluid

Newton's law states that

$$\text{Mass} \times \text{acceleration} = \text{force}.$$

Applied to a fluid, it takes the following form, which will be explained in detail below:

Density × (transient acceleration + convective acceleration)

= −pressure gradient + divergence of normal and shear stresses

+ gravitational force per unit volume.   (1)

Here *density* refers to the mass per unit volume of the fluid, the *transient acceleration* refers to the rate of change of velocity with respect to time at a given place, and the *convective acceleration* refers to the rate of change of velocity of a fluid particle caused by the motion of the particle from one place to another in a nonuniform flow field. *Pressure gradient* has been explained in Section 1.2. The *divergence of normal and shear stresses other than pressure* refers to the rate of change of those stresses arising from fluid viscosity. The word *divergence* is a mathematical term referring to a certain process of differentiation written out in Eq. (13) of Section 2.6 on p. 52. *Gravitational force per unit volume* means weight per unit volume of the fluid, that is, the density of the fluid multiplied by the gravitational acceleration.

Transient and convective accelerations exist in most problems of interest to biomechanics. Figure 1.3:1 shows a record of the velocity of flow in the center of the canine ascending aorta. It is seen that the velocity changes with time. In fact, the velocity record changes from one heart beat to another. An average of a large number of records is called an *ensemble average*. The difference between the velocity history of a specific record and the ensemble average is considered to be *turbulence*. Figure 1.3:1 shows the turbulence velocity of a specified record, and the root mean square values of the turbulence velocity of all records of the ensemble. The *partial derivative* of the velocity with respect to time is the transient acceleration at any given point in a fluid.

Convective acceleration also exists in most problems in bioengineering because interesting flow fields are usually nonuniform (e.g., Figs. 1.5:1 and 2.5:5 for closure of the mitral valve, Fig. 3.2:3 for entry flow in a tube, Fig. 3.7:2 for pulsatile flow, and Fig. 3.17:1 for flow separation from a solid wall). In each case, we can calculate the partial derivatives of the nonuniform velocity components with respect to the spatial coordinates. This set of partial derivatives forms a velocity gradient tensor. The product of the velocity vector and the velocity gradient tensor (by a process called *contraction*) gives us the convective acceleration. Detailed mathematical expressions of these products are given in Fung (1993a, 1993b) listed at the end of this chapter.

Gravitational force is important. If our blood were to stop flowing, then blood pressure would vary with height, just as water pressure in a swimming pool does. The blood pressure at any point in the body is the sum of the static pressure due to gravity and the pressure due to the pumping action of the heart and the frictional loss in the blood vessels. Hence the blood pressure in our body is a continuous variable that changes from place to place and from time to time.

In a soft, distensible organ such as the lung, the effect of gravity is particularly evident. Due to the weight of the blood, the static pressure of blood in the capillary blood vessels at the apex of the lung in the upright posture of a human is less than that at the base by an amount equal to the height of the lung times the gravitational acceleration and the density of the blood. Since pulmonary capillary blood vessels are very compliant, their lumen sizes vary with this static pressure. As a result, blood flow in the lung varies with height. In an upright position there could be virtually no blood flow in the apex region and full flow at the base, whereas in the middle region there is flow limitation. This example points out another basic physical principle of blood flow: Blood vessel dimension is a function of blood pressure, and the feedback influence between these two variables can have dramatic effects at times.

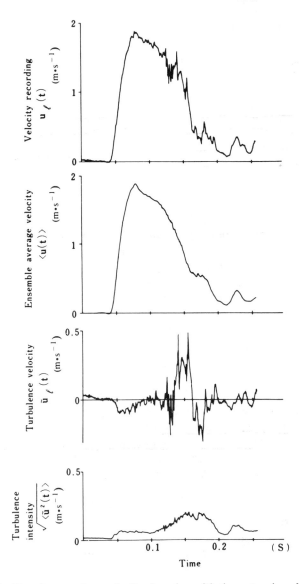

FIGURE 1.3:1. Transient motion: velocity changing with time at a given location. The partial derivative of velocity with respect to time while the location is fixed is the transient acceleration. The example shown here is from a paper by Yamaguchi, T., Kikkawa, S., Yoshikawa, T., Tanishita, K., and Sugawara, M.: Measurement of turbulent intensity in the center of the canine ascending aorta with a hot-film anemometer. *J. Biomech. Eng.* **105**: 177–187, 1983. The first panel shows a recording of blood velocity history. The second panel gives an ensemble average of many experiments. The third panel is a sample of turbulence velocity history. The fourth panel gives the root mean square of the turbulence velocity, or the so-called turbulence intensity.

The other two terms in the equation of motion, Eq. (1), the gradient of pressure and the divergence of normal and shear stresses, have been discussed earlier. The shear stresses are proportional to the coefficient of viscosity of the fluid and the rate of change of strains. Shear stresses are strongly influenced by turbulences, if any exist. In a fluid, turbulences are random motions whose statistical characteristics can be predicted, but for which the exact value of the velocity of any particle at a particular instant of time and at a given point in space is unpredictable.

1.4 Importance of Turbulence

Flow is defined by the field of velocity vectors of all particles in a domain. It is said to be *steady* if the velocity field is independent of time. It is *unsteady* if the velocity field varies with time. It is *turbulent* if the velocity field is stochastic, that is, if the velocity components are random variables described by their statistical properties. A turbulent flow contains small eddies within large eddies, and smaller eddies within small eddies, ad infinitum. A *laminar* flow is one that is not turbulent. Naturally, laminar flows are studied most often because they are easier to understand. Turbulence, on the other hand, is one of the hardest subjects in natural science. It is still poorly understood.

Figure 1.3:1 shows that the flow in the canine ascending aorta is turbulent and unsteady. Blood flow is laminar only in vessels that are sufficiently small. Turbulence dissipates energy. If a laminar flow in a tube turns turbulent, the resistance to the same flow may be increased greatly. The shear stress acting on the blood vessel endothelium may be increased many times when a laminar flow becomes turbulent if the total flow rate is unchanged.

We may think of an axisymmetric flow of blood in a blood vessel as the sliding of a series of concentric tubes of fluid, as illustrated in Figure 1.4:1. Then the differences between laminar and turbulent flows in the tube are twofold. First, the velocity profiles are different: The laminar flow has a parabolic profile, and the turbulent flow has a profile that is much blunter at the central portion of the tube and much sharper at the wall. Hence, the shear strain rate at the tube wall is much higher in the turbulent case as compared with the laminar flow of the same rate of volume flow. Secondly, the coefficients of friction or viscosity between the concentric tubes are different in these two cases. The "effective" or "apparent" coefficient of friction for the turbulent case is much higher than that of the laminar case. This is indicated by smooth shading of the surfaces of the sliding cylinders in the laminar case and a rough appearance in the turbulent case. The viscosity, or friction, in the laminar case is due to molecular motion between the sliding cylinders. The friction in the

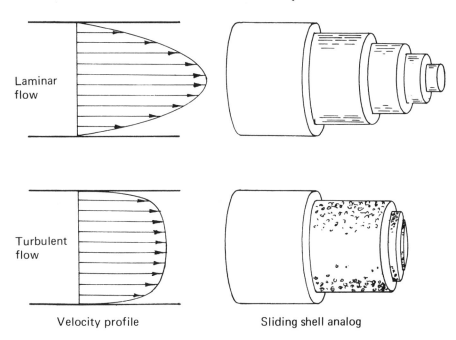

Velocity profile   Sliding shell analog

FIGURE 1.4:1. Schematic representation of fully developed laminar and turbulent flows in a tube at the same flow rate. The time-averaged velocity vectors are plotted in this figure as functions of spatial coordinates.

turbulent case comes from two sources: the molecular transport, which is usually the smaller part, and the convection of the turbulent "eddies," which is often the greater part. Because the shear stress in a fluid is equal to the product of the coefficient of viscosity and the shear strain rate, it is easy to see that the wall friction can become much larger if a laminar flow becomes turbulent. Turbulence is controllable to a certain degree, and the success of many medical devices depends on such control through good engineering design.

Turbulence in blood flow is strongly implicated in atherogenesis. Atherosclerotic plaques are often found at sites of turbulence in the aorta.

Associated with the velocity fluctuations in a turbulent flow are pressure fluctuations. Pressure fluctuations can excite vibrations in the eardrum and cochlea, and can be heard if the frequencies are in the audible range. We can hear the howling of wind because wind is turbulent. We can hear jet noise if the jet flow is turbulent. The Korotkoff sound at systole is the sound of the jet noise of rushing blood. A heart murmur is a turbulent noise. Flow separation at a site of stenosis in an artery often causes turbulence in the separated region, making it possible to detect a stenosis by listening to the noise (bruit) in the blood flow with a stethoscope.

1.5 Deceleration as a Generator of Pressure Gradient

If there is no turbulence and if the gravitational and frictional forces are small enough to be ignored, then Eq. (1) of Section 1.3 becomes

$$\text{Density} \times \text{acceleration} = -\text{pressure gradient}. \qquad (1)$$

Hence if the acceleration is negative (i.e., decelerating), then the left-hand side of Eq. (1) is negative and the equation yields a positive pressure gradient. In other words, in a decelerating fluid the pressure increases in the direction of flow. This mechanism is used effectively in our bodies to operate the heart valves and the valves in the veins and lymphatics.

Principle of Heart Valve Closure

Figure 1.5:1 shows the operation of the mitral valve. The mitral valve is composed of two very flexible, thin membranes. These membranes are pushed open at a stage of diastole when the pressure in the left atrium exceeds that of the left ventricle. Then a jet of blood rushes in from the left atrium into the left ventricle, impinges on the ventricular wall, and is broken up there. Thus, the blood stream is decelerated in its path and a positive pressure gradient is created. Toward the end of diastole, the pressure acting on the ventricular side of the mitral valve membranes becomes higher than that acting on the side of the membranes facing the left atrium. The net force acts to close the valve. In a normal heart, closure occurs *without* any backward flow or regurgitation. The papillary muscles play no role at all in opening and closing of the valve. They serve to generate systolic pressure in the isovolumetric condition by pulling on the membranes, and to prevent inversion of the valves into the atrium in systole.

The same principles applies to the aortic valve (see Fig. 2.5:4) as well as the valves of the veins and lymphatics. Deceleration is the essence, not backward flow.

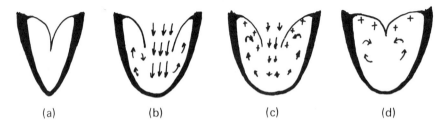

FIGURE 1.5:1. Operation of the mitral valve: (a) End of systole; (b) mitral valve wide open, jet rushing in; (c) toward end of diastole, with the jet being broken. The deceleration creates a pressure gradient that tends to close the valve: (d) Mitral valve closed.

1.6 Pressure and Flow in Blood Vessels—Generalized Bernoulli's Equation

If we know the pressure and velocity of blood at one station in the blood vessel and wish to compute the pressure and velocity at another station, we may integrate Eq. (1) of Sec. 1.3 along a streamline, or we may consider the balance between the changes in kinetic, potential, and internal energies and the work done on the blood by forces acting on it. Consider a system of blood vessels, as illustrated in Figure 1.6:1. Let the distance measured along a blood vessel be denoted by a variable x. Consider blood in the vessel between two stations, 1 and 2, located at $x = 0$ and $x = x$. Then we can derive the following equation from Eq. (1) of Sec. 1.3 (see Sec. 1.11 for details):

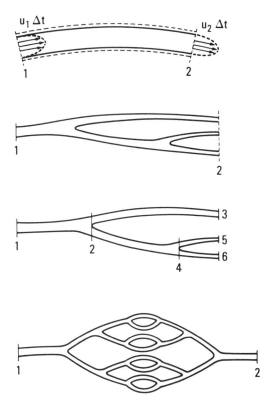

FIGURE 1.6:1. Two arbitrary stations, 1 and 2, in a blood vessel system. The energy equation (1) and that derived in Section 1.11 are applicable to any one of the systems illustrated here. In the uppermost figure the solid lines show the instantaneous boundary of the fluid in the vessel between stations 1 and 2 at an instant of time t; the dotted lines show the boundary at a time Δt later; u_1 and u_2 are the velocities of flow at stations 1 and 2.

Pressure at station 0 – pressure at station x

$$= \tfrac{1}{2} \cdot \text{density} \cdot \left(\text{velocity at } x\right)^2 - \tfrac{1}{2} \cdot \text{density} \cdot \left(\text{velocity at } 0\right)^2$$

$$+ \left(\text{specific weight}\right)\left(\text{height difference of station } x \text{ and station } 0\right)$$

+ rate of change of the kinetic energy of the blood between
stations x and 0

+ integrated frictional loss between stations x and 0. (1)

If the last two lines can be ignored, then we have the famous *Bernoulli's equation*. It says that pressure can be converted from kinetic energy of motion and potential energy of height, provided that the flow is steady and the fluid is inviscid. Pressure rises when the velocity of flow is slowed down, and vice versa. Bernoulli's equation is useful to analyze pressure and flow in a tube under steady conditions, and is used very often to calibrate flow-measuring instruments, pressure transducers, and flow meters.

For flow in blood vessels it is usually necessary to include more terms of Eq. (1) than are in Bernoulli's equation. For flow in normal aorta and vena cava, the last term in Eq. (1) can be neglected, but all the other terms must be retained.

For flow in small blood vessels, the last term of Eq. (1) is not negligible. The smaller the blood vessel, the more significant the frictional loss becomes. In microcirculation, in blood vessels with diameters 100 μm or less, the last term of Eq. (1) becomes the predominant term on the right-hand side of the equation. In the capillary blood vessels, the pressure drop balances the frictional loss exclusively.

1.7 Analysis of Total Peripheral Flow Resistance

If we read the generalized Bernoulli's Equation (1) of the preceding section by putting station $x = 0$ in the aorta at the aortic valve and station x in the vena cava at the right atrium, then the velocities of flow at the two stations are approximately equal and the heights of these two stations are the same, so that the first two lines on the right-hand side of Eq. (1) of Sec. 1.6 vanish. Furthermore, if we make measurements of average pressure and flow over a period of time extended over several cycles of oscillation, then the third line on the right-hand side of Eq. (1) of Sec. 1.6 will average out to zero because the rate of change of the kinetic energy of the blood in this segment oscillates on the positive and negative sides equally. In this case Eq. (1) of Sec. 1.6 becomes

Average pressure at aortic valve – average pressure at right atrium

= integrated frictional loss. (1)

This is often written as

$$\text{Systemic arterial pressure} = \text{flow} \times \text{resistance.} \qquad (2)$$

Here the *systemic arterial pressure* is the difference between the pressure at the aortic valve and that at the vena cava at the right atrium, the flow is the *cardiac output*, and the resistance is the *total peripheral vascular resistance*. Hence, writing in greater detail, we have

Pressure at aortic valve – pressure at right atrium

$$= \left(\text{cardiac output}\right) \times \left(\text{total peripheral vascular resistance}\right). \qquad (3)$$

The last term in Eqs. (1–3) represents the sum of the pressure drops due to the friction loss along all segments of blood vessels in a circuit or pathway. Since there are millions of capillary blood vessels in the body, there are millions of pathways along which one can integrate Eq. (1) of Sec. 1.3 to obtain Eqs. (1) and (2); the final result [Eq. (1)] is the same difference of the pressures at the aortic valve and right atrium no matter which path of integration is used.

Thus, if we define

$$\text{Total peripheral vascular resistance} = \frac{\text{integrated frictional loss}}{\text{cardiac output}}, \qquad (4)$$

then Eq. (2) or (3) is obtained. Hence, the systemic arterial pressure is equal to the product of cardiac output and total peripheral vascular resistance.

The integrated frictional loss is the sum of frictional losses in all segments of vessels of the circuit. In Chapter 3, we derive formulas to compute the friction loss in tube flow according to the Navier–Stokes equation. It is interesting, however, to discuss some major features here. Let us consider first a steady laminar flow (i.e., one that is not turbulent) in a long, rigid, circular, cylindrical tube. Such a flow is governed by the famous *Hagen–Poiseuille law* (or more commonly, *Poiseuille's law*), which states that the flow $\dot{Q}$ (short for "flow rate" or "volume flow rate," i.e., the volume of fluid flowing through a vessel in unit time) is related to the pressure drop Δp by the equation (see Sec. 3.2),

$$\dot{Q} = \frac{\pi d^4}{128} \frac{\Delta p}{\mu L}. \qquad (5)$$

Here d is the diameter of the vessel, μ is the coefficient of viscosity of the fluid, and L is the length of the segment of vessel over which the pressure drop Δp is measured. Equation (5) can be written as

$$\Delta p = \text{laminar resistance in a tube} \times \text{flow in the tube}, \qquad (6)$$

from which we obtain the resistance of a steady laminar flow in a circular cylindrical tube,

$$\text{Laminar resistance in a tube} = \frac{128\,\mu L}{\pi d^4}. \qquad (7)$$

If the nth generation of a vascular tree consists of N identical vessels in parallel, then the

Pressure drop in the nth generation of vessels

= resistance in N parallel tubes × total flow in N tubes

$$= \frac{\text{resistance in one tube}}{N} \times \text{cardiac output.} \tag{8}$$

Note that according to Eq. (7) the laminar flow resistance is proportional to the coefficient of viscosity μ and the length of the vessel L, and *inversely proportional to the fourth power of the diameter, d*. Obviously the vessel diameter d is the most effective parameter to control the resistance. A reduction of diameter by a factor of 2 raises the resistance 16-fold, and hence leads to a 16-fold pressure loss. In peripheral circulation the arterioles are muscular and they control the blood flow distribution by changing the vessel diameters through contraction or relaxation of the vascular smooth muscles.

Equation (7) gives the resistance to a Poiseuillean flow in a pipe and is the minimum of resistance of all possible flows in a pipe. If the flow becomes turbulent, the resistance increases (see Sec. 3.5 and 3.6). If the blood vessel bifurcates, the local disturbance at the bifurcation region raises resistance. In these deviations from the Poiseuillean flow, the governing parameter is a dimensionless number called the *Reynolds number* (in honor of Osborne Reynolds, see Sec. 3.2 and 3.5), which is the ratio of the inertia force to the viscous force in the flow,

$$\text{Reynolds number} = N_R = \frac{\text{inertia force}}{\text{viscous force}}$$

$$= \frac{\text{velocity of flow} \times \text{vessel diameter}}{\text{kinematic viscosity of fluid}}. \tag{9}$$

The kinematic viscosity is defined as the coefficient of viscosity divided by the density of the fluid. The flow in a vessel usually becomes turbulent when the Reynolds number exceeds a critical value of approximately 2300 (with exact value depending on the pulse rate and on whether the flow rate is increasing or decreasing). If a flow is turbulent, then

Resistance of a turbulent flow in a vessel

$$= \left(\text{laminar resistance}\right) \cdot \left(0.005\, N_R^{3/4}\right). \tag{10}$$

Thus, if the Reynolds number is 3000, the resistance of a turbulent flow is over two times that of the laminar resistance. In the ascending and descending aorta of the human and dog, the peak Reynolds number does exceed 3000.

1.8 Importance of Blood Rheology

Blood rheology plays a vital role in circulation because the coefficient of viscosity of blood, μ, figures importantly in Eqs. (5) and (7) of Sec. 1.7. For large blood vessels these equations must be modified by a function of the Reynolds number, as discussed in the preceding section. For smaller blood vessels, Reynolds numbers are smaller. Somewhere at the level of the small arteries and veins, the Reynolds number becomes 1. Microvessels with further reduction of diameter will have Reynolds number less than 1. From the definition of Reynolds number given in Eq. (9) of Sec. 1.7, we see that in these microvessels the viscous force dominates the scene. In capillary blood vessels, the Reynolds number is of the order of 10^{-2} or smaller, the inertia force becomes unimportant, and the flow is controlled almost entirely by the viscous force and pressure.

We have discussed blood rheology quite thoroughly in the book *Biomechanics: Mechanical Properties of Living Tissues* (Fung, 1993). In Chapter 3 of that book, we have shown that the viscosity of blood depends on the protein concentration of the plasma (the part of the blood outside the blood cells), the deformability of the blood cells, and the tendency of the blood cells to aggregate. The viscosity of the whole blood varies with the shear strain rate of the flow. It increases when the shear strain rate decreases. The blood viscosity increases when the hematocrit (i.e., the percentage of the total volume of blood occupied by the cells) increases and when the temperature is decreased. Aggregation and hardening of the cells increase the viscosity. These factors are all affected by various states of health and disease.

For flow in the microvessels, interactions between the cells and the blood vessel wall become very important. As a result of these interactions, the cells are not uniformly distributed in the blood in the microvessels. We speak of the *apparent coefficient of viscosity* of blood to take into account the interactions of blood cells, plasma, and the blood vessel wall. The apparent coefficient of viscosity is the value of μ that keeps Eq. (5) of Sec. 1.7 valid. It is obtained experimentally by measuring $\dot{Q}$, d, Δp, and L, and using Eq. (5) of Sec. 1.7 to compute μ. The apparent coefficient of viscosity of blood in microvessels can be increased under the following conditions:

  a. Existence of large leukocytes or exceptionally large erythrocytes with diameters greater than that of the capillary blood vessel. The flow may be obstructed by these cells.

  b. Smooth muscle in the arterioles or in the sphincters of the capillaries may contract so that the diameters of these vessels are greatly reduced, causing interference with the blood cells. The contraction of the smooth muscle may be initiated by nerves, by metabolites, or by mechanical stimulation. The behavior of vascular smooth muscle is discussed in Sections 7.10 and 7.11.

c. Leukocytes have a tendency to adhere to the blood vessel wall. If they do adhere, they increase resistance to blood flow. If the endothelial cells on the inner wall of the blood vessel are injured, the platelets are activated, causing clotting and increasing resistance.

d. Cell flexibility may be changed. Hardening of the red blood cell, as in sickle cell disease, increases the coefficient of viscosity of the blood.

Effects b and c, besides controlling the apparent coefficient of viscosity, effectively control vessel diameter. Since vessel diameter appears in Eqs. (5) and (7) of Sec. 1.7 in the fourth power, its importance is obvious.

1.9 Mechanics of Circulation

The brief analysis presented in the preceding sections demonstrates how some very important results can be obtained by applying the laws of conservation of mass and momentum to the circulatory system. From a few pieces of information about the anatomy of the heart and blood vessels, we were able to derive a pressure-flow relationship that furnishes the basic principle for the understanding of the perfusion of organs, the control of blood pressure, and the operation of heart valves. These results are simple and quantitative. From these examples we can anticipate that when more specific information about the physical properties of the circulation system is added, more penetrating specific results will be obtained. Some of them will help clarify the physiology and pathology of the organs, and others will lead to diagnostic and clinical tools. The rest of this book develops this theme.

1.10 A Little Bit of History

In Western culture, the concept of blood circulation was established by William Harvey (1578–1657). It was surprisingly late in history. In the early history of Greek physiology, there was a great error in the notion that the heart was the focus of respiration, as well as the center of the vascular system. Thus, it is not a coincidence that the original meaning of the Greek word *arteria* was "windpipe." Aristotle (384–322 B.C.) considered the heart as the focus of the blood vessels, but he made no distinction between arteries and veins, and showed no signs of any knowledge of the cardiac valves. Aristotle's conception was an enormous improvement over the most famous of the earlier treatises of the "Hippocratic" corpus, the *Sacred Disease* (i.e., epilepsy), which regards the head as the starting point of the main blood vessels.

At the beginning of the third century B.C., if not earlier, Praxagoras found that the two separate great trunks of blood vessels—still all called *veins*—

are different. One is about six times thicker than the other. In a corpse the veins collapse if emptied of blood; the arteries do not. Praxagoras insisted that the arteries contain no blood but only air. This mistaken idea of Praxagoras, more than any other single factor, prevented the ancients from arriving at the discovery of the circulation of the blood. How did Praxagoras reach such a conclusion? According to Harris (1980), it was probably based on the observation that when he cut open the chest of the animal that he dissected, the main arteries were found to be empty and full of air. Fåhraeus (1975) explained this phenomenon by the dilation of small arteries after death. Fåhraeus measured the pressures in the carotid or femoral artery in corpses at least 24 hours after death and found them to be negative, varying from a few millimeters to 1 cm Hg below the atmospheric pressure. If one begins an autopsy and opens the thoracic cage at that time, the pressure rises to zero by sucking air into the arteries. Fåhraeus believes that the negative pressure arises due to dilation of the arterial system. He reasoned that this dilation cannot take place in the large arteries; it must therefore be the little arteries that dilate and draw a certain amount of blood from the large arteries. He found that the phenomenon varies with the age of the person. The height of the negative pressure increases sharply from the age of 18 to 70 and then remains relatively constant. In addition, it is not instantaneous at the moment of death. Perhaps this may also explain why the author of *De Corde* found the ventricle, but not the aorta, empty (Harris, 1980). The picture that evolved to the Greeks was that the arteries provide organs with pneuma, veins provide them with blood, and nerves endow muscles with the power of contraction.

In China, the concept of circulation was stated very clearly in one of the oldest books on medicine, the *Nei Jing* (內經), or "Internal Classic." The authorship of this book was attributed to Huang Ti (黃帝) or the *Yellow Emperor* (2697–2597 B.C., according to the dictionary 辭源). But most Chinese scholars believe that it was written by anonymous authors in the *Warring Period* (戰國時代, 475–221 B.C.). It was written in the style of conversations between the Emperor and his officials, discussing medicine and its relationship to heaven, earth, climate, seasons, day, and night. It has 18 chapters; the first 9 are called *Su Wen* (素向), or "Plain Questions," the second 9 are called *Ling Shu* (靈樞), or "Miraculous Pivot." In the chapter 脈要精微論篇 in Su Wen, it states that "the blood vessels are where blood is retained" (夫脈者, 血之府也). In the chapter 五臟生成篇 in Su Wen, it states that "all blood in the vessels originates from the heart" (諸血者, 皆屬於心). In *Ling Shu* 營衛生會第十八, it states that "(the blood and *Chi*) circulate without stopping. In 50 steps they return to the starting point. Yin succeeds Yang, and vice versa, like a circle without an end" (營周不休, 五十而復大會, 陰陽相貫, 如環之無端).

Chi (氣) is a concept not familiar to Western thinking, but it was central to the Chinese idea of life. In the Chinese language Chi means gas; but it may also mean spirit (精氣). When a man would rather die than surrender, you say he has Chi Tsi (氣節). When a person is willing to labor or suffer for his friend without a thought about compensation, you say he has Yi Chi (義氣). The physiological concept of Chi was derived from ancient Chinese experimentation on acupuncture. The Chi circulates along the "meridional lines" (經絡), on which the acupunctural "points" (穴位) are located. The "points" are where the needle should be inserted. What Chi is exactly is not clear even today, but the thought is that it is something that circulates.

1.11 Energy Balance Equation

Equation (1) in Section 1.6 is very important. It is especially useful in studying complex situations, such as blood flow in the aortic arch or air flow in the trachea, in which the pressure gradient in the radial direction can be as large as the pressure gradient in the axial direction. In such a situation an accurate measurement of the pressure field is beyond the state of the art. On the other hand, current technology does permit a detailed measurement of the velocity field. The use of Eq. (1) of Section 1.6 would then enable us to compute the pressure p.

Because of its importance, let us consider the derivation of this equation in detail in order to obtain a deeper insight, and therefore to gain greater confidence in using it.

Consider the blood in a portion of the blood vessel system between two stations, 1 and 2 (see Fig. 1.6:1). Since the rate of gain of energy (the sum of the kinetic, potential, and internal energies) must be equal to the sum of the rate at which work is done on the system and the heat is transported in, the energy balance of the blood may be stated by the following equation:

$$\left(\text{Rate at which pressure force at station 1 does work on blood}\right) \tag{1}$$

$$+\left(\text{rate at which pressure force at station 2 does work on blood}\right) \tag{2}$$

$$+\left(\begin{array}{l}\text{rate at which pressure and shear on vessel wall do work}\\\text{on blood}\end{array}\right) \tag{3}$$

$$+\left(\text{rate at which heat is transported into the system}\right) \tag{4}$$

$$=\left(\text{rate of change of the kinetic energy of blood in the volume}\right) \tag{5}$$

$$+\left(\begin{array}{l}\text{rate at which kinetic energy is carried by particles leaving}\\\text{the vessel at station 2}\end{array}\right) \tag{6}$$

$$-\left(\begin{array}{l}\text{rate at which kinetic energy is carried by particles}\\\text{entering the vessel at station 1}\end{array}\right) \tag{7}$$

$$+\left(\begin{array}{l}\text{rate at which kinetic energy is carried across the blood}\\\text{vessel wall}\end{array}\right) \tag{8}$$

$+\left(\text{rate of gain of potential energy of blood against gravity}\right)$ (9)

$+\left(\text{rate of change of internal energy in the volume}\right).$ (10)

This equation is long, but obvious, except for the signs of the terms (6), (7), and (8). To explain these signs, we must remember that this equation is written for the *fluid particles* that occupy the volume in the blood vessel between stations 1 and 2 instantaneously. But these fluid particles are moving, so that in an infinitesimal time interval Δt later, the boundary of the space occupied by the fluid particles is no longer the original vessel wall and cross sections 1 and 2, but has become the new vessel wall and the curved surfaces, as illustrated in the uppermost part in Figure 1.6:1. Hence the change of kinetic energy (K.E.) of the *fluid particles* is equal to the change of K.E. in the *original volume* (bounded by the solid lines in the figure *plus* the K.E. of the particles that entered into the shaded area in the figure. The term (5) represents the rate of change of K.E. in the original volume; (6) represents the K.E. in the shaded area at the right; (7) represents the K.E. in the shaded area at the left, which does *not* belong to the original group of fluid particles, and hence has to be subtracted. The term (8) represents the K.E. in the shaded area bounding the vessel wall. The sum of (5), (6), (7), (8) is the rate of change of the K.E. of the fluid particles in the vessel between stations 1 and 2.

Let us translate this statement into a mathematical expression. Let us first identify the various rates listed in the equation above. First, the rate at which the pressure force acting on a small area dA does work on a fluid flowing across that surface with velocity u is equal to $pudA$, where u is the velocity component normal to the surface dA. Thus the rate at which the pressure force does work on the fluid passing station 1 is given by an integral over the cross section A_1,

$$\int_{A_1} pudA.$$ (1)′

A similar expression is obtained at station 2, except that a negative sign is needed because the pressure and u act in opposite directions. The expression for the work done on the vessel wall is also similar, and can be written as

$$\int_S \overset{v}{T_i} u_i \, dA,$$ (3)′

where $\overset{v}{T_i}$ is the stress vector acting on the surface of the vessel wall of area dA and normal v (with components $i = 1, 2, 3$ in the directions of a rectangular cartesian coordinate system x_1, x_2, x_3), and u_i is the velocity vector with components u_1, u_2, u_3. The integration is taken over the entire surface of contact between the blood and the vessel wall, S. Repetition of the index i means summation over i from 1 to 3.

The kinetic energy per unit volume of a small fluid element is $\frac{1}{2}\rho q^2$, where ρ is the density of the fluid and q is the speed of the element, that is, $q^2 =$

$u_1^2 + u_2^2 + u_3^2$. The rate at which kinetic energy is changing at any given place is given by its partial derivative with respect to time. Hence the rate of change of the K.E. in the volume V is

$$\int_V \frac{\partial}{\partial t}\left(\frac{1}{2}\rho q^2\right)dv. \tag{5}'$$

The rate at which the K.E. is carried out by particles crossing the instantaneous boundary at station 2 is

$$\int_{A_2} \frac{1}{2}\rho q^2 u\, dA. \tag{6}'$$

That at station 1 is given by a similar integral. The K.E. carried out of the instantaneous boundary by particles at the blood vessel wall is given by an expression similar to Eq. (3)', with $\overset{v}{T_i}$ replaced by $\frac{1}{2}\rho q^2$ and u_i replaced by the component of velocity normal to the vessel wall.

As was explained earlier, the sum of (5), (6), (7) and (8) is the rate of change of the K.E. of all fluid particles that occupy instantaneously the space in the blood vessel between stations 1 and 2. Mathematically, it is represented by the *material derivative* of the kinetic energy and is denoted by the symbol D/Dt,

$$(5)+(6)+(7)+(8) = \frac{D}{Dt}\int \frac{1}{2}\rho q^2\, dA. \tag{11}$$

The potential energy of blood against gravity is ρgh per unit volume, where ρ is the density of the fluid, g is the gravitational acceleration, and h is the height of the fluid element above a fixed plane perpendicular to the vector of gravitational acceleration. The total potential energy of the fluid particles that occupy the volume V instantaneously is

$$G = \int_V \rho gh\, dv, \tag{12}$$

and its rate of change is DG/Dt and can be broken down into four integrals, as in the case of the kinetic energy.

Finally, the change of internal energy is due to the generation of heat through viscosity. It can be shown that this is equal to the scalar product of the stress tensor σ_{ij} and strain rate tensor V_{ij},

$$V_{ij} = \frac{1}{2}\left(\frac{\partial u_i}{\partial x_j} + \frac{\partial u_j}{\partial x_i}\right).$$

Hence, the last term, the rate of dissipation of mechanical energy, is given by

$$\int_V \sigma_{ij}V_{ij}\, dv. \tag{13}$$

See Fung (1993, Sec. 10.8, pp. 220–222) for a rigorous derivation.

Summarizing the above, we obtain the following energy equation:

$$\int_{A_1} pu\,dA - \int_{A_2} pu\,dA + \int_S \overset{v}{T_i} u_i\,dA + \text{heat input}$$

$$= \int_{A_2} \frac{1}{2}\rho q^2 u\,dA - \int_{A_1} \frac{1}{2}\rho q^2 u\,dA + \int_S \frac{1}{2}\rho q^2 u_i v_i\,dA$$

$$+ \int_V \frac{\partial}{\partial t}\left(\frac{1}{2}\rho q^2\right)dv + \frac{D}{Dt}\int_V \rho gh\,dv$$

$$+ \int_V \sigma_{ij} V_{ij}\,dv. \tag{14}$$

For the blood flow problem, usually some of the terms may be neglected. The heat input is often small. The deformation of the vessel wall may be so small that both the work done by the wall force and the kinetic energy crossing the vessel wall are negligible. Then the third and fourth terms on the left-hand side of the equation and the third term on the right can be omitted, and the gravitational potential term is simplified into $(\rho gh_2 - \rho gh_1)$ times the flow rate, with h_1 and h_2 being the heights of stations 1 and 2, respectively.

The final equation is simplifed if we assume ρ to be constant, and define a characteristic pressure $\hat{p}$ and a characteristic square of velocity $\hat{q}^2$ as follows:

$$\hat{p} = \frac{1}{Q}\int pu\,dA, \qquad \hat{q}^2 = \frac{1}{Q}\int q^2 u\,dA, \tag{15}$$

where Q is the volume flow rate,

$$Q = \int u\,dA. \tag{16}$$

Then on dividing Eq. (14) by Q and writing $\mathscr{D}$ for the dissipation function

$$\mathscr{D} = \int \sigma_{ij} V_{ij}\,dv, \tag{17}$$

we have

$$\hat{p}_1 - \hat{p}_2 = \frac{1}{2}\rho\hat{q}_2^2 - \frac{1}{2}\rho\hat{q}_1^2 + \rho gh_2 - \rho gh_1 + \frac{\mathscr{D}}{Q} + \frac{1}{Q}\int \frac{\partial}{\partial t}\left(\frac{1}{2}\rho q^2\right)dv. \tag{18}$$

This final equation is most useful and important. If it is applied to a stream tube (a tube whose wall is composed of streamlines) of such a small cross section that the velocity and shear stress may be considered as uniform in it, then Eq. (1) of Section 1.6 results. If it is applied to tubes of finite cross section, then the definition of $\hat{p}$, $\hat{q}^2$, and $\mathscr{D}$ must be rigorously observed. For example, in Section 1.6 we discuss the pressure at the aortic valve. But the pressure varies from point to point at the aortic valve, and no single pressure can be assumed at the section. One might suggest to take the mean value, that is, $(\int p\,dA)/A$, as "the" pressure, but that is not necessarily the p required by the energy equation. Similarly, the "square of velocity" would have to be $\hat{q}^2 \cdot \hat{p}, \hat{q}^2$ are "velocity-weighted" averages. If the pressure p is uniform over a cross section, then $p = \hat{p}$. If the velocity u is uniform over the cross section, then $\hat{q}^2 = u^2$. If the velocity profile is parabolic as in Poiseuille flow, then $\hat{u}^2 = 2\bar{u}^2$, where $\bar{u}$ is the average value of u.

Equation (18) is a generalization of Bernoulli's equation to viscous and nonstationary flows. It makes it possible for us to speak of "pressure drop" in terms of quantities that can be calculated from velocity measurements. It can be applied to gas flow in the airway (except sneezing) and become the foundation on which pulmonary ventilation theory is built.

Recapitulation

Although the energy principle is straightforward, the earlier detailed statement is long. Let us recapitulate it in another way. Recall a general formula for the material derivative for any integral of a function Φ over all the fluid particles in a volume V that is bounded by a surface S,

$$\frac{D}{Dt} \int_V \Phi \, dv = \int_V \frac{\partial \Phi}{\partial t} \, dv + \int_S \Phi v_j v_j \, dS, \tag{19}$$

where $\mathbf{v}$, with components v_1, v_2, v_3, is the velocity of the fluid, and $\mathbf{v}$, with components v_1, v_2, v_3, is the unit vector normal to the surface, and the summation convention is used: Repetition of an index means summation over the index. See Fung, (1993, Sec. 10.4, pp. 215–217). Applying this to the kinetic energy K, given by $\frac{1}{2}\rho q^2$ integrated over a volume V of the blood bounded by the wall S and cross sections A_1 and A_2 (see Fig. 1.6:1), we obtain

$$\frac{DK}{Dt} = \frac{D}{Dt} \int_V \frac{1}{2} \rho q^2 \, dv = \int_V \frac{\partial}{\partial t} \left(\frac{1}{2} \rho q^2 \right) dv$$

$$+ \int_{A_2} \frac{1}{2} \rho q^2 u_2 \, dA_2 - \int_{A_1} \frac{1}{2} \rho q^2 u_1 \, dA_1 + \int_S \frac{1}{2} \rho q^2 u_n \, dS, \tag{20}$$

where u_1, u_2, u_n are, respectively, velocity components normal to the cross sections A_1, A_2, and the vessel wall S. The right-hand side of Eq. (20) is the sum $(5) + (6) + (7) + (8)$ above.

Similar expressions are obtained for the gravitational energy G and the internal energy E,

$$G = \int_V \rho g h \, dv, \qquad E = \int_V \rho \mathscr{E} \, dv, \tag{21}$$

where $\mathscr{E}$ is the specific internal energy per unit mass of the fluid. The first law of thermodynamics states that

$$\frac{D}{Dt}(K + G + E) = \dot{W} + \dot{H}, \tag{22}$$

where $\dot{W}$ is the rate at which work is done on the fluid and $\dot{H}$ is the rate at which heat is transported into the fluid mass. $\dot{W}$ is the sum of the work done by the fluid pressure over the cross sections A_1, A_2, and the work done by shear and normal stresses over the vessel wall, Eqs. (1)′, (2)′, (3)′ above, respectively. A similar expression can be written for $\dot{H}$.

Equation given at the beginning of this section is a verbal statement of Eq. (22) or (14).

On substituting Eq. (20), etc. into (22) and making use of the equations of motion and continuity, we can show that the rate of change of internal energy is equal to the dissipation function $\sigma_{ij}V_{ij}$. This completes the derivation. The use of the final result is explained earlier and is illustrated in Section 1.7.

Problems

1.1 Why is it that sometimes you can hear a turbulent flow but you cannot hear a laminar flow?

1.2 To listen to a flow, you can use a stethoscope. Apply a stethoscope to an artery. In what range of the eddying frequencies can you hear?

1.3 Discuss the possible relationships between the eddy size, the velocity of mean flow, the velocity fluctuations, and the frequency of pressure fluctuations in a turbulent flow. To be concrete, think of listening to a stenosis in an artery.

1.4 Discuss the answer to the preceding question more rigorously in terms of the correlation functions of velocity components and the corresponding frequency spectrums. Consult a book on modern fluid mechanics to learn about the correlation functions and frequency spectra of turbulent flow fields.

1.5 In what way is a Venturi tube used in a hydrodynamics laboratory? Explain the function of a Venturi tube on the basis that acceleration corresponds to a negative pressure gradient and deceleration corresponds to a positive pressure gradient, or in terms of the Bernoulli equation, Eq. (1) of Section 1.6.

1.6 Why is a red blood cell so deformable whereas a white blood cell is less so?

1.7 What is the evidence that the hemoglobin in the red blood cell is in a liquid state? When a hemoglobin solution is examined under x-ray, a definite diffraction pattern can be found. Why is the existence of such a crystalline pattern not in conflict with the idea that the solution is in a liquid state?

1.8 What is the source of bending rigidity of a red cell membrane?

1.9 How could evidence be found that the red cell membrane has bending rigidity and in what way is it important?

1.10 What is the evidence that the hydrostatic pressure in a red blood cell is about the same as that outside the cell?

1.11 What are the factors that determine the volume of a red blood cell?

1.12 How can a red blood cell deform without changing its surface area and volume? Can a sphere do this?

1.13 In catheterization, the pressure reading varies with the position of the catheter as the catheter is advanced. Why?

1.14 Explain why flow resistance depends on the Reynolds number, as indicated in Eq. (10) of Section 1.7.

1.15 Consulting a book on anatomy, describe the structure of the vascular system of a limb or organ (such as hand, arm, lung, heart, kidney, or brain). Can you estimate the total peripheral vascular resistance for the organ? What data are needed? Where can they be found?

1.16 Referring to Figure 1.4:1, explain why the resistance to flow is higher in the case of turbulent flow as compared with laminar flow.

1.17 Design an artificial heart valve to replace a diseased one. Discuss the pros and cons of your design.

1.18 When you measure blood pressure by inflating a cuff over the brachial artery and then listen to the sound in the brachial artery below the cuff with a stethoscope, what do you hear? What are the meanings of the various sounds? How are the sounds correlated with the physical phenomena happening in the artery? How are systolic and diastolic pressures defined and decided?

References

Most topics considered in this chapter are discussed in greater detail in the rest of this book. See the Index to locate references. Fundamental equations of fluid and solid mechanics are given in Fung (1993a and b). Basic concepts and equations for the description of finite deformation are presented in the companion volume *Biomechanics: Motion, Flow, Stress and Growth* (Fung, 1990, Springer Verlag). References mentioned in the text are the following:

Fåhraeus, R. (1975). Empty Arteries, Lecture delivered at the 15th International Congress of the History of Medicine, Madrid.

Fung, Y.C. (1993a). *A First Course in Continuum Mechanics for Physical and Biological Engineers and Scientists*. Prentice-Hall, Englewood Cliffs, NJ.

Fung, Y.C. (1993b). *Biomechanics: Mechanical Properties of Living Tissues*. Second edition. Springer Verlag, New York.

Harris, C.R.S. (1980). The arteries in Greco-Roman medicine. In *Structure and Function of the Circulation* (Schwartz, C.T., Werthessen, N.T., and Wolf, S., eds.), Plenum Press, New York.

2
The Heart

2.1 Introduction

The heart is the prime mover of blood. By periodic stimulation of its muscles it contracts periodically and pumps blood throughout the body. How the pump works is the subject of this chapter.

In each cycle the left and right ventricles are first filled with blood from the left and right atria, respectively, in the diastolic phase of the cycle. Then by the deceleration of the blood stream a pressure field is generated, which closes the valves between the atria and the ventricles. The contraction of the heart muscle begins and the pressures in the ventricles rise. When the pressure in the left ventricle exceeds that in the aorta, and the pressure in the right ventricle exceeds that in the pulmonary artery, the aortic valve in the left ventricle and the pulmonary valve in the right ventricle are pushed open, and blood is ejected into the aorta and the lung. This is the systolic phase. The ejection continues until the deceleration of the jets of blood creates pressure fields to close the valves. Then the muscle relaxes, the pressures decrease, and the diastolic phase begins.

Thus, in the left ventricle, the blood pressure fluctuates from a low of nearly zero (i.e., atmospheric) to a high of 120 mm Hg or so. But in the aorta, the pressure fluctuation is much less. How does the aorta do it? How is the large fluctuation of blood pressure in the heart converted to the pressure wave in the aorta, with a high mean value and a smaller fluctuation? The answer was given by Stephen Hales (1733), who credited the feat to the elasticity of the aorta. An analogy was drawn between the heart-and-artery system and the old-fashioned hand-pumped fire engine. In the case of the fire engine, the fireman pumps water into a high-pressure air chamber by periodic injections at a higher pressure. Water is then drained from the air chamber, which has a high mean pressure that drives water out in a steady jet. This analogy was used by Otto Frank (1899) in his theory of the cardiovascular system, and is known as the *Windkessel* (German for air vessel) theory. In this theory, the aorta is represented by an elastic chamber and the peripheral blood vessels are replaced by a rigid tube of constant resist-

FIGURE 2.1:1. The *Windkessel* model of the aorta and peripheral circulation.

ance (see Fig. 2.1:1). Let $\dot{Q}$ be the inflow (cm³/s) into this system from the left ventricle. Part of this inflow is sent to the peripheral vessels and part of it is used to distend the elastic chamber. If p is the blood pressure in the elastic chamber (aorta), then the flow in the peripheral vessel is assumed to be equal to p/R, where R is a constant called *peripheral resistance*. For the elastic chamber, its change of volume is assumed to be proportional to the pressure. The rate of change of the volume of the elastic chamber with respect to time, t, is therefore proportional to dp/dt. Let the constant of proportionality be written as K. Then, on equating the inflow to the sum of the rate of change of volume of the elastic chamber and the outflow p/R, the differential equation governing the pressure p is

$$\dot{Q} = K\left(dp/dt\right) + p/R. \tag{1}$$

The solution of this differential equation is

$$p(t) = \frac{1}{K} e^{-t/(RK)} \int_0^t \dot{Q}(\tau) e^{\tau/(RK)} d\tau + p_0 e^{-t/(RK)}, \tag{2}$$

where p_0 is the value of p at time $t = 0$. This gives the pressure in the aorta as a function of the left ventricle ejection history $\dot{Q}(t)$. Equation (2) works remarkably well in correlating experimental data on the total blood flow $\dot{Q}$ with the blood pressure p, particularly during diastole (McDonald 1974, pp. 11, 310, 423; and Wetterer and Kenner, 1968). Hence, in spite of the severity of the underlying assumptions, it is quite useful.

Turning now to the question of stress in the heart itself, there is another very simple analysis that is quite good. Assume that the left ventricle can be approximated by a thick-walled hemispherical shell (Fig. 2.1:2), with

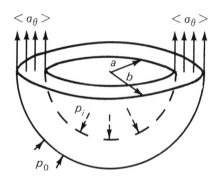

FIGURE 2.1:2. Balance of pressure forces acting on a thick-walled spherical shell.

inner radius a and outer radius b. Let the pressure acting on the inside be p_i (blood pressure) and that acting on the outside be p_o (pressure from pericardium), both assumed to be uniform. Consider the equilibrium of forces in the vertical direction. The force acting downward is equal to p_i times the projected area πa^2. The forces acting upward are $p_o \pi b^2$ and the product of the average wall stress $\langle \sigma_\theta \rangle$ and the area on which it acts, $\pi(b^2 - a^2)$. At equilibrium, these forces are balanced:

$$\langle \sigma_\theta \rangle \left(b^2 - a^2 \right) = p_i a^2 - p_o b^2 . \tag{3}$$

Hence the average circumferential wall stress in the heart is

$$\langle \sigma_\theta \rangle = \frac{p_i a^2 - p_o b^2}{b^2 - a^2} . \tag{4}$$

The highest stress is obtained when p_i is equal to the maximum systolic pressure. The outer pressure p_o is normally close to the pleural pressure, which is subatmospheric (negative), but may become positive in disease states such as cardiac tamponade (a large accumulation of pericardial fluid). The material of the heart wall is incompressible, so that the volume of the heart wall, $(\frac{2}{3})\pi(b^3 - a^3)$, remains a constant. Hence b is a function of a, and we see from Eq. (4) that the maximum stress acting in the heart wall is essentially proportional to $p_i a^2$: the higher the systolic pressure, and the larger the heart (radius a), the larger the stress. From this we can make two deductions:

1. Abnormal enlargement of the heart and increase in pressure will increase the wall stress and may lead to hypertrophy of the heart. The radius a is principally controlled by the diastolic pressure of the heart.

2. The systolic pressure p_i is determined by the wall stress $\langle \sigma_\theta \rangle$, which is due to muscle contraction. The maximum tensile stress that can be generated in an isometric contraction of a cardiac muscle varies with the length of the sarcomere. See the length–tension curve in Figure 2.1:3, and the explanation in Fung (1993b, chapter 10). If a heart normally operates at a sarcomere length marked by the point A in the figure, then when the sarcomere is lengthened the maximum muscle tension will increase, and as a consequence the systolic pressure p_i will increase. Now the number of sarcomeres in a heart muscle is fixed, hence the sarcomere length is proportional to the heart radius a. Thus, if the radius of the heart a is increased, the muscle tension will increase, and so will be the systolic blood pressure. This is known as *Starling's law of the heart*.

Starling's law of the heart works as long as the operating point A lies on the upward-sloping leg of the curve shown in Figure 2.1:3. It ceases to be valid when A moves off this leg.

As mentioned before, the size of the left ventricle in the isovolumetric phase is determined by how well the left ventricle is filled at the end dias-

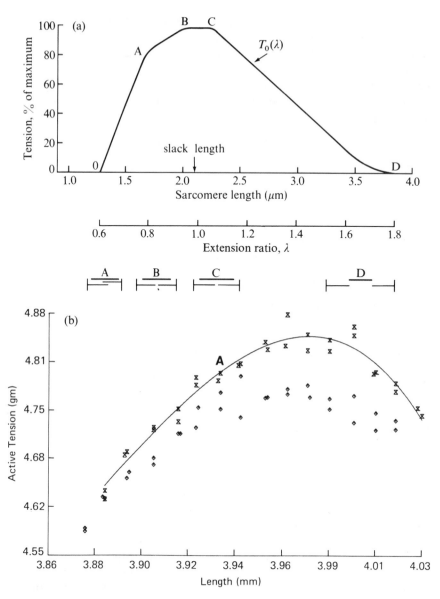

FIGURE 2.1:3. (a) The "length–tension" curve of a skeletal muscle. The sarcomere length is plotted on the abscissa. The maximum tension achieved in isometric contraction at the length specified is plotted on the ordinate. (b) The maximum active tension in single isometric twitches of the papillary muscle of the rabbit obtained by Paul Patitucci in the author's lab. Note that the maximum tension is not a unique function of the muscle length: Its value depends on whether in successive experiments the length is increased (symbol $\mathbf{X}$) or decreased (symbol $\diamondsuit$). Thus, there is hysteresis in active tension. The solid curve represents a fitted Fourier series.

tolic condition. This, in turn, is determined by the balance between p_i, the diastolic pressure, p_o, the pericardial pressure, and $\langle \sigma_\theta \rangle$, the elastic stress in the ventricular wall when it is relaxed while the muscle is in the refractive period between successive stimulations. The same formula (3) applies. The major determinants are the diastolic pressure and the elasticity of the heart muscle in the resting state. By controlling the diastolic pressure, that is, the filling of the heart, the size of the left ventricle, and hence the length of the sarcomeres in the isovolumetric phase, can be controlled. Thus, by controlling the diastolic pressure, the location of the point A on the upward leg of Figure 2.1:3 can be controlled. In this way a physician or surgeon can make use of Starling's law to deal with some clinical problems.

This brief discussion shows the critical role played by biorheology in understanding the function of the organs. Biorheology is described in *Biomechanics: Mechanical Properties of Living Tissues* (Fung, 1993b), which will be referred to frequently in the present work. To know more about the heart we must know its geometry, its materials of construction and their mechanical properties, and the electric, chemical, and nervous events in each cycle of contraction, as well as the way the heart is coupled with the vascular system. On the foundation of this information we should be able to predict the function of the heart by the method of continuum mechanics.

2.2 Geometry and Materials of the Heart

The adult human heart has four chambers: two thin-walled atria separated from each other by an interatrial septum, and two thick-walled ventricles separated by an interventricular septum. As is shown schematically in Figure 2.2:1, the venous blood flows into the right atrium, through the tricuspid valve into the right ventricle, and then is pumped into the pulmonary artery and the lung, where the blood is oxygenated. The oxygenated blood then flows from the pulmonary veins into the left atrium, and through the mitral valve into the left ventricle, whose contraction pumps the blood into the aorta, and then to the arteries, arterioles, capillaries, venules, veins, and back to the right atrium.

The four valves are seated in a plane, as shown in Figure 2.2:1. The mitral and tricuspid valves, which are opened in order to fill the ventricles with blood when the blood pressure is low and velocity is small, are relatively large in area. The aortic and pulmonary valves, used in ventricular systole to pump blood out of the ventricles at high velocity, are smaller. The mitral and triscuspid valves are attached to papillary muscles, which contract in systole, pull down the valves to generate systolic pressure rapidly, and prevent the valves from any danger of inversion into the atrium. The aortic and pulmonary valves have no strings attached. The closing and opening of all valves are operated by blood itself through hydrodynamic forces.

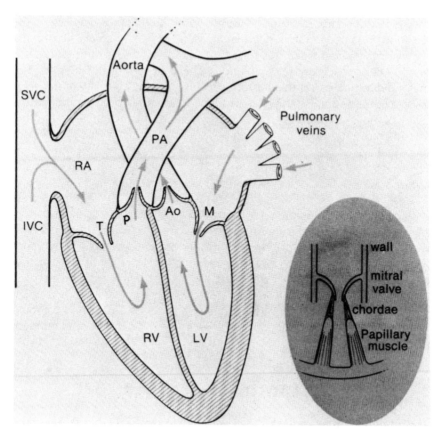

FIGURE 2.2:1. Blood flow through the heart. The arrows show the direction of blood flow. SVC = superior vena cava; IVC = inferior vena cava; RA = right atrium; RV = right ventricle; PA = pulmonary artery; LV = left ventricle; T = tricuspid; P = pulmonary; AO = aortic; M = mitral. From Folkow and Neil (1971) *Circulation*, Oxford Univ. Press, New York, p. 153, by permission.

The heart is a muscle. The muscles of the atria and the ventricles are joined to a skeleton of fibrous tissue on which the rings of the four valves are seated. The *bundle of His* constitutes the only *muscle* connection between the atria and ventricles. The electric pacemaking activity of the right atrium can pass to the ventricles only via the bundle of His.

The muscle fibers in the heart are systematically oriented, see Le Grice et al. (1995), Smail and Hunter (1991), Streeter and Hanna (1973), Streeter (1979), Streeter et al. (1969). Figure 2.2:2 shows the fiber orientation in the left ventricular wall of the dog. On the epicardium, the muscle fibers are oriented from the apex to the base, or, if we borrow the language used in geography in describing features on the globe, we say that the fibers are arranged in the direction of the longitudes. Away from the epicardium the

FIGURE 2.2:2. The orientation of muscle fibers in the left ventricular wall of the dog. From Streeter et al. (1969), by permission.

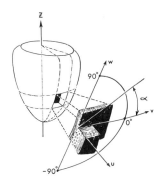

muscle fiber orientation changes continuously. At the midwall the fibers are oriented parallel to the base, that is, in the direction of the latitudes. The rotation continues until the fibers become longitudinal again in the endocardium. If the angle between a muscle fiber and a parallel circle is denoted by α, then the relationship between α and the depth through the ventricular wall is as shown in Figure 2.2:3.

The valves are collagen membranes. They are thin and flexible. A membranous structure of similar flexibility is the pericardium, which encloses the entire heart. The pericardium is attached to arterial trunks above and the central tendon of the diaphragm below. It is a sac containing a small amount of pericardial fluid that serves as a lubricant, and limits the excursions of the heart.

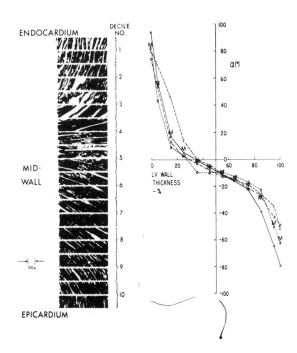

FIGURE 2.2:3. Variation of the inclination of the muscle fibers in the left ventricular wall of the dog from the apex-to-base (longitudinal) direction. From Streeter et al. (1969), by permission.

The reader might enjoy looking over the beautifully illustrated book by Frank Netter, *Heart* (1969). Details of the anatomy of the heart can be found in the *Handbook of Physiology* (Berne and Sperelakis, 1979). A tremendous store of information is in Braunwald (1988).

2.3 Electric System

Since every contraction of the heart muscle needs an electric stimulation, the heart must have a center that generates a periodic electric signal which is conducted to every muscle cell. The main electric generating station of the heart lies in the *sinoatrial node* (S-A node; see Fig. 2.3:1). The S-A node is a pale, narrow structure. For humans it is approximately 25 mm long, 3 to 4 mm wide, and 2 mm thick. It contains two types of cells: (a) the small, round *P cells*, which have few organelles and myofibrils, and (b) the slender, elongated *transitional cells*, which are intermediate in appearance between the *P* and the ordinary myocardial cells. The *P* cells are the dominant pacemaker cells of the heart. They exhibit rhythmicity very early in fetal life. (In the human fetus rhythmic contraction of the heart tube begins on the 19th or 20th day.) Electrophysiological data show that the S-A node is

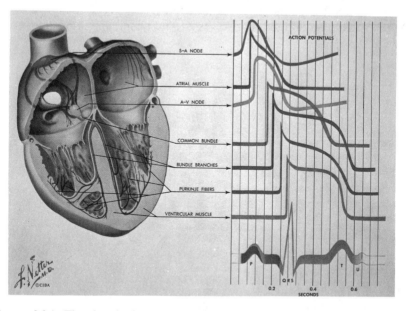

FIGURE 2.3:1. The electric system of the heart and the action potentials at various locations in the heart. From Frank Netter (1969). © Copyright 1969, CIBA Pharmaceutical Company, Division of CIBA-GEIGY Corporation. Reprinted with permission from THE CIBA COLLECTION OF MEDICAL ILLUSTRATIONS illustrated by Frank H. Netter, M.D. All rights reserved.

the first region of the heart to display electric activity during each cardiac cycle.

From the S-A node, the cardiac electric signal spreads radially throughout the right atrium along ordinary myocardial fibers at a conduction velocity of approximately 1 m/s. In the meantime, a special bundle of fibers carries the signal directly from the S-A node to the left atrium, and three other bundles conduct the signal directly from the S-A to the A-V node. These bundles consist of a mixture of ordinary myocardial cells and specialized conducting fibers similar to those that exist in the ventricles.

The *atrioventricular* (A-V) *node*, is a substation of the signal transmission (see Fig. 2.3:1). This node has dimensions $3 \times 10 \times 22$ mm in humans. It contains the same two types of cells as the S-A node, but the *P* cells are more sparse and the transitional cells preponderate. When the signal from S-A node reaches the A-V node, it is delayed for a certain period of time, and then is passed to the ventricles via the *atrioventricular bundle*, or *bundle of His*. It is the only muscular tissue connecting the atria to the ventricles. In humans it is about 12 mm long before it branches.

The delay of the electric signal transmission at the A-V node allows optimal ventricular filling during atrial contraction. It is this delay that is responsible for the interval between the *P wave* and the *QRS complex* in the electrocardiogram. Detailed studies attribute the delay to the small fibers in the junctional region between the atrial myocardium and the A-V node. It is known that the conduction velocity is directly related to fiber diameter, and in these small fibers the velocity is 0.05 m/s as compared with 1 m/s in the main body of the A-V node. The cells in the A-V node have been found to be significantly less excitable than other cardiac cells, and the relative refraction period persists for an appreciable period of time after repolarization.

If the A-V delay were prolonged or if some or all of the atrial excitations were prevented from reaching the ventricles, then the conduction is said to be *blocked*. Pathological blocks may be provoked by nervous, inflammatory, circulatory, or drug factors, such as acute rheumatic fever or digitalis.

The bundle of His passes down the right side of the interventricular septum and then divides into the right and left *bundle branches*. From these further branches, called *Purkinje fibers*, spread over both ventricles (see Fig. 2.3:1). Electric signal propagates fast in the Purkinje fibers at a speed of 1 to 4 m/s.

The last stage of electric transmission is done by the cardiac muscle itself, transmitting from one cell to the next. There is a semblance of *syncytium* between cardiac muscle cells, which make such excitation possible. The speed of signal transmission in the myocardium is 0.3 to 0.4 m/s.

Figure 2.3:2 shows an example of the sequence of initiation of excitation spreading in the ventricles of humans. Different parts of the heart are excited in a definite sequence of time. Such a map is important not only for

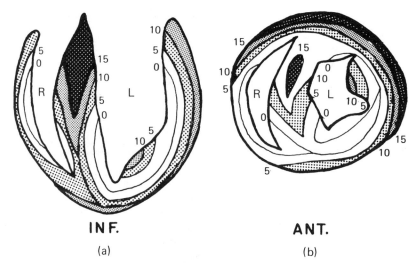

FIGURE 2.3:2. Excitation sequence of the myocardium as determined by miniature electrodes. (a) Meridional section. (b) Latitudinal section. The numbers indicate the intervals in millseconds after the earliest excitation. The shading increases in darkness with increasing length of intervals. From W.G. Guntheroth, (1965) *Pediatric Electrocardiography*, W.B. Saunders Co., Philadelphia, on the basis of data in Scher and Young (1956), by permission.

the interpretation of the electrocardiogram, but also for the dynamics of the movement of the heart.

So far we have outlined the layout of the electric system and the sequence of spreading of the electric signal in the normal condition. However, apparently some cells in the walls of all four cardiac chambers are capable of initiating beats. Regions of the heart, other than the S-A node, that initiate beats under special circumstances are called *ectopic foci* or *ectopic pacemakers*. The special circumstances include block of conduction pathways, enhancement of rhythmicity of the ectopic foci, or depression of the rhythmicity of the S-A node.

The electric activity of the heart is of great clinical significance, particularly because it is revealed in the electrocardiogram (ECG). The ECG thus becomes a powerful clinical tool. The reader should consult chapters in *Handbook of Physiology* (edited by Berne and Sperelakis, 1979), especially by Scher and Spach (1979), to learn something about the electrophysiology of the heart.

From the point of view of mechanics, it is probably sufficient to know the layout of the electric system and to know that the *action potential* of myocardial cells in different parts of the heart has somewhat different courses in time. The action potential of a cardiac muscle cell is recorded by inserting microelectrodes into the interior of the cell. A typical example of a record of potential changes occurring in a ventricular cell is shown in Figure 2.3:3. When two electrodes are put next to a strip of quiescent cardiac muscle in an electrolyte solution, there is no potential difference between the two electrodes (in the period of time from point A to point B in the figure). At time B one of the microelectrodes is inserted into a cell. Immediately the galvanometer records a potential difference across the cell membrane, indicating that the potential of the interior of the cell is about 90 mV lower than that of the surrounding medium. At point C a propagated action potential is transmitted to the cell impaled with the microelectrode (assuming that the cell lies in a strip of ventricular muscle in which the action potential is propagated). Very rapidly the cell membrane becomes depolarized and the potential difference reversed, so that the potential of the interior of the cell exceeds that of the exterior by about 20 mV. Immediately following the upstroke there is a brief rapid change of potential in the direction of repolarization. Repolarization then decelerates (phase 2), and again accelerates (phase 3), until finally it decelerates asymptotically to the resting potential (phase 4). The period C–D is the effective refractory period, in which the muscle cell will not respond to any additional electric stimulation. The period D–E is the relative refractory period, in which a graded action potential can be initiated in the muscle. In the supernormal period E–F, an action potential can be initiated by electric stimulation, for example, by depolarization waves transmitted in the muscle.

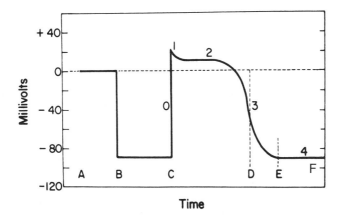

FIGURE 2.3:3. Changes in electric potential recorded by an intracellular microelectrode. From time A to B the electrode was outside the cell. At B the cell was impaled by the electrode. At C an action potential begins in the impaled cell. Time C–D represents the effective refractory period. Time D–E is the relative refractory period, while E–F is the supernormal period. For the meaning of the refractory period and the relationship between the action potential and the muscle contraction, see the text. From Berne, R.M. and Levy, M.N. (1972) *Cardiovascular Physiology*, 2nd edn. C.V. Mosby Co., Saint Louis, p. 6, by permission.

Mechanically, the contraction of the muscle cell begins after the action potential changes the polarization of the cell membrane at time C. In the period of time C–D, active tension in the muscle cell increases if the muscle length was fixed. The tension reaches the maximum at the point D, then it decreases gradually in the period D–E until it becomes zero at point F. If the muscle tension is maintained at a constant level, then the muscle length shortens in the period C–F.

Figure 2.3:1 shows that the time course of the action potential at different parts of the heart is somewhat different. These changes affect the rate of tension generation in the isometric condition and the speed of shortening of the muscle in the isotonic condition. Electrochemical studies of the muscle cell show that the various phases of the cardiac action potential are associated with changes in the ion channels of Na^+, K^+, and Ca^{2+} in the membrane (see Fung, 1990, chapter 8). The Ca^{2+} is involved in the excitation–contraction coupling of the muscle. Hence, the variation in the shape of the action potential history curve affects the muscle contraction history.

Of mechanical significance is the difference in the duration of action potentials in the subendocardial cells from that in the subepicardial cells. As a result, although the wave of depolarization in the ventricles proceeds from the endocardium to epicardium, the wave of repolarization travels in the opposite direction. It has been postulated that the compressive strain in the ventricular muscle during systole retards the repolarization process more in the subendocardial than in the subepicardial region.

2.4 Mechanical Events in a Cardiac Cycle

The electric events described in Section 2.3 determine the sequence in which the cardiac muscle cells contract. This contraction, operating in coordination with a set of valves, circulates the blood. Figure 2.4:1 shows the pattern of the motion of the ventricular walls when the heart is ejecting blood: The mitral and tricuspid valves are closed, the aortic and pulmonary valves are open, the muscles of both ventricles are contracting, the left ventricle ejects blood into the aorta, and the right ventricle ejects blood into the lung. The pressure in the left ventricle is much higher than the pressure in the right ventricle. Hence, the left ventricle remains almost ellipsoid in shape, whereas the right ventricle is bellow-shaped. The common wall shared by the two ventricles—the interventricular septum—bulges into the right ventricles, as shown in Figure 2.4:1. The transmural pressure acting on the interventricular septum = (left ventricular pressure − right ventricular pressure). The transmural pressure acting on the free wall of the left ventricle = (left ventricular pressure − pericardial pressure). The transmural pressure acting on the right ventricular free wall = (right ventricular pressure − pericardial pressure). The heart is wrapped in a thin collagenous

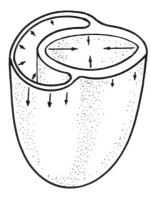

FIGURE 2.4:1. Patterns of ventricular contraction. Right ventricular ejection is accomplished primarily by compression of the right ventricular cavity, but also by downward displacement of the tricuspid valve ring (shortening of the free wall). Left ventricular ejection is accomplished primarily by constriction of the left ventricular chamber with only a minor contribution of shortening of the long axis. From Rushmer R. (1976) *Cardiovascular Dynamics*, W.B. Saunders, Philadelphia, p. 92, by permission.

membrane called the *pericardium*. The space between the pericardium and the heart is filled with a layer of fluid. The pressure in the pericardial fluid is the pericardial pressure, the value of which is of the same order of magnitude as the pleural pressure between the lung and the chest wall, but its exact value depends on the relative motion of the heart and pericardium, and the amount of the pericardial fluid that is present. Under these pressure loadings, and with the sarcomere lengths of the muscle cells defined by the end-diastolic condition of the heart, the heart muscle contracts and induces motion of the ventricles, as indicated by the velocity vectors in Figure 2.4:1. The motion of the walls of the ventricles is mainly radial, and is larger at the left ventricle and smaller at the right ventricle. But the radial motion is accompanied by some shortening in the longitudinal direction.

The fluid dynamics of blood ejection from the left and right ventricles into the aorta and the lung, respectively, must be analyzed by solving the Navier–Stokes equations, with the motions of the solid walls of the heart, the heart valves, the aorta, and the pulmonary arteries as boundary conditions. On the other hand, the motion of the heart, heart valves, aorta, and pulmonary arteries must be determined by solving Navier's equations with the pressures of the blood, the pericardial fluid, and the body fluid as external loading and the anatomical structure as geometric constraints. Thus, as is usual in circulation research, fluid mechanics and solid mechanics are coupled together. Furthermore, the heart, lung, and aorta are coupled together. Computational methods are available for detailed analysis. In this

TABLE 2.4:1. Pressures and Volumes in the Normal Human Heart*.

Left atrial pressure, mean ≤12 mm Hg
Left ventricular pressure
    Peak systolic, 100–150 mm Hg in adults
    End diastolic, ≤12 mm Hg
Aortic pressure
    Systolic, 100–150 mm Hg in adults
    Diastolic, 60–100 mm Hg in adults
Right atrial pressure, mean ≤6 mm Hg
Right ventricular pressure
    Peak systolic, 15–30 mm Hg
    End diastolic, ≤6 mm Hg
Pulmonary arterial pressure
    Systolic, 15–30 mm Hg
    Diastolic, 4–12 mm Hg
Left ventricular end-diastolic volume, at rest, 70–100 ml/m^2 body surface area
Left ventricular end-systolic volume, at rest, 25–35 ml/m^2 body surface area
Stroke volume, at rest, 40–70 ml/m^2 body surface area
Ejection fraction at rest, (stroke volume divided by end-diastolic volume), 0.55–0.80
Cardiac index, 2.8–4.2 l/m^2/min
Systemic vascular resistance, 770–1,500 dyne s cm^{-5}
Pulmonary vascular resistance, 20–120 dyne s cm^{-5}

Body surface area is given approximately by (weight in kg + height in cm − 60)/100 m^2.
* Grossman, W. (ed.) (1974) *Cardiac Catherization and Angiography*. Lea & Febiger, Philadelphia; Grossman, Brodie, Mann, and McLaurin (1977) Effects of sodium nitroprusside on left ventricular diastolic pressure–volume relations. *J. Chin. Inves.* **59**: 59–68.

section, however, we look only at the general features to see how the heart works.

Typical values of pressures and volumes of the chambers in the normal human heart are listed in Table 2.4:1. Volume changes and wall motion of the heart are affected significantly by posture (supine or erect), anesthetization, and surgery (open chested or not, pericardium open or not). Measurement of shape and volume in an intact heart remains a challenging problem. Some experiments have shown that the strain is nonuniform in the left ventricle during systole: The shortening is greater in the neighborhood of the apex than that at the base. Opening of the pericardium has an important effect on the shape and size of the heart. All this is quite expected from the mechanical point of view, because any change in boundary conditions changes the stresses and strains in the heart.

Correlation with Electric Events

Figure 2.4:2 shows the relationship between the mechanical and electric events in the heart. The electrocardiogram, or ECG, is shown at the bottom. It shows the characteristic P wave induced by the atrial electric activity, the QRS complex induced by the ventricular depolarization, and the T wave representing ventricular repolarization. The valve motion is indicated in the middle of the figure. The opening O and closing C of each of the valves (A = aortic, T = tricuspid, P = pulmonic, M = mitral) are marked in the figure. Thus, TC means tricuspid valve closing, etc.

The curve above the ECG represents the displacement of the apex of the heart, which can usually be felt through the chest wall when a person lies horizontally on the left side. Starting from the left, the *a* wave coincides with the atrial contraction at the end of filling of the left ventricle. IC represents isovolumic contraction, ending at E the most outward position. The apex

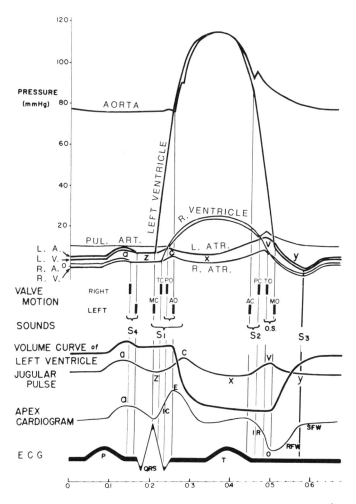

FIGURE 2.4:2. Correlation of various hemodynamic measurements in a single cardiac cycle. The time scale reference at the bottom is in tenths of a second. From Hurst, J.W., and Logue R.B. (1970) *The Heart*, 2nd edn., McGraw-Hill, New York, p. 76, by permission.

then moves away from the chest wall during the period of ejection of blood after E. In isovolume relaxation (IR) the apex moves further inward. At O the mitral valve opens. Then a rapid filling wave (RFW) follows due to the initial rush of blood into the left ventricle. This is followed by a slow-filling wave (SFW) during mid-diastole, up to the *a* wave contraction at end diastole.

The jugular vein pulse shows a similar *a* wave due to right atrial contraction at end diastole. The *c* wave reflects bulging backward of the tricuspid valve into the right atrium during right ventricular systole. The *x* wave is due to downward displacement of the base of the ventricles during systole and continued atrial relaxation. The *xv* wave is due to filling of the right atrium. The *vy* wave follows the opening of the tricuspid valve.

The right atrial pressure wave (R.A.) is essentially the same as the jugular vein pulse. The volume curve of the left ventricle shows an *a* wave due to the contraction of the left atrium, followed by an isovolumic period of ventricular contraction. As the aortic valve opens, blood is ejected into the aorta and the left ventricular volume is rapidly decreased. This continues until end systole. The blood volume remains constant during the isovolumic relaxation phase until the mitral valve opens to admit fresh blood.

The pressure curve of the left ventricle (L.V.) is shown in the uppermost part of the figure. When the ventricular pressure exceeds the pressure in the aorta, the aortic valve opens and blood is ejected into the aorta. Later, when the ventricular pressure becomes smaller than that in the aorta, the ejecting jet decelerates. At the instant marked AC the aortic valve closes.

The aortic pressure curve reflects the closure of the aortic valve with a *dicrotic notch*. The right ventricular pressure curve (R.V.) is similar to that of the left ventricle, but at a lower level. The pulmonary artery pressure curve is similar to that of the aorta but also at a lower level.

A certain sound is associated with each of these events. The timing of the four heart sounds—S_1, S_2, S_3, and S_4—is indicated in the figure. They are associated with movement of the valves, and acceleration and deceleration of blood, and movement and vibration of the heart muscle, and their transmission through the thorax. If fully understood, the heart sounds can become as useful a tool to the clinician as seismic waves are to geologists.

Ventricular Function Curve

A ventricular function curve is one that relates a certain measure of preload (which determines the length of the heart muscle fibers in a relaxed state before contraction), such as the end-diastolic pressure, with some measure of cardiac performance, such as stroke volume or stroke work (see Fig. 2.4:3). Depending on the choice of the "measures," many different ventri-

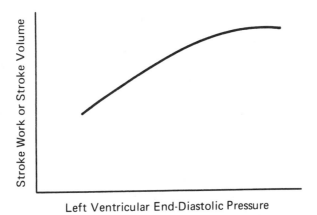

FIGURE 2.4:3. Schematic diagram of a left ventricular function curve. A measure of ventricular performance, such as stroke work or stroke volume, is plotted on the ordinate as a function of a measure of preload, such as left ventricular end-diastolic pressure, on the abscissa.

cular function curves can be plotted. The choice is usually based on considerations of clinical relevance. For example, as the filling pressure of the ventricle decreases, symptoms of dyspnea appear; and as the stroke work decreases signs of insufficient peripheral perfusion occur; hence, a figure such as Figure 2.4:3 is useful. It tells a clinician a lot about a patient, especially with respect to therapeutic interventions that may shift the ventricular function curve up or down, left or right.

Pressure–Volume Loop

If we plot the pressure and volume of a left ventricle throughout a cardiac cycle on a plane with respect to a set of coordinates, as shown in Figure 2.4:4, with volume as the abscissa and pressure as the ordinate, we see a clear representation of the four phases of the cardiac cycle. The area within the loop describes the external work (stroke work) done by the left ventricle as it contracts.

The lower curve of the loop, phase 1 (the arc AB in Fig. 2.4:4), represents the ventricular filling. During this period the heart is in diastole and the muscle is resting. The curve AB, therefore, should be the same as the pressure–volume (P–V) curve of a resting heart. Some deviation of the arc AB from the P–V curve of a resting heart occurs at the corners A and B owing to the dynamic events in the heart. The mitral valve opens at A; it closes at B.

Points on the straight line segment BC represent the isovolumetric systolic phase of the cardiac cycle. In this phase, both the mitral and aortic valves are closed and the blood volume of the left ventricle remains con-

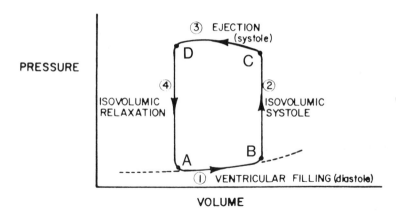

FIGURE 2.4:4. A plot of the pressure and volume in the left ventricle during a cardiac cycle. Phase 1 represents ventricular filling during diastole. Phase 2 represents isovolumic pressure development. Phase 3 represents ejection of blood into the aorta. Phase 4 represents the isovolumic relaxation period.

stant. Tension develops in the heart muscle, and the blood pressure in the left ventricle rises. At the point C, the aortic valve opens. Points on the curve CD represent the ejection phase of the cardiac cycle. Blood is ejected from the left ventricle into the aorta. At the point D, the aortic valve closes. Then the muscle tension decreases while both the aortic and mitral valves are closed. The relaxation is again isovolumetric.

Recalling Figure 2.1:3, one may ask whether the point D in Figure 2.4:4 corresponds to the maximum isometric tension in the heart muscle at the volume specified. To answer this question, Suga et al. (1973) made an experiment on a dog's heart in which the aorta was transiently occluded at specified volumes of the heart and the heart muscle was allowed to go through cycles of contraction during which the heart could develop pressure but not eject blood. The maximum and minimum pressures developed in such cycles were noted, and were plotted against the volume. Such a curve of the maximum pressure versus volume is called an *isovolumic pressure curve*, and was found to be approximately a straight line. When the isovolumic pressure line was plotted onto the P–V curves of Figure 2.4:4, it was found that the corner D fell in the neighborhood of the isovolumic pressure line, as shown in Figure 2.4:5. Thus, the point D does seem to correspond to the maximum isometric tension of heart muscle at the volume specified. Of course, we should object to the use of the word *isometric*, because isovolumic contraction does not guarantee constant length of the muscle fibers. Indeed, shape change of the ventricles in the isovolumic contraction period, as shown in Figure 2.4:1, implies the lack of isometry in the muscle fibers. But, roughly, isovolumic condition is an approximation to isometric condition.

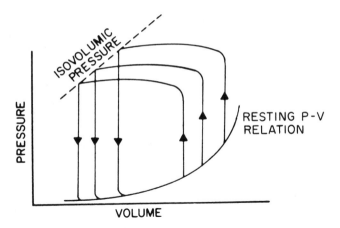

FIGURE 2.4:5. Relationship between the "isovolumic pressure line" and the cardiac cycle. The isovolumic pressure line represents the capacity of the ventricle to develop pressure at each initial end-diastolic volume if it were not allowed to eject blood. The three ejection cycles illustrated represent cardiac contractions at different initial end-diastolic volumes and aortic pressures.

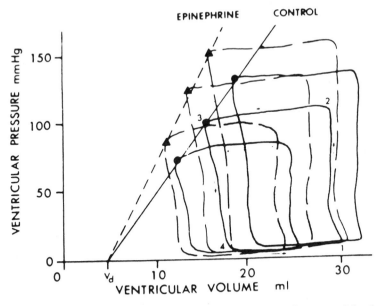

FIGURE 2.4:6. Left ventricular pressure–volume loops from a denervated dog heart. Mean aortic pressure was fixed at three different levels, while cardiac output was kept constant during both the control (solid loops) and enhanced (2 μg/kg/min epinephrine infusion) (broken loops) contractile states. The upper left-hand corners of the pressure–volume loops form a line that intersects the volume axis at a point V_d of about 5 ml. From Suga et al. (1973), by permission.

The relationship plotted in Figure 2.4:5 suggests that the upper left-hand corner of each loop approximates the isovolumic pressure line, independent of the initial end-diastolic volume and the aortic pressure against which the heart is working. The isovolumic pressure line, when extrapolated to the volume axis, intercepts that axis at approximately 5 ml (for the dog). If the contractility of the ventricle is increased by an inotropic drug such as epinephrine, the isovolumic pressure line is shifted upward and to the left, although the volume intercept remains approximately the same, as shown in Figure 2.4:6. Thus, under these circumstances, an increase in stroke volume can be produced at the same end-diastolic pressure and arterial pressure, since the end-systolic point of contraction will approximate the new isovolumic pressure line. A reverse situation would occur during the process of heart failure, when the isovolumic pressure line would be shifted downward and to the right. Much more information can be found in Parmley and Talbot (1979).

2.5 How Are the Heart Valves Operated?

The most interesting fluid-dynamical events in the heart are the filling and ejecting of blood in the ventricles, which are associated with the motion of the heart valves. The human heart has four valves: the tricuspid (T), pulmonic (P), aortic (AO), and mitral (M) (see Fig. 2.2:1). They lie essentially in a plane.

The aortic valve consists of three thin, crescent-shaped cusps (thus the name *semilunar*), which in the open position are displaced outward toward the aorta. In the closed position the three cusps come together to seal the aortic orifice. Behind the cusps there are outpouchings of the aortic root, called the *sinuses of Valsalva*, which play a role in the closure of the valve. The pulmonic valve has a similar structure.

The mitral valve consists of two thin membranous cusps of roughly trapezoidal shape that originate from the slightly elliptical mitral ring. In the open position these membranes form a scalloped, conelike structure. The distal margins of the two cusps have an irregular appearance because they are pulled by the chordae tendinea, which originate from the papillary muscle of the ventricular wall. The cusp adjacent to the aortic valve is called the *anterior* or *aortic cusp*, the other is called the *posterior* or *mural cusp*. In closed condition the free edges of the cusps are pressed together. The tricuspid valve structure is similar, except that it has three cusps.

The muscle bundles of the heart are organized into a unified whole, and one part cannot move without affecting other parts. The aortic and mitral valves, covering the base of the left ventricle, are an integral unit. Opening of the mitral valve occurs when the pressure above the aortic valve is low, so an enlargement of the mitral valve orifice coincides with a reduction of the orifice of the aortic valve. On the other hand, during systole, when the aortic valve is opened and its orifice is distended, the mitral valve is closed and reduced in area. The principle of opening and closing of the valves has been discussed in Section 1.5 of the preceding chapter. We now review the mechanism in greater detail.

The history of heart research must mention Leonardo da Vinci. Leonardo (1452–1519) was born 126 years before Harvey (1578–1657). He left 20 beautiful paintings and 5,000 pages of notes, including hundreds of anatomical drawings. His drawings of coronary blood vessels and heart valves are accurate. His studies of fluid motion, waves, vortices, and circulation were original. It is hard to believe that he did not see the function of the heart as a pump. Nevertheless, the first correct fluid mechanial interpretation of the mechanism of closing of the heart valves was probably that of Henderson and Johnson (1912). They performed several experiments. In the first experiment they used a tube connected to a reservoir of dye and dipped into a tank of water [Fig. 2.5:1(a)]. When they stopped the flow [Fig. 2.5:1(b)], the jet was seen to conserve its forward motion, while clear water was drawn into the wake of the jet in the vicinity of the tube opening. In the second experiment a section of curved tube was attached to the midportion of a straight tube to form a "D" configuration [Figure 2.5:2]. A mem-

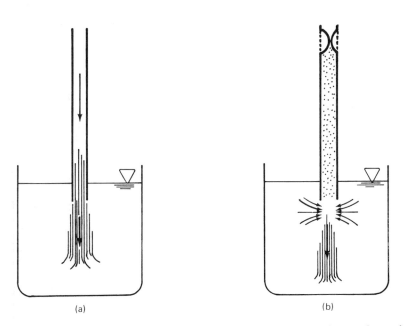

(a)   (b)

FIGURE 2.5:1. Schematic of Henderson and Johnson's first experiment drawn by Parmley and Talbot (1979). (a) A tube feeds a jet flowing into a cup of water. (b) Tube flow stopped. Jet breaks away. Fluid rushes in toward the end of the tube.

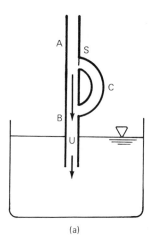
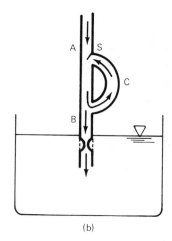

FIGURE 2.5:2. Henderson and Johnson's second experiment. A flexible valve is installed at the D-shaped side tube at point S. (a) Flow through. Valve does not move. (b) Flow interrupted. Valve moves to left. See text for explanation.

brane valve was installed at a point marked S. Flow directed down the straight tube into a water-filled tank caused little motion in the curved part of the D as shown in Figure 2.5:2(a). The flow was then halted by occluding the tube, thereupon the fluid began to circulate in the curved portion of D, as indicated in Figure 2.5:2(b), and caused the valve at S to close against the forward flow coming from A.

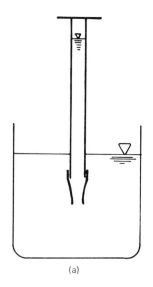
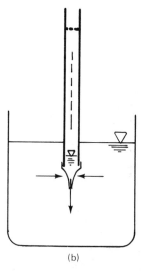

FIGURE 2.5:3. Henderson and Johnson's third experiment. A flexible sleeve is installed at the end of the verticle tube. (a) Downward flow in tube. (b) Sleeve closes when the water level in the tube is lower than that in the cup. See text.

In another of their experiments, a glass tube fitted with a flexible rubber sleeve was dipped into a tank of water (Fig. 2.5:3) and the column of water within the tube was raised to a level above that of the tank. Upon release of the column of fluid, it was observed that when the level in the tube fell below that of the tank, the fluid within the tank moved inward to collapse the sleeve and seal the tube.

Lee and Talbot (1979) explained Henderson and Johnson's experiment shown in Figure 2.5:2 as follows: Consider the phenomenon described in Figure 2.5:2. On assuming that the flow velocity U is uniform across the tube between A and B, and that there is initially little flow in the curved section C, and neglecting gravity and viscous effects, the equation of motion for the fluid contained in AB is

$$\frac{\partial U}{\partial t} + U\frac{\partial U}{\partial x} = -\frac{1}{\rho}\frac{\partial p}{\partial x}, \tag{1}$$

where U is the velocity in the x (flow) direction, ρ is the fluid density, p is the fluid pressure, and t is time. However, since the section AB is of constant area, there is no velocity gradient along the tube (according to the principle of conservation of mass), and the convective acceleration $U(\partial U/\partial x)$ vanishes. We are thus left with

$$\frac{\partial U}{\partial t} = -\frac{1}{\rho}\frac{\partial p}{\partial x}. \tag{2}$$

Now, when the flow in AB is caused to decelerate, $\partial U/\partial t$ takes on a negative value. Accordingly, the pressure gradient $\partial p/\partial x$ must take on a positive value. Thus, the pressure at B becomes higher than that at A. This causes the fluid initially at rest in the curved section C to be set in motion, in the direction shown in Figure 2.5:2. Other experiments sketched in Figures 2.5:1 and 2.5:3 can be explained similarly.

Operation of the Aortic Valve

An aortic valve with the sinus of Valsalva behind it is sketched in Figure 2.5:4. According to model experiments by Bellhouse and Bellhouse (1969, 1972), the flow issuing from the ventricle immediately upon opening of the valve at the inception of systole is split into two streams at each valve cusp, as shown in the figure. Part of the flow is directed into the sinus behind the valve cusp, where it forms a vortical flow before re-emerging, out of the plane of the figure, to rejoin the main stream in the ascending aorta.

When the aortic pressure rises sufficiently so that deceleration of the flow occurs, an adverse pressure gradient is produced—p_2 at the valve cusp tip exceeds the pressure p_1 at a station upstream. The higher pressure p_2 causes a greater flow into the sinus, which carries the cusp toward apposition. The peak deceleration occurs just before the valve closure. The vortical motion established earlier upon the opening of the valve has the merit of prevent-

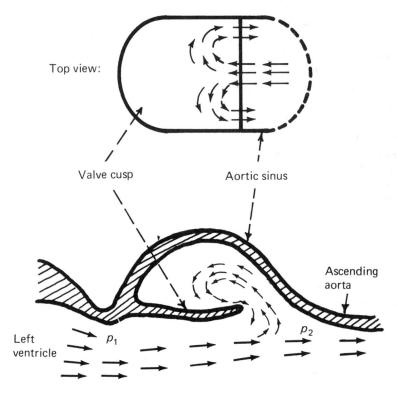

Top view:

Valve cusp   Aortic sinus

Ascending
aorta

Left
ventricle p_1 p_2

FIGURE 2.5:4. Flow pattern within the sinus of Valsalva.

ing the valve cusp from bulging outward to contact the walls of the sinuses.
The open sinus chamber thus can be supplied with fluid to fill the increasing volume behind the valve cusps as they move toward closure.

Operation of the Mitral Valve

The flow through the mitral valve can be illustrated by the photographs of
a model by Lee and Talbot (1979) shown in Figure 2.5:5. The model consists of two freely hinged rigid cusps, with a simulated flexible ventricle and
an atrium. In Figure 2.5:5(a), at $t = 450$ ms after the onset of diastole, the
valve has opened. The fluid motion in the ventricle is predominantly
outward. At $t = 800$ ms, Fig 2.5:5(b), a vortex motion can be seen behind the
cusps. At $t = 1,100$ ms and $1,600$ ms [Fig. 2.5:5(c) and (d)] the fully developed diastolic flow pattern is seen, which continues until the period of active
diastole terminates at $t = 1,650$ ms. It can be seen that the vortices behind
the valve cusps have decayed in strength, and there is very little motion in
this region. In Figure 2.5:5(e), at $t = 1,800$ ms an early stage of valve closure
is seen. A strong jet through the valve is still evident, although the fluid in
the upper portion of the valve is moving much less rapidly. At $t = 1,850$ ms

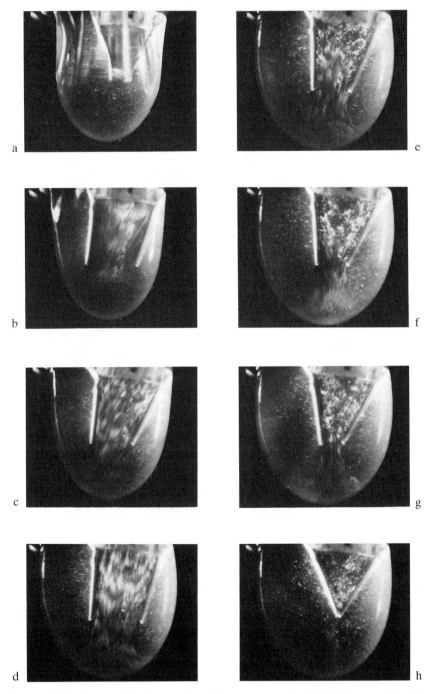

FIGURE 2.5:5. Sequential photographs of flow patterns inside model ventricle with mitral valve. (a) $t = 450\,ms$, (b) $t = 800\,ms$, (c) $t = 1,100\,ms$, (d) $t = 1,600\,ms$, (e) $t = 1,800\,ms$, (f) $t = 1,850\,ms$, (g) $t = 1,900\,ms$, (h) $t = 1,950\,ms$. Time t was measured with respect to onset of ventricular diastole. The flow pattern was made visible by hydrogen bubbles evolved from fine wires. From Lee and Talbot (1979), by permission.

47

[Fig. 2.5:5(f)] a stagnation point has formed in the flow within the valve, and the fluid in the upper portion of the valve has begun to move in the reverse direction. We see here the "breaking of the jet" mentioned earlier. Also, a circulating motion with the ventricle behind the valve cusps is evident. At $t = 1,900$ ms [Fig. 2.5:5(g)], although the valve is nearly closed, there is still a narrow jet issuing from the valve, and the circulatory motion within the ventricles has become stronger. Closure is complete at $t = 1,950$ ms, 50 ms before the onset of ventricle systole.

Lee and Talbot's model uses water as the circulating fluid, and the *Reynolds number* and *Strouhal number* of the heart are simulated. The Strouhal number is defined as $L/(U_m T)$, where L is the mitral valve cusp length, U_m is the maximum velocity of flow through the mitral valve, and T is the time interval of one heartbeat. It is also called the *reduced frequency*. The Reynolds number and Strouhal number together define the dynamic similarity for model testing.

The flow patterns in Figure 2.5:5 show that the deceleration of the jet and the associated adverse pressure gradient are the mechanism responsible for valve closure during diastole. The deceleration causes the pressure within the valve to fall below that of the surrounding fluid in the ventricle, and an inward motion of the ventricular fluid results, which closes the valve.

Based on these observations, Lee and Talbot formulated a mathematical theory to calculate the valve cusp motion from a knowledge of the velocity–time history of the flow through the valve, under the assumptions that the valve cusps are massless and passive, and the velocity in the valve cross section is uniform. Using the theory in reverse, they also showed how to calculate the velocity of flow from the known motion of the valve cusps. This theory is useful because valve cusp motion can be measured non-invasively by echocardiography.

What Are the Papillary Muscles for?

The operation of the mitral and tricuspid valves does not need the help of the papillary muscles, although these muscles do pull on the edges of the membranes of these valves (see Figs. 2.2:1 and 2.3:1). In the human heart the papillary muscles constitute about 10% of the total heart mass. What are they for?

To understand the function of the papillary muscles we should go back to the law of Laplace (see Fung, 1993b, p. 14), which states that for a curved surface with principal curvatures $1/r_1$ and $1/r_2$, and principal membrane stress resultants T_1 and T_2, resisting an internal pressure p_i and an external pressure p_e, the equation of equilibrium is

$$p_i - p_0 = \frac{T_1}{r_1} + \frac{T_2}{r_2}. \tag{3}$$

Now, during isovolumic contraction the papillary muscles are not restrained in length, and they can shorten, pulling on the valve membrane, increasing its tension T_1, T_2 in the direction of the papillary muscles, and decreasing its radii of curvature r_1, r_2, so that the ventricular pressure p_i is increased. Thus the papillary muscle may be regarded as the controller of the ventricular pressure when the mitral or tricuspid valve is closed. When the membranes touch and the valves are closed, the stress and strain in the membrane are supported by the papillary muscles. These muscles prevent the membranes from inverting like an umbrella in a strong wind. Hence the papillary muscles function as structural extensions of the membranes.

2.6 Equations of Heart Mechanics

Basic equations of heart mechanics are assembled below. These equations are needed for the analysis of stress and strain in the heart muscles, flow of blood in the heart chambers and coronary blood vessels, flow of interstitial fluid and lymph, motion and stress in heart valves, nervous and electric control of the heart, heat transfer, homeostatic and disturbed conditions, tissue growth and remodeling, and disease states. In this brief presentation, it is assumed that the reader is familiar with the elements of continuum mechanics at the level of the book *First Course in Continuum Mechanics* (Fung, 1993a). In particular, it is assumed that the reader has thought about the applicability of the concept of continuum to complex tissues such as the heart muscle. In Sections 1.5–1.10 of the above mentioned reference, the hierarchy of the structure of the living tissues at different scales of magnification (observations made by bare eyes, or optical microscopes, or electron microscopes) are discussed. We define a series of mathematical copies of the tissue according to the hierarchy of the structure, and obtain a hierarchy of continuum models. Each model is specified by a minimum linear scale (or the largest magnification). The materials' mass per unit volume (density), force per unit area (stress), and deformation measures (displacements, deformation gradients, and strains) are defined in scales above the assumed minimum, with acceptable variabilities. The laws of conservation of mass, momentum, and energy, and thermodynamic laws are then expressed in the form of equations governing the mathematical model. When these equations are applied to real materials, they are understood to be stochastic, valid only within the limits defined by the acceptable variability of the basic concepts of density, stress and strain.

Eulerian and Lagrangian Descriptions

Let (x, x_2, x_3) or (x, y, z) be a rectangular Cartesian frame of reference fixed in space. The motion of a fluid in this space is described by the velocity vector $\mathbf{v}$ (with components v_1, v_2, v_3 or u, v, w in the directions of coordinate axes x_1, x_2, x_3 respectively) as a function of x_1, x_2, x_3. The deformation of a solid in this space is described by the displacement of every material particle. If the particle is located at **X** (with components X_1, X_2, X_3) at time $t = 0$, and moved to **x** (with components x_1, x_2, x_3) at time t, then the displacement vector is **u** = **x** − **X** which is function of (X_1, X_2, X_3). These two descriptions, **v**(**x**) and **u**(**X**), are called Eulerian and Lagrangian descriptions, respectively. They can be used for both fluids and solids.

Conservation of Mass, Momentum, and Energy

The index notation is used below. Unless stated otherwise, all indices range over 1, 2, 3. Thus, the stress tensor σ_{ij} has components ($\sigma_{11}, \sigma_{12}, \sigma_{13}, \sigma_{21}, \sigma_{22}, \sigma_{23}, \sigma_{31}, \sigma_{32}, \sigma_{22}$). The summation convention is used: repetition of an index means summation over 1, 2, 3.

Consider the integral $I(t)$

$$I(t) = \int_V A(\mathbf{x}, t)dV, \tag{1}$$

where $A(\mathbf{x}, t)$ is a continuously differentiable function of spatial coordinates $\mathbf{x}(x_1, x_2, x_3)$ and time t, V is a region which is bounded by a surface S that consists of a finite number of parts whose field of unit outward pointing normal vectors $\mathbf{v}(v_1, v_2, v_3)$ are continuous. V is occupied by a given set of material particles. The integration is extended over a given set of particles, so that when time changes and the particles change their position, the spatial region V changes with them. The rate at which $I(t)$ changes with respect to t is defined as the *material derivative of I* and is denoted by DI/Dt:

$$\frac{DI}{Dt} = \lim_{dt \to 0} \frac{1}{dt} \left[\int_{V'} A(\mathbf{x}, t + dt)dV - \int_V A(\mathbf{x}, t)dV \right], \tag{2}$$

where V' is the volume occupied by the same set of particles at time $t + dt$. By means of the Gauss theorem which states that

$$\int_V \frac{\partial A}{\partial x_i} dV = \int_S A v_i dS \qquad (i = 1, 2, 3), \tag{3}$$

where A is a scalar, a vector, or a tensor, continuously differentiable in a convex region V which is bounded by a surface S whose unit outer normal is $\mathbf{v}(v_1, v_2, v_3)$, we can derive the following (see Fung, 1993a, p. 217)

$$\frac{D}{Dt} \int_V A dV = \int_V \frac{\partial A}{\partial t} dV + \int_S A v_j v_j dS$$

$$= \int_V \left(\frac{\partial A}{\partial t} + v_j \frac{\partial A}{\partial x_j} + A \frac{\partial v_j}{\partial x_j} \right) dV$$

$$= \int_V \left(\frac{DA}{Dt} + A \frac{\partial v_j}{\partial x_j} \right) dV, \tag{4}$$

where

$$\frac{DA}{Dt} = \left(\frac{\partial A}{\partial t}\right)_{\mathbf{x}=\text{const.}} + v_1 \frac{\partial A}{\partial x_1} + v_2 \frac{\partial A}{\partial x_2} + v_3 \frac{\partial A}{\partial x_3}, \tag{5}$$

is the *material derivative of A*; i.e., the rate at which the quantity A associated with a particle is seen changing as the particle moves about in a velocity field. v_i is the velocity field, of course.

If A in the equation above is the velocity vector with components u, v, w, then DA/Dt is the acceleration vector, and $\partial A/\partial t$ is the transient acceleration, whereas the last three terms in Eq. (5) are the convective acceleration. Thus, the components of transient acceleration are

$$\frac{\partial u}{\partial t}, \quad \frac{\partial v}{\partial t}, \quad \frac{\partial w}{\partial t};$$

whereas the three components of the convective acceleration are

$$u\frac{\partial u}{\partial x} + v\frac{\partial u}{\partial y} + w\frac{\partial u}{\partial z}, \quad u\frac{\partial v}{\partial x} + v\frac{\partial v}{\partial y} + w\frac{\partial v}{\partial z}, \quad u\frac{\partial w}{\partial x} + v\frac{\partial w}{\partial y} + w\frac{\partial w}{\partial z}.$$

Consider the mass of a material of density $\rho(\mathbf{x})$ in a region V:

$$M = \int_V \rho dV. \tag{6}$$

Refer to Eq. (1), identify A with ρ and I with M. The law of conservation of mass states that $DM/Dt = 0$ if the group of particles within V is fixed. Applying Eq. (4) we obtain the following *equation of continuity*, which expresses the law of conservation of mass:

$$\frac{\partial \rho}{\partial t} + \frac{\partial \rho v_j}{\partial x_j} = 0. \tag{7}$$

If the fluid density ρ is a constant, then the material is said to be *incompressible*. The *equation of continuity of an incompressible fluid* is

$$\frac{\partial v_j}{\partial x_j} = 0 \quad \text{or} \quad \frac{\partial u}{\partial x} + \frac{\partial v}{\partial y} + \frac{\partial w}{\partial z} = 0. \tag{8}$$

In the heart chambers and blood vessels, blood is incompressible and Eq. (8) applies. In the walls of the heart, blood in the coronary vessels cannot be restricted within a boundary surface (S) of a region V. If $\dot{m}$ is the rate at which the coronary blood increases in the ventricular wall per unit wall volume per unit time, then, in the ventricular wall, the equation of continuity is

$$\frac{\partial \rho}{\partial t} + \frac{\partial \rho v_j}{\partial x_j} = \dot{m} \tag{9}$$

A similar situation exists in the blood vessel wall, in which the flow in the vasa vasorum cannot be constrained locally.

The momentum of a material of density ρ and velocity $\mathbf{v}$ in V is:

$$\mathcal{P}_i = \int_V \rho v_i dV. \tag{10}$$

Newton's law of motion states that the material rate of change of momentum is equal to the force acting on the body, $\mathbf{F}(F_1, F_2, F_3)$:

$$\frac{D}{Dt}\mathcal{P}_i = F_i. \tag{11}$$

If the body is subjected to surface tractions $\overset{v}{T_i}$ and body force per unit volume X_i, the resultant force acting on the body is

$$F_i = \int_S \overset{v}{T_i}\, dS + \int_V X_i dV. \tag{12}$$

Now, if σ_{ij} is the stress tensor and v_i is the unit outer normal of S, then $\overset{v}{T_i} = \sigma_{ji}v_j$, and Eqs. (1)–(4), and (10)–(12) yield the *Euler's equation of motion*

$$\rho\frac{Dv_i}{Dt} = \frac{\partial\sigma_{ij}}{\partial x_j} + X_i. \tag{13}$$

If there is no heat input into the system and no heat source in the region, the equation for balance of energy is the same as balance of mechanical energy and work: it leads to no new independent equation.

When these equations are combined with the constitutive equation of the material, we obtain the basic equations of mechanics. We shall illustrate this by several examples below.

Navier–Stokes Equations for an Incompressible Newtonian Fluid

Consider an incompressible, Newtonian viscous fluid. Let us use the same notations for pressure, velocity, and coordinates for position as in the previous section. Let μ be the coefficient of viscosity which is a constant for a Newtonian fluid. Then the stress–strain rate relationship is given by

$$\sigma_{ij} = -p\delta_{ij} + \lambda V_{kk}\delta_{ij} + 2\mu V_{ij} \tag{14}$$

where σ_{ij} is the stress tensor, δ_{ij} is Kronecker delta, p is pressure,

$$V_{ij} = \frac{1}{2}\left(\frac{\partial v_i}{\partial x_j} + \frac{\partial v_j}{\partial x_i}\right) \tag{15}$$

is the *strain rate* tensor, and λ and μ are two material constants. Since the fluid is assumed to be incompressible, the condition of incompressibility

$$\frac{\partial v_i}{\partial x_i} = 0 \quad \text{or} \quad \frac{\partial u}{\partial x} + \frac{\partial v}{\partial y} + \frac{\partial w}{\partial z} = 0 \tag{16}$$

reduces Eq. (14) to the form

$$\sigma_{ij} = -p\delta_{ij} + 2\mu V_{ij}. \tag{17}$$

Substituting these into Eq. (13), we obtain the Navier–Stokes equations

$$\frac{\partial u}{\partial t} + u\frac{\partial u}{\partial x} + v\frac{\partial u}{\partial y} + w\frac{\partial u}{\partial z} = X - \frac{1}{\rho}\frac{\partial p}{\partial x} + \frac{\mu}{\rho}\nabla^2 u, \tag{18a}$$

$$\frac{\partial v}{\partial t} + u\frac{\partial v}{\partial x} + v\frac{\partial v}{\partial y} + w\frac{\partial v}{\partial z} = Y - \frac{1}{\rho}\frac{\partial p}{\partial y} + \frac{\mu}{\rho}\nabla^2 v, \tag{18b}$$

$$\frac{\partial w}{\partial t} + u\frac{\partial w}{\partial x} + v\frac{\partial w}{\partial y} + w\frac{\partial w}{\partial z} = Z - \frac{1}{\rho}\frac{\partial p}{\partial z} + \frac{\mu}{\rho}\nabla^2 w, \tag{18c}$$

The ratio μ/ρ is the *kinematic viscosity* of the fluid, and ∇^2 is the *Laplacian operator*

$$\nabla^2 = \frac{\partial^2}{\partial x^2} + \frac{\partial^2}{\partial y^2} + \frac{\partial^2}{\partial z^2}. \tag{19}$$

Equations (16) and (18) comprise four equations for the four variables u, v, w, and p occurring in an incompressible viscous flow. To solve these equations, we need to specify appropriate boundary conditions. *If the fluid is in contact with a solid, the boundary condition is that there is no relative motion between the solid and the fluid. The fluid adheres to the solid, whether the surface is wettable or not.* The justification of this condition is discussed at some length in Fung (1993a, p. 233). If the fluid is in contact with another fluid, then at the interface the boundary conditions must be consistent with the *interfacial surface tension, surface viscosity*, and *conditions of cavitation* or its absence. See Fung (1993a, p. 235).

Navier–Stokes equations have applications to blood flow when the non-Newtonian features of blood can be ignored and when the blood vessel diameter is much larger than the dimensions of individual blood cells. These equations are not accurate enough for flow in the neighborhood of a stagnation point, or flow in a capillary blood vessel, or flow with blood clots.

Effect of Blood Rheology

Blood is a non-Newtonian incompressible viscoplastic fluid. As it is discussed in *Biomechanics: Mechanical Properties of Living Tissues* (Fung, 1993b), blood viscosity can be described in three regimes:

(a) Elastic regime, blood not flowing. This regime is defined by the yield condition. Since yielding is due to distortion and is unaffected by pressure or mean stress, we make use of the *stress deviation* tensor $\boldsymbol{\sigma}'$ with components σ'_{ij} defined by the equation

$$\sigma'_{ij} = \sigma_{ij} - \tfrac{1}{3}\sigma_{kk}\delta_{ij} \tag{20}$$

to describe the yield condition. The mean stress of σ'_{ij} is zero. The yield condition is stated in terms of the second invariant of the stress deviation tensor:

$$J'_2(\boldsymbol{\sigma}') = \tfrac{1}{2}\sigma'_{ij}\sigma'_{ij}. \tag{21}$$

The material yields if

$$J'_2(\boldsymbol{\sigma}') = K. \tag{22}$$

It remains elastic if

$$J'_2(\boldsymbol{\sigma}') < K. \tag{23}$$

For blood, the value of K is of the order of $4 \times 10^{-4}\,\mathrm{dyne^2/cm^4}$ or 4×10^{-6} $\mathrm{N^2 m^{-4}}$, with exact number depending on the hematocrit. When Eq. (23) applies, blood obeys Hooke's law.

(b) If $J'_2(\boldsymbol{\sigma}') \geq K$, then flow ensues, and the second invariant of the strain rate tensor $J_2(v) \neq 0$. If $J_2(v) > c$, a certain constant which depends on the hematocrit, then blood obeys the Newtonian viscosity law, Eq. (14).

Here

$$J_2(v) = \tfrac{1}{2}V_{ij}V_{ij}, \tag{24}$$

$$V_{ij} = \frac{1}{2}\left(\frac{\partial v_i}{\partial x_j} + \frac{\partial v_j}{\partial x_i}\right). \tag{25}$$

(c) If $J'_2(\boldsymbol{\sigma}') \geq K$, and $J_2(v) \leq c$, then blood obeys the following constitutive equation:

$$\sigma_{ij} = -p\delta_{ij} + \mu(J_2)\left(\frac{\partial v_i}{\partial x_j} + \frac{\partial v_j}{\partial x_i}\right). \tag{26}$$

where

$$\mu(J_2) = \left[\left(\eta^2 J_2\right)^{1/4} + 2^{-1/2}\tau_y^{1/2}\right]^2 J_2^{-1/2} \tag{27}$$

and η and τ_y are constants known as the *Casson viscosity* and *yielding stress*, respectively. Hence, in the flow regime, we obtain, on substituting Eqs (14), and (24)–(27) into Eq. (13),

$$\rho\frac{Dv_i}{Dt} = X_i - \frac{\partial p}{\partial x_i} + \frac{\partial}{\partial x_k}\left(\mu\frac{\partial v_k}{\partial x_i}\right) + \frac{\partial}{\partial x_k}\left(\mu\frac{\partial v_i}{\partial x_k}\right). \tag{28}$$

Differentiating and using Eq. (8) under the assumption that blood is incompressible, we obtain

$$\rho \frac{Dv_i}{Dt} = X_i - \frac{\partial p}{\partial x_i} + \mu \frac{\partial^2 v_i}{\partial x_k \partial x_k} + \frac{\partial \mu(J_2)}{\partial x_k}\left[\frac{\partial v_k}{\partial x_i} + \frac{\partial v_i}{\partial x_k}\right].$$ (29)

Equations (8) and (29) comprise four equations for the four variables u, v, w, and p. What are the boundary conditions? In blood vessels, the endothelium of the blood vessel may be regarded as a solid surface, and the boundary conditions are *no-slip* and *continuity*, i.e., *there are no relative tangential and normal velocities between the blood and the wall*. The normal velocities of the fluid and wall are the same because the fluid and the wall move together. The no-slip condition is gathered from general experience of fluid mechanics.

Navier's Equation for an Isotropic Hookean Elastic Solid

If the material is an isotropic elastic body obeying Hooke's law, the stress–strain relationship can be expressed either as

$$\sigma_{ij} = \lambda e_{\alpha\alpha} \delta_{ij} + 2G e_{ij}$$ (30)

or as

$$e_{ij} = \frac{1+\nu}{E}\sigma_{ij} - \frac{\nu}{E}\sigma_{\alpha\alpha}\delta_{ij}.$$ (31)

Here e_{ij} ($i, j = 1, 2, 3$) is the strain tensor, σ_{ij} is the stress tensor, λ, G, E, ν are elastic constants. λ and G and called *Lamé's constants*, G is called the *shear modulus* or the *modulus of rigidity*, E is called the *Young's modulus* or *modulus of elasticity*, and ν is called the *Poisson's ratio*.

On substituting Eq. (30) into the equation of motion (13), we obtain

$$\rho \frac{Dv_i}{Dt} = \lambda \frac{\partial}{\partial x_i} e_{\alpha\alpha} + 2G \frac{\partial e_{ij}}{\partial x_j} + X_i.$$ (32)

To proceed further we need to express Dv_i/Dt and e_{ij} in terms of the displacements of material particles. An elastic body has a "natural" state of zero stress and zero strain. We measure elastic displacements of every point in the body relative to the natural state with respect to a set of inertial, rectangular, Cartesian coordinates. Let the displacement of a point located at x_i at time t be $u_i(x_1, x_2, x_3, t)$, $i = 1, 2, 3$. The acceleration Dv_i/Dt and the strain e_{ij} can be expressed in terms of u_i, but the expressions are non-linear and complex if u_i is finite and if the velocity is high. Simplicity can be achieved if we can assume u_i to be infinitesimal. Since there is a large class of important practical problems in which this infinitesimal displacement assumption holds well, we shall use this assumption to obtain some simple results.

If $u_i(x_1, x_2, x_3, t)$ is infinitesimal, then, on neglecting small quantities of higher order, we have

$$e_{ij} = \frac{1}{2}\left(\frac{\partial u_i}{\partial x_j} + \frac{\partial u_j}{\partial x_i}\right), \tag{33}$$

$$v_i = \frac{\partial u_i}{\partial t}, \quad \frac{Dv_i}{Dt} = \frac{\partial^2 u_i}{\partial t^2}. \tag{34}$$

To the same order of approximation, the material density is a constant:

$$\rho = \text{const.} \tag{35}$$

On substituting Eqs. (33) and (34) into (32), we obtain the well-known *Navier's equation*:

$$G\nabla^2 u_i + (\lambda + G)\frac{\partial e}{\partial x_i} + X_i = \rho\frac{\partial^2 u_i}{\partial t^2}, \tag{36}$$

where e is the divergence of the displacement vector u_i:

$$e = \frac{\partial u_j}{\partial x_j} = \frac{\partial u_i}{\partial x_i} + \frac{\partial u_2}{\partial x_2} + \frac{\partial u_3}{\partial x_3}. \tag{37}$$

∇^2 is the Laplace operator, see Eq. (19).

If we introduce the Poisson's ratio as in Eq. (31),

$$v = \frac{\lambda}{2(\lambda + G)}, \tag{38}$$

we can write Navier's equation (26) as

$$G\left(\nabla^2 u_i + \frac{1}{1 - 2v}\frac{\partial e}{\partial x_i}\right) + X_i = \rho\frac{\partial^2 u_i}{\partial t^2}. \tag{39}$$

Navier's equation is the basic field equation of the linearized theory of elasticity. It must be solved with appropriate initial and boundary conditions.

Unfortunately, most biological soft tissues are subject to finite deformation in normal function. Therefore, very often the linearization cannot be justified. In heart mechanics, Navier's equations are used only for preliminary studies because Hooke's law is not sufficiently accurate to represent the mechanical properties of the myocardium.

More Realistic Constitutive Equations

For further progress we need the stress–strain–strain rate-history relationships of the materials involved, that is, the constitutive equations of living tissues.

First, let us consider deformations large enough so that the strains are not infinitesimal. When deformation is large, we must distinguish the location of material particles before and after deformation, and define finite strain and stress components in a manner consistent with tensor analysis and physics. There are at least three definitions of strain tensors: Cauchy's, Almansi's, and Green's strains; and three finite stress tensors: Cauchy's, Lagrange's, and Kirchhoff's. I believe that the simplest introduction to the subject is given in *Biomechanics: Motion, Flow, Stress and Growth*, Chapter 10, (Fung, 1990). More advanced treatises are listed in the above reference. It is known that when the coordinates of material particles are denoted by (a_1, a_2, a_3) initially, and (x_1, x_2, x_3) after deformation, the Green's strain is defined as

$$
\begin{aligned}
E_{ij} &= \frac{1}{2}\left(\delta_{\alpha\beta} \frac{\partial x_\alpha}{\partial a_i} \frac{\partial x_\beta}{\partial a_j} - \delta_{ij} \right) \\
&= \frac{1}{2}\left(\frac{\partial u_j}{\partial a_i} + \frac{\partial u_i}{\partial a_j} + \frac{\partial u_\alpha}{\partial a_i} \frac{\partial u_\alpha}{\partial a_j} \right),
\end{aligned}
\tag{40}
$$

where δ_{ij} is the Kronecker delta. The displacement vector is

$$
u_i = x_i - a_i.
\tag{41}
$$

If the material is elastic or pseudoelastic in finite deformation, a strain energy function $\rho_o W(E_{11}, E_{12}, \ldots)$ exists whose derivative yields the Kirchhoff's stress tensor S_{ij}:

$$
S_{ij} = \frac{\partial (\rho_o W)}{\partial E_{ij}}.
\tag{42}
$$

The Kirchhoff stress S_{ij} is related to the Cauchy stress σ_{ij} by the relation

$$
\sigma_{ij} = \frac{\rho}{\rho_0}\left[S_{ij} + \left(\delta_{i\beta} \frac{\partial u_j}{\partial a_\alpha} + \delta_{j\alpha} \frac{\partial u_i}{\partial a_\beta} + \frac{\partial u_i}{\partial a_\alpha} \frac{\partial u_j}{\partial a_\beta} \right) S_{\alpha\beta} \right].
\tag{43}
$$

Here ρ and ρ_0 are the density of the material in the deformed and initial states, respectively. The equation of motion Eq. (13) is valid with σ_{ij} representing Cauchy stress:

$$
\rho \frac{Dv_i}{Dt} = \frac{\partial \sigma_{ij}}{\partial x_j} + X_i.
\tag{44}
$$

The particle acceleration is given by the material derivative of the velocity:

$$
\frac{Dv_i}{Dt} = \frac{\partial v_i}{\partial t} + v_j \frac{\partial v_i}{\partial x_j}.
\tag{45}
$$

The particle velocity is given by the material derivative of the displacement:

$$
v_i = \frac{\partial u_i}{\partial t} + v_j \frac{\partial u_i}{\partial x_j}.
\tag{46}
$$

Taking Viscoelasticity into Account

Most linearly viscoelastic bodies have the following constitutive equation:

$$\sigma_{ij}(\mathbf{x},t) = \int_{-\infty}^{t} G_{ijkl}(t-\tau)\frac{\partial}{\partial \tau}E_{kl}(\mathbf{x},\tau)d\tau, \tag{47}$$

where G_{ijkl} is the tensor of relaxation functions, σ_{ij} is the stress tensor, E_{kl} is the strain tensor, t and τ are time, and the integration over τ is from the beginning of motion to the time t. The equation of motion is obtained by substituting Eq. (47) into Eq. (13).

Biological materials usually have very complex viscoelastic properties. Most living soft tissues, however, may be represented approximately by a *quasi-linear* constitutive equation, as follows:

$$\sigma_{ij}(\mathbf{x},t) = \int_{-\infty}^{t} G_{ijkl}(t-\tau)\frac{\partial \sigma_{kl}^{(e)}(\mathbf{x},\tau)}{\partial \tau}d\tau, \tag{48}$$

where $\sigma^{(e)}$ is the *pseudoelastic stress*, see *Biomechanics* (Fung, 1993b, Section 7.6, pp. 277–293), and is a function of the strain, which in turn is a function of $\mathbf{x}$ and t. If a *pseudo-strain-energy-function* $\rho_o W$ exists, then

$$\sigma_{kl}^{(e)} = \frac{\partial \rho_o W}{\partial E_{kl}}, \tag{49}$$

where $\rho_o W$ is a function of the strain components E_{11}, E_{22}, $E_{12}, \ldots$, symmetric with respect to the symmetric shear strains $E_{ij} = E_{ji}$. Appropriate forms of $\rho_o W$ are discussed below. If we let $E_1, E_2, \ldots, E_6$ represent $E_{11}, E_{22}, E_{33}, E_{12}, E_{23}, E_{31}$, respectively, and $\sigma_1, \sigma_2, \ldots \sigma_6$ for σ_{11}, $\sigma_{22}, \sigma_{33}, \sigma_{12}, \sigma_{23}, \sigma_{31}$, nespectively, then, on substituting (49) into (48), we obtain

$$\sigma_i(\mathbf{x},t) = \int_{-\infty}^{t} G_{im}(t-\tau)\frac{\partial^2(\rho_o W)}{\partial E_n \partial E_m}\frac{\partial E_n(\mathbf{x},\tau)}{\partial \tau}d\tau. \tag{50}$$

A substitution of Eq. (50) into Eq. (13) yields the equation of motion. If the strains were finite, we recognize E_{ij} to be the Green's strains, and $\sigma_{kl}^{(e)}$ in Eq. (49) to be the Kirchhoff's stress, which is denoted by S_{ij} in Eq. (42), and the σ in Eq. (48) is also Kirchhoff's stress.

Myocardium in Relaxed State

For soft tissues in general, including the heart and blood vessels, the following strain energy function is recommended (Fung et al., 1993):

$$\rho_o W = \frac{c}{2}(e^Q - Q - 1) + \frac{q}{2} \tag{51a}$$

where c is a constant, and q and Q are quadratic forms of the Green's strains:

$$Q = a_1 E_{11}^2 + a_2 E_{22}^2 + a_3 E_{33}^2 + 2a_4 E_{11}E_{22} + 2a_5 E_{22}E_{33} + 2a_6 E_{33}E_{11}$$
$$+ a_7 E_{12}^2 + a_8 E_{23}^2 + a_9 E_{31}^2 \tag{51b}$$

$$q = b_1 E_{11}^2 + b_2 E_{22}^2 + b_3 E_{33}^3 + 2b_4 E_{11}E_{22} + 2b_5 E_{22}E_{33} + 2b_6 E_{33}E_{11}$$
$$+ b_7 E_{12}^2 + b_8 E_{23}^2 + b_9 E_{31}^2 \tag{51c}$$

in which $a_1, a_2, \ldots, a_9, b_1, b_2, \ldots, b_9$ are material constants. The unit of $c, b_1,$ $b_2, \ldots, b_9$ are those of stress, whereas $a_1, a_2, \ldots, a_9$ are nondimensional.

If the constant c vanishes, then the strain energy function $\rho_o W = q$ defines a linear Hookean elastic body. Its use has a long history, and is still much used (see the journal *Applied Mechanics Reviews* for new articles). Strain energy functions of finite strains of crystalline materials have been examined exhaustively by Green and Adkins (1960). In the simplest case of initially isotropic elastic body, Green and Adkins show that $\rho_o W$ must be a polynomial of the form

$$W = W(I_1, I_2, I_3), \tag{52}$$

where I_1, I_2, I_3 are the first, second, and third invariants of the strain tensor. If the material is incompressible, so that $I_3 = 1$, then the strain energy is a function of $I_1,$ and I_2 only. The linear form

$$W = C_1(I_1 - 3) + C_2(I_2 - 3), \tag{53}$$

where C_1 and C_2 are constants, has been found valuable in the study of large deformations of rubber. A material that obeys Eq. (53) is known as a Mooney–Rivlin material, in recognition of the contributions of M. Mooney who proposed it in 1940 on shear experiments on rubber, and R. Rivlin who produced many exact mathematical solutions, see References in Green and Adkins (1960). No biological tissue is known to obey Eq. (53).

The strain energy function proposed by Fung (1972, 1973) for blood vessels and skin was in the form of

$$\rho_o W = q + ce^Q \tag{54}$$

where q and Q are polynomials of the strain components. Fung and his associates have examined Q as a combination of strain invariants $I_1, I_2,$ and a third order polynomial. Q was put into the simpler form shown in Eq. (54) by Fung et al. (1979) for blood vessels on recognizing that the material is anisotropic, and that the inclusion of the third power terms in Q is unnecessary according to Tong and Fung (1976). The final form of Eq. (51) was tested for blood vessels by Zhou (1992). It differs from (54) only in the inclusion of the terms $-1 - Q$ in the parenthesis of Eq. (51). This crucial little step is the result of recognizing the experimental fact that in the neighborhood of the zero-stress state the stress–strain relationship of most biological tissues is linear. Since 1983, when the significance of the zero-stress state of living tissues was recognized, many experiments have been done on living tissues beginning with the zero-stress state. It was found that nonlinear features appear at larger stains. Equation (51) adds nonlinear corrections at larger strains without affecting the results of linear elasticity measurements made at small strains.

Since the 1970s, the strain energy function given in Eq. (54) has been used by many authors for many tissues, and was found to be reasonably satisfactory. For papillary heart muscle at relaxed state Pinto and Fung (1973) have presented an extensive set of experimental results including the viscoelastic relaxation and creep functions. The results of Pinto and Fung are consistent with the exponential form of the strain energy function, and the quasi-linear viscoelasticity Eq. (48) proposed by Fung (1971, 1973).

For myocardium, various choices of the polynomial Q have been made by authors. Smail and Hunter (1991) used the strain energy

$$W = C_1\left(e^{Q_1} - 1\right) + C_4\left(e^{Q_2} - 1\right) \tag{55a}$$

where

$$Q_1 = C_2 E_{11}^2 + C_3\left(E_{12}^2 + E_{13}^2\right) \tag{55b}$$

and

$$Q_2 = C_5\left(E_{22} + E_{33}\right)^2 + C_6\left(E_{22}E_{33} + E_{23}^2\right) \tag{55c}$$

which is a transversely isotropic law with x_1 aligned with the fiber axis. McCulloch and Omens (1991) and Guccione et al. (1991) used

$$W = \frac{C}{2}\left(e^{Q} - 1\right) - \frac{1}{2}p\left(I_3 - 1\right) \tag{56a}$$

where

$$Q = b_1\left(E_{11} + E_{22} + E_{33}\right) + b_2 E_{11}^2 + b_3\left(E_{22}^2 + E_{33}^2 + E_{23}^2 + E_{32}^2\right)$$
$$+ b_4\left(E_{12}^2 + E_{21}^2 + E_{13}^2 + E_{31}^2\right) \tag{56b}$$

Humphrey et al. (1991) on the other hand, chose a polynomial form for the relaxed myocardium:

$$W = C_1\left(\alpha - 1\right)^2 + C_2\left(\alpha - 1\right)^3 + C_3\left(I_1 - 3\right) + C_4\left(I_1 - 3\right)\left(\alpha - 1\right) + C_5\left(I_1 - 3\right)^2 \tag{57}$$

in which C_1, C_2, C_3, C_4, C_5 are constants, I_1 is the first invariant of the Green strain tensor, and α is the stretch ratio in a preferred direction.

Many other forms of strain energy functions and stress-strain relations have been proposed by Blatz et al. (1969), Hoeltzel et al. (1992), Kenedi et al. (1964), Morgan (1960), Ridge and Wright (1964), Rivlin (1947), Rivlin and Saunders (1951), Takamizawa and Hayashi (1987), Valanis and Landel (1967), Veronda and Westmann (1970), Vidik (1966), Werthein (1847) etc. Reviews of these papers and references are given in Fung (1993b). Generally speaking, heart muscles, skeletal muscles, and blood vessels are obviously anisotropic so that those formulations that are based on isotropic concept have questionable applicability. In the case of blood vessels, claims of transverse isotropy are often not valid upon close examination, see Debes and Fung (1995), Deng et al. (1994), and Zhou (1992). Many ad hoc proposals are written for uniaxial stretch, and it is not clear how the formulas can be generalized to three dimensions. Finally, virtually all nonlinear stress-strain relationships are written for stress in terms of strain, i.e., to compute stress when strains are known. Most of these nonlinear equations cannot be inverted to solve for strains in terms of stresses. Most strain energy functions $\rho_o W(\mathbf{E})$ for materials of nonlinear elasticity cannot be inverted to complementary energy functions W_c (σ), which are analytic functions of stresses so that $E_{ij} = \partial W_c/\partial \sigma_{ij}$. An exception is the exponential strain energy functions, Eqs. (51), (54), (55), for which the complementary energy are known, and the stress-strain relationship can be inverted, see Fung (1979), or Fung (1993b, p. 307).

Derivation of Constitutive Equation from Microstructure

In optical and electron microscopes, the myocardium and blood vessel tissues are seen as consisting of muscle cells and connective tissues. The structure may be separated into various "compartments" as illustrated in Fig. 2.6:1 in order to visualize how the mechanical properties of the whole tissue are related to the properties of the individual compartments. Each compartment is multiphasic, that is, consists of fluids, solids, and ions. A multiphasic analysis will shed light on the mechanical properties of each compartment, and so on. Linking the mechanical properties at successive hierachial levels of structures is not only intellectually satisfying, but also important practically. For the myocardium and the blood vessel, the view of Fig. 2.6:1 suggests separating their mechanical properties into two parts: one part is contributed by the active contraction of the muscle cells; the other part is that of the tissue in which the muscle cells are relaxed. Whether these two parts can be determined by separate series of experiments is debatable. Nevertheless, a separation is a great convenience.

Lanir (1983) has focused on deriving constitutive equation from microstructure of connective tissues. Horowitz (1991) has outlined the Lanir approach to the heart muscle, showing how collagen fibers in the interstitial space between muscle cells contribute to the mechanical properties of the heart. Hunter and his associates, Le Grice et al. (1995) have documented the connective tissue architecture in dog heart.

Whatever theoretical constitutive equations being proposed for a living tissue at whatever levels of hierarchy, one must validate the constitutive equations experimentally and determine the material constants. It is nice to explain the constitutive equation of one level of hierarchy with those of the lower levels; but it is impractical to hope that the constitutive equations at lower levels of hierarchy in biology is simpler than those at the higher levels. In fact, at the most fundamental level of quantum mechanics, the biological theory is the most complex. Continuum models are great simplifications of the atomic and molecular models.

Active Contraction of Muscle Cells

The equations describing the active contraction of the heart muscle and the generation of tensile force are presented in Section 2.7.

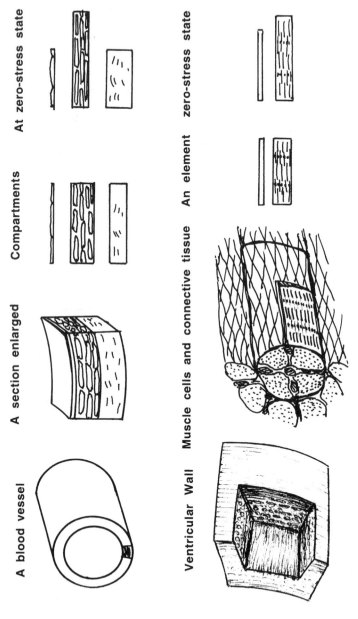

FIGURE 2.6:1. Schematic drawings of the tissues of a blood vessel and a left ventricular wall showing a breakdown into compartments and the configurations of the compartments in vivo and at zero-stress state. Each compartment is a multiphasic material.

Tissue Remodeling Under Stress

In the first edition of this book, the author raised the question whether the zero-stress states of the ventricles and blood vessels are the same as the no-load states of these organs, and presented some photographs that showed the difference. Since then the zero-stress states of heart and blood vessels have been studied extensively, and used to document tissue remodeling under stress. The subject is discussed in the books *Biomechanics: Motion, Flow, Stress, and Growth,* (Fung, 1990) and *Biomechanics: Mechanical Properties of Living Tissues,* (Fung, 1993b), which contain lists of references to many articles. Since living cells respond to stress and strain by division, or growth, or produce proteins, the materials in our bodies change with stress and strain; the structure of the tissues and organs change dynamically, and the mechanical properties change with them. Thus, all the constitutive equations presented above must be regarded as transient, the "constants" in these formulas are functions of time, and are functionals of the stress and strain history. The geometry, the materials, the structure, and the mechanical properties change under the loading, stress and strain. All the governing equations listed above have to be modified to reflect this fact.

Using blood vessel as an example, the phenomenon of tissue remodeling may be described in the following way. Let $\mathbf{Y}$, with components Y_1, Y_2, $Y_3, \ldots$, be a matrix describing the features of a blood vessel, (e.g., $Y_1 =$ opening angle, $Y_2 =$ lumen diameter, $Y_3 =$ thickness of smooth muscle layer, ...). At homeostasis, $\mathbf{Y}$ has the value $\mathbf{Y_0}$. When the homeostasis is disturbed, $\mathbf{Y}$ becomes $\mathbf{Y_0} + \mathbf{X}$, where X is the perturbation of Y, with components $X_1, X_2. \ldots$ $\mathbf{X}$ and $\mathbf{Y}$ are functions of time t, blood pressure P, oxygen tension pO_2, blood flow Q, concentrations of chemicals $c_1, c_2 \ldots$, physical parameters φ_1, $\varphi_2 \ldots$, and the location of the vessel on the vascular tree, x. Our hypothesis is that $\mathbf{X}$ **is a functional** of the following form:

$$\mathbf{X}(x,t) = \text{Functional of} \left[\mathbf{Y_0}, \mathbf{X}(t), P(t), pO_2(t), Q(t), \right.$$
$$\left. c_1, c_2, \varphi_1, \varphi_2, \cdots, t, \tau, x \right] \tag{58}$$

in which τ is time prior to the instant of time t, and the term **Functional** means a function of all the past histories of the variables enclosed in the bracket, up to the instant of time t. In practice, the existence of a stable equilibrium state in life (homeostasis) is a fundamental hypothesis in biology, and most of the variables in the brackets of Eq. (58) have stable homeostatic values. To investigate, one attempts to vary one parameter at a time. If we focus our attention on the effect of blood pressure on tissue remodeling, we may write,

$$\mathbf{X}(x,t) = \text{Functional of} \left[\mathbf{Y_0}, \mathbf{X}(t), t, \tau, x, P(t) \right] \tag{59}$$

To seek further simplification we may standardize the pressure perturbation into a step function:

$$P(t) = P_0 + \Delta P_0 \mathbf{1}(t) \tag{60}$$

where P_0 and ΔP_0 are constants, and $\mathbf{1}(t)$ has the value 1 when $t \geq 0$, and 0 when $t < 0$. Then the histories of all the components of $\mathbf{X}(t)$ are definite, and Eq. (58) can be simplified to

$$\mathbf{X}(x,t) = \Delta P_0 \cdot \text{Function of} \left[\mathbf{Y_0}, \mathbf{X}(t), t, x, P_0, \Delta P_0 \right] \tag{61}$$

Here the word **Functional** has been replaced by $\Delta P_0 \cdot$ **Function**. The **Function** on the right-hand side of Eq. (61) is the **Indicial Function**. If the system is **linear** with respect to ΔP_0, then the function in Eq. (61) is independent of ΔP_0. The functions $X_i(t)$ on both sides of Eq (61) can be unscrambled to yield

$$X_i(x,t) = \Delta P_0 \cdot \text{Indicial function of} \left[Y_0, t, x, P_0, \Delta P_0 \right] \tag{62}$$

If the experimental data cannot make ΔP_0 disappear from the indicial function, then the system is **nonlinear** with respect to ΔP_0. Data on tissue remodeling can be expected to come in the form of indicial functions.

For heart or any organ, let m be the mass per unit volume of a tissue, L be a characteristic dimension, E_{ij} be the Green's strain measured relative to the zero-stress state, t be time, and τ be time between 0 and t. Then the stress constitutive equation may be written as

$$\sigma_{ij}(t) = \text{Functional of} \left[E_{ij}(\tau), m(\tau), L(\tau), 0 < \tau < t, x \right] \tag{63}$$

The values of m(t) and L(t) are obtained by integrating the rate of growth:

$$m(t) = m(0) + \int_0^t \dot{m}(\tau) d\tau \tag{64a}$$

$$L(t) = L(0) + \int_0^t \dot{L}(\tau) d\tau \tag{64b}$$

and $\dot{m}$, $\dot{L}$ are given by growth laws, or indicial functions. Thus, the constitutive and the growth equations are coupled. All the equations listed in this section are affected by these new equations. The boundary conditions are obviously affected by the characteristic dimension $L(t)$.

The realm of cell response to stress and strain belongs to cell and molecular biology. The growth law is part of the laws of gene expression. The future of biomechanics is clearly connected to molecular biology. In the hierarchy of biomechanics the molecular level is basic.

A Perspective

All these equations are condensations of observations, thoughts, and experiments in mathematical form. One can then use mathematics to make deductions, to formulate boundary-value problems and solve them.

It is a biological axiom that a living organism has a stable equilibrium state (homeostasis). When a homeostatic state is disturbed, tissue remodeling takes place. Would the remodeling destroy the tissue? Would a new homeostatic state evolve? Is the new state good or bad for the individual? In what sense? Equations (1) to (64) are needed to formulate definitive questions and to find answers.

2.7 Active Contraction of Heart Muscle

Mechanical properties of heart muscle in active contraction have been described by Hill's three-element model shown in Figure 2.7:1 (A.V. Hill, 1939). In this model, the parallel element represents the connective tissues and the quiescent muscle cells, such as an unstimulated papillary muscle in a testing machine. The contractile element represents the active contraction machinery, the actin–myosin sliding mechanism. The series element represents the elasticity of the actin and myosin fibers and crossbridges in action. The separation of the active and quiescent functions of the muscle cell into

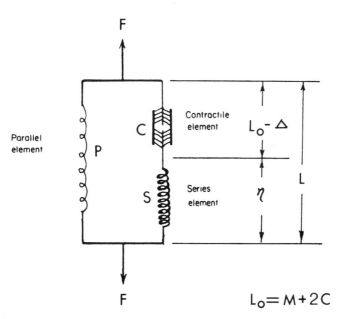

FIGURE 2.7:1. Schematic diagram of the action of the parallel, series, and contractile elements in a heart muscle based on Hill's model. F = force acting on a muscle specimen; P = force in the parallel element; S = force acting in the contractile element; L = length of the specimen; M = length of myosin fibers in the specimen; C = length of actin filaments; Δ = overlap between actin and myosin filaments; η = extension of the series elastic element.

two elements has no clear foundation. Hence, the model is an extra working hypothesis. The model allows one to proceed in the following manner: First, on a quiescent specimen, perform the length–tension experiments, such as periodic stretching and releasing in various modes and frequencies, step stretching and relaxation afterwards, step load and creep measurements, etc.; from the data extract a constitutive equation to represent the mechanical properties of the parallel element. Then stimulate the muscle to twitch and measure the tension in isometric condition at various muscle lengths, the velocity of contraction in the isotonic condition at various loads, the conditions of isometric–isotonic changeover at specified afterloads, and the movement of the muscle in quick release from one tension to another. From the measured tension in these twitch experiments, subtract the tension in the parallel element at the same strain and call the difference the *active tension* in the contractile element. From the data extract formulas that describe the mechanical behavior of the contractile element and the series elastic element. The details of these experiments and the formulas extracted are described in *Biomechanics: Mechanical Properties of Living Tissues* (Fung, 1993b, Chapters 9 and 10). Briefly, the total force, F, acting in the muscle is

$$F = P + S, \tag{1}$$

where P and S represent the forces in the parallel and series elements, respectively. P is a function of the length L (or the history of L if viscoelasticity is emphasized). In discussing the stress–strain relationship, no generality is lost if we consider L to be the length of one sarcomere. Sarcomere length consists of one myosin fiber of length M and two actin fibers of length C each. Hence if the fibers were lined up end-to-end the length would be $M + 2C$. But they overlap by an amount Δ; so the length of an unstressed contractile element is $M + 2C - \Delta$. In the unstressed state the series element extension (i.e., the contractile element's elastic extension) is zero. When the tension is S the series elastic element extension becomes η. Hence in the stressed state the sarcomere length is

$$L = M + 2C - \Delta + \eta. \tag{2}$$

By differentiation, we have

$$\frac{dL}{dt} = -\frac{d\Delta}{dt} + \frac{d\eta}{dt}, \tag{3}$$

which is a basic kinematic relation connecting the rate of muscle length change with the rate at which the actin–myosin overlap changes, and the velocity of extension of the series element.

Experimental data on resting heart muscles show that (Pinto and Fung, 1973a)

$$\frac{dP}{d\lambda} = \alpha\left(P + \beta\right) \tag{4}$$

or

$$P = -\beta + ce^{\alpha\lambda}, \tag{5}$$

where λ is the stretch ratio of the muscle (measured from the state of zero tension in resting state), and α, β are material constants. c is an integration constant. To determine c, let a point be chosen on the curve, so that $P = P^*$ when $\lambda = \lambda^*$; then

$$P = \left(P^* + \beta\right)e^{\alpha(\lambda - \lambda^*)} - \beta. \tag{6}$$

Experimental results on series element show that (Edman and Nilsson, 1972; Parmley and Sonnenblick, 1967; Parmley et al., 1969; Sonnenblick, 1964; Sonnenblick et al., 1964, 1967)

$$\frac{dS}{d\eta} = \alpha'\left(S + \beta'\right), \tag{7}$$

where α' and β' are also constants. Hence S is also an exponential function of the extension η:

$$S = \left(S^* + \beta'\right)e^{\alpha'(\eta - \eta^*)} - \beta'. \tag{8}$$

Experimental data on the papillary muscles show that the contractile element shortening velocity may be represented by the following modified Hill's equation (Fung, 1970):

$$\frac{d\Delta}{dt} = \frac{b\left[S_o f(t) - S\right]}{a + S}, \tag{9}$$

in which the constants a and b are functions of muscle length L, S_o is the peak tensile stress reached in an isometric contraction at length L, t is the time after the stimulus, and $f(t)$ is a function that may be represented as

$$f(t) = \sin\left[\frac{\pi}{2}\left(\frac{t + t_o}{t_{ip} + t_o}\right)\right]. \tag{10}$$

Here t_o is a phase shift related to the initiation of the active state at stimulation, and t_{ip} is the time to reach the peak isometric tension after the instant of stimulation. The constant a as a function of L is known empirically to be proportional to S_o, which is also a function of L, and can be written as

$$a(L) = \gamma S_o(L). \tag{11}$$

The value of γ is of the order of 0.45. The relationship between S_o and L is shown in Figure 2.1:3.

If active tension history $S(t)$ is known, then from Eq. (7) we can compute the series element velocity,

$$\frac{d\eta}{dt} = \frac{1}{\alpha'\left(S + \beta'\right)} \frac{dS}{dt}. \tag{12}$$

Using Eqs. (9), (11), and (3), we can obtain the velocity of muscle contraction dL/dt. If the length history $L(t)$ is known, then from Eqs. (3) and (9) we can compute $d\eta/dt$, and then by an integration obtain η. Then Eq. (8) yields S, Eq. (6) yields P, and Eq. (1) yields the tension in the muscle. In some important problems, however, one would have to find both $S(t)$ and $L(t)$ from specified boundary conditions.

Although Hill's model and the above-mentioned formulas represent the state of the art, the basic arbitrariness of lumping the actin–myosin interaction of an unstimulated muscle cell in the parallel element clouds the issue. If the unstimulated muscle cell is infinitely soft compared with the collagen tethers, then we can identify the muscle cell as the sole contractile element. But Brady (1984) has shown that the passive stiffness of isolated intact single rat cardiac muscle cell is comparable with that of the papillary muscle. Hence, the actin–myosin complex in the passive state contributes a major share to the stiffness of the parallel elements. It follows that the parallel and contractile elements cannot be assigned to different phases in different compartments of Figure 2.7:1.

A better constitutive equation for the heart muscle needs to be formulated that takes into account the sliding-element theory, and the flux of calcium ion and ATP, and distinguishes the influences of shear stress and strain from those of normal stress and strain. Theories have been developed by Guccione and McCulloch, 1993; Huntsman et al., 1983; Lacker and Peskin, 1986; Panerai, 1980; ter Keurs, 1983; Tözeren, 1985. Using the data of Hunter et al. (1983) and ter Keurs et al. (1980) on the calcium flux rate, Guccione and McCulloch (1993) have shown that the equations yield results that fit the experimental data well. Without going into details, we cite one of their results in Figure 2.7:2, which shows the force–velocity relationship of rat trabeculae muscle at the instant of time of developing the peak isometric tension and then released to isotonic condition of specified force. The experimental data were Krueger et al.'s (1988). Hill's equation is represented by the solid curve with constants determined by Daniels et al. (1984). Guccione and McCulloch's calculated results are represented by the dashed curve.

Still an important piece of information is missing: the elasticity of an active muscle in directions perpendicular to the muscle fibers. This information is needed to write down the equation balancing the forces acting at the border of coronary blood vessels and cardiac muscle cells. The equation is required to determine how much the coronary capillaries are squeezed by the muscle cells during systole. See Sections 7.9, 7.13, and 7.14.

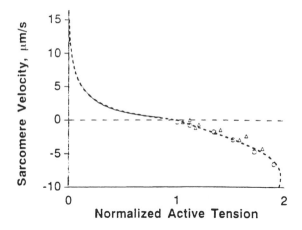

FIGURE 2.7:2. The Relationship between force and sarcomere shortening velocity of a heart muscle in isometric contraction and suddenly released to a lower or higher tension at a time close to the instant of peak isometric tension. The tensile force is normalized by the isometric value at the instant of release. Solid curve: Hill's equation. Dotted curve: from the model by choosing the cross-bridge detachment rate constants to fit the data. Data and curve for normalized tension >1 were given by Krueger and Tsujioka. Deltas (Δ) were measured at 90 ms after stimulation. Circles (0) were measured at 190 ms. From Guccione and McCulloch (1993), by permission.

2.8 Fluid Mechanics of the Heart

To analyze the motion of the heart and the coupling of the heart with the lung and aorta, we must deal with the fluid mechanics of blood flow and the solid mechanics of the heart and blood vessel walls. The mathematical equations have been collected in Section 2.6. The boundary-value problems are complex, in general.

Many problems have been solved with heavy calculations. Instead of providing a survey, an allegory of the ejection phase of a heart cycle by Robert T. Jones (1969, 1972) is presented here to show that a great deal can be learned from simplified thoughts. Jones made no attempt to simulate the exact geometry or mechanical properties of a ventricle. He considered blood as a nonviscous ideal incompressible fluid. Hence, his governing equation is the "potential" equation. Under this hypothesis, any solution of the potential equation is a possible velocity potential of a flow. Jones picked a velocity potential to see what kind of boundary conditions and pressure distribution it leads to. If the boundary looks somewhat like a pulsating bulb, then the solution tells us something about the pressure and velocity distribution in an idealized heart in the ejection phase.

Consider a velocity potential ϕ given by

$$\phi = \alpha x^2 + \beta y^2 + \gamma z^2, \tag{1}$$

where α, β, γ are functions of time, t, with

$$\alpha(t) + \beta(t) + \gamma(t) = 0 \tag{2}$$

and x, y, z are Cartesian coordinates. The velocity components u, v, w in the directions of x, y, z are given by the derivatives of ϕ:

$$u = \frac{\partial \phi}{\partial x}, \quad v = \frac{\partial \phi}{\partial y}, \quad w = \frac{\partial \phi}{\partial z}, \tag{3}$$

They are functions of time and linear in x, y, z. Hence flows represented by Eq. (1) have the property that initially plane surfaces convected with the fluid remain plane. Successive changes in the shape of a bulb are thus related to an (arbitrary) initial shape by affine transformation. If x, y, z are the initial coordinates of a particle of fluid (or of the boundary), we have

$$\zeta = 2x \int \alpha dt, \quad \eta = 2y \int \beta dt, \quad \xi = 2z \int \gamma dt, \tag{4}$$

for the Lagrangian coordinates. This is an affine transformation between (x, y, z) and (ξ, η, ζ).

If the time dependence in Eq. (1) is represented by a single function, as in

$$\phi = f(t) [ax^2 + by^2 + cz^2], \tag{5}$$

where a, b, c are constants with

$$a + b + c = 0, \tag{6}$$

then the streamlines are stationary though the flow is time dependent. The streamlines for a flow of this type, with $a = 2, b = -1, c = -1$, is shown in Figure 2.8:1. In this case the flow is axially symmetric with a stagnation point

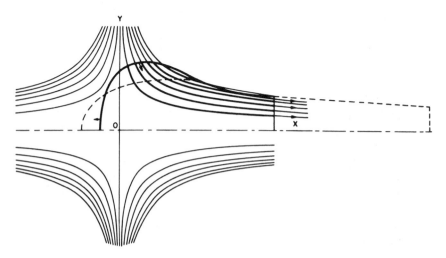

FIGURE 2.8:1. Streamlines for a collapsing bulb. From Jones (1972), by permission.

FIGURE 2.8:2. Isobaric surfaces of Jones's solution. From Jones (1972), by permission.

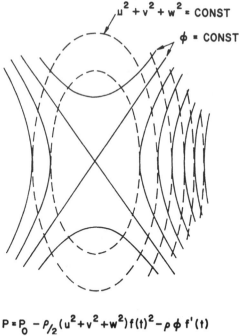

$$u^2 + v^2 + w^2 = \text{CONST}$$

$$\phi = \text{CONST}$$

$$P = P_0 - \rho/2\,(u^2 + v^2 + w^2)\,f(t)^2 - \rho\,\phi\,f'(t)$$

at the origin. Starting with a bulb of arbitrary shape, successive shapes are obtained by a simple stretching transformation. Figure 2.8:1 illustrates this process. Here a bulb is drawn in heavy line; its mouth opens to the right. At a later moment the bulb's shape is changed into the one outlined by the dotted curve.

With the effect of viscosity omitted, the pressure at points within the bulb is given by the generalized Bernoulli's equation (see Sec. 1.11)

$$p = p_o(t) - \frac{\rho}{2}\left(u^2 + v^2 + w^2\right) - \rho\frac{\partial\phi}{\partial t}, \tag{7}$$

where ρ is the density of the fluid. Equation (7) contains an arbitrary function, $p_o(t)$, which must be determined by the impedance into which the bulb works.

For flows with stationary streamlines the isobaric surfaces given by $u^2 + v^2 + w^2$ and ϕ in Eq. (5) are also stationary. Figure 2.8:2 shows these isobaric surfaces for flows with axial symmetry. Jones called the component associated with $\partial\phi/\partial t$ the *acceleration pressure*, while $(\rho/2)\,(u^2 + v^2 + w^2)$ is called the *Bernoulli pressure*. In an oscillatory flow the latter component will oscillate at twice the frequency of the former.

Figure 2.8:3 shows pressures computed by the Eq. (7) for the initial instant of contraction of the bulb shown. Here $u^2 + v^2 + w^2 = 0$ and the pressures are entirely due to acceleration of the flow ($p = p_o - \partial\phi/\partial t$). For an

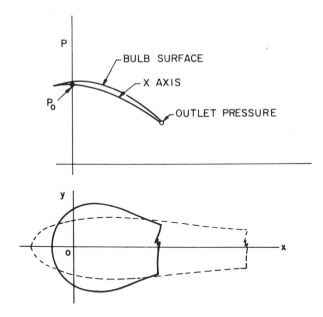

FIGURE 2.8:3. Initial distribution of pressure within a collapsing bulb. From Jones (1972), by permission.

ejection curve of the type shown in Figure 2.4:4, the pressure variation within the bulb amounts to 3 to 5 mm Hg, and the pressures due to the acceleration of flow within the heart are of the same order as the Bernoulli pressures.

More complete analysis of fluid motion in the heart can be done with numerical methods [see, e.g., Peskin (1977), Peskin and Wolfe (1978)].

2.9 Solid Mechanics of the Heart

To determine the stress and strain in the heart, we must use the equations listed in Section 2.6 and the boundary conditions on the surfaces of the wall, then devise an algorithm to solve these equations, and perform experiments to validate the solution. Every step is nontrivial.

An analogy is to treat a left ventricle as a spherical shell made of a homogeneous isotropic material that obeys Hooke's law (see Figs. P2.1 and P2.3, associated with Problems 2.1 to 2.3). The spherical shell has an inner radius a and outer radius b. It is subjected to a uniform internal pressure p_i and external pressure p_o (see Fig. 2.1:2). Assume that when p_i and p_o are zero the shell is stress free. Under these hypotheses, the stress distribution was found by Lamé (1852) a long time ago (see Fung, *Foundations of Solid Mechanics*, 1965, p. 191).

The classical solution gives the radial stress σ_r:

$$\sigma_r = \frac{p_o b^3 \left(r^3 - a^3\right)}{r^3 \left(a^3 - b^3\right)} + \frac{p_i a^3 \left(b^3 - r^3\right)}{r^3 \left(a^3 - b^3\right)} \tag{1}$$

and the circumferential stress σ_θ:

$$\sigma_\theta = \frac{p_o b^3 \left(2r^3 + a^3\right)}{2r^3 \left(a^3 - b^3\right)} - \frac{p_i a^3 \left(2r^3 + b^3\right)}{2r^3 \left(a^3 - b^3\right)}. \tag{2}$$

If $p_o = 0$, the greatest circumferential stress is at the inner surface, at which

$$\left(\sigma_\theta\right)_{max} = \frac{p_i}{2} \frac{2a^3 + b^3}{b^3 - a^3}. \tag{3}$$

If $a = 1$, $b = 2$, then $(\sigma_\theta)_{max} = (5/7)p_i$, which is 2.14 times the mean circumferential stress [see the Laplace formula Eq. (4) in Sec. 2.1] and 3.33 times the circumferential stress at the outer wall.

Here we found an answer that we do not like. This is why: Eq. (3) tells us to expect strain concentration at the inner wall of the left ventricle. If the sarcomere length of all muscle cells is the same in the no-load state, then the sarcomere length at the end-diastolic condition will be longer at the endocardium and shorter at the epicardium. Then at the end of the iso-volumic contraction period, the peak tensile stress in the myocytes will be much higher in the endocardial region and much lower in the epicardial region. The myocytes in the endocardial side will do much more work than those in the epicardial region. Parenthetically, there must be an unusually large ATP supply, calcium flux, and coronary capillary blood flow in the endocardial region. These theoretical concentrations have not been seen.

Janz et al. (1974) and Wong and Rautaharju (1968) relaxed the spherical shell hypothesis to an ellipsoidal shape. Mirsky (1973, 1979), Janz and Grimm (1973), and Janz and Waldron (1976) relaxed Hooke's law to nonlinear stress–strain relationships. The stress concentration persisted.

Searching and Finding the Residual Stresses

In 1982, the author thought there might be residual stresses in the myocardium at the no-load state. A former student, Paul Patitucci, and the author cut up an unloaded left ventricle of a rabbit, and found the results shown in Figure 2.9:1(a) and 2.9:1(b). This revealed the existence of residual strain. Omens and Fung (1989) obtained similar results with greater quantitative details. Then we collected similar data on blood vessels. Independently, Vaishnav and Vossoughi (1983) also found residual stress in blood vessels. Since then, many authors have contributed to this subject, and the zero-stress state question has become well understood. The results are summarized in *Biomechanics* (Fung, 1990, 1993b). Its application to tissue engineering became obvious (Fung, 1988).

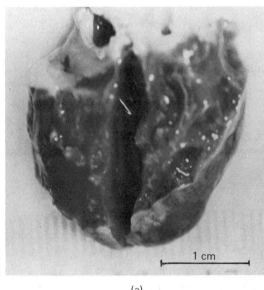

(a)

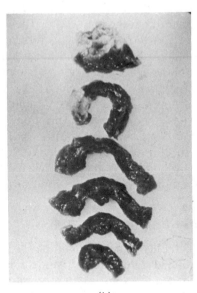

(b)

FIGURE 2.9:1. Change of shape of the left ventricle of the rabbit when it was cut open. (a) Appearance of the left ventricle after a longitudinal cut was made through the interventricular septum (with the right ventricle removed). (b) The cut left ventricle was further sliced into six strips. A further large deformation was seen, revealing large residual stresses. Photograph courtesy of Paul Patitucci.

Principle of Optimal Operation

In the first edition of this book, and also in (Fung, 1983), the author formulated the "principle of optimal operation" to explain the residual stress. This hypothesis states that each organ operates in such a manner as to achieve optimal performance. If the optimal performance of the left ventricle is that all muscle cells develop approximately the same the peak tensile stress at end systole consistent with muscle length, then the sarcomere length of all muscle cells must be approximately the same at end diastole. This implies a uniform tensile stress in myocytes at end diastole and a uniform sarcomere length in the zero-stress state. From this condition, we can compute the geometry of the zero-stress state of the left ventricle if we know the geometry of the end-diastolic state *in vivo*. The same theme was developed independently by Takamizawa and Hayashi (1987) and Takamizawa and Matsuda (1990), who proposed the "uniform strain" theory.

The sarcomere length distribution in the left ventricle was studied by Rodriquez et al. (1993) who obtained two equatorial slices of a rat heart that had been potassium arrested, as done by Omens and Fung (1989). As shown in Figure 2.9:2, one slice was kept in the no-load state, whereas the other was cut open to the zero-stress state. Such large slices were fixed and embedded in plastic, and then sectioned to obtain histological measurements of sarcomere length. Sections were cut along planes perpendicular to the radial direction and were collected from each heart at 16 equidistant transmural sites, as shown in Figure 2.9:3. The results are shown in Figure 2.9:4. The average sarcomere length for the stress-free tissue was 1.84 ± 0.05 (SD) μm and for the unloaded tissue was 1.83 ± 0.06 (SD) μm. However, whereas the sarcomere length was uniform across the wall in the stress-free

Equatorial slices   No-load state   Stress-free state

FIGURE 2.9:2. Two equatorial slices (rings) were cut from unloaded, arrested, isolated, and perfused rat left ventricle. One slice was kept in intact unloaded state. The other slice was cut radially, which sprang open immediately with a measurable opening angle. From Rodriquez et al. (1993), with permission.

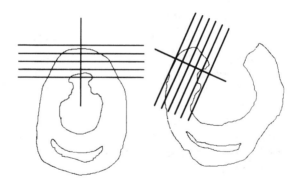

FIGURE 2.9:3. Embedded slices were sectioned to obtain histological measurements of sarcomere length. Sections were cut along planes perpendicular to the radial direction. From Rodriquez et al. (1993), with permission.

state, with a mean gradient of -0.014 ± 0.044 (SD) μm/total wall thickness, the sarcomere length decreased from the epicardium to the endocardium in the intact unloaded tissue with a slope of -0.114 ± 0.054 (SD) μm/total wall thickness. This supports the hypothesis that the sarcomere length of myocytes is uniform in the zero-stress state. The no-load state has nonuniform sarcomere length, and the nonuniformity is such that it tends to make sarcomere length uniform in the end-diastolic state.

Is there evidence that the sarcomere length is approximately uniform throughout the ventricular wall in the end-diastolic condition? In a crude way, yes. Figure 2.9:5 shows some data on the distribution of sarcomere

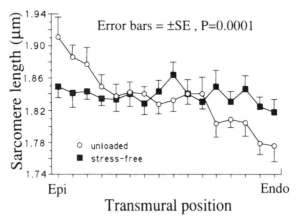

FIGURE 2.9:4. Transmural distributions from epicardium (Epi) to endocardium (Endo) of sarcomere length in 2 states of stress under consideration: unloaded and stress-free. Average distributions from 11 hearts are shown from measurements at 16 equidistant transmural sites. Sarcomere length distribution is uniform across the wall in the stress-free state but is nonuniform in the no-load state. From Rodriquez et al. (1993), with permission.

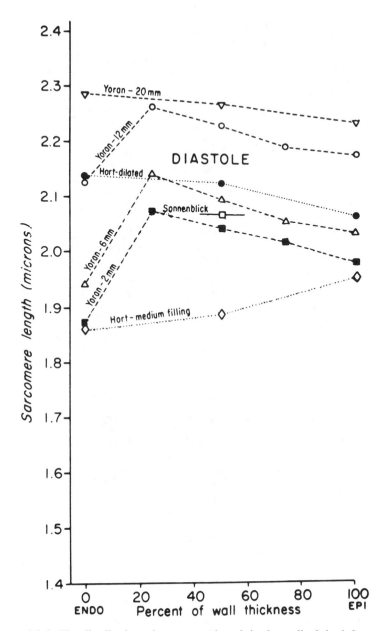

FIGURE 2.9:5. The distribution of sarcomere length in the wall of the left ventricle of the dog in the diastolic condition as given by Hort (1960), Sonnenblick et al. (1967), and Yoran et al. (1973).

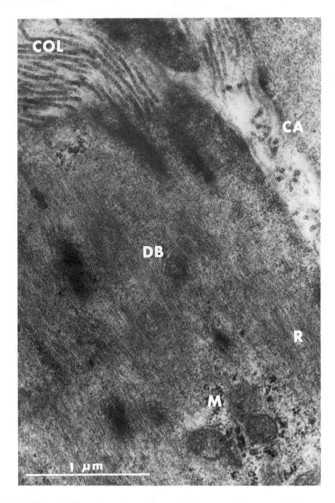

FIGURE 2.9:6. *View 1.* Vascular smooth muscle in a small artery of the mesentery of the rabbit, at a magnification of 53,185, showing the dense bodies (DB), collagen (COL) caveoli (CA), ribosomes (R), and mitochondria (M). Electron micrograph by John Hardy in Los Angeles County-USC Cardiovascular Research Lab., Sid Sobin, Director.

length in the wall of the left ventricle of the dog. The data are taken from the papers of Hort (1960), Sonnenblick et al. (1967), and Yoran et al. (1973). While none of them shows exactly uniform distribution of the sarcomere length at end diastole (at some finite ventricular pressure), they are all far different from what one would have predicted from the hypothesis that the sarcomere length is uniform when the left ventricle is load-free (with zero transmural pressure). In fact, the sarcomere length is most nonuniform near the no-load state (the curve Yoran 2 mm Hg in Fig. 2.9:5), and the non-uniformity is exactly of the type expected (short at the endocardium in the no-load condition).

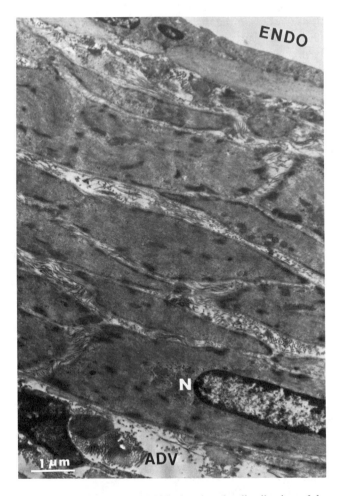

FIGURE 2.9:6. *View 2*. Same artery. ×10,725, showing the distribution of dense bodies in the muscle layers, and endothelium (ENDO), nucleus (N), and adventitia (ADV). Blood pressure = 100 mm Hg. Inner diameter of blood vessel = 320 μm.

If the principle of optimal operation were valid for other organs, then we could apply the same argument to the blood vessel. The optimal condition is that at homeostasis every vascular smooth muscle cell does the same amount of work per cycle of pulsation. This implies that the strain energy, and hence the strain, in the muscle cells is uniform in homeostasis. Now, the blood vessel does not have sarcomeres to serve as natural grid lines to measure the change of strains in a muscle. But the dense bodies of the smooth-*muscle* cells (Fung, 1993b, p. 469) can also serve as markers, although they are not as evenly spaced as the Z-lines of the sarcomeres of the heart muscle. The dense bodies are the smooth muscle's equivalent of the Z-lines of the skeletal and heart muscles. Figure 2.9:6 shows the distribution of dense bodies in the mesenteric artery of the rabbit. It can be seen that the dense bodies are either attached to the cell membranes or are free floating. By measuring the number of dense bodies per unit length of the cell membrane as seen in these electron micrographs, we can obtain a measure of the spacing of the dense bodies. Our results are shown in Figure 2.9:7(a), which shows the spacing between neighboring dense bodies attached to the cell membranes in longitudinal cross sections of the smooth muscle. It is seen that the spacing is statistically uniform throughout the vessel wall when the blood pressure is 20 mm Hg and above. Figure 2.9:7(b) shows some results calculated under the assumption that the stress is uni-

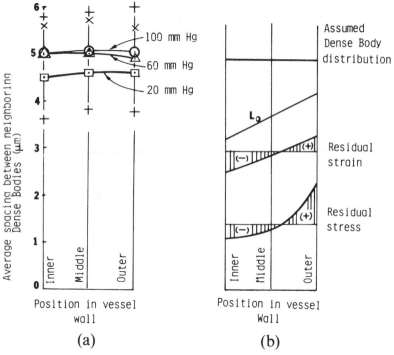

FIGURE 2.9:7. (a) The distribution of dense bodies on the cell membranes of the smooth muscles in the small mesenteric arteries of the rabbit (inner diameter 300–425 μm). Spacing between neighboring dense bodies averaged over inner, middle, and outer one third of the vessel wall, in micrometers. Mean ± SD (+ for 20 and 100 mm Hg, × for 60 mm Hg). (b) Probable distribution of residual circumferential stress and strain in arterial wall calculated under the assumption that the circumferential stress and strain are distributed uniformly in the wall at 60 mm Hg blood pressure. See text.

formly distributed at a blood pressure of 60 mm Hg. L_0 is the reference length of wall material at zero stress. The residual circumferential strain and stress in the vessel wall when the blood pressure is reduced to zero are calculated on the basis of a nonlinear stress–strain relationship. At zero pressure the inner wall is compressed and the outer wall is in tension. If the wall is cut longitudinally, relief of the circumferential stress will cause the artery to spring open (see Fung, 1990, Chapter 11).

2.10 Experimental Strain Analysis

Every good method to measure strain in vivo has clinical applications. The nuclear magnetic resonance imaging (NMRI) method, being completely noninvasive, will one day become the most powerful instrument to measure the strains in the heart. Current methods using sound (phonocardiography, echophonocardiography, echocardiography) x-ray (radiological and angiographic methods), electricity (electrocardiography), catheterization (coronary arteriography), and nuclear physics (scintillation camera, tomography, radiopharmaceuticals, position-emission tomography) do not have the precision required for strain analysis. These methods are described concisely in the book *Heart Disease* (Braunwald, 1988).

In vivo strain measurements in heart have been done by Dieudonne (1969) with strain rosette, and by Arts et al. (1980, 1982), McCulloch et al. (1987, 1991), Meier et al. (1980), Prinzen et al. (1984), and Waldman et al. (1985, 1991) with embedded markers visualized with x-ray, and electromagnetic sensors. Accuracy and error estimates are discussed by Waldman and McCulloch (1993). The methods, results, and applications are illustrated briefly here.

Figure 2.10:1 shows schematically a set of radio-opaque markers of small gold or lead balls implanted in a canine left ventricle. X-ray ciné photographys of these markers in a beating heart were taken in two directions simultaneously. From the photographs, the coordinates of the markers were determined. Every subset of four markers forms a tetrahedron with six edges. Let a set of rectangular cartesian coordinates x_1, x_2, x_3 be chosen, and denote the coordinates of a point in a reference configuration by x_{i0} and those in a deformed state by x_i, ($i = 1, 2, 3$). Denote the distance between two points in the deformed position by ds and that in the reference state by ds_0. Then according to the definition of Green's strains, E_{ij}, we have (Fung, 1993a)

$$ds^2 - ds_0^2 = 2E_{ij}\, dx_{i0}\, dx_{j0}. \tag{1}$$

This is a differential equation. Treating it as a difference equation in the sense of a finite element, then each edge of the tetrahedron satisfies an

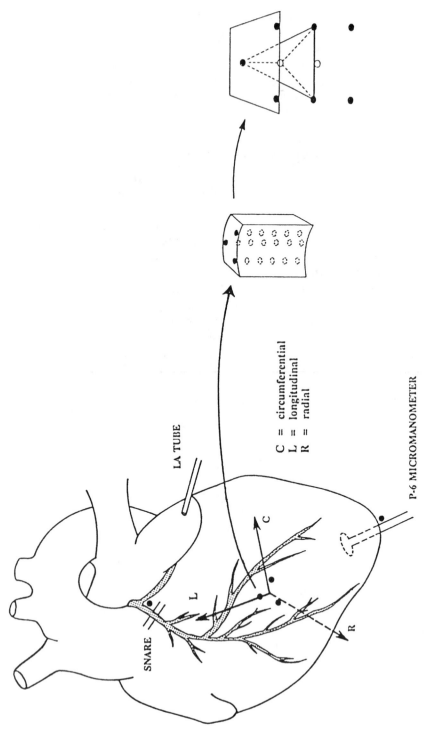

FIGURE 2.10:1. Schematic diagram of the ventricular site in the anterior free wall where the bead columns were inserted, the coordinate system to which the strains are referred, and the site of coronary occlusion (snare). The other two diagrams show an expanded view of the transmural markers and a representative tetrahedron. LA, left atrial. From Villarreal et al. (1991), by permission.

equation like Eq. (1). The six edges of a tetrahedron provide six such equations, from which the six strain components—E_{11}, E_{22}, E_{33}, $E_{12} = E_{21}$, $E_{13} = E_{31}$, $E_{23} = E_{32}$—can be computed. These are the values of the strain in the tetrahedron. If the strains are homogeneous in the tetrahedron, then this calculation is exact. If the strains are variable in space, then the calculated values are a kind of average over the tetrahedron. McCullcoh and Omens (1991) have improved the accuracy and interpretation of the averages by fitting a finite element to a group of markers and then using known mathematical methods (see Sec. 2.12). But errors can be minimized if care was taken to avoid tetrahedrons that are too large, or too long, or too small, or too flat. In addition, triplets of sonomicrometers sewn onto the heart were used, which yielded dimensional data quickly without film processing.

Figure 2.10:2 shows a set of data obtained in a canine left ventricle at a subendocardial location. The six components of the strain tensor relative to the end-diastolic state are plotted for one cardiac cycle, with the initial end-diastolic frame as the reference configuration. The x_1, x_2, x_3 axes lie in the circumferential, longitudinal, and radial directions. Substantial shortening strains are seen in the circumferential direction, $E_{11} < 0$. Large wall thickening ($E_{33} > 0$) is seen throughout the ejection process, and even in the isovolumic periods. Simultaneously, all three components of shear occur. Hence, vectors normal to the local epicardial surface at end diastole do not remain normal during contraction. The transverse shear is significant and Kirchhoff's hypothesis, often used in the theory of thin elastic shells, does not apply. This is not surprising because left ventricles are not "thin."

The three principal strains—E_1, E_2, and E_3—computed from E_{ij}, are plotted in the right-hand column of Figure 2.10:2. Note the large peaks of E_1 and E_3. The shear strains contributed to the differences between E_1 and E_{11}, and between E_3 and E_{33}.

Figure 2.10:3 illustrates the strain changes in the left ventricle when acute ischemia occurs. During acute ischemia caused by occluding the left anterior descending artery, systolic shortening deteriorates within several beats and is replaced by lengthening and wall thinning. In the experiment, coronary occlusion was imposed with a snare at the location shown in Figure 2.10:1. Control data are shown in open circles in Figure 2.10:3, and data after 10 min of ischemia are shown in closed circles. The strains were computed from the coordinate data of a tetrahedron whose centroid was located at a 50% depth in the left ventricular wall. The left ventricular end-diastolic pressure was 4 mm Hg, the end-diastole was at time 0 and end-systole was at 265 ms. After 10 min of ischemia the shortening in the circumferential and longitudinal directions became lengthening in both directions, and the strain in the radial direction became negative. Large changes of the shear strains E_{13} and E_{23} are also seen, with a change of sign.

With data of this kind, Omens et al. (1991) studied the transmural distribution of strain in the isolated arrested canine left ventricle, and found that

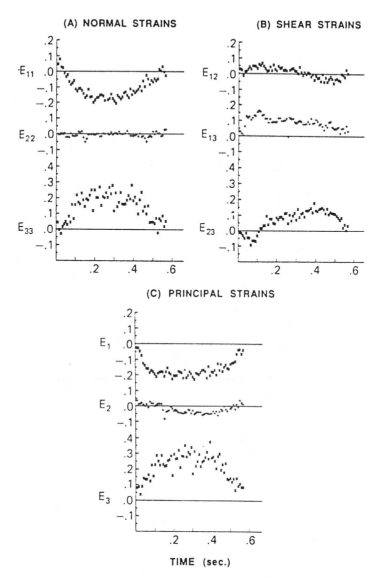

FIGURE 2.10:2. Time series (120 frames per sec) of strains relative to the end-diastolic state near the endocardium of the ventricular wall observed during a cardiac cycle. (A) Normal strains E_{11} (circumferential), E_{22} (longitudinal), and E_{33} (radial). (B) Shear strains E_{12} (longitudinal – circumferencial), E_{13} (circumferential-radial), and E_{23} (longitudinal-radial). (C) Principal strains computed from six strain components shown in (A) and (B). From Waldman (1991), by permission.

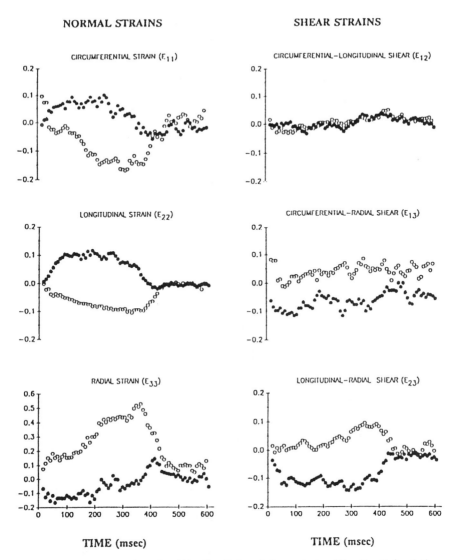

FIGURE 2.10:3. Normal strains (E_{11}, E_{22}, E_{33}) and shear strains (E_{12}, E_{13}, E_{23}) relative to the end-diastolic state at a midwall site during control (open circles) and 10 minutes of ischemia (closed circles) as functions of time. Data shown are from one experiment for one full cardiac cycle, with end diastole occurring at 0 ms and end systole at 265 ms. From Villarreal et al. (1991), by permission.

the fiber strain has no significant transmural variation ($p = 0.57$) from the no-load condition to 20 to 30 mm Hg left ventricular pressure. The principal axis of the greatest strain was close to the fiber orientation on the epicardium, but was closer to the crossfiber direction near the endocardium.

Holmes et al. (1994) studied scar remodeling and transmural deformation after infarct, and Ono et al. (1995) studied the effect of coronary artery reperfusion on cardiac tissue remodeling. Holmes et al. found that significant passive systolic wall thickening and large shears occurred in the scar tissue 3 weeks after infarction when the scar tissue had become entirely collagenous. They concluded that infarct expansion and scar shrinkage may be controlled by different factors, and that systolic wall thickening cannot be taken to be recovery. Ono et al. found that at 3 months after reperfusion, the tissue volume and function of the outer wall recovered completely.

The use of surface marker can also yield lots of information. Hashima et al. (1993) used an array of lead markers sewn onto the epicardium of the free wall of the left ventricle to study the change of surface strain distribution during acute myocardial ischemia in the dog. Here the mathematical trick is to fit a curved surface to the markers. Similarly, Van Leuven et al. (1994) used radio-opaque markers to study the gradients of epicardial strain across the perfusion boundary during acute myocardial ischemia. Young (1991) and associates studied epicardial surface strains using the bifurcation points of the coronary blood vessels as natural markers. Undoubtedly these techniques will find applications in other organs.

2.11 Constitutive Equations of the Materials of the Heart

The efforts over many years to determine the constitutive equations of the materials of the heart have not been rewarded with the desired success. We still do not have a handbook listing the formulas and values of the material constants. It is easy to list the difficulties: The structure is very complex, it is composed of a number of different materials, dissection causes injuries whose effects are difficult to evaluate, and cutting off blood flow changes the tissue. For active muscle, the scheme of separating the structure of the heart tissue into parallel, contractile, and series elements is nonunique and cannot identify the elements with physical entities. The remodeling of tissue under stress, changes of composition, structure, zero-stress state, and possible edema further complicates the picture.

The conclusion is simple: We should do as many in vivo tests as possible. We realize, however, that it is not only difficult to measure strains in vivo, but we cannot measure stress because a stress gauge does not exist. The only information available is the stress boundary condition. Hence, the evaluation of stress in tissue must be done by solving boundary-value problems with hypothetical constitutive equations and assumed material constants, and then identifying the constants against experimental results. This is a herculean task, but it is gradually being accomplished.

New data are fast arriving. Guccione and McCulloch (1991), Humphrey and Yin (1989), and Humphrey et al. (1990, 1991), have provided the mathematical formulation, experimental validation, and data showing the differences in the constitutive behavior of the epicardium, myocardium, and endocardium. MacKenna et al. (1994) and Whittaker et al. (1994) presented morphometric and mechanics data on collagen in ventricular wall. Blum et al. (1995) reported on the mechanical behavior of single myocytes. Rogers and McCulloch (1994) presented a model of cardiac action potential propagation. Omens et al. (1991, 1993, 1994) have obtained data on strain distribution in vivo and have analyzed the stress in the myocardium. Leaving the details to these references and the bibliographies contained therein, the author will state three personal views.

First, with regard to experiments on organs in vivo, more frequent use of concentrated loads of one kind or another are recommended to bend the ventricular wall, as well as measurement of the applied forces and the deflection surface of the epicardium. This is because we need high quality data on stress and strain over a wide range of positive and negative values of strain in order to determine the material constants in different parts of the ventricle. Such data can be obtained by local loading and significant bending. Some examples in the case of blood vessels are given in Fung and Liu (1995), Xie et al. (1995), Yu et al. (1993).

Secondly, with regard to the choice of theoretical forms of the stress–strain relationship for any material in the ventricular wall, or for the wall as a whole, we should remember the following basic requirements: (a) The expression should contain a minimum number of material constants. This is especially important if the stress–strain relationship is nonlinear, if the deflection is so large that the deflection–strain relationship is nonlinear, or if the direction of the load changes with the local slope of the deflection surface. It has been a common experience that decreasing the number of unknown material constants by one may produce a world of difference with regard to the stability of the entire set of constants (see Fung, 1993b, Sec. 8.6). Remember that it is the *form* of the function that determines how well the empirical formula fits the data, not the number of unknown constants. (b) The form should related stress tensor to strain tensor, and not just one component of each. (c) The stress–strain relationship should be *invertible* in the sense that one should be able to compute stress from a known strain, and strain from a known stress. (d) One should be very careful about claiming *isotropy*, or constructing strain energy function with strain invariants alone. For example, consider a fabric. Let the warp and weave be made of the same fiber material. On testing the fabric with loading in the direction of the warp, one determines Young's modulus, E, and Poisson's ratio, v. Repeat testing with loading in the direction of the weave yields the same modulus and ratio. If an author claims isotropy, then he should show that the shear modulus obeys the relation

$$G = \frac{E}{2(1+v)}. \tag{1}$$

But in the case of the fabric, G will depend on how the warp and weave fibers were put together: Are they cemented or are they lubricated? It is thus unlikely that Eq. (1) will hold.

Thirdly, how should the residual stress be handled? Currently, two developments are taking place in this regard. In one, the traditional method, the no-load state of an organ is regarded as the proper starting point of mechanical analysis. One takes care of the residual stress by developing new constitutive equations in terms of strains defined with reference to the no-load state of the organ. In the other development, one pays attention to the zero-stress state of the tissues, defines strains with reference to the zero-stress state of the tissues, and uses them in the constitutive equations. In this approach, the no-load state of the organ could be ignored altogether, and the residual stress and strain are computed only if one wants to know them, but in fact they are of little interest. The biology of the tissue—its growth, resorption, and remodeling—are expressed by the change in the zero-stress state of the tissues. These two approaches, however, may be regarded as modeling an organ at two levels of hierarchies: the whole organ and the tissue. Each tissue can be modeled further to successively lower levels of hierarchy. At each level there are compartments and phases (see, e.g., Fig. 2.6:1), each of which has its own constitutive equations referred to its own zero-stress state. Experimental determination of the zero-stress states of all the compartments is the quickest way to learn about the biology of the tissue at each specific level. Relating the constitutive equations and zero-stress states at successive levels of hierarchy of an organ then reveals the biological structure and function from genes and cells, up to the organ level.

2.12 Stress Analysis

There is no stress gauge to measure stress in the heart, but we can measure the pressure and shear acting on the heart and the restrictions imposed on its movement. We know the laws of conservation of mass, momentum, and energy. If we know also the stress–strain relationship of the materials of the heart and the laws governing the active contraction of the heart muscle, then we can calculate the stress from the boundary conditions and field equations. This is stress analysis.

For a body as complex as the heart, stress analysis needs the most powerful computational methods. The currently available methods include finite elements, finite differences, boundary integrals, and the Ritz–Galerkin. A specialist would have to be an expert in at least one of these methods. It is not the objective of this book to discuss computational methods. In the following, we limit ourselves to the general view of the finite element analysis, which has yielded excellent results in heart analysis in the hands of McCulloch, Omens, Guccione, Rogers, Costa, and others. The author will add some comments on how to take the zero-stress state into account in the analysis. We begin by considering the question: How detailed does a stress analysis need to be?

Stress Analysis and Hierarchical Considerations of the Structure

Plans for any analysis must be drawn according to the objective of the effort. In Table 2.12:1 are listed some objectives and associated details about the structures and materials of the tissues to be considered. It is seen that the refinement with which one should look at the structure depends on the level of detail one wishes to know about the stress. For rough overall estimates, we may look at the tissue macroscopically, often as a body of homogenized material, but when we want to know details about the living tissue, then we have to take a microscopic view. A clear understanding of the hierarchy of the structure is thus essential.

Let us consider the last entry in Table 2.12:1 and use the coronary blood vessel as an example. At the level of the length scale between micrometers and centimeters, we need to know the endothelial cells, vascular smooth muscle cells, fibroblasts, collagen, elastin, and ground substances in the vessel wall. For a bioengineering study of the physiology and pathology of the coronary blood vessels, it is sufficient to consider the vessel as made up of three parallel compartments: the intima, media, and adventitia layers. Each compartment is a continuum, and each has its own zero-stress state.

TABLE 2.12:1. Relationship Between the Objective of Stress Analysis and the Hierarchical Considerations of the Structure, the Zero-Stress State of Each Compartment, and the Constitutive Equations

Objective	Final results expected	Material hypothesis	Zero-stress state information
Rough estimation	Averages of Laplace eq. style	Homogeneous	Same as no-load
Stress/strain distribution	Pressure/flow and stress relationship	Constitutive eq. includes residual stress effect	Hidden
Local stress, ischemia, infarct, surgery	Local stress distribution	Geographic nonuniformity	May be important
Tissue remodeling	Precise stress and strain at site	Multiphasic compartments	Essential
Identifying constitutive eqs. of material	Precise stress and strain in compartments	Multiphasic compartments	Essential

In the zero-stress state there may be discontinuities between these compartments. Yet these compartments can be deformed by internal residual stress and strain into a connected continuous body without external load. They can be deformed also into a continuous body in the in vivo state that is subjected to an external load. The no-load state has no physiological significance. To establish a relationship between the in vivo body and the zero-stress states of the compartments, it is unnecessary to bring in the no-load state as an intermediate step. The constitutive equations of the materials in the three compartments are simplest if strains in each compartment are referred to their respective zero-stress state. The three compartments can be analyzed as three separate bodies and then cemented together into the in vivo state. While the displacements are continuous over the entire organ in the in vivo state, the strains are discontinuous at the borders of the compartments because the zero-stress states of the compartments are different. This approach is feasible both in theory and in experimental practice.

Coordinate Systems

Consider an *element* of the heart in vivo, shown as the body $\overline{\mathbf{B}}$ in Figure 2.12:1. $\overline{\mathbf{B}}$ may represent a piece of myocardium or a compartment within it. The forces of interaction between the material in $\overline{\mathbf{B}}$ and materials surrounding it are revealed by stresses acting on the surface of $\overline{\mathbf{B}}$. If all the stresses were reduced to zero, the shape of $\overline{\mathbf{B}}$ would be changed to $\mathbf{B}_0$, the zero-stress state. The heart is represented by a contiguous set of elements $\overline{\mathbf{B}}$. The corresponding set of $\mathbf{B}_0$ may be discontinuous, as we have explained earlier. However, each $\mathbf{B}_0$ is a continuum. The no-load state, often called the *undeformed body*, will be denoted by $\mathbf{B}$, (not shown in Fig. 2.12:1).

Costa et al. (1996) used several coordinate systems, as shown in Figure 2.12:1. A point with position vector $\mathbf{R}$ in $\mathbf{B}$ or $\mathbf{R}_0$ in $\mathbf{B}_0$ moves to a point with position vector $\mathbf{r}$ in $\overline{\mathbf{B}}$. The coordinates of a material point in the body at zero-stress state $\mathbf{B}_0$ are denoted by capital letters Y_1^0, Y_2^0, Y_3^0; those at the no-load state $\mathbf{B}$ are denoted by capital letters Y_1, Y_2, Y_3 when referred to a rectangular cartesian frame of reference, or by $\Theta_1, \Theta_2, \Theta_3$ when referred to a set of curvilinear coordinates. The corresponding coordinates of the same material point in the deformed body or body in vivo, $\overline{\mathbf{B}}$, are usually denoted by lower case letters, y_1, y_2, y_3 in cartesian coordinates, or $\theta_1, \theta_2, \theta_3$ in curvilinear coordinates. One also uses a set of local orthonormal (cartesian) coordinates X_1, X_2, X_3, with X_1 axis parallel to the muscle fiber to describe the location of a material point in the no-load state $\mathbf{B}$, or the zero-stress state $\mathbf{B}_0$, and x_1, x_2, x_3 in lower case when the body is in vivo, or deformed. Finally, for interpolation in a finite element, we use a set of locally normalized coordinates ξ_1, ξ_2, ξ_3, each with a range of $[0, 1]$ for points in the finite element. We use (ξ_1, ξ_2, ξ_3) coordinates in the Lagrangian sense that each set of numbers (ξ_1, ξ_2, ξ_3) refer to a material point, whether it is in $\overline{\mathbf{B}}, \mathbf{B}$, or $\mathbf{B}_0$. Some examples follow.

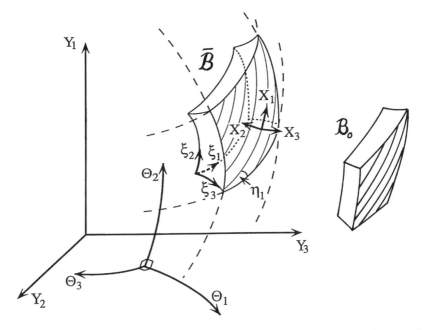

FIGURE 2.12:1. Sketch of a physical element and its zero-stress state, and the use of four coordinate systems to describe their geometry. See text. Adapted from Costa et al. (1996), by permission.

Example of Zero-Stress Element

Costa et al. (1996) have analyzed three-dimensional residual stress in the canine left ventricle. They implanted three columns of radio-opaque beads across the anterior wall of a canine left ventricle and recorded their motion. The deformations measured were used in a finite element to compute the stress. Surface markers locating the element are shown in Figure 2.10:1. The positions of the beads shown in Figure 2.12:2(a) were recorded by biplane fluoroscopy. A block of tissue was then cut out [Fig. 2.12:2(b)], was considered to be stress-free, and the position of the beads in it were again recorded. This illustrates the idea shown in Figure 2.12:1. The fitting of these markers into a finite element is illustrated in Figure 2.12:3. The tissue was then fixed to obtain the morphometric data on the muscle fibers relative to the beads.

Example of Finite Element Interpolation

According to McCulloch and Omens (1991), the coordinates of each marker is identified by a set of local coordinates (ξ_1, ξ_2, ξ_3), each spanning the internal [0, 1], as shown in Figure 2.12:1. With three columns of beads,

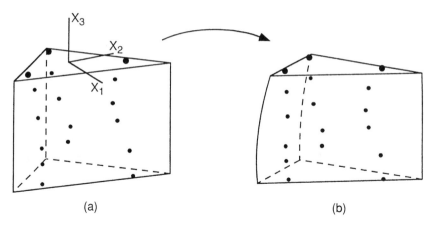

FIGURE 2.12:2. A three-dimensional finite element. (a) An element containing beads. (b) An updated deformed element. Adapted from Costa et al. (1996), by permission.

the ξ_1, ξ_2 axes form the "base" of the finite element and the ξ_3 axis is "transmural." A bilinear Lagrangian interpolation formula is used in the (ξ_1, ξ_2) plane, but cubic Hermite polynomials are used for interpolation in the ξ_3 direction. Each coordinate (e.g., x) is interpolated from eight nodal values $(x_1, x_2, x_3, x_4, x_5, x_6, x_7, x_8)$ at the vertices of the element and eight nodal derivatives with respect to ξ_3 at the vertices by the following formula:

$$
\begin{aligned}
x(\xi_1,\xi_2,\xi_3) = {} & L_1(\xi_1)L_1(\xi_2)H_1^0(\xi_3)x_1 + L_2(\xi_1)L_1(\xi_2)H_1^0(\xi_3)x_2 \\
& + L_1(\xi_1)L_2(\xi_2)H_1^0(\xi_3)x_3 + L_2(\xi_1)L_2(\xi_2)H_1^0(\xi_3)x_4 \\
& + L_1(\xi_1)L_1(\xi_2)H_2^0(\xi_3)x_5 + L_2(\xi_1)L_1(\xi_2)H_2^0(\xi_3)x_6 \\
& + L_1(\xi_1)L_2(\xi_2)H_2^0(\xi_3)x_7 + L_2(\xi_1)L_2(\xi_2)H_2^0(\xi_3)x_8 \\
& + L_1(\xi_1)L_1(\xi_2)H_1^1(\xi_3)\left(\frac{\partial x}{\partial \xi_3}\right)_1 + L_2(\xi_1)L_1(\xi_2)H_1^1(\xi_3)\left(\frac{\partial x}{\partial \xi_3}\right)_2 \\
& + L_1(\xi_1)L_2(\xi_2)H_1^1(\xi_3)\left(\frac{\partial x}{\partial \xi_3}\right)_3 + L_2(\xi_1)L_2(\xi_2)H_1^1(\xi_3)\left(\frac{\partial x}{\partial \xi_3}\right)_4 \\
& + L_1(\xi_1)L_1(\xi_2)H_2^1(\xi_3)\left(\frac{\partial x}{\partial \xi_3}\right)_5 + L_2(\xi_1)L_1(\xi_2)H_2^1(\xi_3)\left(\frac{\partial x}{\partial \xi_3}\right)_6 \\
& + L_1(\xi_1)L_2(\xi_2)H_2^1(\xi_3)\left(\frac{\partial x}{\partial \xi_3}\right)_7 + L_2(\xi_1)L_2(\xi_2)H_2^1(\xi_3)\left(\frac{\partial x}{\partial \xi_3}\right)_8,
\end{aligned}
$$

$$(1)$$

where the one-dimensional linear Lagrange basis functions L and cubic Hermite basis functions H are given by

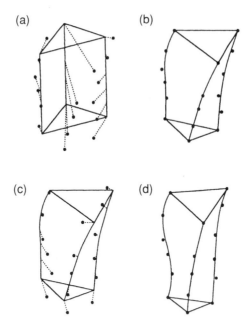

FIGURE 2.12:3. Steps in nonhomogeneous deformation analysis showing 3-D perspective of bilinear—cubic finite element and measured bead positions (top, epicardium; bottom, endocardium). (a) Beads in the end-diastolic reference state projected onto initial element edges. (b) Undeformed element with updated nodal parameters fitted by least squares to end-diastolic bead coordinates. (c) Beads in deformed (end-systolic) configuration projected onto the *same* material points in the undeformed element from (b). (d) Deformed configuration of the element with nodal parameters fitted to end-systolic bead coordinates. From McCulloch and Omens (1991), and Waldman *et al.* (1985), by permission.

$$L_1(\xi) = 1 - \xi, \quad L_2(\xi) = \xi, \tag{2}$$

and

$$H_1^0(\xi) = 1 - 3\xi^2 + 2\xi^3, \quad H_1^1(\xi) = \xi(\xi - 1)^2.$$
$$H_2^0(\xi) = \xi^2(3 - 2\xi), \quad H_2^1(\xi) = \xi^2(\xi - 1). \tag{3}$$

Example of Least-Square Error Fitted Element

In Section 2.10 it was shown that Waldman et al. have made a practice of implanting three columns of radio-opaque beads into the ventricular wall to measure strains in vivo. Using the change of lengths of the six edges of every tetrahedron formed by four beads, we can compute the strains in the tetrahedron based on the hypothesis that the strains are homogeneous. Now, if the beads can be fitted into a finite element with adequate interpolation formulas, then the homogeneity hypothesis can be removed and the accuracy can be improved.

McCulloch and Omens (1991) did this using the steps shown in Figure 2.12:3. Figure 2.12:3(a) shows a perspective drawing of the centers of the beads at end diastole. The reconstructed x-ray coordinates were transformed into a local, rectangular cartesian system (X_1, X_2, X_3), with the X_3 axis parallel to the columns of beads. The dotted lines show the "projections" from each bead center onto the edges of an initial three-dimensional finite element, having six global nodes, one at each vertex. With the bilinear-cubic Hermite interpolation formula (1), let a triangular base be (ξ_1, ξ_2) plane and a column be in the (ξ_3) direction. The location of the bead projections (ξ_d) on the element were chosen so that each column of beads corresponds to a different edge of the element. With respect to an initial finite element, the ξ values were computed for each bead, and the coordinates of the dth bead $X(\xi_d)$ are computed from Eq. (1). The coordinate of the dth data point is X_d. The least squares error function to be minimized with respect to the nodal parameters $x_1, x_2, \ldots x_8, (\partial x/\partial \xi_3)_1, (\partial x/\partial \xi_3)_2, \ldots (\partial x/\partial \xi_3)_8$ is

$$E(X) = \sum_d \gamma_d \left| X(\xi_d) - X_d \right|^2, \tag{4}$$

where γ_d are weighting factors to reflect the relative confidence associated with the dth measurement.

With successful minimization, the updated end-diastolic element is shown in Figure 2.12:3(b). Then the data on the beads locations at end systole are projected onto the end-diastolic fit, as Figure 2.12:3(c) shows. This time, the beads were projected onto the same local coordinates of the finite element as in the end-diastolic state. In Figure 2.12:3(d) the deformed element is fitted to the end-systolic bead coordinates using the least-squares fitting method. Now the position vectors X and x are defined throughout the element. The deformation gradients and the strains can be computed accurately.

Examples of Efficient Elements

The left ventricle is well approximated by a truncated ellipsoid of revolution, as shown in Figure 2.12:4. Its geometry can be defined by the major and minor radii of the two surfaces representing the endocardium and the epicardium. The prolate spherical (elliptic-hyperbolic-polar) coordinates (λ, μ, θ) system can be used for such a body. The cartesian coordinates of a point are given in terms of its prolate spherical coordinates by

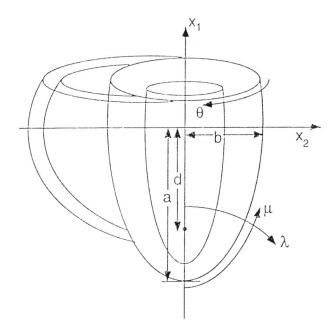

FIGURE 2.12:4. Truncated ellipsoid representation of ventricular geometry, showing major left ventricular radius (a), minor radius (b), focal length (d), and prolate spheroidal coordinates (λ, μ, θ). From McCulloch (1995), by permission.

$$x_1 = d \cosh \lambda \cos \mu,$$
$$x_2 = d \sinh \lambda \sin \mu \cos \theta,$$
$$x_3 = d \sinh \lambda \sin \mu \sin \theta. \tag{5}$$

Here d is the focal length, and $d^2 = a^2 - b^2$ (see Fig. 2.12:4). Two limiting cases are well known: the spherical-polar coordinates and the cylindrical-polar coordinates. If a problem can be solved by using these coordinates, it certainly will be advantageous to use them. Many examples can be found in the literature.

Problems

2.1 Consider a blood vessel (Fig. P2.1). Let us make the following assumptions: (1) When the internal and external pressures are both zero and the length is fixed, it is a uniform long circular cylindrical tube with inner radius R_i and outer radius R_o. (2) In this state the circumferential stress is zero everywhere. (3) The vessel wall material is incompressible. (4) The relationship between the circumferential stress T and the circumferential stretch ratio λ is given by Eq. (5) on p. 328 of *Biomechanics* (Fung, 1993b):

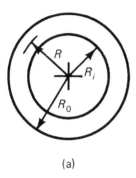
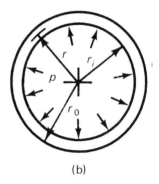

(a)   (b)

FIGURE P2.1. Cross section of a blood vessel when it is (a) unloaded and (b) loaded.

$$T = \left(T^* + \beta\right)e^{\alpha\left(\lambda - \lambda^*\right)} - \beta, \tag{1}$$

where α, β are material constants and (T^*, λ^*) represent a point on the curve, $T = T^*$ when $\lambda = \lambda^*$. The values of α and $E_0 = \alpha\beta$ are given in Figure 8.3:2, p. 330 of that book.

Now let the vessel be subjected to an internal pressure p and an external pressure of zero. The inner radius then becomes r_i. Under assumption (3) (incompressibility), how large would the outer radius r_o be? What would be the stress distribution in the wall? Consider all the stress and strain components. Derive the generalized Navier's equation. Attempt a rigorous solution.

2.2 In Problem 2.1, change the second hypothesis to (2′): At an internal pressure of p, an outer pressure of zero, and an inner radius of r_i, the circumferential stress T is uniform throughout the vessel wall. λ is, therefore, a constant. Now, let the internal pressure be reduced to zero. The blood vessel diameter will be reduced. Calculate the new stress distribution. What assumptions must be made in order to get a unique answer? Use the generalized Navier's equation.

2.3 Consider an approximate theory of the heart (Fig. P2.3). Let the left ventricle be approximated by a spherical shell of uniform wall thickness. In the *end-diastolic* condition the internal pressure is p_d, the active tension in the muscle is zero, and the inner and outer radii are R_i and R_o, respectively. Let us *assume that in the end-diastolic condition the circumferential stress in the wall is uniformly distributed in the wall. Now let the valves be closed, the muscle be stimulated, and the left ventricle be contracted isometrically. Assume that the active muscle tension is also uniformly distributed throughout the wall in this isometric condition.* When the systolic pressure becomes p_s, what would the active tensile stress be?

You may analyze the problem with algebraic symbols. When you have finished, try the following numerical values and find the values of the passive and active tensile stresses:

$$R_i = 3\,\text{cm}, \quad R_o = 3.6\,\text{cm},$$
$$p_d = 11\,\text{cm}\,H_2O \doteq 11 \times 10^3 \,\text{dyne}/\text{cm}^2,$$
$$p_s = 150\,\text{cm}\,H_2O \doteq 150 \times 10^3 \,\text{dyne}/\text{cm}^2.$$

Assume that the active tension corresponds to the point A on the curve of length–tension diagram shown in Figure 2.1:3. What is the sarcomere length of this heart in the end-diastolic condition?

In the second stage of systolic contraction, the aortic valve is open and blood is ejected. At the end of systole the inner and outer radii become r_i, r_o, respectively, and the aortic valve is closed again. Assume that the wall material is uniform and incompressible, that the passive property of the muscle obeys the stress–strain relationship given by

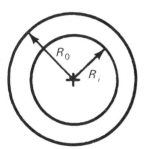

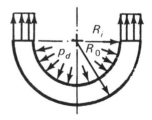

Diastolic condition

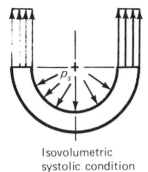

Isovolumetric
systolic condition

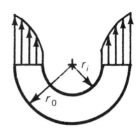

End-systolic
condition

FIGURE P2.3. Simplified ventricular analysis.

Eq. (1) of Problem **2.1**, and that the total stress is the sum of the passive stress and the active stress, as in Hill's model. In this contracted condition, what is the strain (or stretch ratio) in the ventricular wall? What is the passive stress distribution in the wall? What is the active stress distribution in the wall? Derive rigorous equations in terms of stress and strain tensors. Attempt an approximate solution.

2.4 Formulate a mathematical analysis of the blood flow in the opening process of the aortic and mitral valves. According to experimental results, it is allowable to assume velocity of flow as uniform across the cross section of the valve at any time. Use of this assumption will simplify the analysis. An example can be found in Lee and Talbot (1979).

2.5 Stenotic heart valves have increased resistance to flow. Use dimensional analysis [by comparing the physical dimensions of various terms, for example, pressure $(ML^{-1}T^{-2})$, where M, L, T, stand for mass, length, and time; velocity (LT^{-2}); and the so-called Π-theorem], find the relationship between the pressure drop across a heart valve, the average velocity of flow through the valve, the cross-sectional area of the valve, and the density of the fluid. An empirical constant will be involved, which can be determined by experiments. Find a formula for the determination of the cross-sectional area of the valve.

Note: For dimensional analysis, see *Biomechanics*, (Fung, 1993b, pp. 168–172). For experiments on stenotic heart valves, see Gorlin and Gorlin (1951).

2.6 Write down a full set of basic equations and boundary conditions governing the flow of blood in the left ventricle [see e.g., Peskin (1977)].

2.7 Write down a full set of field equations and boundary conditions governing the deformation of the myocardium.

2.8 Write down a full set of equations governing blood flow through the heart valves in the process of closing.

2.9 The flow through heart valves was analyzed by Lee and Talbot (1979) using a cone with straight walls to simulate the aortic valve and a pair of rigid flaps (see Fig. 2.5:5) to simulate the mitral valve. As a result of this simplifying simulation, a singularity exists at the instant of valve closing: The velocity becomes infinitely large and the valves cannot close without regurgitation (backward flow). To remove this difficulty, the valve leaflets should be assumed flexible, with a curved wall. Formulate an improved theory.

2.10 At the end of the diastolic state, the left ventricular wall may be considered as a layered orthotropic pseudoelastic material. One of the material axes of symmetry must coincide with the direction of muscle fibers. Since fiber direction changes systematically throughout the ventricular wall, the directions of the material axes of symmetry change in different layers. Assume a pseudo–strain-energy function for the temporarily resting mycoardium in a form similar to those presented in Chapters 7 and 8 of *Biomechanics* (Fung, 1993b, Sec. 7.10, and pp. 302–304, 345). Use it to specialize the general equations obtained in Problem 2.7.

2.11 In many publications on stress analysis of the myocardium, the mechanical property of the material is assumed to be isotropic. Discuss qualitatively the effect of this hypothesis on the stress distribution in the myocardium (Mirsky, 1979).

2.12 Consider the question of stress concentration in a thick-walled spherical shell subjected to internal pressure. Under the hypotheses that (1) the initial stress is zero when the shell is unloaded and (2) the material obeys Hooke's law, we have the classical solution by Lamé, Eqs. (1) and (2) of Section 2.9. Now, retain the assumption (1) but replace assumption (2) by a more realistic exponentially stiffening stress-strain law, for example, by one of incompressible material with a pseudo–strain-energy function,

$$\rho_0 W = C \exp\left[a_1 E_{\theta\theta}^2 + a_2 E_{\phi\phi}^2 + a_4 E_{\theta\theta} E_{\phi\phi}\right],$$

where $E_{\theta\theta}$, $E_{\phi\phi}$ are strains in the circumferential (θ) and longitudinal (ϕ) directions, respectively, and C, a_1, a_2, a_4 are constants. Will the stress concentration be increased or decreased? Give an estimate.

Note: This pseudo–strain-energy function is discussed in detail in *Biomechanics* (Fung, 1993b, Section 8.6, p. 345).

2.13 The finite-element method should be suitable for the analysis of the stress distribution in the myocardium. For an analysis of myocardium at the end-diastolic state considered in Problem 2.7, how can a finite element be formulated?

2.14 The finite-element method should be useful also for the analysis of the contraction process of the heart. In formulating a finite element of the heart muscle in active contraction, what basic information is needed? For the whole heart, what additional information is needed? How can the coupling of the circulation in the heart to that in the lung and aorta be described mathematically? (The last question is discussed in Waldman, 1983.)

2.15 In Section 2.5, we explained the function of the papillary muscles in the control and stability of the mitral and tricuspid valves. The aortic and pulmonary valves do not have the help of papillary muscles. How do aortic and pulmonary valves resist the transmural pressure acting on them when they are closed? The same Laplace equation (2.5:3) must apply. How are the membrane tensions T_1, T_2 and the radii of curvature r_1, r_2 distributed in the aortic and pulmonary valves when they are closed? Formulate a mathematical theory of aortic valve operation. Derive the governing differential equations and boundary conditions. Obtain a solution in a simplified case.

2.16 Design an experiment to test the no-slip boundary condition of blood flow in a blood vessel.

2.17 Blood drips at the end of a pipette connected to a reservoir as in a blood transfusion apparatus. Let the pipette be vertical and let the drops form under gravity. It is important to know the size of each drop and the rate at which it is formed. A controlling factor is the surface tension. Measuring the drop size is a way of measuring the surface tension between blood and air. Describe mathematically the boundary conditions that apply to the surface of the drop. Make a mathematical analysis of the droplets.

2.18 The inner wall of a blood vessel (endothelium) is easily injured by mechanical stress. Platelets are activated by the injured endothelium, white blood cells begin to attach to the endothelium, and a thrombus is gradually formed. Derive the governing field equations for the blood and the thrombus, and all the relevant boundary conditions. Formulate a theory of vaso-occlusion and find the critical condition.

2.19 A plastic surgeon cuts out a piece of skin and sutures the remainder together. In analyzing the stresses in the skin as a result of the surgery, what are the boundary conditions at the suture? What are the field equations for the skin? What is the stress and strain distribution?

2.20 In a membrane blood oxygenator, blood flows between two sheets of silastic membranes which are exposed to concentrated oxygen. Design a possible oxygenator, write down all the equations necessary in order to analyze the flow of blood in the oxygenator, and solve them.

2.21 Bone is an orthotropic elastic solid. Derive the field equations and boundary conditions for a diarthrodial joint.

2.22 In the derivation of the equations in Section 2.7 the fluid is assumed incompressible. How should these equations be modified to account for fluid compressibility so that they are applicable to a gas?

2.23 Write down all the field equations and boundary conditions necessary for solving the problem of propagation of pulse wave in arteries.

2.24 Do the same for the ventilation in airways.

2.25 Think of some problems, in biology in which gas–fluid interfaces exist and in which surface tension on the interfaces is important. Formulate the problem mathematically, and discuss its solution.

2.26 Think of and formulate mathematically some problems, real or hypothetical, in which surface viscosity plays an important role.

2.27 Consult a book on anatomy; sketch a knee joint. It is filled with viscoelastic synovial fluid. The ends of bone are covered with articulate cartilage, which is a porous, compressible tissue. Give a mathematical formulation of the problem of joint lubrication as one walks or runs.

2.28 A surgeon has to excise a certain dollar-sized circular patch of skin from a patient with a skin tumor. A circular patch was cut off and the edges sewn together. What would be the stress and strain distribution in the skin after surgery? What is the tension in the suture? Formulate an approximate theory for the solution.

Note: In the theory of functions of a complex variable, a transformation from $z = x + iy$ to $w = u + iv$ by

$$w = z + \frac{1}{z}$$

transforms a circle to a straight line segment. Would this knowledge be helpful?

2.29 Consider the skin surgery problem of Problem 2.28 again. There will be a large stress (and strain) concentration at the ends of the line of suture if the surgery was done as stated in Problem 2.28. To minimize the stress concentration, a diamond-shaped area of skin is excised and the edges sewn together. Formulate the mathematical problem and explain why is the stress concentration reduced?

2.30 Consider the skin problem again. Design a better way of excising a piece of diseased skin from the point of view of minimizing stress concentration. In skin surgery there is a technique called **z**-plasty. Explain why is it a good technique.

2.31 In some soft tissues there is a natural direction along which collagen fibers are aligned. Healing will be faster if a cut is made in such a direction as compared with a cut perpendicular to this direction. Explain this qualitatively and formulate a mathematical theory for this.

References

Arts, T., and Reneman, R.S. (1980). Measurements of deformation of canine epicardium in vivo during cardiac cycle. *Am. J. Physiol.* **239**: H432–H437.

Arts, T., Reneman, R.S., and Veenstra, P.C. (1982). Epicardial deformation and left ventricular wall mechanics during ejection in the dog. *Am. J. Physiol.* **243**: H379–H390.

Bellhouse, B.J., and Bellhouse, F.H. (1969). Fluid mechanics of model normal and stenosed aortic valves. *Circ. Res.* **25**: 693–704.

Bellhouse, B.J., and Bellhouse, F.H. (1972). Fluid mechanics of a model mitral valve and left ventricle. *Cardiovasc. Res.* 6: 199–210.

Berne, R.M., and Sperelakis, N. (eds.) (1979). *Handbook of Physiology.* Sec. 2. *The Cardiovascular System*, Vol. 1. *The Heart.* American Physiological Society, Bethesda, MD.

Blum, W.F., McCulloch, A.D., and Lew, W.Y.W. (1995). Active force in rabbit ventricular myocytes. *J. Biomech.* **28**: 1119–1122.

Brady, A.J. (1984). Passive stiffness of rat cardiac myocytes. *J. Biomech. Eng.* **106**: 25–30.

Braunwald, E. (ed.) (1988). *Heart Disease.* 3rd Edition, Saunders Co., Philadelphia, PA.

Costa, K.D., Hunter, P.J., Rogers, J.M., Guccione, J.M., Waldman, L.K., and McCulloch, A.D. (1996). A three-dimensional finite element method for large elastic deformations of ventricular myocardium: Part 1 cylindrical and spherical polar coordinates. *J. Biomech. Eng.* Submitted.

Daniels, M., Noble, M.I.M., ter Keurs, H.E.D.J., and Wohlfart, B. (1984). Velocity of sarcomere shortening in rat cardiac muscle: relationship to force, sarcomere length, calcium and time. *J. Physiol.* **355**: 367–381.

Debes, J.C., and Fung, Y.C. (1995). Biaxial mechanics of excised canine pulmonary arteries, *Am. J. Physiol.* **269**: H433–H442.

Deng, S.X., Tomioka, J., Debes, J.C., and Fung, Y.C. (1994). New experiments on shear modulus of elasticity of arteries. *Am. J. Physiol.* **266**: H1–H10.

Dieudonné, J.M. (1969). La determination experimentale des contraintes myocardiques. *J. Physiol. (Paris)* **61**: 199–218.

Edman, K.A.P., and Nilsson, E. (1972). Relationship between force and velocity of shortening in rabbit papillary muscle. *Acta Physiol. Scand.* **85**: 488–500.

Frank, O. (1899). Die grundform des arteriellen pulses. Erste Abhandlung, Mathematische Analyse. *Z. Biol.* **37**: 483–526.

Fung, Y.C. (1965). *Foundations of Solid Mechanics.* Prentice-Hall, Englewood Cliffs, NJ.

Fung, Y.C. (1970). Mathematical representation of the mechanical properties of the heart muscle. *J. Biomech.* **3**: 381–404.

Fung, Y.C. (1971). Stress-strain-history relation of soft tissues in simple elongation, In *Biomechanics: Its Foundation and Objectives.* (Fung, Y.C., Perrone, N., and Anliker, M., eds.), Prentice-Hall, Englewood Cliffs, NJ, pp. 181–208.

Fung, Y.C. (1973). Biorheology of soft tissues, *Biorheology*, **19**: 139–155.

Fung, Y.C. (1979). Inversion of a class of nonlinear stress-strain relationships of biological soft tissues. *J. Biomech. Eng.*, **101**: 23–27.

Fung, Y.C. (1983). What principle governs the stress distribution in living organisms, In *Biomechanics in China, Japan, and USA.* (Fung, Y.C., Fukada, E., and Wang, J.J., eds.), Science Press, Beijing, pp. 1–13.

Fung, Y.C. (1988). Cellular growth in soft tissues affected by the stress level in service, In *Tissue Engineering.* (Skalak, R., and Fox, C.F., eds.), Alan Liss, Inc., New York, pp. 45–50.

Fung, Y.C. (1990). *Biomechanics: Motion, Flow, Stress and Growth*, Springer-Verlag, New York.

Fung, Y.C. (1993a). *A First Course in Continuum Mechanics*, 3rd Edition, Prentice-Hall, Englewood Cliffs, NJ, pp. 165–180.

Fung, Y.C. (1993b). *Biomechanics: Mechanical Properties of Living Tissues*, 2nd Edition, Springer-Verlag, New York.

Fung, Y.C., and Liu., S.Q. (1995). Determination of the mechanical properties of the different layers of blood vessels in vivo. *Proc. U.S. Natl. Acad. Sci.* **92**: 2169–2173.

Fung, Y.C., Fronek, K., and Patitucci, P. (1979). Pseudoelasticity of arteries and the choice of its mathematical expression, *Am. J. Physiol.* **237**, H620–H631.

Fung, Y.C., Liu, S.Q., and Zhou, J. (1993). Remodeling of the constitutive equation while a blood vessel remodels itself under stress, *J. Biomech. Eng.* **115**: 453–459.

Glass, L., McCulloch, A., and Hunter, P. (eds.) (1991). *Theory of Heart*. Springer-Verlag, New York.

Gorlin, R., and Gorlin, S.G. (1951). Hydraulic formula for calculation of the area of the stenotic mitral valve, other cardiac valves, and central circulatory shunts. *Am. Heart J.* **41**: 1–29.

Green, A.E., and Adkins, J.E. (1960). *Large Elastic Deformations and Non-linear Continuum Mechanics*. Oxford Univ. Press, London.

Guccione, J.M., and McCulloch, A.D. (1991). Finite element modeling of ventricular mechanics. In *Theory of Heart*, pp. 124–144, see Glass et al. (1991).

Guccione, J.M., and McCulloch, A.D. (1993). Mechanics of active contraction in cardiac muscle: Part 1—constitutive relations for fiber stress that describe deactivation. *J. Biomech Eng.* **115**: 72–81.

Guccione, J.M., Costa, K.D., and McCulloch, A.D. (1995). Finite element stress analysis of left ventricular mechanics in the beating heart. *J. Biomech.* **28**: 1167–1177.

Guccione, J.M., McCulloch, A.D., and Waldman, L.K. (1991). Passive material properties of intact ventricular myocardium determined for a cylindrical model. *J. Biomech Eng.* **113**: 42–55.

Hales, S. (1733). *Statical Essays: II. Haemostaticks*. Innays and Manby, London. Reprinted by Hafner, New York.

Hashima, A.R., Young, A.A., McCulloch, A.D., and Waldman, L.K. (1993). Nonhomogeneous analysis of epicardial strain distributions during acute myocardial ischemia in the dog. *J. Biomech.* **26**: 19–35.

Henderson, Y., and Johnson, F.E. (1912). Two modes of closure of the heart valves. *Heart* **4**: 69–82.

Hill, A.V. (1939). The heat of shortening and the dynamic constants of muscle. *Proc. R. Soc. London (Biol.) B* **126**: 136–195.

Holmes, J.W., Yamashita, H., Waldman, L.K., and Covell, J.W. (1994). Scar remodeling and transmural deformation after infarction in the pig. *Circulation* **90**: 411–420.

Horowitz, A. (1991). Structural considerations in formulating material laws for the myocardium. In *Theory of Heart*, pp. 31–58, see Glass et al. (1991).

Hort, W. (1960). Makroskopische und mikrometrische untersuchungen am Myokard verschieden stark gefullter linker kammern. *Virchows Arch. Path. Anat.* **333**: 523–564.

Humphrey, J.D., and Yin, F.C.P. (1989a). Biomechanical experiments on excised myocardium: theoretical considerations. *Am. J. Physiol.* **22**: 377–383.

Humphrey, J.D., and Yin, F.C.P. (1989b). Constitutive relations and finite deformations of passive cardiac tissue II: stress analysis in the left ventricle. *Circ. Res.* **65**: 805–817.

Humphrey, J.D., Strumpf, R.K., and Yin, F.C.P. (1990). Biaxial mechanical behavior of excised ventricular epicardium. *Am. J. Physiol.* **259**: H101–H108.

Humphrey, J.D., Strumpf, R.K., Halperine, H., and Yin, F. (1991). Toward a stress analysis in the heart. In *Theory of Heart*, pp. 59–75, see Glass et al. (1991).

Hunter, W.C., Janicki, J.S., Weber, K.T., and Noordergraaf, A. (1983). Systolic mechanical properties of the left ventricle: effects of volume and contractile state. *Circ. Res.* **52**: 319–327.

Huntsman, L.L., Rondinone, J.F., and Martyn, D.A. (1983). Force-length relations in cardiac muscle segments. *Am. J. Physiol.* **244**: H701–H707.

Janz, R.F., and Grimm, A.F. (1973). Deformation of the diastolic left ventricle. I. Nonlinear elastic effects. *Biophys. J.* **13**: 689–704.

Janz, R.F., and Waldron, R.J. (1976). Some implications of a constant fiber stress hypothesis in the diastolic left ventricle. *Bull. Math. Biol.* **38**: 401–413.

Janz, R.F., Grimm, A.F., Kubert, B.R., and Moriarty, T.F. (1974). Deformation of the diastolic left ventricle. II. Nonlinear geometric effects. *J. Biomech.* **7**: 509–516.

Jones, R.T. (1969). Blood flow. In *Annual Review of Fluid Mechanics.* (Sears, W.R., and van Dyke, M., eds.), Annual Reviews, Palo Alto, CA.

Jones, R.T. (1972). Fluid dynamics of heart assist devices. In *Biomechanics: Its Foundations and Objectives.* (Fung, Y.C., Perrone, N., and Anliker, M., eds.), Prentice-Hall, Englewood Cliffs, NJ, Chapter 1, pp. 549–565.

Krueger, J.W., Tsujioka, K., Okada, T., Peskin, C.S., and Lacker, H.M. (1988). A "give" in tension and sarcomere dynamics in cardiac muscle relaxation. *Adv. Exp. Med. Biol.* **226**: 567–580.

Lacker, H.M., and Peskin, C.S. (1986). A mathematical method for unique determination of crossbridge properties from steady-state mechanical and energetic experiments on macroscopic muscle. In *Some Mathematical Questions in Biology—Muscle Physiology.* (Miura, R.M., ed.), American Mathematics Society, Providence, RI, pp. 121–153.

Lamé, E. (1852). *Leçons sur la Théorie de l'Elasticité.* Paris.

Lanir, Y. (1983). Constitutive equation for fibrous connective tissue. *J. Biomech.* **16**: 1–12.

Lee, C.S.F., and Talbot, L. (1979). A fluid mechanical study on the closure of heart valves. *J. Fluid Mech.* **91**: 41–63.

LeGrice, I.J., Smail, B.H., Chai, L.Z., Edgar, S.G., Gavin, J. B., and Hunter, P.J. (1995). Laminar structure of the heart: ventricular myocyte arrangement and connective tissue architecture in the dog. *Am. J. Physiol.* **269**: H571–H582.

MacKenna, D.A., Omens, J.H., McCulloch, A.D., and Covell, J.W. (1994). Contribution of collagen matrix to passive left ventricular mechanics in isolated rat hearts. *Am. J. Physiol.* **266**: H1007–H1018.

McCulloch, A.D. (1995). Cardiac mechanics. In *Biomedical Engineering Handbook.* (Bronzino, J.D., ed.), Chapter 31, pp. 418–439. CRC Press, Inc. Boca Raton, FL.

McCulloch, A.D., and Omens, J.H. (1991). Factors affecting the regional mechanics of the diastolic heart. In *Theory of Heart*, pp. 87–119, see Glass et al. (1991).

McCulloch, A.D., and Omens, J.H. (1991). Non-homogeneous analysis of three-dimensional transmural finite deformation in canine ventricular myocardium. *J. Biomech.* **24**: 539–548.

McCulloch, A.D., Smail, B.H., and Hunter, P.J. (1987). Left ventricular epicardial deformation in isolated arrested dog heart. *Am. J. Physiol.* **252**: H233–H241.

McDonald, D.A. (1974). *Blood Flow in Arteries.* Williams & Wilkins, Baltimore, MD.

Meier, G.D., Bove, A.A., Santamore, W.P., and Lynch, P.R. (1980). Contractile function in canine right ventricle. *Am. J. Physiol.* **239**: H794–H804.

Mirsky, I. (1973). Ventricular and arterial wall stresses based on large deformation analysis. *Biophys. J.* **13**: 1141–1159.

Mirsky, I. (1979). Elastic properties of the myocardium: A quantitative approach with physiological and clinical applications. In *Handbook of Physiology, Sec. 2, Vol. 1. The Heart.* (Berne, R.M., and Sperelakis, N., eds.), American Physiological Society, Bethesda, MD., pp. 497–531.

Netter, F. (1969). *The Ciba Collection of Medical Illustrations*, Vol. 5, *Heart*, CIBA Publications Dept., Summit, NJ.

Omens, J.H., and Covell, J.W. (1991). Transmural distribution of myocardial tissue growth induced by volume-overload hypertrophy in the dog. *Circulation* **84**: 1235–1245.

Omens, J.H., and Fung, Y.C. (1989). Residual strain in the rat left ventricle. *Circ. Res.* **66**: 37–45.

Omens, J.H., Mac Kenna, D.A., and McCulloch, A.D. (1993). Measurement of strain and analysis of stress in resting rat left ventricular myocardium. *J. Biomech.* **26**: 665–676.

Omens, J.H., May, K.D., and McCulloch, A.D. (1991). Transmural distribution of three-dimensional strain in the isolated arrested canine left ventricle. *Am. J. Physiol.* **261**: H918–H928.

Omens, J.H., Rockman, H.A., and Covell, J.W. (1994). Passive ventricular mechanics in tight-skin mice. *Am. J. Physiol.* **266**: H1169–H1176.

Ono, S., Waldman, L.K., Yamashita, H., Covell, J.W., and Ross, Jr., J. (1995). Effect of coronary artery reperfusion on transmural mycoardial remodeling in dogs. *Circulation* **91**: 1143–1153.

Parmley, W.W., and Sonnenblick, E.H. (1967). Series elasticity of heart muscle: Its relation to contractile element velocity and proposed muscle models. *Circ. Res.* **20**: 112–123.

Parmley, W., and Talbot, L. (1979). Heart as a pump. In *Handbook of Physiology. Sec. 2. The Cardiovascular System, Vol. 1, The Heart.* (Berne, R.M., and Sperelakis, N., eds.), American Physiological Society, Bethesda, MD, pp. 429–460.

Parmley, W.W., Brutsaert, D.L., and Sonnenblick, E.H. (1969). The effects of altered loading on contractile events in isolated cat papillary muscle. *Circ. Res.* **24**: 521–532.

Panerai, R.B. (1980). A model of cardiac muscle mechanics and energetics. *J. Biomech.* **13**: 929–940.

Peskin, C.S. (1977). Numerical analysis of blood flow in the heart. *J. Comput. Phys.* **25**: 220–252.

Peskin, C.S., and Wolfe, A.W. (1978). The aortic sinus vortex. *Fed. Proc.* **37**: 2784–2792.

Pinto, J.G., and Fung, Y.C. (1973a). Mechanical properties of the heart muscle in the passive state. *J. Biomech.* **6**: 597–616.

Pinto, J.G., and Fung, Y.C. (1973b). Mechanical properties of stimulated papillary muscle in quick-release experiments. *J. Biomech.* **6**: 617–630.

Prinzen, F.W., Arts, T., Van der Vusse, G.J., Comans, W.A., and Reneman, R.S. (1986). Gradients in fiber shortening and metabolism across the left ventricler wall. *Am. J. Physiol.* **250**: H255–H264.

Rodriquez, E.K., Hoger, A., and McCulloch, A.D. (1994). Stress-dependent finite growth in soft elastic tissues. *J. Biomech.* **27**: 455–467.

Rodriquez, E.K., Omens, J.H., Waldman, L.K., and McCulloch, A.D. (1993). Effect of residual stress on transmural sarcome length distributions in rat left ventricle. *Am. J. Physiol.* **264**: H1048–H1056.

Rogers, J.M., and McCulloch, A.D. (1994). A collocation-Galerkin finite element model of cardiac action potential propagation. *IEEE Trans. Biomed. Eng.* **41**: 743–757.

Scher, A.M. and Spach, M.S. (1979). Cardiac depolarization and repolarization and the electrocardiogram. In *Handbook of Physiology, Sec. 2, Vol. 1, The Heart.* (Berne, R.M., and Sperelakis, N., eds.), American Physiological Society, Bethesda, MD, pp. 357–392.

Smail, B.H., and Hunter, P.J. (1991). Structure and function of the diastolic heart. In *Theory of Heart*, pp. 1–30, see Glass et al. (1991).

Sonnenblick, E.H. (1964). Series elastic and contractile elements in heart muscle: Changes in muscle length. *Am. J. Physiol.* **207**: 1330–1338.

Sonnenblick, E.H., Ross, Jr. , Covell, J.W., Spontnitz, H.M., and Spiro, D.   (1967). Ultrastructure of the heart in systole and diastole: Changes in sarcomere length. *Circ. Res.* **21**: 423–431.

Streeter, Jr., D. (1979). Gross morphology and fiber geometry of the heart. In *Handbook of Physiology, Sec. 2, Cardiovascular System. Vol. 1. The Heart.* (Berne, R.M., and Sperelakis, N., eds.), American Physiology Society, Bethesda, MD, pp. 61–112.

Streeter, Jr., D., and Hanna, W.T. (1973). Engineering mechanics for successive states in canine left ventricular myocardium. I. Cavity and wall geometry. II. Fiber angle and sarcomere length. *Circ. Res.* **33**: 639–655(I), 656–664(II).

Streeter, D., Jr., Spotnitz, H.M., Patel, D.J., Ross, Jr., J., and Sonnenblick, E.H. (1969). Fiber orientation in the canine left ventricle during diastole and systole. *Circ. Res.* **24**: 339–347.

Suga, H., Sagawa, K., and Shoukas, A.A. (1973). Load independence of the instantaneous pressure-volume ratio of the canine left ventricle and effects of epinephrine and heart rate on the ratio. *Circ. Res.* **32**: 314–322.

Takamizawa, K., and Hayshi, K. (1987). Strain energy density function and uniform strain hypothesis for arterial mechanics. *J. Biomech.* **20**: 7–17.

Takamizawa, K., and Matsuda, T. (1990). Kinematics for bodies undergoing residual stress and its applications to the left ventricle. *J. Appl. Mech.* **57**: 321–329.

ter Keurs, H.E.D.J. (1983). Calcium in contractility. In *Cardiac Metabolism*, (Drake-Holland, A.J., and Noble, M.I.M., eds.), Wiley, New York, pp. 73–99.

ter Keurs, H.E.D.J., Rijnsburger, W.H., Van Heuningen, R., and Nagelsmit, M.J. (1980). Tension development and sarcomere length in rat cardiac trabeculate: evidence of length-dependent activation. *Circ. Res.* **46**: 703–713.

Tong, P., and Fung, Y.C. (1976). The Stress-Strain Relationship for the Skin. *J. Biomech.* **9**: 649–657.

Tözeren, A. (1985). Continuum rheology of muscle contraction and its application to cardiac contractillity. *Biophys. J.*, **47**: 303–309.

Vaishnav, R.N., and Vossoughi, J. (1983). Estimation of the residual strains in aortic segments. In *Biomedical Engineering*, II, *Recent Developments.* (Hall, C.W., ed.), Pergamon Press, New York, pp. 330–333.

Van Leuven, S.L., Waldman, L.K., McCulloch, A.D., and Covell, J.W. (1994). Gradients of epicardial strain across the perfusion boundary during acute myocardial ischemia. *Am. J. Physiol.* **267**: H2348–H2362.

Villarreal, F.J., Waldman, L.K., and Lew, W.Y.W. (1988). A technique for measuring regional two-dimensional finite strains in canine left ventricle. *Circ. Res.* **62**: 711–721.

Villarreal, F.J., Lew, W.Y.W., Waldman, L.K., and Covell, J.W. (1991). Transmural myocardial deformation in the ischemic canine left ventricle. *Circ. Res.* **68**: 368–381.

Waldman, L.K. (1983). On the Mechanical Coupling of the Heart to the Circulation. Ph.D. thesis. University of California, San Diego, CA.

Waldman, L.K. (1991). Multidimensional measurements of regional strains in the intact heart. In *Theory of Heart*, pp. 145–174, see Glass et al. (1991).

Waldman, L.K., and Covell, J.W. (1987). Effects of ventricular pacing on finite deformation in canine left ventricle. *Am. J. Physiol.* **252**: H1023–H1030.

Waldman, L.K., and McCulloch, A.D. (1993). Nonhomogeneous ventricular wall strain: Analysis of errors and accuracy. *J. Biomech. Eng.* **115**: 497–502.

Waldman, L.K., Fung, Y.C., Covell, J.W. (1985). Transmural myocardial deformation in the canine left ventricle: normal in vivo three-dimensional finite strains. *Circ. Res.* **57**: 152–163.

Waldman, L.K., Nosan, D., Villarreal, F.J., and Covell, J.W. (1988). Relation between transmural deformation and local myofiber direction in canine left ventricle. *Circ. Res.* **63**: 550–562.

Wetterer, E., and Kenner, T. (1968). *Die Dynamik des Arterien-Pulses*. Springer-Verlag, Berlin.

Whittaker, P., Kloner, R.A., Boughner, D.R., and Pickering, J.G. (1994). Quantitative assessment of myocardial collagen with picrosirius red staining and circularly polarized light. *Basic Res. in Cardiol.* **89**: 397–410.

Wong, A.Y.K., and Rautaharju, P.M. (1968). Stress distribution within the left ventricular wall approximated as a thick ellipsoidal shell. *Am. Heart J.* **75**: 649–662.

Xie, J.P., Zhou, J., and Fung, Y.C. (1995). Bending of blood vessel wall: Stress-strain laws of the intima-media and adventitial layers. *J. Biomech. Eng.* **117**: 136–145.

Yoran, C., Covell, J.W., and Ross, Jr., J. (1973). Structural basis for the ascending limb of left ventricular function. *Circ. Res.* **32**: 297–303.

Young, A. (1991). Epicardial deformation from coronary cinéangiogranis. In *Theory of Heart*, pp. 175–207, see Glass et al. (1991).

Yu, Q., Zhou, J.B., and Fung, Y.C. (1993). Neutral axis location in Bending and Young's modulus of different layers of arterial wall. *Am. J. Physiol.* **265**: H52–H60.

Zhou, J. (1992). *Theoretical Analysis of Bending Experiments on Aorta and Determination of Constitutive Equations of materials in Different Layers of Arterial Walls*. Doctoral Dissertation, University of California, San Diego, CA.

Zienkiewicz, O.C., and Morgan, K. (1982). *Finite Elements and Approximation*. Wiley, New York.

3
Blood Flow in Arteries

3.1 Introduction

The larger systemic arteries, shown in Figure 3.1:1, conduct blood from the heart to the peripheral organs. Their dimensions are given in Table 3.1:1. In humans, the aorta originates in the left ventricle at the aortic valve, and almost immediately curves about 180°, branching off to the head and upper limbs. It then pursues a fairly straight course downward through the diaphragm to the abdomen and legs. The aortic arch is tapered, curved, and twisted (i.e., its centerline does not lie in a plane). Other arteries have constant diameter between branches, but every time a daughter branch forks off the main trunk the diameter of the trunk is reduced. Overall, the aorta may be described as tapered. In the dog, the change of area fits the exponential equation,

$$A = A_0 e^{(-Bx/R_0)},$$

where A is the area of the aorta, A_0 and R_0 are, respectively, the area and radius at the upstream site, x is the distance from that upstream site, and B is a "taper factor," which has been found to lie between 0.02 and 0.05. Figure 3.1:2 shows a sketch of the dog aorta.

   If there is a fluid of sufficiently large quantity in static condition outside a blood vessel so that the blood vessel may be considered as an isolated tube bathed in a large reservoir, then at a given blood pressure the stress in the blood vessel wall depends on the radius and wall thickness of the vessel. These quantities change considerably with age (see, e.g., Fig. 3.1:3). Associated with these geometric changes are changes in elastic properties. In the thoracic aorta, at a physiological pressure of $1.33 \times 10^4 \, Nm^{-2}$ (100 mm Hg), the incremental Young's modulus E increases steadily with age; but in more peripheral vessels there is either no change or a fall [Fig. 3.1:4(a) and (b)]. The explanation for this appears to be that the diameter of the thoracic aorta increases with age, whereas that of the iliac and femoral arteries either decreases or changes little with age (see Fig. 3.1:3);

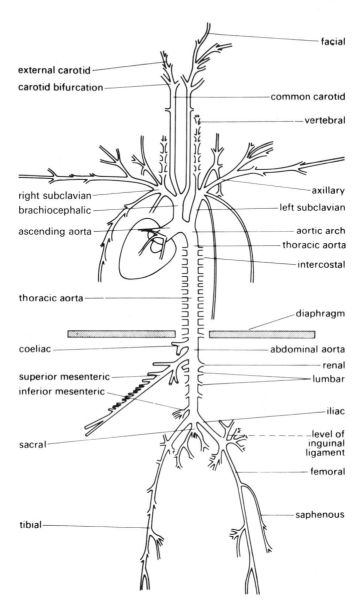

FIGURE 3.1:1. Major branches of the canine arterial tree. From McDonald (1974), by permission.

thus, at the same transmural pressure the stress in the thoracic aorta of the old is greater than that of the young, whereas the reverse is true for the iliac or femoral arteries. A glance at the nonlinear stress–strain relationship, as shown in Figure 7.5:1 of *Biomechanics: Mechanical Properties of Living*

TABLE 3.1:1. Normal Values for Canine Cardiovascular Parameters

Site	Units	Ascending aorta	Descending aorta	Abdominal aorta	Femoral artery	Carotid artery	Arteriole	Capillary	Venule	Inferior vena cava	Main pulmonary artery
Internal diameter d_i	cm	1.5 / 1.0–2.4	1.3 / 0.8–1.8	0.9 / 0.5–1.2	0.4 / 0.2–0.8	0.5 / 0.2–0.8	0.005 / 0.001–0.008	0.0006 / 0.0004–0.0008	0.004 / 0.001–0.0075	1.0 / 0.6–1.5	1.7 / 1.0–2.0
Wall thickness h	cm	0.065 / 0.05–0.08		0.05 / 0.04–0.06	0.04 / 0.02–0.06	0.03 / 0.02–0.04	0.002	0.0001	0.0002	0.015 / 0.01–0.02	0.02 / 0.01–0.03
h/d_i		0.07 / 0.055–0.084		0.06 / 0.04–0.09	0.07 / 0.055–0.11	0.08 / 0.053–0.095	0.4	0.17	0.05	0.015	0.01
Length	cm	5	20	15	10	15 / 10–20	0.15 / 0.1–0.2	0.06 / 0.02–0.1	0.15 / 0.1–0.2	30 / 20–40	3.5 / 3–4
Approximate cross-sectional area	cm^2	2	1.3	0.6	0.2	0.2	2×10^{-5}	3×10^{-7}	2×10^{-5}	0.8	2.3
Total vascular cross-sectional area at each level	cm^2	2	2	2	3	3	125	600	570	3.0	2.3
Peak blood velocity	cm s^{-1}	120 / 40–290	105 / 25–250	55 / 50–60	100 / 100–120		0.75 / 0.5–1.0	0.07 / 0.02–0.17	0.35 / 0.2–0.5	25 / 15–40	70
Mean blood velocity	cm s^{-1}	20 / 10–40	20 / 10–40	15 / 8–20	10 / 10–15					15	15 / 6–28
Reynolds number (peak)		4500	3400	1250	1000		0.09	0.001	0.035	700	3000
α (heart rate 2 Hz)		13.2	11.5	8	3.5	4.4	0.04	0.005	0.035	8.8	15
Calculated wave speed c_o	cm s^{-1}	580		770	840	850				100	350
Measured wave speed c	cm s^{-1}	500 / 400–600		700 / 600–750	900 / 800–1030	800 / 600–1100				400 / 100–700	250 / 200–330
Young's modulus E	Nm$^{-2} \times 10^5$	4.8 / 3–6		10 / 9–11	10 / 9–12	9 / 7–11				0.7 / 0.4–1.0	6 / 2–10

An approximate average value, and then the range, is given where possible.
From Caro, Pedley, and Seed (1974). Reproduced by permission.

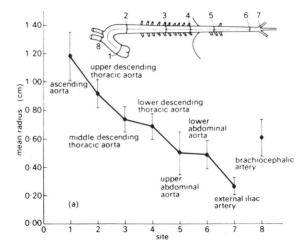

FIGURE 3.1:2. A sketch of the dog's aorta from data measured at physiological pressure in 10 large dogs. From Fry, Griggs, Jr., and Greenfield, Jr. (1964) In vivo studies of pulsatile blood flow. In *Pulsatile Blood Flow*, Attinger, (ed.). McGraw-Hill, New York, p. 110, by permission.

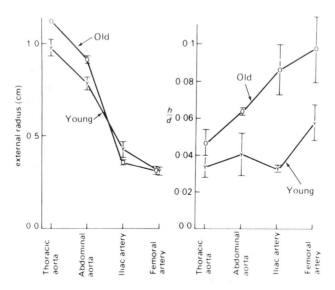

FIGURE 3.1:3. The radius and wall thickness of human arteries for young (Y) and old (O) persons. From Learoyd and Taylor, (1966) Alterations with age in the viscoelastic properties of human arterial walls. *Circ. Res.* **18**: 278–292, by permission of the American Heart Association.

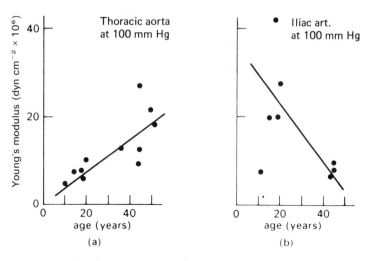

FIGURE 3.1:4. Incremental modulus of elasticity of arteries of normal young and old persons at a pressure of 100 mm Hg. (a) Thoracia aorta. (b) Iliac artery. From Learoyd and Taylor, (1966) *Circ. Res.* **18**: 278–292, by permission of the American Heart Association.

Tissues (Fung, 1993b, p. 270), or Figure 3.1:5 of this book, tells us that at higher stress the incremental Young's modulus is larger.

A detailed discussion of the mechanical properties of arteries is given in Chapter 8 of *Biomechanics: Mechanical Properties of Living Tissues* (Fung, 1993b). In the present chapter we consider the flow of blood in these elastic vessels. The basic equations of hemodynamics are presented in Section 2.6. In most organs, the tissue outside the blood vessel cannot be considered to be a static fluid reservoir. In the skeletal muscle and myocardium, the blood vessels are closely integrated with muscle cells. In the lung, the pulmonary arteries and veins are tethered by interalveolar septa. The interaction between a blood vessel and its surrounding tissues then becomes a major feature of the blood flow in organs. This will be illustrated in Chapters 6 to 8.

We proceed from the simple to the complex. First, we give a solution to the problem of steady flow in a uniform rigid pipe. It is interesting to see that this simple solution has important applications. We also see that even this simple case has some very difficult aspects, for example, the questions of stability and turbulence. We then proceed to study aspects of flow in elastic tubes, first steady flow and then wave propagation. Reflection and transmission of waves in branching vessels is a subject of major interest. This brings us to pulsatile flow in the arteries, nonlinear effects, flow separation, entrance flow, messages carried in the pulse waves, and finally, mechanics of atherogenesis.

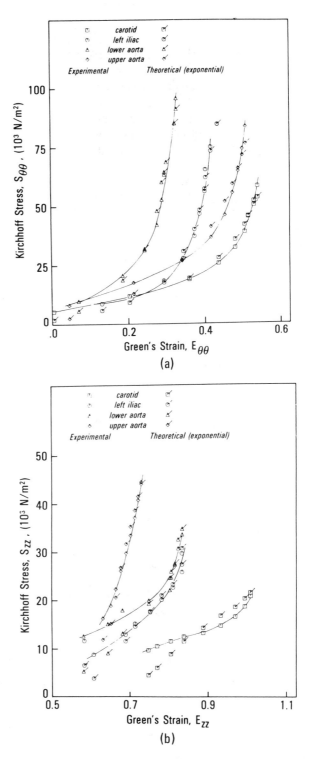

FIGURE 3.1:5. The stress–strain relationship of the thoracic aorta of the rabbit. From Fung, Fronek, and Patitucci (1979), by permission.

3.2 Laminar Flow in a Channel or Tube

Consider first a steady flow of an incompressible Newtonian fluid in a rigid, horizontal channel of width $2h$ between two parallel planes, as shown in Figure 3.2:1. The channel is assumed horizontal so that the gravitational effect (a body force) may be ignored. The walls are assumed to be so rigid that their geometry is uninfluenced by the flow. The coordinate system is shown in the figure, with x parallel to the wall.

With (u, v, w) denoting velocity, μ denoting the coefficient of viscosity, p the pressure, τ the shear stress, γ the shear strain rate, we search for a uniaxial flow with u the only nonvanishing velocity which is a function of y:

$$u = u(y), \quad v = 0, \quad w = 0, \tag{1}$$

The equation of continuity [Eq. (8) of Section 2.6],

$$\frac{\partial u}{\partial x} + \frac{\partial v}{\partial y} + \frac{\partial w}{\partial z} = 0$$

is satisfied by Eq. (1). The equations of motion [Eq. (13) of Sec. 2.6] which have been reduced to the Navier-Stokes equations [Eq. (18) of Sec. 2.6], now become

$$0 = -\frac{\partial p}{\partial x} + \mu \frac{d^2 u}{dy^2}, \tag{2}$$

$$0 = \frac{\partial p}{\partial y}, \tag{3}$$

$$0 = \frac{\partial p}{\partial z}. \tag{4}$$

The no-slip conditions on the boundaries $y = \pm h$ are

$$u(h) = 0, \quad u(-h) = 0. \tag{5}$$

Equations (3) and (4) show that p is a function of x only. If we differentiate Eq. (2) with respect to x and use Eq. (1), we obtain $\partial^2 p/\partial x^2 = 0$. *Hence $\partial p/\partial x$ must be a constant.* Equation (2) then becomes

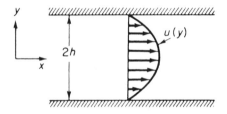

FIGURE   3.2:1. Laminar   flow   in   a channel.

$$\frac{d^2u}{dy^2} = \frac{1}{\mu}\frac{dp}{dx}, \tag{6}$$

which has the solution

$$u = A + By + \frac{1}{\mu}\frac{y^2}{2}\frac{dp}{dx}. \tag{7}$$

The two constants A and B can be determined by the boundary conditions (2) to yield the final solution

$$u = -\frac{1}{2\mu}\left(h^2 - y^2\right)\frac{dp}{dx}. \tag{8}$$

Thus, the velocity profile is a parabola.

A corresponding problem is the flow through a horizontal circular cylindrical tube of radius a (Fig. 3.2:2). We search for a solution, as follows:

$$u = u(y, z), \quad v = 0, \quad w = 0.$$

In analogy with Eq. (6), the Navier–Stokes equation becomes

$$\frac{\partial^2 u}{\partial y^2} + \frac{\partial^2 u}{\partial z^2} = \frac{1}{\mu}\frac{dp}{dx}, \tag{9}$$

where dp/dx is a constant. For convenience we will use cylindrical polar coordinates x, r, θ, with $r^2 = y^2 + z^2$, instead of the cartesian coordinates x, y, z. Then Eq. (9) becomes

$$\frac{\partial^2 u}{\partial y^2} + \frac{\partial^2 u}{\partial z^2} = \frac{1}{r}\frac{\partial}{\partial r}\left(r\frac{\partial u}{\partial r}\right) + \frac{1}{r^2}\frac{\partial^2 u}{\partial \theta^2} = \frac{1}{\mu}\frac{dp}{dx}. \tag{10}$$

Let us assume that the flow is symmetric so that u is a function of r only; then $\partial^2 u/\partial\theta^2 = 0$, and the equation

$$\frac{1}{r}\frac{d}{dr}\left(r\frac{du}{dr}\right) = \frac{1}{\mu}\frac{dp}{dx} \tag{11}$$

can be integrated immediately to yield

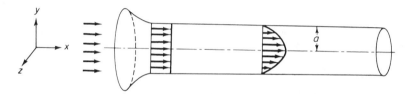

FIGURE 3.2:2. Laminar flow in a circular cylindrical tube.

$$u = \frac{1}{\mu}\frac{r^2}{4}\frac{dp}{dx} + A\log r + B. \tag{12}$$

The constants A and B are determined by the conditions of no-slip at $r = a$ and symmetry on the center line, $r = 0$:

$$u = 0 \quad \text{at} \quad r = a, \tag{13}$$

$$\frac{du}{dr} = 0 \quad \text{at} \quad r = 0. \tag{14}$$

The final solution is

$$u = -\frac{1}{4\mu}\left(a^2 - r^2\right)\frac{dp}{dx}. \tag{15}$$

This is the famous parabolic velocity profile of the *Hagen–Poiseuille flow*; the theoretical solution was worked out by Stokes. The velocity profile is sketched in Figure 3.2:2.

From the solution (15) we can obtain the *rate of flow* through the tube by an integration,

$$\dot{Q} = 2\pi \int_0^a ur\,dr. \tag{16}$$

This leads to the *Poiseuille formula*,

$$\dot{Q} = -\frac{\pi a^4}{8\mu}\frac{dp}{dx} = -\frac{\pi a^4}{8\mu}\frac{\Delta p}{L}. \tag{17}$$

Here, in the last term, Δp represents the pressure drop in a segment of blood vessel of length L. Δp varies linearly with L because dp/dx is a constant. Dividing the rate of flow by the cross-sectional area of the tube yields the *mean velocity of flow* in the laminar, Poiseuillean case,

$$u_m = -\frac{a^2}{8\mu}\frac{dp}{dx}. \tag{18}$$

Finally, the shear stress at the tube wall is given by $-\mu(\partial u/\partial r)$ at $r = a$. Using Eqs. (15) and (17), we obtain

$$\text{shear stress on the tube wall} = -\frac{a}{2}\frac{dp}{dx} = -\frac{a}{2}\frac{\Delta p}{L}$$

$$= \frac{4\mu}{\pi a^3}\dot{Q} = 4\mu\frac{u_m}{a}. \tag{19}$$

If we divide the shear stress by the mean dynamic pressure $\frac{1}{2}\rho u_m^2$ the ratio is called the *skin friction coefficient*. Denoting the skin friction coefficient by C_f, we obtain, for a laminar Poiseuillean flow,

$$C_f = \frac{\text{shear stress}}{\text{mean dynamic pressure}} = \frac{-\mu\left(\partial u/\partial r\right)_{r=a}}{\frac{1}{2}\rho u_m^2} = \frac{16}{N_R}, \tag{20}$$

where N_R is the Reynolds number,

$$N_R = 2au_m/\nu. \tag{21}$$

The formula for shear stress on the wall in a laminar Poiseuillean flow is then

$$\text{shear stress} = C_f \tfrac{1}{2}\rho u_m^2. \tag{22}$$

Hagen and Poiseuille obtained Eq. (13) through experimental measurements; it was their empirical formula. The theoretical derivation was due to Stokes. Equation (13) is not valid near the entrance or exit section of a tube. It is satisfactory at a sufficiently large distance from the ends, but is again invalid if the tube is too large or if the velocity is too high. The difficulty at the entry or exit region is due to the transitional nature of the flow in that region, so that our assumption $v = 0$, $w = 0$ is not valid. The difficulty with too large a Reynolds number, however, is of a different kind: The flow becomes turbulent!

Osborne Reynolds demonstrated the transition of a laminar flow to a turbulent flow in a pipe by a classical experiment in which he examined the flow in a small outlet from a large water tank. He used a stopcock at the end of the tube to control the speed of water flow through the tube. The junction of the tube with the tank was nicely rounded, and a filament of colored fluid was introduced at the mouth. When the speed of water was slow, the filament remained distinct through the entire length of the tube. When the speed was increased, the filament broke up at a given point and diffused throughout the cross section (Fig. 3.2:3). Reynolds identified the governing parameter $u_m d/\nu$—the Reynolds number—where u_m is the mean velocity, d is the diameter, and ν is the kinematic viscosity. The region in which the colored filament diffuses to the whole tube is the transition zone from laminar to turbulent flow in the tube. Reynolds found that transition occurred at Reynolds numbers between 2,000 and 13,000, depending on the smoothness of the tube wall and the shape of the entry condition. When extreme care is taken, the transition can be delayed to Reynolds numbers as high as 40,000. On the other hand, a value of 2,000 appears to be about the lowest value obtainable on a rough entrance. This is interesting, but hard to understand. Indeed, turbulence is one of the most difficult problems in fluid mechanics.

The theoretical solution can be modified to account for the non-Newtonian rheological properties of blood, which have been discussed in Sections 3.1 and 3.2 in *Biomechanics: Mechanical Properties of Living Tissues* (Fung, 1993b). Steady flow of blood in circular cylindrical tubes is discussed in Section 3.3 of that book. It is shown that the effect of nonlin-

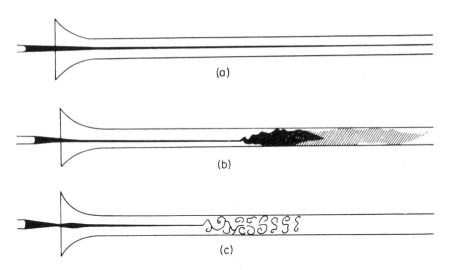

FIGURE 3.2:3. Reynolds' turbulence experiment: (a) laminar flow; (b) and (c) transition from laminar to turbulent flow. After Reynolds, O. (1883): An experimental investigation of the circumstances which determine whether the motion of water shall be direct or sinuous, and of the law of resistance in parallel channels. *Phil. Trans. Roy. Soc.* **174**: 935–982.

ear blood rheology on the resistance of blood flow in arteries is relatively minor but its effect on flow separation can be great.

Historically, the great significance of Poiseuille's contribution is at least fourfold: (a) The great precision of his results. (b) By using tubes of very small diameters, he made sure that the flow was laminar. (c) Stokes and others have regarded the agreement of Poiseuille's empirical formula with theoretical prediction based on the Navier–Stokes equation as a proof of the *no-slip condition on the solid boundary* mentioned in Section 2.6. The importance of the no-slip condition is paramount; the theoretical derivation of this condition is forever fascinating. (d) Poiseuille made this study for the explicit purpose of laying the foundation of biomechanics.

3.3 Applications of Poiseuille's Formula: Optimum Design of Blood Vessel Bifurcation

Poiseuille's formula has many uses. It tells us that the most effective factor controlling blood flow is the radius of the blood vessel. For a given pressure drop, a 1% change in vessel radius will cause a 4% change in blood flow. Conversely, if an organ needs a certain amount of blood flow to function, then the pressure difference needed to send this flow through depends on the vessel radius. For a fixed flow a 1% decrease in vessel radius will cause a 4% increase in the required pressure difference. This is seen from Eq. (3.2:17) as follows:

a. If Δp, μ, and L are constant, then by taking the logarithm on both sides of the equation and differentiating, we obtain

$$\frac{\delta \dot{Q}}{\dot{Q}} = 4 \frac{\delta a}{a}. \tag{1}$$

b. If $\dot{Q}$, μ, and L are constant, then by differentiation and rearranging terms, we obtain

$$\frac{\delta(\Delta p)}{\Delta p} = -4 \frac{\delta a}{a}. \tag{2}$$

Hence an effective way of controlling blood pressure is to change the vessel radius. Hypertension (high blood pressure) can be caused by narrowing of blood vessels, and can be reduced by relaxing the smooth muscle tension that controls the blood vessel radius. Reducing blood viscosity is another way of reducing the resistance to blood flow, and hemodilution is sometimes used in surgery.

Now let us consider a different application. We know that arteries bifurcate many times before they become capillaries. Can we guess at a design principle of the blood vessel bifurcation? To be more concrete, let us consider three vessels, AB, BC, and BD, connecting three points, A, C, and D, in space (Fig. 3.3:1). There is a flow $\dot{Q}_0$ coming through A into AB. The flow is divided into $\dot{Q}_1$ in BC and $\dot{Q}_2$ in BD. Let the points A, C, D be fixed, but the location of B and the vessel radii are left for the designer to choose. Is there an optimal position for the point B?

By asking such a question we are seeking a principle of optimum design. Some *cost function* is assumed, and the design parameters are chosen so that the cost function is minimized. Some of the great theories of physics and chemistry are based on such principles. One may recall the principle

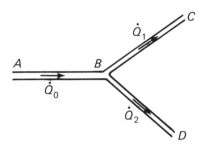

FIGURE 3.3:1. Bifurcation of a blood vessel AB into two branches BC and BD, supplying blood at a rate of $\dot{Q}_0$ (cm^3/sec) from point A to points C and D, with outflow of $\dot{Q}_1$ at C and $\dot{Q}_2$ at D.

of minimum potential energy in elasticity, the principle of minimum entropy production in irreversible thermodynamics, the Fermat principle of least time of travel in optics, Maupertius' principle of least action, Hamilton's principle in physics, and so on. The potential energy, entropy production, travel time, action, and the Hamiltonian are the cost functions in these cases.

For blood vessels, Murray (1926) and Rosen (1967) proposed a cost function that is the sum of the rate at which work is done on the blood and the rate at which energy is used up by the blood vessel by metabolism. The former is the product of $\dot{Q}\Delta p$. The latter is assumed to be proportional to the volume of the vessel $\pi a^2 L$, with a proportional constant K. Hence

$$\text{Cost function for blood vessels} = \dot{Q}\Delta p + K\pi a^2 L. \tag{3}$$

With Eq. (3.2:17) we can write

$$\text{Cost function} = \frac{8\mu L}{\pi a^4}\dot{Q}^2 + K\pi a^2 L. \tag{4}$$

The cost function of the entire system of blood vessels is the sum of the cost functions of individual vessel segments. Hence, each vessel must be optimal and the system must be put together optimally. For a given vessel of length L and flow $\dot{Q}$, there is an optimal radius a, which can be calculated by minimizing the cost function with respect to a. At the optimal condition, the following derivative must vanish:

$$\frac{\partial}{\partial a}\left(\text{cost function}\right) = -\frac{32\mu L}{\pi}\dot{Q}^2 a^{-5} + 2K\pi La = 0. \tag{5}$$

This yields the solution

$$a = \left(\frac{16\mu}{\pi^2 K}\right)^{1/6}\dot{Q}^{1/3}. \tag{6}$$

Hence, the optimal radius of a blood vessel is proportional to $\dot{Q}$ to the 1/3 power. On substituting (6) into (4), we obtain the minimum value of the cost function:

$$\text{Min. cost function} = \frac{3\pi}{2}KLa^2. \tag{7}$$

Bifurcation Pattern

Now consider the bifurcation problem. Since the cost functions of all vessels are additive, we see at once that the vessels connecting $A, C,$ and D in Figure 3.3:1 should be straight and lie in a plane (because this minimizes the length, L, when other things are fixed.) To find out the details let the geometric parameters be specified as shown in Figure 3.3:2. The three branches

FIGURE 3.3:2. Geometric parameters of the branching pattern. Theory shows that B should lie in the plane of ACD.

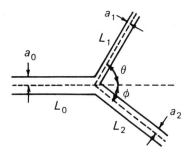

will be denoted by subscripts $0, 1, 2$. The total cost function will be denoted by P:

$$P = \frac{3\pi K}{2}\left(a_0^2 L_0 + a_1^2 L_1 + a_2^2 L_2\right). \tag{8}$$

The lengths L_0, L_1, L_2 are affected by the location of the point B, and the radii a_0, a_1, a_2 are related to the flows $\dot{Q}_0, \dot{Q}_1, \dot{Q}_2$ through Eq. (6). Let us now minimize P by properly choosing the location of the bifurcation point B.

Since a small movement of B changes P by

$$\delta P = \frac{3\pi K}{2}\left(a_0^2 \delta L_0 + a_1^2 \delta L_1 + a_2^2 \delta L_2\right), \tag{9}$$

an optimal location of B would make $\delta P = 0$ for arbitrary small movement of B. Let us consider three special movements of B. First, let B move to B′ in the direction of AB, as shown in Figure 3.3:3. In this case

$$\delta L_0 = \delta, \quad \delta L_1 = -\delta\cos\theta, \quad \delta L_2 = -\delta\cos\phi,$$

$$\delta P = \frac{3\pi K}{2}\delta\left(a_0^2 - a_1^2\cos\theta - a_2^2\cos\phi\right). \tag{10}$$

The optimum is obtained when

$$a_0^2 = a_1^2\cos\theta + a_2^2\cos\phi. \tag{11}$$

Next, let B move to B′ in the direction of CB, as shown in Figure 3.3:4. Then

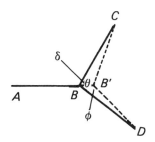

FIGURE 3.3:3. A particular variation of $\delta L_0, \delta L_1, \delta L_2$ by a small displacement of B in the direction of AB.

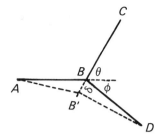

FIGURE 3.3:4. Another particular variation of δL_0, δL_1, δL_2 by a displacement of B to B' along BC.

$$\delta L_0 = -\delta \cos\theta, \quad \delta L_1 = \delta, \quad \delta L_2 = \delta \cos(\theta + \phi),$$

$$\delta P = \frac{3\pi K \delta}{2}\left[-a_0^2 \cos\theta + a_1^2 + a_2^2 \cos(\theta + \phi)\right], \tag{12}$$

and the optimal condition is

$$-a_0^2 \cos\theta + a_1^2 + a_2^2 \cos(\theta + \phi) = 0. \tag{13}$$

Finally, let B move a short distance δ in the direction of DB (Fig. 3.3:5). Then the optimal condition is obviously,

$$-a_0^2 \cos\phi + a_1^2 \cos(\theta + \phi) + a_2^2 = 0. \tag{14}$$

Solving Eqs. (11), (13), and (14) for $\cos\theta$, $\cos\phi$, and $\cos(\theta + \phi)$, we obtain

$$\cos\theta = \frac{a_0^4 + a_1^4 - a_2^4}{2a_0^2 a_1^2},$$

$$\cos\phi = \frac{a_0^4 - a_1^4 + a_2^4}{2a_0^2 a_2^2},$$

$$\cos(\theta + \phi) = \frac{a_0^4 - a_1^4 - a_2^4}{2a_1^2 a_2^2}. \tag{15}$$

The equation of continuity (conservation of mass) is

$$Q_0 = Q_1 + Q_2. \tag{16}$$

By Eq. (6), this is

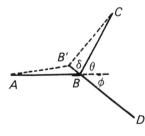

FIGURE 3.3:5. A third variation caused by a displacement of B to B' along BD.

$$a_0^3 = a_1^3 + a_2^3, \tag{17}$$

which is often referred to as Murray's law. Thus Eq. (15) can be reduced to

$$\cos\theta = \frac{a_0^4 + a_1^4 - \left(a_0^3 - a_1^3\right)^{4/3}}{2a_0^2 a_1^2}, \quad \text{etc.} \tag{18}$$

Ghassan Kassab has collected an extensive set of data on the coronary arteries of the pig (see Chapter 7; Kassab et al., 1993), and Kassab and Fung (1995) tested Murray's law, Eq. (17), against the experimental data. They found excellent agreement of Murray's law and experimental data from control and hypertensive hearts. Oka (1974) has proposed to improve Murray's cost function by adding a term of metabolic cost proportional to the volume of the blood vessel wall. Oka's cost function is a modification of Eq. (4):

$$\text{Cost } P = \frac{8\mu L}{\pi a^4} \dot{Q}^2 + K\pi a^2 L + K_w 2\pi a h L \tag{19}$$

where K_w and h are the metabolic constant and thickness of the vessel wall, respectively, and are assumed to be constant. Kassab and Fung (1995) found Oka's modified Murray's law does not improve the agreement.

Applications of these formulas are illustrated in the following Problems.

Problems

3.1 Show that, according to Murray's cost function, if $a_1 = a_2$, then $\theta = \phi$. Thus, if the radii of the daughter branches are equal, the bifuracting angles are equal.

3.2 Show that if $a_2 > a_1$, then $\theta > \phi$.

3.3 Show that if $a_2 \gg a_1$, then $a_2 \doteq a_0$ and $\phi \doteq \pi/2$.

3.4 When $a_1 = a_2$, show that $a_1/a_0 = 2^{-1/3} = 0.794$, and $\cos\theta = 0.794$. Thus $\theta \doteq 37.5°$.

3.5 The cost function specified in Eq. (4) is somewhat arbitrary. Develop some other cost functions and deduce the consequences, such as

(a) Minimum total surface area of the blood vessels,
(b) Minimum total volume of the blood vessels,
(c) Minimum power for the blood flow,
(d) Minimum total shear force on the vessel wall.

See Kamiya and Togawa (1972), Murray (1926), and Zamir (1976, 1977).

The results of Problems 3.1–3.4 are in reasonable agreement with empirical observations. The result of Problem 3.4 is especially interesting.

Let a_0 denote the radius of the aorta, and assume equal bifurcation in all generations. Then the radius of the first generation is 0.794 a_0, that of the second generation is $(0.794)^2 a_0$, and, generally, that of the nth generation is

$$a_n = (0.794)^n a_0. \tag{20}$$

If a capillary blood vessel has a radius of 5×10^{-4} cm and the radius of the aorta is $a_0 = 1.5$ cm, then Eq. (20) yields $n \doteq 30$. Thus 30 generations of equal bifurcation are needed to reduce that aorta to the capillary dimension. Since each generation multiplies the number of vessels by 2, the total number of blood vessels is $2^{30} \doteq 10^9$. But these estimates cannot be taken too seriously, because arteries rarely bifurcate symmetrically (as required by the hypothesis $a_1 = a_2$). There is one symmetric bifurcation of the arteries of humans; there is none in the dog.

The Uniform Shear Hypothesis

Problem 3.5 is very significant. It turns out that several cost functions lead to almost the same results. Zamir (1976) deduced that the shear stress on blood vessel wall is uniform throughout the arterial system according to his principle of minimum total shear force. The hypothesis of uniform shear stress or shear strain gradient and constant coefficient of viscosity on the blood vessel wall has been supported by several studies. Kamiya and Togawa (1980) surgically constructed an arteriovenous shunt from the common carotid artery to the external jugular vein, causing an increase of blood flow in one segment of the artery and a decrease of flow in another. They then showed that 6 to 8 months after the operation, the segment with increased flow dilated, while the segment with decreased flow atrophied to a smaller diameter, just enough so that the shear strain rate remained almost constant if the change of flow was within four times of the control. Liebow (1963), Thoma (1893) and others, on observing embryologic vascular development and studying arteriovenous fistulas and collateral circulation, have shown that increased flow induces vessel growth, reduced flow leads to atrophy. Rodbard (1975) collected clinical evidence of the same. Then Friedman and Deters (1987), Giddens et al. (1990) and Kamiya et al, (1984) collected data from literature and their own research and concluded that the arterial wall shear stress on dog's peripheral and coronary arterioles, arteries, and aorta lies in a remarkably narrow range of 10–20 dynes/cm².

Kassab and Fung (1995) showed that if the Poiseuille formula given in Eq. (19) of Section 3.2 is substituted into the equation of continuity Eq. (16), one obtains the relation

$$(a_0^3/\mu_0)\tau_{w0} = (a_1^3/\mu_1)\tau_{w1} + (a_2^3/\mu_2)\tau_{w2} \tag{21}$$

or

$$a_0^3 \gamma_0 = a_1^3 \gamma_1 + a_2^3 \gamma_2 \qquad (22)$$

Where τ_w denotes shear stress and γ denotes shear strain rate, $\gamma = \partial u / \partial y$, on the blood vessel wall, with γ_0, γ_1, γ_2, τ_{w0}, τ_{w1}, τ_{w2}, and μ_0, μ_1, μ_2 refer to the values of γ, τ_w, and μ at the boundaries of the tubes 0, 1, 2, respectively. Now, if one introduces the hypothesis that the shear strain rate or the coefficients of viscosity and the wall shear stresses are the same in all three vessels, then Eqs. (21) and (22) become Murray's law, Eq. (17). On the other hand, if we assume Murray's law as an empirical fact, then the coexistence of Eqs. (17), (21) and (22) implies that

$$\tau_{w0} = \tau_{w1} = \tau_{w2}, \quad \gamma_0 = \gamma_1 = \gamma_2 \qquad (23)$$

Thus, Murray's law and Poiseuille's law imply uniform shear; and vice versa, uniform shear and Poiseuille's law imply Murray's law.

Finally, the rate of viscous dissipation per unit volume of blood is equal to the product of shear stress and strain rate. At the blood vessel wall, this is:

$$\text{Volumetric rate of viscous dissipation} = \tau_w \gamma_w. \qquad (24)$$

Thus the uniformity of shear implies a uniform energy dissipation throughout the arteriolar walls.

Stress Distribution on Blood Vessel Wall and in Endothelial Cells

The wall shear stress discussed in Poiseuille flow is defined at the length scale hierarchy level of the blood vessel diameter. If we go one level lower, to the scale of a single endothelial cell, a very different picture exists. At level, the endothelium surface is wavy, with hills and valleys. The no-slip condition must be applied on the cell membranes facing the blood. The hills and valleys will cause nonuniform shear stress distribution, even if the shear flow far above the surface is uniform. Inside the individual endothelial cells, there is another world of structures and materials. Intracellular mechanics must be investigated at smaller and smaller scales. Finally, there is also a class of problems concerned with the long-range variation of certain features in individual cells. An example is the longitudinal variation of tensile and shear stresses in the endothelial cell membranes along the length of the aorta, as discussed by Fung and Liu (1993) and Liu et al. (1994).

3.4 Steady Laminar Flow in an Elastic Tube

As another application of Poiseuille's formula, let us consider the flow in a circular cylindrical elastic tube (Fig. 3.4:1). The flow is maintained by a pressure gradient. The pressure in the tube is, therefore, nonuniform—higher at

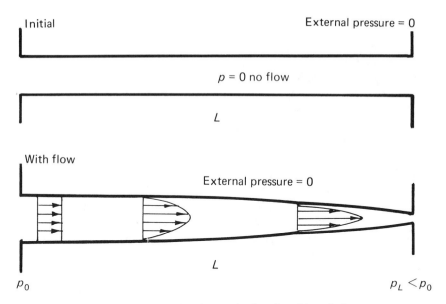

FIGURE 3.4:1. Flow in an elastic tube of length L.

the entry end and lower at the exit end. Because the tube is elastic, the high-pressure end distends more than the low-pressure end. The diameter of the tube is, therefore, nonuniform (if it were uniform originally), and the degree of nonuniformity depends on the flow rate.

If we wish to determine the pressure–flow relationship for such a system, we may break down the problem into two familiar components. This is illustrated in Figure 3.4:2. In the lower block, we regard the vessel as a rigid conduit with a specified wall shape. For a given flow, we compute the pressure distribution. This pressure distribution is then applied as loading on the elastic tube, represented by the upper block. We then analyze the deformation of the elastic tube in the usual manner of the theory of elasticity. The result of the calculation is then used to determine the boundary shape of the hydrodynamic problem of the lower block. Thus, back and forth, until a consistent solution is obtained, the pressure distribution corresponding to a given flow is determined.

Let us put this in mathematical form. Assume that the tube is long and slender, that the flow is laminar and steady, that the disturbances due to entry and exit are negligible, and that the deformed tube remains smooth and slender. These assumptions permit us to consider the solution given in Section 3.2 as valid (a good approximation) everywhere in the tube. Assuming a Newtonian fluid, we have (Eq. (17) of Sec. 3.2)

$$\frac{dp}{dx} = -\frac{8\mu}{\pi a^4}\dot{Q}. \tag{1}$$

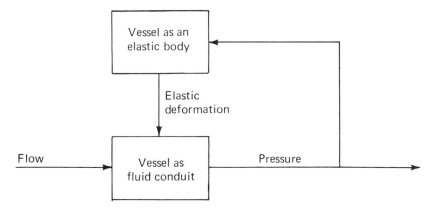

FIGURE 3.4:2. A hemoelastic system analyzed as a feedback system of two functional units: an elastic body and a fluid mechanism.

Here $\dot{Q}$ is the volume–flow rate. In a stationary, nonpermeable tube $\dot{Q}$ is a constant throughout the length of the tube. The tube radius is a, which is a function of x because of the elastic deformation. An integration of Eq. (1) yields

$$p(x) = p(0) - \frac{8\mu}{\pi}\dot{Q}\int_0^x \frac{1}{[a(x)]^4}\,dx. \tag{2}$$

The integration constant is $p(0)$, the pressure at $x = 0$. The exit pressure is given by Eq. (2) with $x = L$. L is the length of the tube.

Now let us turn our attention to the calculation of the radius $a(x)$. Let the tube be initially straight and uniform, with a radius a_0. Assume that the tube is thin walled, and that the external pressure is zero (Fig. 3.4:3). (If the external pressure was not zero, we would replace p in Eq. (3) by the difference of internal and external pressures.) Then a simple analysis yields the average circumferential stress in the wall:

$$\sigma_{\theta\theta} = \frac{p(x)a(x)}{h}, \tag{3}$$

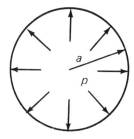

FIGURE 3.4:3. Distension of an elastic tube due to internal pressure.

where h is the wall thickness (Fung, 1993a, p. 25). Let the axial length and tension be constant, and assume that the material obeys Hooke's law. Then the circumferential strain is

$$e_{\theta\theta} = \frac{\sigma_{\theta\theta}}{E},\tag{4}$$

where E is the Young's modulus of the wall material. (Srictly, the right-hand side of Eq. (4) should be

$$\frac{1}{E}\left(\sigma_{\theta\theta} - v\sigma_{rr}\right),$$

where v is the Poisson's ratio. But σ_{rr} is, in general, much smaller than $\sigma_{\theta\theta}$ for thin-walled tubes.) The strain $e_{\theta\theta}$ is equal to the change of radius divided by the original radius, a_0:

$$e_{\theta\theta} = \frac{a(x) - a_0}{a_0} = \frac{a(x)}{a_0} - 1.\tag{5}$$

Combining (5), (4), and (3), we obtain

$$a(x) = a_0\left[1 - \frac{a_0}{Eh}p(x)\right]^{-1}.\tag{6}$$

Substituting (6) into (1), we may write the result as

$$\left(1 - \frac{a_0}{Eh}p\right)^{-4}dp = -\frac{8\mu}{\pi a_0^4}\dot{Q}dx.\tag{7}$$

Recognizing the boundary conditions $p = p(0)$ when $x = 0$ and $p = p(L)$ when $x = L$, and integrating Eq. (7) from $p(0)$ to $p(L)$ on the left and 0 to L on the right, we obtain the pressure–flow relationship:

$$\frac{Eh}{3a_0}\left\{\left[1 - \frac{a_0}{Eh}p(L)\right]^{-3} - \left[1 - \frac{a_0}{Eh}p(0)\right]^{-3}\right\} = -\frac{8\mu}{\pi a_0^4}L\dot{Q},\tag{8}$$

which shows that the flow is not a linear function of pressure drop $p(0) - p(L)$.

Another Solution

The solution obtained in Eq. (8) is based on the assumption of Hooke's law. Most blood vessels do not obey Hooke's law, their zero-stress states are open sectors, and their constitutive equations are nonlinear (see Section 2.6, pp. 56–60, and Section 3.8, p. 145).

A simple result can be obtained if we assume the pressure–radius relationship to be linear:

$$a = a_0 + \alpha p/2.\tag{9}$$

Here a_0 is the tube radius when the transmural pressure is zero. α is a compliance constant. Equation (9) is a good representation of the pulmonary blood vessels (see Sec. 4.10, Fig. 4.10:2, p. 257 and Sec. 6.7, Fig. 6.7:4).

Using Eq. (9), we have

$$\frac{dp}{dx} = \frac{dp}{da}\frac{da}{dx} = \frac{2}{\alpha}\frac{da}{dx}. \tag{10}$$

On substituting Eq. (10) into Eq. (1) and rearranging terms, we obtain

$$a^4\frac{da}{dx} = \frac{1}{5}\frac{da^5}{dx} = -\frac{4\mu\alpha}{\pi}\dot{Q}. \tag{11}$$

Since the right-hand side term is a constant independent of x, we obtain at once the integrated result

$$\left[a(x)\right]^5 = -\frac{20\mu\alpha}{\pi}\dot{Q}x + \text{const.} \tag{12}$$

The integration constant can be determined by the boundary condition that when $x = 0$, $a(x) = a(0)$. Hence the constant $= [a(0)]^5$. Then, by putting $x = L$, we obtain from Eq. (12) the elegant result

$$\frac{20\mu\alpha L}{\pi}\dot{Q} = \left[a(0)\right]^5 - \left[a(L)\right]^5. \tag{13}$$

The pressure–flow relationship is obtained by substituting Eq. (9) into Eq. (13). Thus the flow varies with the difference of the fifth power of the tube radius at the entry section ($x = 0$) minus that at the exit section ($x = L$). If the ratio $a(L)/a(0)$ is 1/2, then $[a(L)]^5$ is only about 3% of $[a(0)]^5$, and is negligible by comparison. Hence when $a(L)$ is one half of $a(0)$ or smaller, the flow varies directly with the fifth power of the tube radius at the entry, whereas the radius (and the pressure) at the exit section has little effect on the flow.

Problems

3.6 If the elastic deformation is small,

$$\frac{a_0 p(0)}{Eh} \ll 1, \quad \frac{a_0 p(L)}{Eh} \ll 1,$$

show that the pressure–flow relationship Eq. (8) or (13) then becomes approximately linear.

3.7 Plot curves to show the flow–pressure relationship given by Eqs. (8) and (13), and discuss the results.

3.8 The actual relationship between the pressure and radius in peripheral blood vessels is nonlinear. See Chapter 8 of *Biomechanics: Mechanical*

Properties of Living Tissues (Fung, 1993b). Outline a theory that will take into account the nonlinear pseudo-elastic stress–strain relationship in deriving the pressure–flow relationship of the blood vessel.

3.9 Outline further a theory that will take into account the viscoelastic behavior of the blood vessel in deriving the pressure–flow relationship of the blood vessel.

3.5 Dynamic Similarity. Reynolds and Womersley Numbers. Boundary Layers

Consider first a blood flow in which the shear rate is sufficiently high so that the blood has a constant coefficient of viscosity. Then the Navier–Stokes equations presented in Section 2.6 apply. This equation is

$$
\rho \frac{\partial u_i}{\partial t} + \rho \left(u_1 \frac{\partial u_i}{\partial x_1} + u_2 \frac{\partial u_i}{\partial x_2} + u_3 \frac{\partial u_i}{\partial x_3} \right)
$$

$$
= X_i - \frac{\partial p}{\partial x_i} + \mu \left(\frac{\partial^2}{\partial x_1^2} + \frac{\partial^2}{\partial x_2^2} + \frac{\partial^2}{\partial x_3^2} \right) u_i. \tag{1}
$$

Here u_i denotes the velocity vector, with the index i ranging over $1, 2, 3$, so that the components of u_i are u_1, u_2, u_3, or u, v, w, x_i, with components x_1, x_2, x_3 or x, y, z, is a position vector referred to a rectangular cartesian frame of reference. ρ is the density or mass per unit volume of the fluid. X_i is the body force per unit volume. p is pressure. μ is the coefficient of viscosity of the fluid. μ/ρ is called *kinematic viscosity*, and is designated by a Greek symbol v which will be used later.

Equation (1) represents the balance of four kinds of forces. Term by term, they are

transient	convective	body	pressure	viscous
inertia	inertia	force	force on	force on
force per +	force per	= per	+ sides of	+ sides of .
unit vol	unit vol	unit	unit	unit
		vol	control	control
			vol	vol

Let us put the Navier–Stokes equation in dimensionless form. Choose a characteristic velocity V, a characteristic frequency ω and a characteristic length L. For example, if we investigate the flow in the aorta, we may take V to be the average speed of flow, ω to be the heart rate, and L to be the blood vessel diameter. Having chosen these characteristic quantitites, we introduce the dimensionless variables

$$x' = \frac{x}{L}, \quad y' = \frac{y}{L}, \quad z' = \frac{z}{L}, \quad u' = \frac{u}{V},$$

$$v' = \frac{v}{V}, \quad w' = \frac{w}{V}, \quad p' = \frac{p}{\rho V^2}, \quad t' = \omega t, \tag{2}$$

and the parameters

$$\text{Reynolds number} = N_R = \frac{VL\rho}{\mu} = \frac{VL}{v}, \tag{3}$$

$$\text{Stokes number} = N_S = \frac{\omega L^2}{v}, \tag{4}$$

$$\text{Womersley number } \alpha = N_w = \sqrt{N_S} = L\sqrt{\left(\frac{\omega}{v}\right)}. \tag{5}$$

On substituting Eqs. (2) to (5) into Eq. (1), omitting body force, and dividing through by $\rho V^2/L$, we obtain

$$\frac{N_S}{N_R}\frac{\partial u'}{\partial t'} + u'\frac{\partial u'}{\partial x'} + v'\frac{\partial u'}{\partial y'} + w'\frac{\partial u'}{\partial z'} = -\frac{\partial p'}{\partial x'} + \frac{1}{N_R}\left(\frac{\partial^2 u'}{\partial x'^2} + \frac{\partial^2 u'}{\partial y'^2} + \frac{\partial^2 u'}{\partial z'^2}\right) \tag{6}$$

and two additional equations obtainable from Eq. (6) by changing u' into v', v' into w', w' into u' and x' into y', y' into z', z' into x'. The body force is ignored. The $\partial\mu/\partial x_k$ term is dropped because μ is a constant. The equation of continuity (Eq. 8 of Sec. 2.6) can also be put in dimensionless form:

$$\frac{\partial u'}{\partial x'} + \frac{\partial v'}{\partial y'} + \frac{\partial w'}{\partial z'} = 0. \tag{7}$$

Since Eqs. (6) and (7) constitute the complete set of field equations for an incompressible fluid, it is clear that only two physical parameters, the Reynolds number N_R, and the Womersley number N_w, enters into the field equations of the flow.

To solve these equations for a specific problem we must consider the boundary equations. Consider two flows in two geometrically similar vessels. The vessels have the same shape but different sizes. The boundary conditions are identical (no-slip). Then the two flows will be identical (in the dimensionless variables) if the Reynolds numbers and the Womersley numbers for the two flows are the same, because two geometrically similar bodies having the same Reynolds number and Womersley number will be governed by identical differential equations and boundary conditions (in dimensionless form). Therefore, flows about geometrically similar bodies at the same Reynolds and Womersley numbers are completely similar in the sense that the functions $u'(x', y', z', t')$, $v'(x', y', z', t')$, $w'(x', y', z', t')$, $p'(x', y', z', t')$ are the same for the various flows. Thus the Reynolds and Womersley numbers are said to govern the dynamic similarity.

The Reynolds number expresses the ratio of the convective inertia force to the shear force. In a flow the inertial force due to convective acceleration arises from terms such as $\rho u \partial u/\partial x$, whereas the shear force arises from terms such as $\mu \partial^2 u/\partial y^2$. The orders of magnitude of these terms are, respectively,

$$\text{Convective inertia force: } \rho V^2/L$$
$$\text{Shear force: } \mu V/L^2$$

The ratio is

$$\frac{\text{Convective inertia force}}{\text{Shear force}} = \frac{\rho V^2/L}{\mu V/L^2} = \frac{\rho VL}{\mu} = \text{Reynolds number.} \quad (8)$$

A large Reynolds number signals a preponderant convective inertia effect. A small Reynolds number signals a predominant shear effect.

Similarly, the Womersley number expresses the ratio of the transient or oscillatory inertia force to the shear force. The transient inertia force is given by the first term of Eq. (1). If the frequency of oscillation is ω and the amplitude of velocity is V, the order of magnitude of the first term of Eq. (1) is $\rho \omega V$. The order of magnitude of the last term of Eq. (1) is, as before, $\mu V/L^2$. Thus

$$\text{Transient inertia force: } \rho \omega V$$
$$\text{Shear force: } \mu V/L^2 \,.$$

The ratio is

$$\frac{\text{Transient inertia force}}{\text{Shear force}} = \frac{\rho \omega L^2}{\mu} = \frac{\omega L^2}{\nu} = \text{Stokes number}$$
$$= \left(\text{Womersley Number}\right)^2. \quad (9)$$

If the Womersley number is large, the oscillatory inertia force dominates. If N_W is small, the viscous force dominates. Typical values of Reynolds and Womersley numbers in blood vessels at normal heart rate are given in Table 3.1:1, p. 110. The Womersley number is usually denoted by α.

Boundary Layers and Their Thicknesses

The concept of boundary layer was presented by Ludwig Prandtl (1875–1953) in a brief but truly epoch-making paper (1904). It can be understood by comparing the significance of various terms of the governing equation, Eq. (1). If the coefficient of viscosity of the fluid is zero, $\mu = 0$, then the fluid is said to be *ideal*, and the last term of Eq. (1) vanishes. At the solid wall, an ideal fluid must not penetrate the solid, but its tangential velocity is unrestricted. For a viscous fluid, however, the no-slip condition must apply, no matter how small the viscosity is. Prandtl's idea is that for a fluid with small μ the influence of the no-slip condition and the last term in Eq. (1) is limited to a thin layer next to the solid wall, whereas in the bulk of the fluid the influence of no-slip is small and the last term in Eq. (1) can be dropped. If this was true, then the thickness of the boundary layer, denoted by δ, can be estimated by comparing proper terms in Eq. (1). Consider first an oscillatory velocity field of frequency ω and amplitude U. The first term of Eq. (1) shows that the transient inertia force is of the order of magnitude $\rho\omega U$. The last term in Eq. (1) shows that the order of magnitude of the viscous force is of the order of $\mu U/\delta_1^2$, where a subscript 1 is added to indicate that this boundary layer is associated with the transient acceleration. In the transient boundary layer, these two terms are of equal importance. Hence,

$$\rho\omega U = \frac{\mu U}{\delta_1^2} \tag{10}$$

or

$$\delta_1 = \sqrt{\frac{\mu}{\rho\omega}} = \sqrt{\frac{\nu}{\omega}}. \tag{11}$$

In a tube flow, let the characteristic length be the radius, L. Then the ratio of L to δ_1 is

$$\frac{L}{\delta_1} = L\sqrt{\frac{\omega}{\nu}} = \sqrt{N_\mathrm{S}} = N_\mathrm{W} \tag{12}$$

Hence, if the Womersley number N_W or Stokes number N_S is large, the transient boundary layer is very thin compared with the tube radius.

Next, consider the convective inertia force given by the second group of terms in Eq. (1). At a convective boundary layer thickness δ_2, the order of magnitude of the second term is $\rho U^2/L$; whereas that of the viscous force term is $\mu U^2/\delta_2^2$. In the convective boundary layer these two terms are equally important. Hence,

$$\rho U^2/L = \mu U^2/\delta_2^2 \tag{13}$$

or

$$\delta_2 = \sqrt{\frac{\mu L}{\rho U}} \tag{14}$$

The ratio of L to δ_2 is

$$\frac{L}{\delta_2} = \sqrt{\frac{\rho UL}{\mu}} = \sqrt{N_\mathrm{R}} \tag{15}$$

i.e., the square root of the Reynolds number. Hence, when the Reynolds number is large, the convective boundary layer is very thin. In a tube flow, at a distance from the wall much larger than δ_1 and δ_2, the fluid may be regarded as ideal.

These estimates have many applications in the following sections. In Section 3.18, it is shown that the boundary layer thickness increases with the distance from the entrance section; and there are interactions between the transient and convective boundary layers. Exact calculations can be found in Schlichting (1968).

3.6 Turbulent Flow in a Tube

In Section 3.2 we mentioned that when the Reynolds number exceeds a certain critical value the flow becomes turbulent. Turbulence is marked by random fluctuations. With turbulence the velocity field can no longer be predicted with absolute precision, but its statistical features (mean velocity, root mean square velocity, mean pressure gradient, etc.) are perfectly well defined. If a steady flow in a straight, long pipe changes from laminar to turbulent, two important changes will occur: (a) The profile of the mean velocity will become much more blunt at the center of the pipe, and (b) the shear gradient will become much greater at the wall. This is shown in Figure 1.4:1, p. 7. As a consequence of this change of velocity profile, the resistance to flow is greatly increased.

The best way to see how the resistance to flow changes with turbulence is to study the *friction coefficient*, C_f, defined in Eq. (3.2:21):

$$\text{Shear stress on pipe wall} = C_f\left(\tfrac{1}{2}\rho U_m^2\right). \tag{1}$$

Here ρ is the fluid density, U_m is the mean velocity over the cross section of the tube, capitalized here to show that this velocity is not only averaged over space, but also over a sufficiently long period of time so that the random fluctuations of turbulence are averaged out. C_f is a function of the Reynolds number (based on tube radius and the mean velocity of flow, U_m) and the roughness of the tube surface. Roughness influences the position of transition in the entrance zone of a flow into a pipe at which the flow in the boundary layer is changed from laminar to turbulent. It affects also the skin-friction drag on that portion of the surface over which the layer is turbulent.

The experimental results of Nikuradse are shown in Figure 3.6:1. The surface of the tube was sprinkled with sand of various grain sizes, which were expressed in the ratio a/ε in the figure, where a is the radius of the tube and ε is the mesh size of the screen through which the sand will just pass. The dashed straight line on the left refers to a fully developed laminar flow [Eq. (3.2:19)],

$$C_f = 16\left(\frac{2aU_m}{v}\right)^{-1} = \frac{16}{N_R}, \tag{2}$$

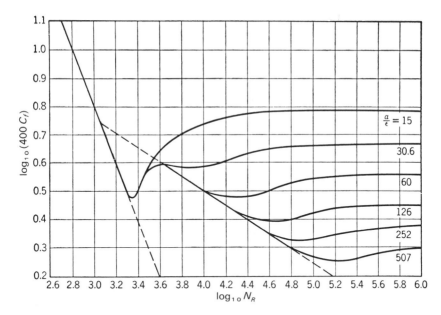

FIGURE 3.6:1. Resistance coefficient for fully developed flow through a tube of radius a with various sizes of roughness elements on the wall. ε is roughly the diameter of sand grain sprinkled on the wall. The solid curves represent the average experimental results by Nikuradse. The dashed line on the left represents a theoretical result for laminar flow. The dashed line on the right is the Blasius empirical formula for turbulent flow in a smooth tube. Based on Nikuradse, J. (1933) *Strömungsgesetze in Rauhen Rohren*, Forschungsheft 361, Ver. deutsch. Ing.

where N_R is the Reynolds number based on tube diameter and mean speed of flow. The dashed line on the right is an empirical formula given by Blasius for turbulent flow in smooth pipes,

$$C_f = 0.0655\left(\frac{U_m a}{\nu}\right)^{-1/4} = \frac{0.0779}{\left(N_R\right)^{1/4}}. \tag{3}$$

The solid curves represent the mean experimental results. It is clear that at large Reynolds numbers the friction coefficient of turbulent flow is much greater than that of laminar flow. For example, at a Reynolds number of 4,000 (i.e., $\log_{10} N_R \doteq 3.6$), a rough pipe with $a/\varepsilon = 30$ will have a skin friction about the same as that in a smooth pipe if the flow were turbulent, but it would be 2.51 times larger than that of a laminar flow if laminar flow were possible. At $N_R = 10^5$ the skin friction of a smooth pipe with turbulent flow would be 27 times larger than that given by Eq. (2), and for a rough pipe with $a/\varepsilon = 30$ the skin friction would be increased again 2.77-fold.

It seems natural to expect that natural selection in the animal world would favor laminar flow in the blood vessels so that energy is not wasted in turbulence. Furthermore, turbulence is implicated in atherogenesis. To avoid turbulent flow in aorta, the Reynolds number should be kept below a certain critical value. Let the cardiac output (volume flow per unit time) be $\dot{Q}$, and the radius of the aorta be R_a. Then the cross-sectional area of the aorta is πR_a^2, and the mean velocity of flow is

$$U_m = \frac{\dot{Q}}{\pi R_a^2}. \tag{4}$$

The Reynolds number is

$$N_R = \frac{2U_m R_a}{v} = \frac{2\dot{Q}}{\pi v R_a}. \tag{5}$$

Rosen (1967) plotted the radius of the aorta of animals versus the cardiac output, and obtained a regression line

$$R_a = 0.013\dot{Q}. \tag{6}$$

On substituting R_a from Eq. (6) into Eq. (5), we obtain a Reynolds number $2/(0.013\pi v)$, which is 1224 if $v = 0.04$ and 1632 if $v = 0.03$. These values are fairly close to but somewhat lower than the transition Reynolds number in steady flow, which seems to mean that animal aortas are designed for laminar flow, but are fairly close to the borderline of transition to turbulence.

So far our discussion of turbulence is based on steady mean flow. Pulsatile flow makes the phenomenon of laminar–turbulence transition much more complex, as is shown presently.

3.7 Turbulence in Pulsatile Blood Flow

Reynolds' experiment (see Fig. 3.2:3) shows that in pipe flow the entry region remains laminar even though turbulence develops downstream when the Reynolds number exceeds the critical value. This shows that turbulence must develop gradually in a laminar flow. It takes time for some unstable modes of motion in a flow to grow into turbulence. We may apply this concept to the pulsatile blood flow in the arteries. The flow velocity changes with time. The Reynolds number, $2aU/v$, based on the instantaneous velocity of flow averaged over the cross section, varies with time. Figure 3.7:1 shows a record of velocity of flow versus time. In a period of rising velocity the Reynolds number increases slowly until it reaches a level marked by the dotted line ($N_R = 2{,}300$), at which the flow could be expected to become turbulent if it were steady. But an accelerating flow is more stable than a steady flow, because turbulence cannot develop instantaneously. So, when the turbulence finally sets in, the velocity and N_R are much higher than the dotted line level. On the other hand, in a

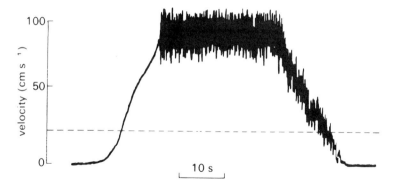

FIGURE 3.7:1. A turbulent flow velocity-versus-time record made with a hot-film probe in a pipe in which the flow rate was slowly increased until turbulence occurred, and later stopped. Peak Reynolds number was 9,500. The dotted line corresponds to Reynolds number 2,300. From Nerem, R., and Seed, W.A. (1972) An in vivo study of aortic flow disturbances, *Cardiovasc. Res.* **6**: 1–14, by permission.

period of decreasing velocity the disappearance of turbulence occurs at a level of velocity considerably below the dotted line. This is partly because decelerating flow is inherently less stable than steady flow, and partly because existing eddies take a finite time to decay. Thus, the *critical Reynolds number* of laminar–turbulent transition depends on the rate of change of velocity, as well as on the eddies upstream and the roughness of the pipe wall.

The experiment corresponding to Figure 3.7:1 was designed to show the transition from laminar to turbulent flow and vice versa. Figure 3.7:2 shows a record of velocity waves from the upper descending aorta of an anesthetized dog. Turbulence is seen during the deceleration of systolic flow. Hot-film anemometry was used to obtain such records.

Quantitative studies of the laminar–turbulent transition may seek to express the critical Reynolds number as a function of the Womersley number. Experimental results are plotted in Figure 3.7:3. The ordinate is the peak Reynolds number. The stippled area indicates the conditions under which the flow is stable and laminar. In the experiments, the wide variations of velocity and heart rate were obtained with drugs and nervous stimuli in anesthetized dogs. In normal, conscious, free-ranging dogs the peak Reynolds number usually lies in an area high above the stippled area of Figure 3.7:3. This suggests that some turbulence is generally tolerated in deceleration of systolic flow in the dog.

Turbulence in blood flow implies a fluctuating pressure acting on the arterial wall and an increased shear stress. These stresses are implicated in murmurs, poststenotic dilatation, and atherogenesis. Experimental methods are described in Deshpande and Giddens (1980), and Nerem et al. (1972).

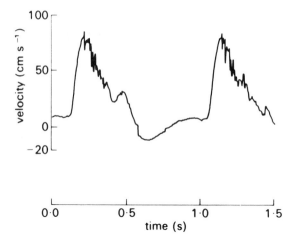

FIGURE 3.7:2. A record of velocity-versus-time of blood flow in the upper descending aorta of a dog, showing turbulence during the deceleration of systolic flow. From Seed, W.A., and Wood, N.B. (1971) Velocity patterns in the aorta. *Cardiovasc. Res.*, **5**: 319–333, by permission.

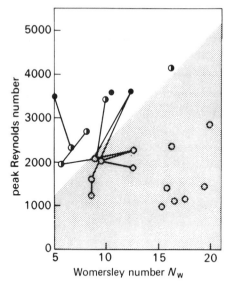

FIGURE 3.7:3. The stability of blood flow in the descending aorta of anesthetized dogs as influenced by the peak Reynolds number and the Womersley number. Points joined by the lines refer to the same animal. Open circles, laminar flow; filled circles, turbulent flow; half-filled circles, transiently turbulent flow. From Nerem, R.M., and Seed, W.A. (1972) An in vivo study of aortic flow disturbances. *Cardiovasc. Res.* **6**: 1–14, by permission.

For pulsatile flow in a tube, when the Womersley number is large, the effect of the viscosity of the fluid does not propagate very far from the wall. In the central portion of the tube the transient flow is determined by the balance of the inertial forces and pressure forces as if the fluid were non-viscous. We therefore expect that when the Womersley number is large the velocity profile in a pulsatile flow will be relatively blunt, in contrast to the parabolic profile of the Poiseuillean flow, which is determined by the balance of viscous and pressure forces. That this is indeed the case can be seen from Figures 3.7:4 and 3.7:5. In Figure 3.7:4 the velocity profiles constructed from time-mean measurements at several sites along the aorta of the dog are shown. They are seen to be quite blunt in the central portion of the aorta. Similar profiles constructed from instantaneous measurements show that this is true throughout the flow cycle.

Figure 3.7:5 shows the theoretical velocity profiles computed for a straight circular cylindrical tube in which a sinusoidally oscillating pressure gradient acts. As the Womersley number increases from 3.34 to 6.67, the

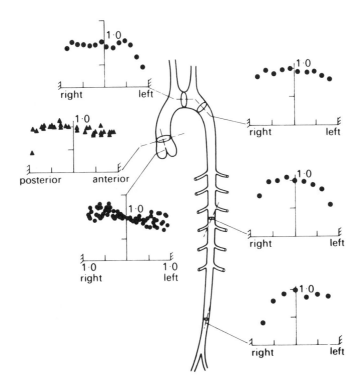

FIGURE 3.7:4. Normalized mean velocity profiles in dog aorta. The mean velocity at each site is normalized by dividing through by the centerline mean velocity. From Schultz, D.L. (1972) Pressure and flow in large arteries. In *Cardiovascular Fluid Dynamics* Bergel, D.H. (ed.). Vol. 1. Academic Press, New York, by permission.

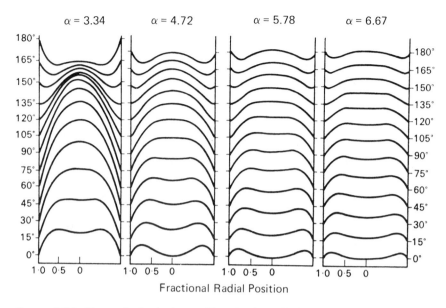

$\alpha = 3.34 \qquad \alpha = 4.72 \qquad \alpha = 5.78 \qquad \alpha = 6.67$

Fractional Radial Position

FIGURE 3.7:5. Theoretical velocity profiles of a sinusoidally oscillating flow in a pipe, with the pressure gradient varying as cos ωt. α is the Womersley number. Profiles are plotted for phase angle steps of $\Delta \omega t = 15°$. For $\omega t > 180°$, the velocity profiles are of the same form but opposite in sign. From McDonald (1974), by permission.

profiles are seen to become flatter and flatter in the central portion of the tube.

Problems

3.10 The total volume rate of flow in all generations of blood vessels is the same. In which vessels is the Reynolds number the largest in the human and dog?

3.11 If the diameter of the aorta of a person is unusually small, would the blood flow be more likely to be laminar or turbulent? If cardiac output is the same but the heart rate is increased, would the blood flow be more likely to become turbulent?

3.12 Estimate the difference between the peak Reynolds number and the mean Reynolds number of blood flow in the aorta of the dog.

3.8 Wave Propagation in Blood Vessels

Before taking up the full complexity of pulse-wave propagation in arteries, let us consider some idealized cases and learn a few basic facts. Let us consider first an infinitely long, straight, isolated, circular, cylindrical, elastic tube containing a homogenous, incompressible, and nonviscous liquid. When this tube is disturbed at one place, the disturbance will be propagated as waves along the tube at a finite speed. The problem is to determine this speed.

Let us impose some further simplifications.* Let the wave amplitude be small and the wave length be long compared with the tube radius, so that the slope of the deformed wall remains $\ll 1$ at all times. Under these conditions we can introduce an important hypothesis that the flow is essentially one dimensional, with a longitudinal velocity component $u(x, t)$, which is a function of the axial coordinate x and time t. In comparison with u, other velocity components are negligibly small. Then the basic equations can be obtained from the general equations listed in Section 2.6. They are the equation of continuity (or conservation of mass),

$$\frac{\partial A}{\partial t} + \frac{\partial}{\partial x}(uA) = 0, \tag{1}$$

and the equation of motion,

$$\frac{\partial u}{\partial t} + u\frac{\partial u}{\partial x} + \frac{1}{\rho}\frac{\partial p_i}{\partial x} = 0. \tag{2}$$

Here $A(x, t)$ is the cross-sectional area of the tube and $p_i(x, t)$ is the pressure in the tube. The relationship between p_i and A may be quite complex. For simplicity we introduce another hypothesis, that A depends on the transmural pressure, $p_i - p_e$, alone,

$$p_i - p_e = P(A), \tag{3}$$

where p_e is the pressure acting on the outside of the tube. Equation (3) is a gross simplification. In the theory of elastic shells we know that the tube deformation is related to the applied load by a set of partial differential equations and that the external load includes the inertial force of the tube wall (see Eqs. (4) and (5) of Sec. 3.15). Hence Eq. (3) implies that the mass of the tube is ignored and that the partial differential equations are replaced by an algebraic equation. By assuming Eq. (3) the dynamics of the tube is replaced by statics. The viscoelasticity of tube wall is ignored.

In the theoretical development, the derivative of the function $P(A)$ is very important, particularly in the following combination,

$$c = \sqrt{\frac{A}{\rho}\frac{dP}{dA}}. \tag{4}$$

We shall see later that c is the velocity of propagation of progressive waves.

These equations are not difficult to solve since Georg Riemann (1826–1866) has shown the way. But before solving these equations we shall

---

* In subsequent sections we shall relax these assumptions and evaluate their effects.

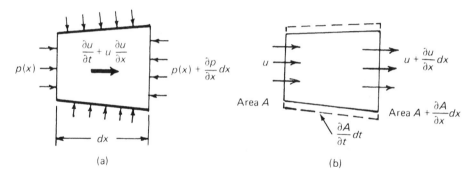

FIGURE 3.8:1. Free-body diagram of an arterial element, showing pressure, velocity, and wall displacement.

derive them once more from elementary considerations to make sure that we know them well.

Consider first the balance of forces acting in the axial direction on a fluid element of length dx and cross-sectional area A. A free-body diagram is shown in Figure 3.8:1(a). Since the fluid is nonviscous there is no shear stress acting on it. The force acting on the left end due to the pressure is pA toward the right; that acting on the right end is $[p + (\partial p/\partial x)dx][A + (\partial A/\partial x)dx]$ toward the left. The pressure acting on the lateral sides contributes an axial force $p(\partial A/\partial x)dx$ toward the right. Therefore, on neglecting the second-order term, the net pressure force is $A(\partial p/\partial x)dx$ acting toward the left. The mass is $\rho A dx$, with ρ being the density of the blood. According to Newton's law the net force will cause an acceleration $\partial u/\partial t + u\partial u/\partial x$. On equating the force with mass times acceleration, we obtain Eq. (2).

Next, consider the conservation of mass in a segment of the tube of length dx, as illustrated in Figure 3.8:1(b). In a unit time the mass influx at the left end is equal to ρuA; the efflux at the right is $\rho\{uA + [\partial(uA)/\partial x]dx\}$. In the mean time the volume of the element is increased by $(\partial A/\partial t)dx$. The law of conservation of mass then leads to Eq. (1).

Next, consider the elasticity of the tube. If the tube behaves like a pulmonary artery or vein, then the situation is simple. The pulmonary arterial diameter $2a_i$ is linearly proportional to the blood pressure in the vessel p_i (see Sec. 6.7):

$$2a_i = 2a_{i0} + \alpha p_i, \tag{5}$$

where a_{i0} and α are constants that depend on the pleural pressure p_{PL} and the airway pressure p_A, but are independent of blood pressure p_i. α is the compliance constant of the vessel, and a_{i0} is the radius when $p_i = 0$. Differentiation of Eq. (5) then yields the relationship

$$da_i = \frac{\alpha}{2} dp_i. \tag{6}$$

Equations (1), (2), and (6) govern the wave propagation phenomenon. Let us first solve a linearized version of these equations. Consider small disturbances in an initially stationary liquid-filled circular cylindrical tube. In this case u is small and the second term in Eq. (2) can be neglected. Hence

$$\frac{\partial u}{\partial t} + \frac{1}{\rho}\frac{\partial p_i}{\partial x} = 0. \tag{7}$$

The area A is equal to πa_i^2. Substituting πa_i^2 for A in Eq. (1), remembering the hypothesis that the wave amplitude is much smaller than the wave length, so that $\partial a_i/\partial x \ll 1$, then, on neglecting small quantities of the second order, we can reduce Eq. (1) to the form

$$\frac{\partial u}{\partial x} + \frac{2}{a_i}\frac{\partial a_i}{\partial t} = 0. \tag{8}$$

Combining Eqs. (8) and (6), we obtain

$$\frac{\partial u}{\partial x} + \frac{\alpha}{a_i}\frac{\partial p_i}{\partial t} = 0. \tag{9}$$

Differentiating Eq. (7) with respect to x and Eq. (9) with respect to t, subtracting the resulting equations, and neglecting the second order term (α/a_i^2) $(\partial a_i/\partial t)(\partial p_i/\partial t)$, we obtain

$$\frac{\partial^2 p_i}{\partial x^2} - \frac{1}{c^2}\frac{\partial^2 p_i}{\partial t^2} = 0, \tag{10}$$

where

$$c^2 = \frac{a_i}{\rho\alpha}. \tag{11}$$

Equation (10) is the famous *wave equation*. The quantity c is the *wave speed*,

$$c = \sqrt{\frac{a_i}{\rho\alpha}}. \tag{12}$$

The derivation of Eq. (12) is simple because the pressure-diameter relationship Eq. (5) is simple. The derivation of wave speed for blood vessels that obey more complex pressure-diameter relationships is given below.

Wave Speed in Thin-Walled Elastic Tube

If the tube is thin walled and the material obeys Hooke's law, then for a small change in radius da_i the circumference is changed by $2\pi da_i$ and the circumferential strain is $2\pi da_i/2\pi a_i = da_i/a_i$. If E is the Young's modulus of the wall material the circumferential stress is changed by the amount $E da_i/a_i$. If the wall thickness is h, the tension in the wall is changed by $Eh da_i/a_i$. This increment of tension is balanced by the change of pressure

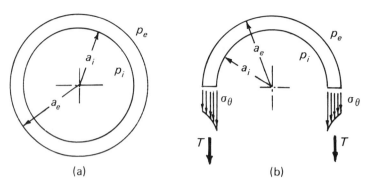

FIGURE 3.8:2. The balances of forces in an arterial wall.

dp_i. According to the condition of equilibrium of the forces acting on a free body shown in Figure 3.8:2(b), we have

$$\frac{Ehda_i}{a_i} = a_i dp_i. \tag{13}$$

This equation is of the same form as Eq. (6) with

$$\frac{\alpha}{2} = \frac{a_i^2}{Eh}. \tag{14}$$

The wave speed in such a tube is, therefore,

$$c = \sqrt{\frac{Eh}{2\rho a_i}}. \tag{15}$$

This formula was first derived by Thomas Young in 1808, and is known as the Moens–Korteweg formula, because it was popularized and modified by Korteweg (1878), Moens (1878).

Note that if the thin-wall assumption is not made the accuracy of the result can be improved by computing the strain on the midwall of the tuber, $da_i/(a_i + h/2)$. Then the wave speed is

$$c = \sqrt{\frac{Eh}{2\rho(a_i + h/2)}}. \tag{16}$$

Wave Speed in Arteries with Nonlinear Elasticity

More realistic constitutive equations of arteries are given in Eqs. (40)–(51) of Section 2.6; and are discussed in greater detail in (Fung, 1993b, Chap. 8). Let the internal and external radii of the vessel be a_i and a_e, respectively, and the corresponding pressures be p_i and p_e (see Fig. 3.8:2(a)). Let the radii be a_{i0} and a_{e0} when the pressures p_i and p_e are zero. The condition of equilibrium of the forces acting on a free body shown in Figure 3.8:2(b) yields the average circumferential stress,

$$\langle \sigma_\theta \rangle = \left(p_i a_i - p_e a_e \right) / \left(a_e - a_i \right).$$ (17)

Let us define the stretch ratio λ_θ and strain $E_{\theta\theta}$ on the midwall by the formulas

$$\lambda_\theta = \frac{a_i + a_e}{a_{i0} + a_{e0}}, \quad E_{\theta\theta} = \frac{1}{2}\left(\lambda_\theta^2 - 1\right).$$ (18)

Then, if $\rho_0 W^{(2)}$ denotes the strain energy function in the arterial wall expressed as a function of the strains $E_{\theta\theta}$ and E_{zz} (the longitudinal strain), we have (Fung, 1993b, Sec. 7.11 and 8.6)

$$\langle \sigma_\theta \rangle = \lambda_\theta^2 \frac{\partial \left(\rho_0 W^{(2)} \right)}{\partial E_{\theta\theta}}.$$ (19)

Combining Eqs. (17) and (19) we have

$$p_i a_i - p_e a_e = \left(a_i - a_e \right) \lambda_\theta^2 \frac{\partial \left(\rho_0 W^{(2)} \right)}{\partial E_{\theta\theta}}.$$ (20)

The function $\rho_0 W^{(2)}$ is given by Fung, Fronek, Patitucci (1979) and in Fung (1993b, Sec. 8.6, Eq. (3)):

$$\rho_0 W^{(2)} = \frac{1}{2} C' \exp\left[a_1 E_{\theta\theta}^2 + a_2 E_{zz}^2 + 2a_4 E_{\theta\theta} E_{zz} \right],$$ (21)

where C', a_1, a_2, a_4 are constants. The radii a_i and a_e are related by the condition of incompressibility of the wall,

$$\pi \left(a_e^2 - a_i^2 \right) = \pi \left(a_{e0}^2 - a_{i0}^2 \right).$$ (22)

On computing $\partial a_e / \partial a_i$ from Eq. (22) and using it in an equation obtained by differentiating Eq. (20), we obtain

$$a_i \, dp_i + p_i \, da_i - p_e \frac{a_i}{a_e} da_i = \lambda_\theta^2 \frac{\partial \left(\rho_0 W^{(2)} \right)}{\partial E_{\theta\theta}} \left(1 - \frac{a_i}{a_e} \right) da_i$$

$$+ \left(a_i - a_e \right) \frac{\partial}{\partial \lambda_\theta} \left[\lambda_\theta^2 \frac{\partial \left(\rho_0 W^{(2)} \right)}{E_{\theta\theta}} \right] \frac{d\lambda_\theta}{da_i} da_i.$$ (23)

This can be put in the form of Eq. (6) if we identify

$$\frac{2}{\alpha} = -\frac{p_i}{a_i} + \frac{p_e}{a_e} + \left(\frac{1}{a_i} - \frac{1}{a_e} \right) \lambda_\theta^2 \frac{\partial \left(\rho_0 W^{(2)} \right)}{\partial E_{\theta\theta}}$$

$$+ \left(\frac{a_i}{a_e} - \frac{a_e}{a_i} \right) \frac{\partial}{\partial \lambda_\theta} \left[\lambda_\theta^2 \frac{\partial \rho_0 W^{(2)}}{\partial E_{\theta\theta}} \right].$$ (24)

The compliance α varies obviously with p_i, p_e, and a_i. If only infinitesimal disturbances da_i, dp_i, dp_e are considered, then the quantity on the right-hand side of Eq. (24) can be evaluated at the steady state and used as a constant in Eq. (11). In that case the linearized wave equation (10) applies.

Solution of the Wave Equation

To understand the nature of the phenomenon described by differential Eq. (10), let us take the following mathematical approach. Let $f(z)$ be an arbitrary function of z, which is differentiable at least twice and whose second derivative is continuous for a certain prescribed region of z. Let z be a function of two variables x and t,

$$z = x - ct, \tag{25}$$

where x represents the coordinate of a point on a straight line and t represents time. Now, by the rules of differentiation, we have

$$\frac{\partial f}{\partial x} = \frac{df}{dz}\frac{\partial z}{\partial x} = \frac{df}{dz}, \quad \frac{\partial f}{\partial t} = \frac{df}{dz}\frac{\partial z}{\partial t} = -c\frac{df}{dz},$$

$$\frac{\partial^2 f}{\partial x^2} = \frac{d^2 f}{dz^2}, \quad \frac{\partial^2 f}{\partial t^2} = c^2 \frac{d^2 f}{dz^2}. \tag{26}$$

The last line shows that the function $f(x - ct)$ satisfies the differential equation

$$\frac{\partial^2 f}{\partial x^2} - \frac{1}{c^2}\frac{\partial^2 f}{\partial t^2} = 0,$$

which is exactly Eq. (10). Thus Eq. (10) is solved by $p = f(x - ct)$.

Now suppose that a disturbance occurs at time $t = 0$ over a segment of the vessel as illustrated in Figure 3.8:3. The amplitude of the disturbance is represented by $f(x)$ at $t = 0$. At a time t_1 later, the same disturbance will appear translated to the right. The value of the disturbance $f(x - ct)$ will remain constant as long as $x - ct$ has the same value; hence an increase in t requires an increase in $x = ct$. Thus the function $f(x - ct)$ represents a wave propagating to the right (in the direction of increasing x) with a speed c. In exactly the same manner, we can shown that $f(x + ct)$ satisfies the wave equation and represents a wave moving in the negative x direction with a speed c.

Flow Velocity and Wall Displacement Waves

Equation (10) shows that the pressure in the elastic vessel is governed by a wave equation. Because the axial velocity u is linearly related to p through Eq. (7), and small change of the radius, a, is linearly related to changes in p through Eq. (6), (13), or (23), we see that u and a are governed by the

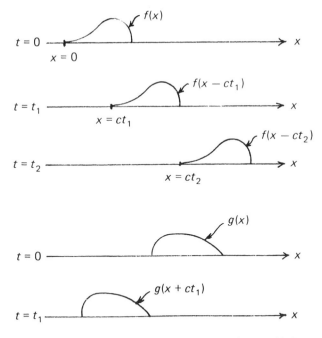

FIGURE 3.8:3. Wave propagation to the right and left.

same wave equation with the same wave speed. In other words, the p in Eq. (10) can be replaced by u and a. (Verify this by direct differentiation.) Thus disturbances in velocity and radius of the vessel are propagated by waves of speed c, in association with the pressure wave.

We use pulse waves in arteries of the wrist, ankle, or temple to determine the heart rate. If we press very gently on the artery, we feel the pulsation of the radius of the artery. If we press harder, so that an area of the artery under the finger is flattened, we should feel the pressure wave in the artery (Fung, 1993b, p. 20, Prob 1.5). With a Doppler ultrasound flow meter, you can detect the velocity waves.

Our derivation of the wave equation is subjected to many simplifying assumptions. All the factors ignored in this derivation have some effect on real wave propagation in the arteries. We discuss them in due course.

Relationship Between the Pressure and Velocity Waves

We have argued that the pressure and velocity satisfy the same wave equation. We can show that the wave equation is satisfied by

$$p = p_o f(x - ct) + p'_o g(x + ct),$$
$$u = u_o f(x - ct) + u'_o g(x + ct), \tag{27}$$

and that by Eq. (7) or (9) the *amplitudes p_o and u_o are related by the simple relationship*

$$p_o = \rho c u_o \tag{28}$$

for a wave that is moving in the positive x direction, and

$$p'_o = -\rho c u'_o \tag{29}$$

for a wave that moves in the negative x direction.

The proof is very simple. On substituting Eq. (27) into Eq. (7), carrying out the differentiation and cancelling the common factor df/dz, we obtain Eq. (28) or (29).

This important relationship shows *that the amplitude of the pressure wave is proportional to the product of wave speed, velocity disturbance, and the fluid density, and nothing else.* This conclusion holds for progressive waves in long tubes without reflection. This, incidentally, is a general result for one-dimensional longitudinal waves, which may occur, for example, in a car crash, or in a plane compressional wave in the earth during earthquake.

Problems on Series Representation of Waves (Fig. 3.8:4)

3.13 Consider a half-sine pulse,

$$f(x) = \sin\frac{\pi x}{L} \quad \text{for} \quad 0 \le x \le L,$$
$$f(x) = 0 \quad\quad \text{for} \quad x < 0, \quad x > L,$$

propagating to the right at speed c. Sketch the wave after 1 s.

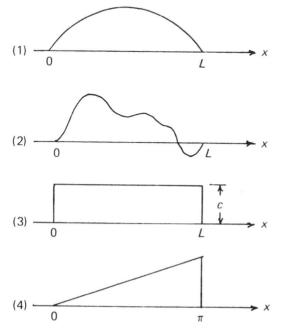

FIGURE 3.8:4. Several wave forms.

3.14 If at time $t = 0$ a wave is observed to have a spatial distribution

$$f(x) = \sum_{n=1}^{\infty} a_n \sin \frac{n\pi x}{L} \quad \text{for} \quad x \text{ in } (0, L) \tag{30}$$

and $f(x) = 0$ outside this interval, show that at time t the solution of wave Eq. (14) is

$$f(x \pm ct) = \sum_{n=1}^{\infty} a_n \sin \frac{n\pi}{L}(x \pm ct) \tag{31}$$

for $x \pm ct$ in $(0, L)$, and $f(x \pm ct) = 0$ elsewhere. The $\pm$ sign is chosen according to whether the direction of propagation is to the left (+) or right (−).

If a wave described by Eq. (31) propagating to the right is observed at a fixed station $x = 0$, then the time sequence is

$$f(-ct) = -\sum_{n=1}^{\infty} a_n \sin \frac{n\pi c}{L} t. \tag{32}$$

Each term in Eq. (32) is a *harmonic* of the wave. The nth is called the *nth harmonic*. The factor $n\pi c/L$ is the *frequency* (or more precisely, *circular frequency*) of the nth harmonic of the pulse wave.

3.15 Show that a square wave of amplitude c in the region $0 < x < L$ can be represented by

$$f(x) = c = \frac{4c}{\pi}\left(\sin \frac{\pi x}{L} + \frac{1}{3} \sin \frac{3\pi x}{L} + \frac{1}{5} \sin \frac{5\pi x}{L} + \cdots \right), \tag{33}$$

whereas one in the region $-L/2 < x < L/2$ can be represented by

$$f(x) = c = \frac{4c}{\pi}\left(\cos \frac{\pi x}{L} - \frac{1}{3} \cos \frac{3\pi x}{L} + \frac{1}{5} \cos \frac{5\pi x}{L} - \cdots \right). \tag{34}$$

Both of these formulas hold for the open intervals indicated. At the ends $x = 0$ and L, the Gibbs phenomenon occurs: The value represented by the series oscillates about c.

3.16 Show that a triangular wave $f(x) = x$ in $-\pi < x < \pi$ can be represented as

$$f(x) = x = 2\left(\sin x - \frac{\sin 2x}{2} + \frac{\sin 3x}{3} - \frac{\sin 4x}{4} + \cdots \right) \tag{35}$$

whereas in $0 \leq x \leq \pi$, inclusive, we have

$$f(x) = x = \frac{\pi}{2} - \frac{4}{\pi}\left(\cos x + \frac{\cos 3x}{3^2} + \frac{\cos 5x}{5^2} + \frac{\cos 7x}{7^2} + \cdots \right). \tag{36}$$

Note the difference in the rate of convergence of the series that represents the same function.

Mathematical Choices of Series to Represent a Function

The choice of the series expansions, as illustrated in Problems 3.15 and 3.16, seems arbitrary, but there are beautiful theorems, such as the following. If one chose to represent a function in the range $-1 \le x \le 1$ by a series of *ultraspherical polynomials* (Szegö, 1939), which includes powers of x, Legendre polynomials, Chebyshev polynomials, and others, then, as Lanczos (1952) has shown, *if the series is truncated at n terms, the estimated error of an expansion into Chebyshev polynomials is smaller than that of any other expansion into ultraspherical polynomials. While the expansion into powers of x (Taylor series) gives the slowest convergence, the expansion into Chebyshev polynomials gives the fastest convergence.*

In case the reader is not familiar with the Chebyshev polynomial, remember that it is nothing but the simple trigonometric function $\cos k\theta$, but expressed in the variable

$$x = \cos \theta.$$

Thus, the Chebyshev polynomial $T_k(x)$ is

$$T_k(x) = \cos(k \text{ arc cos } x). \tag{37}$$

What is meant by this theorem is that if a function $f(x)$ of bounded variation is expanded into a series

$$f(x) = \tfrac{1}{2}c_0 + c_1 T_1(x) + c_2 T_2(x) + \cdots + c_n T_n(x) + \eta_n(x), \tag{38}$$

then the maximum value of the remainder η_n is smaller than that of any other expansions in which $T_k(x)$ is replaced by other ultraspherical polynomials. If the expansion (38) is rearranged into an ordinary power series of the form

$$f(x) = b_0 + b_1 x + b_2 x^2 + \cdots + b_n x^n + \eta_n'(x), \tag{39}$$

then the coefficients b_i decrease slower than the coefficients c_i as i increases and the maximum of the remainder $\eta_n'(x)$ is greater than that of $\eta_n(x)$. In fact, the convergence of the power series in Eq. (39) is the slowest among all expansions in ultraspherical polynomials.

This theorem shows that an orthogonal expansion of $f(x)$ into the polynomials $T_k(x)$ yields an expansion that for the same number of terms represents $f(x)$ with greater accuracy than the expansion into any other sets of orthogonal functions (this includes the Legendre polynomials, which give a better average error but a worse maximum error in the given range).

An Example of Harmonic Analysis of Pulse Waves

In Figure 3.8:5, experimental data on pressure and flow in the ascending aorta of a dog are shown by dotted curves. These curves are analyzed into a Fourier series with a constant term and 10 harmonics (with frequencies

FIGURE 3.8:5. An example of Fourier series representation of pressure and flow waves in the ascending aorta. The experimental wave form is analyzed into a Fourier series with 10 harmonics. The series is then summed and plotted, showing good agreement with experimental data. From McDonald (1974), by permission.

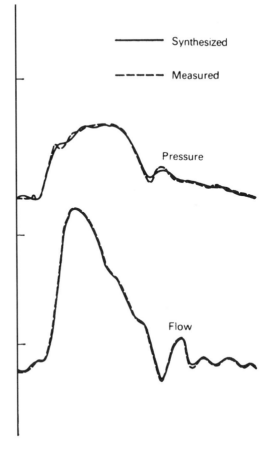

up to about 20 Hz). The solid curves represent the Fourier series. It is shown that the accuracy of the 10-harmonic approximation is acceptable. Further away from the heart, the wave forms are smoother and can be adequately described by fewer harmonics.

3.9 Progressive Waves Superposed on a Steady Flow

The results derived so far apply to a straight, cylindrical, elastic tube filled with a nonviscous liquid that is not flowing. Now, we shall continue to assume the fluid to be nonviscous, but let it have a steady flow to the right. Since the fluid is nonviscous, the no-slip condition on the solid wall does not apply. The velocity profile can be uniform. Then we can show that all equations of Section 3.8 are applicable provided that we adopt a coordinate system that moves with the undisturbed flow, and interpret u as the perturbation velocity superposed on the steady flow and c as the speed of perturbation wave relative to the undisturbed flow. The proof is as follows:

Let U be the velocity of the undisturbed flow and u the small perturbation superposed on it. Treating u as an infinitesimal quantity of the first order, we see that the equation of motion, Eq. (3.8:2), can be linearized into

$$\frac{\partial u}{\partial t} + U \frac{\partial u}{\partial x} = -\frac{1}{\rho} \frac{\partial p_i}{\partial x}. \tag{1}$$

This can be reduced to Eq. (3.8:7) by introducing a transformation of variables from x, t to x', t':

$$x' = x - Ut, \quad t' = t. \tag{2}$$

From Eq. (2) we have

$$\frac{\partial}{\partial t} = \frac{\partial}{\partial t'} \frac{\partial t'}{\partial t} + \frac{\partial}{\partial x'} \frac{\partial x'}{\partial t} = \frac{\partial}{\partial t'} - U \frac{\partial}{\partial x'},$$

$$\frac{\partial}{\partial x} = \frac{\partial}{\partial t'} \frac{\partial t'}{\partial x} + \frac{\partial}{\partial x'} \frac{\partial x'}{\partial x} = \frac{\partial}{\partial x'}. \tag{3}$$

Hence, a substitution into Eq. (1) reduces it to

$$\frac{\partial u}{\partial t'} = -\frac{1}{\rho} \frac{\partial p}{\partial x'}, \tag{4}$$

which is exactly Eq. (3.8:7) in the new coordinates.

The equation of continuity, Eq. (3.8:1), now becomes

$$\frac{\partial a_i}{\partial t} + U \frac{\partial a_i}{\partial x} + \frac{a_i}{2} \frac{\partial u}{\partial x} = 0 \tag{5}$$

when πa_i^2 is substituted for A, with a_i being the inner radius of the tube, and $U + u$ is substituted for u and the equation is linearized for small perturbations. Under the transformation Eq. (2), and using Eq. (5), Eq. (6) becomes

$$\frac{\partial a_i}{\partial t'} + \frac{a_i}{2} \frac{\partial u}{\partial x'} = 0, \tag{6}$$

which is exactly Eq. (3.8:8).

The pressure–radius relationship—Eq. (3.8:5), (3.8:6), (3.8:13), or (3.8:23) —is independent of reference coordinates; thus Eq. (3.8:9) is unchanged when t is replaced by t'. Thus all the basic equations are unchanged. Equations (4) and (6) govern the fluid and Eq. (3.8:9) governs the tube and fluid interaction, that is, the boundary conditions. By eliminating u, the same wave equation (3.8:10) is obtained, except that the independent variables are replaced by x' and t'. But x' and t' are the distance and time measured in the moving coordinates that translate with the undisturbed flow. Thus what we set out to prove is done.

Can Boundary Layers Save the Ideal Fluid Theory?

The wave theory of Section 3.8 is an ideal fluid theory. The superposition of a uniform velocity is valid for an ideal fluid only, not for blood which is viscous. A viscous fluid must obey the no-slip condition. The question is: Could the boundary layer theory discussed in Section 3.5 save the ideal fluid solution for the bulk of blood in the vessel, leaving the boundary layer to adjust to the no-slip condition on vessel wall? Heuristically, the answer is "yes," if the Reynolds and Womersley numbers are large and the vessel is not too long, so that the boundary layers are very thin compared with the tube radius. The short length requirement is related to the boundary layer thickness growth discussed in Section 3.18.

For blood flow in large arteries, in which the Reynolds and Womersley numbers are $\gg 1$, pulse wave analysis of ideal fluid flow provides a good approximation. Hence, we continue to use the ideal fluid hypothesis in the study of wave propagation, reflection, and refraction in large arteries in Sections 3.10 to 3.13. For waves in small arteries and arterioles, in which either the Womersley number, or the Reynolds number, or both approach 1 or <1, we must take viscosity into account, as is done in Section 3.15.

Experimental Validation

Experimental evidence of the theoretical result is shown in Figure 3.9:1. Anliker et al. (1968) installed two electromagnetic wave generators at two stations along a dog aorta and recorded the pressure fluctuations at two points between the two wave generators. A short train of high-frequency waves generated by the upstream wave generator propagates downstream with a theoretical velocity

$$c^D = c + U, \tag{7}$$

which can be determined experimentally by the arrival times of the wave train at the two recording stations. On the other hand, if the wave train is generated by the downstream generator and propagated upstream, the theoretical wave speed is

$$c^U = c - U, \tag{8}$$

which again can be determined experimentally. From Eqs. (7) and (8) we have

$$U = \tfrac{1}{2}\left(c^D - c^U\right). \tag{9}$$

In Figure 3.9:1, c^D, c^U, and U are shown during a cardiac cycle. The flow velocity U can also be measured by a flow gauge, and as Anliker stated, a good agreement is obtained.

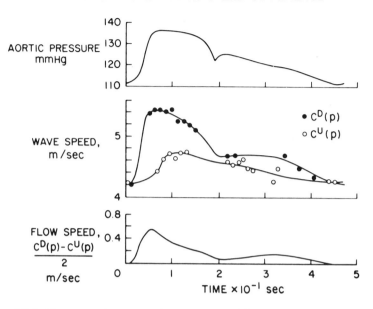

FIGURE 3.9:1. Wave speeds measured upstream and downstream in the aorta of a dog. *Top:* Natural pulse wave. *Middle:* Upstream wave speed (open symbols) and downstream wave speed (closed symbols) measured at different instants of the cardiac cycle. The upstream and downstream data correspond to two heartbeats a few seconds apart, but with matching pressure patterns. *Bottom:* Mean flow velocity U. From Anliker, M. (1972), by permission.

3.10 Nonlinear Wave Propagation

A more general solution of Eqs. (1), (2), and (3) of Section 3.8 is given by Riemann's method of characteristics. This method is explained most clearly in Lighthill (1978), and Yih (1977). Adding $\pm c/A$ times Eq. (1) of Section 3.8 to Eq. (2) of Section 3.8, one can show that on the characteristic curves defined by

$$dx/dt = u \pm c, \tag{1}$$

the quantities (Riemann invariants)

$$R_\pm = \frac{1}{2}\left[u \pm \int_{A_0}^{A} \frac{c}{A} dA \right] \tag{2}$$

are constants, where A_0 is the undisturbed area and c is the velocity

$$c^2 = \frac{A}{\rho} \frac{dp}{dA}. \tag{3}$$

Thus nonlinear waves are propagated in the $\pm x$ directions with speeds $u \pm c$. The linearized theory presented in Section 3.8 results if the condition $c \gg u$ is imposed.

The general solution, Eq. (2), can be used to investigate the effect of some of the simplifying assumptions used in the preceding section. It has been used by Pedley (1980, pp. 79–87) to investigate the formation of shock waves in blood vessels.

The method of characteristics is one of the most important devices to investigate nonlinear wave propagation. See Lighthill (1978), and Skalak (1966, 1972) for in-depth reviews of this subject. Lambert (1958) averaged the equations of motion and continuity over the arterial cross section to obtain uniaxial equations. Van der Werff (1973) introduced a special method to handle periodic conditions. Atabek (1980) combined the characteristics method with Ling and Atabek's (1972) "local flow" analysis to predict velocity profiles of the flow and waves in a segment from known pressure and pressure gradient at the proximal end of the segment. Atabek's detailed comparison between calculated results and those from animal experiments shows the importance of the effects of nonlinearity from various sources (see Sec. 3.16 in this book); he concludes that these effects are not yet fully understood.

Problem

3.17 We know that the blood vessel wall does not obey Hooke's law. Use the information on the pseudo-elasticity and viscoelasticity of the arteries presented in *Biomechanics: Mechanical Properties of Living Tissues* (Fung, 1993b, chapter 8) to derive an expression for the wave speed in arteries.

Devise a theory of your own to handle the viscoelasticity of the blood vessel wall in the problem of pulse wave propagation. Discuss the effect of viscoelasticity in detail.

3.11 Reflection and Transmission of Waves at Junctions of Large Arteries

Thus far we have discussed propagation of uniaxial disturbances in an infinitely long, straight, cylindrical, elastic tube filled with an incompressible nonviscous liquid. Our results are simple and interesting, but they are true only if all the idealizing qualifiers hold. Real arteries do not obey these qualifiers: They are short, tapered, branching, and filled with a non-Newtonian viscous fluid. They are sometimes curved. Their walls are nonlinearly viscoelastic. It turns out that in the large arteries the effect of nonlinear viscoelasticity on wave propagation is not so severe; neglecting the blood viscosity in the tube outside the boundary layer next to the wall is often acceptable for the wave propagation problem because the frequency para-

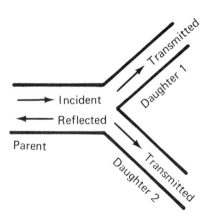

FIGURE 3.11:1. A bifurcating artery.

meter (Sec. 3.7) α and the Reynolds number N_R are sufficiently large that the boundary layer is very thin (we will discuss these factors later), but the "infinitely long" assumption must be removed.

A tube of finite length must have two ends. When waves of pressure and velocity reach an end, they must conform to the end conditions. As a result the waves will be modified. To clarify the situation, consider first a single junction, as shown in Figure 3.11:1, where a tube branches into two daughters. A wave traveling down the parent artery will be partially reflected at the junction and partially transmitted down the daughters. Now, at the junction, the conditions are as follows:

a. The pressure is a single-valued function.
b. The flow must be continuous.

To express this mathematically, let p_I denote the oscillatory pressure associated with the incident wave, p_R that associated with the reflected wave, and p_{T_1} and p_{T_2} those associated with the transmitted waves in the two daughter tubes; then according to (a) we must have

$$p_I + p_R = p_{T_1} = p_{T_2}. \tag{1}$$

Similarly, let $\dot{Q}$ denote the volume-flow rate, and let the subscripts $I, R, T_1,$ T_2 refer to the various waves as before; then, according to (b) we must have

$$\dot{Q}_I - \dot{Q}_R = \dot{Q}_{T_1} + \dot{Q}_{T_2}. \tag{2}$$

The left-hand side of Eq. (2) represents the flow out of the parent tube, and the right-hand side represents the flow into the daughters. But $\dot{Q}$ is just the product of the cross-sectional area A and the mean velocity u. We have already learned the relationship between u and p in Section 3.8. Hence, using Eq. (3.8:28) and Eq. (3.8:29) we obtain the flow–pressure relationship,

$$\dot{Q} = Au = \pm \frac{A}{\rho c} p. \tag{3}$$

Here ρ is the density of the blood and c is the wave speed. The $+$ sign applies if the wave goes in the direction of positive x-axis; the $-$ sign applies if the wave goes the other way. The quantity $\rho c/A$ is an important characteristic of the artery and is called the *characteristic impedance* of the tube, and is denoted by the symbol Z,

$$Z = \frac{\rho c}{A}. \tag{4}$$

Z is the ratio of oscillatory pressure to oscillatory flow when the wave goes in the direction of positive x-axis,

$$Z = \frac{p}{Q}, \quad Z\dot{Q} = p, \tag{5}$$

analogous to the resistance in an electric circuit,

$$R = \frac{V}{I}, \quad RI = V, \tag{6}$$

connecting the voltage V and current I. Z has the physical dimensions $[ML^{-4}T^{-1}]$ and can be measured in the units $\mathrm{kg\,m^{-4}\,sec^{-1}}$. With the Z notation, Eq. (2) can be written as

$$\frac{p_I - p_R}{Z_0} = \frac{p_{T_1}}{Z_1} + \frac{p_{T_2}}{Z_2}. \tag{7}$$

Solving Eqs. (1) and (7) for the p's, we obtain

$$\frac{p_R}{p_I} = \frac{Z_0^{-1} - \left(Z_1^{-1} + Z_2^{-1}\right)}{Z_0^{-1} + \left(Z_1^{-1} + Z_2^{-1}\right)} = \mathscr{R} \tag{8}$$

and

$$\frac{p_{T_1}}{p_I} = \frac{p_{T_2}}{p_I} = \frac{2Z_0^{-1}}{Z_0^{-1} + \left(Z_1^{-1} + Z_2^{-1}\right)} = \mathscr{T}. \tag{9}$$

The right-hand sides of Eqs. (8) and (9) shall be denoted by $\mathscr{R}$ and $\mathscr{T}$, respectively. Hence the amplitude of the reflected pressure wave at the junction is $\mathscr{R}$ times that of the incident wave, the amplitude of the transmitted pressure waves at the junction is $\mathscr{T}$ times the incident wave. The amplitude of the reflected velocity wave is, however, equal to $-\mathscr{R}$ times that of the incident velocity wave, because the wave now moves in the negative x-axis direction, and according to Eqs. (3.8:28) and (3.8:29), there is a sign change in the relation between u and p depending on whether the waves move in the $+$ or $-$ x-axis direction.

The meaning of $\mathscr{R}$ and $\mathscr{T}$ can be clarified further by considering the transmission of energy by pressure waves. Imagine a cross section of the tube. The normal stress acting on this section is the pressure p. The force is

p times the area, A. The fluid pushed by this pressure moves at a velocity u. The rate at which work is done is therefore the product pAu. But $Au = \dot{Q}$ and $\dot{Q} = p/Z$. Therefore the rate of work done is

$$\dot{W} = p\dot{Q} = p^2/Z. \tag{10a}$$

This is the rate of transmission of mechanical energy through the cross section. Now, at the junction of a bifurcating vessel, the rate of energy transmission of the incident wave is p_I^2/Z_0, whereas that of the reflected wave is

$$\frac{p_R^2}{Z_0} = \frac{(\mathscr{R}p_I)^2}{Z_0} = \mathscr{R}^2\,\frac{p_I^2}{Z_0}. \tag{10b}$$

Hence the ratio of the rate of energy transmission of the reflected wave to that of the incident wave is $\mathscr{R}^2$. For this reason $\mathscr{R}^2$ is called the *energy reflection coefficient*. Similarly, the rate of energy transfer in the two transmitted waves, compared with that in the incident wave, is

$$\frac{Z_1^{-1} + Z_2^{-1}}{Z_0^{-1}}\,\mathscr{T}^2, \tag{10c}$$

which is called the *energy transmission coefficient*.

We can express the waves more explicitly as follows. Let the incident wave be

$$p_I = p_0 f\!\left(t - x/c_0\right). \tag{11}$$

Let the junction be located at $x = 0$, so that x is negative in the parent tube and positive in the daughter tubes; then at the junction, $x = 0$, the pressure of the incident wave is

$$p_I = p_0 f(t).$$

The reflectional and transmitted waves are, therefore,

$$p_R = \mathscr{R}p_0 f\!\left(t + x/c_0\right),$$
$$p_{T_1} = \mathscr{T}p_0 f\!\left(t - x/c_1\right),$$
$$p_{T_2} = \mathscr{T}p_0 f\!\left(t - x/c_2\right). \tag{12}$$

Here c_0, c_1, c_2 are the wave speeds in the respective tubes. Note that $p_{T_1} = p_{T_2}$ at the junction, $x = 0$ [see Eq. (1)], but c_1 may be different from c_2. The resultant disturbance in the parent tube is

$$p = p_I + p_R = p_0 f\!\left(t - x/c_0\right) + \mathscr{R}p_0 f\!\left(t + x/c_0\right). \tag{13}$$

The corresponding flow disturbance in the parent tube is, according to Eq. (3) and taking the direction of propagation into account,

$$\dot{Q} = \frac{Ap_0}{\rho c_0} f\!\left(t - x/c_0\right) - \mathscr{R}\frac{Ap_0}{\rho c_0} f\!\left(t + x/c_0\right). \tag{14}$$

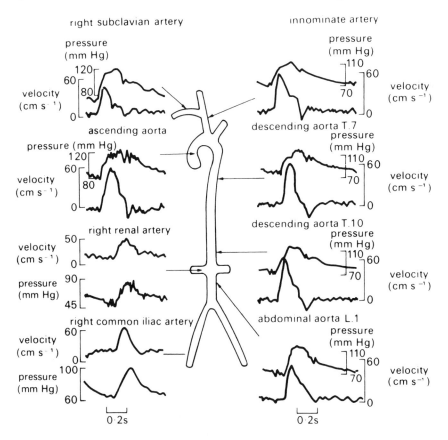

FIGURE 3.11:2. Pressure and flow waves in human arterial tree. From Mills et al. (1970) Pressure-flow relationships and vascular impedance in man. *Cardiovascular Res.* **4**: 405–417, by permission.

A comparison of Eqs. (13) and (14) shows that with reflection, the pressure and flow wave forms are no longer equal.

Inequality of pressure and flow wave forms is a common feature of pulse waves in arteries (Fig. 3.11:2), indicating the effect of reflection at branches.

Problems

3.18 Consider the case in which a parent tube gives rise to three daughter tubes at a junction. Show that $\mathscr{R}$ and $\mathscr{T}$ are given by expressions similar to Eqs. (8) and (9), except that $Z_1^{-1} + Z_2^{-1}$ should be replaced by $Z_1^{-1} + Z_2^{-1} + Z_3^{-1}$ (Fig. P3.18).

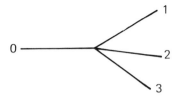

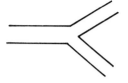

FIGURE P3.19. Matched impedance.

FIGURE P3.18. Trifurcation.

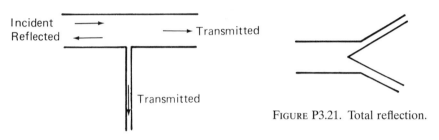

FIGURE P3.21. Total reflection.

FIGURE P3.20. A small daughter branch.

3.19 Under what condition is the reflected wave zero? State the condition $\mathcal{R} = 0$ in terms of the physical parameters of the tubes.

Note: When $\mathcal{R} = 0$ the junction is said to be one at which the *impedances* are matched (Fig. P3.19).

A parent tube gives out a small daughter branch (Fig. P3.20). What are reflection and transmission characteristics ($\mathcal{R}$ and $\mathcal{T}$) at the junction?

Under what condition would a wave be totally reflected ($\mathcal{R} = 1$) (Fig. P3.21)?

If the impedance of a parent tube is perfectly matched to the daughter tubes at a junction so that $\mathcal{R} = 0$, show that $\mathcal{T} = 1$ and the transmission coefficient given in Eq. (10c) is 1. Show that $1 + \mathcal{R} = \mathcal{T}$.

3.20 Consider a bifurcating artery (see Fig. 3.11:1). The functions $\mathcal{T}$ [Eq. (9)] and $\mathcal{R}$ are called, respectively, the *transmission* and the *reflection coefficient* (without the word *energy*). In the special case in which the two daughter branches are of equal size and the wave speed c is the same in the parent and daughter branches, $\mathcal{R}$ and $\mathcal{T}$ are functions of the "area ratio," $(A_1 + A_2)/A_0$, that is, the ratio of the combined area of the branches to the area of the parent tube. Derive expressions of $\mathcal{R}$ and $\mathcal{T}$ in terms of the area ratio and sketch curves to show the variation of $\mathcal{R}$ and $\mathcal{T}$ with the area ratio.

Note: Cf. Atabek (1980), p. 302, in which the viscosity of the blood and viscoelasticity of the vessel wall are considered (see Sec. 3.15 *infra*), while the motion is limited to be simple harmonic (see p. 161).

Our results, Eqs. (4) to (10c), are derived for inviscid fluid in an elastic tube. Whereas the wave speed c is a real number in Eq. (4), it is a complex number in Atabek (see Sec. 3.15). Figure 7.16 of Atabek (1980) shows that there is a minor dependence of the magnitude of $\mathcal{R}$ and $\mathcal{T}$ on the Womersley number (α), and a sudden change of the phase angle of $\mathcal{R}$ (from 0 to 180°) when the area ratio exceeds about 1.2 to 1.4.

3.21 Design an instrument to measure pulse waves noninvasively at some conveniently located arteries, such as the radial artery at the wrist. What can you measure? Pressure? Force? Velocity? What significant use can be made of such measurements? (See Sec. 3.20.)

3.22 There are several machines in clinical use that apply pressure or vacuum on arteries of the arms or legs in a suitable periodic manner to serve as heart assist devices. One machine works on veins to reduce the threat of thrombosis. Invent one yourself, and explain why is it good.

Harmonic Waves

Oscillations that are sinusoidal in time and space are called *harmonic waves*. For example, a pressure wave,

$$p = p_0 \cos\left\{\omega\left(t - \frac{x}{c_0}\right)\right\} = p_0 \cos\left(\omega t - \frac{2\pi x}{\lambda}\right)$$

$$= p_0 \cos\left\{\frac{2\pi}{\lambda}(x - c_0 t)\right\}, \tag{15}$$

is a harmonic progressive wave. Here ω is the *circular frequency* (unit, rad/sec), $\omega/2\pi$ is the *frequency* (unit, Hz), and λ is the *wave length*, (unit, m). They are related by

$$\lambda = \frac{c_0}{(\omega/2\pi)}, \quad \frac{\omega}{2\pi} = \frac{c_0}{\lambda}. \tag{16}$$

Thus the wave length is the wave speed divided by frequency, or the distance traveled per cycle. The wave speed is the product of frequency and wave length.

For harmonic waves, a convenient mathematical device is the *complex representation*. This is based on the relation

$$e^{iz} = \cos z + i \sin z, \tag{17}$$

where $i = \sqrt{-1}$, e is the exponential function, and z is a real variable. Thus $\cos z$ is the real part of e^{iz} and $\sin z$ is the imaginary part of e^{iz}. We can write Eq. (15) as

$$p = \mathcal{Re}\left\{p_0 e^{i\omega(t-x/c_0)}\right\}. \tag{18}$$

The symbol $\mathcal{Re}$ means the real part of the complex quantity. A great advantage of the complex representation is that in Eq. (18) p_0 does not have to be limited to a real number. If p_0 is a complex number,

$$p_0 = a + ib = Pe^{i\phi},$$

$$P = \sqrt{a^2 + b^2}, \quad \phi = \tan^{-1}\frac{b}{a};$$

then Eq. (18) means

$$p = a\cos\left\{\omega\left(t - x/c_0\right)\right\} - b\sin\left\{\omega\left(t - x/c_0\right)\right\}$$
$$= P\cos\left\{\omega\left(t - x/c_0\right) + \phi\right\}. \tag{19}$$

Hence P is the *amplitude* and ϕ is the *phase angle* of the wave. Similar expressions can be written for the flow rate and for waves traveling in the opposite direction. It is conventional to omit the symbol $\mathcal{Re}$, so that whenever a complex number is used to represent a physical quantity, it is assumed that its real part is being used.

We shall use this method to discuss multiple reflections later.

Problem

3.23 Consider energy transmission. We have shown in Eq. (10) that the rate of energy transmission in a progressive wave is

$$W = Ap \cdot u = p \cdot \dot{Q} = p^2/Z.$$

If p is a harmonic wave, show that this is

$$W = \mathcal{Re}\,p \cdot \mathcal{Re}\,\dot{Q} = \left(\mathcal{Re}\,p\right)^2/Z$$

and is not equal to $\mathcal{Re}(p^2)/Z$. This important example shows that one has to be careful in using the complex representation.

Show that if p is given by Eq. (15) the average value of W over a period is

$$W = \tfrac{1}{2}p_0^2/Z.$$

Multiple Reflections

Waves in more complex systems of tubes can be analyzed by repeated application of the results presented earlier. For example, in the double junction illustrated in Figure 3.11:3, a wave reflected once at junction B is reflected a second time at junction A, and so on. The amplitudes of the reflected and transmitted waves on each occasion are determined by the characteristics of the junction.

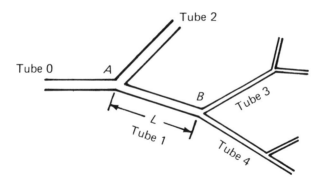

FIGURE 3.11:3. Multiple reflection sites of an artery with two branching junctions.

To see what is going on, let us consider a continuous harmonic excitation and write out in detail the perturbations in the segment AB. Let the origin of the x-axis be taken at A. Let the first wave transmitted through A be

$$p_1 e^{i\omega(t-x/c_0)}. \tag{20}$$

At B, where x = L, the pressure due to this wave is

$$p_1 e^{i\omega(t-L/c_0)}.$$

Here the wave is reflected. Let the reflection parameter be denoted by $\mathscr{R}_{10}$. Then the reflected wave is

$$\mathscr{R}_{10} p_1 e^{i\omega[t-L/c_0+(x-L)/c_0]}. \tag{21}$$

When this wave reaches A, the pressure is

$$\mathscr{R}_{10} p_1 e^{i\omega(t-2L/c_0)}.$$

This wave is reflected at A. To calculate the reflection parameter we must treat the segment AB as the parent tube and tubes 0 and 2 as daughters. Let the reflection parameters be denoted by $\mathscr{R}_{01}$. Then the reflected wave is

$$\mathscr{R}_{01}\mathscr{R}_{10} p_1 e^{i\omega(t-2L/c_0-x/c_0)}.$$

The process continues. The pressure perturbation in the tube AB is the sum of all these waves.

But the story cannot end here. At the ends A and B, the waves do not just bounce back and forth; they are also transmitted into the vessels beyond them, to segments 0, 2, 3, 4, etc. These transmitted waves will be reflected at the junctions further away and will come back to segment AB. The total picture will not be known until the entire system is accounted for. In practice, if the impedance is reasonably well matched, the series converges rapidly.

Standing Waves

The reflection and trapping of waves are related to the phenomenon of resonance. Consider a particular condition in which the tube AB is closed at both ends so that $\mathscr{R}_{10}$ and $\mathscr{R}_{13}$ are both equal to 1. In this case the sum of the first two waves, Eqs. (20) and (21), becomes, with Eq. (16),

$$e^{i\left(\omega t - 2\pi x/\lambda\right)} + e^{i\left[\omega t - 2\pi L/\lambda + 2\pi\left(x-L\right)/\lambda\right]}.$$

The sum of every two succeeding terms is similar, differing only in phase angle. In the special case in which the tube length is equal to the half-wave length,

$$L = \frac{\lambda}{2}, \tag{22}$$

the sum above becomes (because $e^{-i2\pi}$ is equal to 1)

$$e^{i\omega t}\left\{e^{-i\pi x/L} + e^{i\pi x/L}\right\} = 2e^{i\omega t}\cos\frac{\pi x}{L}. \tag{23}$$

Thus, in this case the oscillation is a standing wave, and the motion is a *resonant vibration*. This occurs at a frequency of

$$\frac{\omega}{2\pi} = \frac{c_0}{\lambda} = \frac{c_0}{2L} = \frac{1}{2L}\sqrt{\frac{Eh}{2\rho a}}, \tag{24}$$

which is said to be the *fundamental frequency* of natural vibration; Eq. (23) is said to be the *fundamental mode*. If a system is excited at a resonance frequency, the amplitude of vibration can only be limited by damping. Higher modes are obtained if $L = \lambda/(2n)$, where n is an integer, in which case the mode shape is $\cos 2\pi n x/L$ and the frequency is n times the fundamental.

Real use of this concept is limited. The vibration mode and natural frequency depend on the end conditions. Any change of the end conditions changes the modes. The mode (23) corresponds to a tube with closed ends. Open the ends and the mode is changed.

3.12 Effect of Frequency on the Pressure–Flow Relationship at any Point in an Arterial Tree

The complex branching pattern of the arteries tells us at once that multiple reflections of pulse waves must be a major feature of blood flow. The differences between the pressure and flow profiles shown in Figure 3.11:2 quoted in the preceding section support this statement because, if it were not for the reflections, the pressure and flow waves would have similar profiles. But if reflection is important, then the flow and pressure relationship at any given site in the artery must depend on how the multiple reflections at the bifurcation points are seen at this site, how far away the bifurcation point is, and how long it takes for each wave to travel from a bifurcation point to that site. At any given time, the pressure and flow at a given site are the sums of the newly arrived waves and the retarded waves of reflection from earlier fluctuations. This means that the pressure–flow relationship is frequency dependent.

To express the frequency-dependent characteristics of an arterial tree, it is customary to consider each harmonic of the pulse wave separately and to write, at a given site and a given frequency, the ratio of pressure to flow:

$$\frac{p}{\dot{Q}} = Me^{i\theta} = Z_{\text{eff}}. \tag{1}$$

p and $\dot{Q}$ are represented by complex numbers multiplied by $e^{i\omega t}$. Their ratio is, of course, a complex number, and is called the *input impedance* or *effective impedance*. Its modulus, M, is the ratio of the amplitudes of pressure and flow, whereas its argument, θ, is the phase lag of flow-rate oscillation behind the pressure oscillation.

The input impedance of the human arterial tree can be obtained by analyzing the measured pressure and velocity waves at a given site (e.g., one of those illustrated in Fig. 3.11:2) by Fourier series (e.g., Fig. 3.8:4), and computing the ratio of the corresponding complex-valued harmonics. An example of experimental input impedance measured in the ascending aorta is shown in Figure 3.12:1, which was taken from the same set of measurements as the wave forms shown in Figure 3.11:2. There is a minimum of M at a frequency of 3 Hz, and calculation shows that this implies the presence of a major reflection site roughly at the level of the aortic bifurcation. Measurements at different sites in the aorta lead to the same conclusion.

The input (or effective) impedance is not the same as the characteristic impedance of the tube in which the measurements are made. Don't use the word *impedance* without telling the reader what impedance you mean. The ratio of pressure to flow at any point is called the *effective impedance*. The effective impedance at a point A (see, e.g., Fig. 3.11:3) is called the *input impedance* of the system distal to A.

This terminology comes from electric circuit theory. If a circuit is connected to a voltage source and we want to know if the system can be operated successfully, we often need to know only the input impedance that the circuit offers to the source. Similarly, if we want to couple the arterial system to the heart, we need to know the input impedance of the arterial system at the aortic valve. If we want to know the function of the kidney, we want to know the input impedance of the kidney at the point where the renal artery branches from the abdominal aorta.

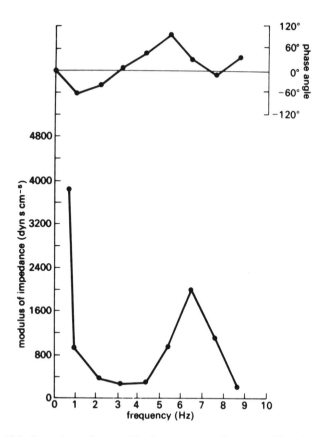

FIGURE 3.12:1. Input impedance of the human ascending aorta. The phase angle and modulus are plotted against the wave frequency. The single minimum of the modulus suggests that there is a single effective reflection site at the level of aortic bifurcation. From Mills et al. (1970) Pressure–flow relationships and vascular impedance in man. *Cardiovasc. Res.* **4**: 405–417, by permission.

Examples

1. *Input Impedance of a Branching Artery* (see Fig. 3.11:3). Consider an artery AB (segment 1) that branches into segments 3 and 4. Let a pressure wave $p_I e^{i\omega t}$ be imposed at the terminal A. A pressure wave $p_I e^{i\omega(t-x/c_1)}$ propagates to the right. When it reaches B at time $t = L/c_1$, it is reflected as a pressure wave,

$$p_R e^{i\omega[t-L/c_1-(L-x)/c_1]},\tag{2}$$

propagating toward A, and is transmitted into segments 3 and 4 as progressive waves

$$p_{T_3} e^{i\omega(t-L/c_1-x_3/c_3)}, \quad p_{T_4} e^{i\omega(t-L/c_1-x_4/c_4)},\tag{3}$$

respectively, where x_3, x_4 are distances measured from point B. Since the pressure at B is single valued, we have, on substituting $x = L$, $x_3 = x_4 = 0$ (at B) and cancelling the factors $e^{i\omega(t-L/c_1)}$ throughout,

$$p_I + p_R = p_{T_3} = p_{T_4}. \tag{4}$$

The flows associated with the incident and reflected waves in segment 1 are obtained by dividing the pressure waves with the characteristic impedance Z_1 of that segment. The flows into branches 3 and 4 are obtained by dividing the pressures at point B by the effective impedances $Z_{3\text{eff}}$ and $Z_{4\text{eff}}$, respectively. Hence, on equating the inflow with outflow at B and again cancelling the factor $\exp{[i\omega(t - L/c_1)]}$,

$$\frac{1}{Z_1}p_I - \frac{1}{Z_1}p_R = \frac{p_{T_3}}{Z_{3\text{eff}}} + \frac{p_{T_4}}{Z_{4\text{eff}}}. \tag{5}$$

These equations are the same as those of Section 3.11, except that Z_3 and Z_4 are replaced by effective impedances. By solving these equations for p_R, p_{T_3}, and p_{T_4}, as before, we obtain

$$\frac{p_R}{p_I} = \mathcal{R}_{\text{eff}}, \quad \frac{p_{T_3}}{p_I} = \frac{p_{T_4}}{p_I} = \mathcal{T}_{\text{eff}}, \tag{6}$$

where

$$\mathcal{R}_{\text{eff}} = \frac{Z_1^{-1} - \left(Z_{3\text{eff}}^{-1} + Z_{4\text{eff}}^{-1}\right)}{Z_1^{-1} + \left(Z_{3\text{eff}}^{-1} + Z_{4\text{eff}}^{-1}\right)}, \quad \mathcal{T}_{\text{eff}} = 1 + \mathcal{R}_{\text{eff}}. \tag{7}$$

Note that this result is the same as that of Section 3.11 except for a change in notation and interpretation. In Section 3.11 we speak of progressive waves going through the bifurcation point B, anticipating the waves to be reflected at other points of bufurcation but discussing the situation at B before any of the reflected waves arrive at B. In the present section we consider periodic oscillations and allow the waves to be reflected and transmitted as the system permits and demands, and find that a progressive wave is reflected at a junction with a complex amplitude ratio $\mathcal{R}_{\text{eff}}$ when the characteristic impedances used in Section 3.11 are replaced by the effective impedances of the downstream branches.

Now, back at point A, where $x = 0$, let us assume that the reflected wave passes through without further reflection. Then the pressure and flow are

$$p_A = p_I e^{i\omega t} + p_R e^{i(\omega t - 2L\omega/c_1)} = p_I e^{i\omega t}\left(1 + \mathcal{R}_{\text{eff}}\, e^{-2i\omega L/c_1}\right), \tag{8}$$

$$\dot{Q}_A = \frac{p_I}{Z_1}e^{i\omega t} - \frac{p_R}{Z_1}e^{i(\omega t - 2L\omega/c_1)} = p_I e^{i\omega t}\frac{1}{Z_1}\left(1 - \mathcal{R}_{\text{eff}}\, e^{-2i\omega L/c_1}\right). \tag{9}$$

Using Eq. (1), we obtain, finally, the ratio of p_A to $\dot{Q}_A$, which is the input impedance at A:

$$\frac{p_A}{Q_A} = Z_{1\text{eff}} = Z_1 \frac{1 + \mathscr{R}_{\text{eff}} e^{-2i\omega L/c_1}}{1 - \mathscr{R}_{\text{eff}} e^{-2i\omega L/c_1}}. \tag{10}$$

We can recast the final result in a different form. On substituting Eq. (7) for $\mathscr{R}_{\text{eff}}$ into Eq. (10), multiplying both the numerator and denominator by $(Z_1^{-1} + Z_{3\text{eff}}^{-1} + Z_{4\text{eff}}^{-1})e^{i\omega L/c_1}$, and noting that for any α,

$$\frac{e^{i\alpha} + e^{-i\alpha}}{2} = \cos\alpha, \quad \frac{e^{i\alpha} - e^{-i\alpha}}{2} = i\sin\alpha, \tag{11}$$

we obtain the important result

$$Z_{1\text{eff}}^{-1} = Z_1^{-1} \frac{\left(Z_{3\text{eff}}^{-1} + Z_{4\text{eff}}^{-1}\right) + iZ_1^{-1}\tan\left(\omega L/c_1\right)}{Z_1^{-1} + i\left(Z_{3\text{eff}}^{-1} + Z_{4\text{eff}}^{-1}\right)\tan\left(\omega L/c_1\right)}. \tag{12}$$

By repeated use of this equation, we can obtain the effective impedance at any point, that is, the relationship between the oscillatory pressure and flow rate at that point, from their values at the distal ends.

The factor $\omega L/c_1$ is equal to $2\pi L/\lambda$, where λ is the wavelength $2\pi c_1/\omega$. If $\omega L/c_1 = n\pi$ (n an integer), then $\tan(\omega L/c_1) = 0$ and

$$Z_{1\text{eff}}^{-1} = Z_{3\text{eff}}^{-1} + Z_{4\text{eff}}^{-1}. \tag{13}$$

Thus, if the arterial length L is much smaller than the wavelength, then $n \to 0$ and Eq. (13) shows that the artery may be considered as part of the junction and there is no change of the pressure–flow relationship in that segment.

On the other hand, if $\omega L/c_1$ is equal to an odd multiple of $\pi/2$, that is, if L is equal to an odd multiple of quarter-wavelengths, then $\tan(\omega L/c_1) = \infty$ and

$$Z_{1\text{eff}}^{-1} = \frac{Z_1^{-2}}{Z_{3\text{eff}}^{-1} + Z_{4\text{eff}}^{-1}}. \tag{14}$$

In this case, if Z_1^{-1} is smaller (or greater) than $Z_{3\text{eff}}^{-1} + Z_{4\text{eff}}^{-1}$, then $Z_{1\text{eff}}^{-1}$ is smaller (or greater) than Z_1^{-1}.

2. *Reverberative Reflections in an Artery.* Consider an artery with two sites of reflection, A and B (see Fig. 3.11:3). A pressure wave $p_o e^{i(\omega t - kx)}$ enters at A. At B it is reflected with a change of amplitude. The reflected wave, on arriving at A, is reflected again, and so on. Let the ratio of the complex amplitude of the reflected wave to that of the incident wave be denoted by $\mathscr{R}_1$ at A and $\mathscr{R}_2$ at B (the subscripts "eff" being omitted for simplicity). Then, at a station at a distance x from A and at time t, the pressure is

$$p(x,t) = p_o e^{i\omega(t - x/c)} + \mathscr{R}_2 p_o e^{i\omega\left[(t - L/c) - (L - x)/c\right]}$$
$$+ \mathscr{R}_1 \mathscr{R}_2 p_o e^{i\omega\left[(t - 2L/c) - x/c\right]}$$
$$+ \mathscr{R}_1 \mathscr{R}_2^2 p_o e^{i\omega\left[(t - 3L/c) - (L - x)/c\right]} + \cdots. \tag{15}$$

We now assemble terms that represent waves going to the right and, separately, those representing waves going to the left. We obtain

$$p(x,t) = p_o e^{i\omega(t-x/c)}\left[1 + \mathcal{R}_1\mathcal{R}_2 e^{-i2\omega L/c} + \cdots\right]$$
$$+ \mathcal{R}_2 p_o e^{i\omega(t-2L/c+x/c)}\left[1 + \mathcal{R}_1\mathcal{R}_2 e^{-i2\omega L/c} + \cdots\right]. \tag{16}$$

Using the summation formula

$$1 + \alpha + \alpha^2 + \alpha^3 + \cdots = \frac{1}{1-\alpha}, \tag{17}$$

for whatever α, we obtain an important formula

$$p(x,t) = \frac{p_o e^{i\omega(t-x/c)} + \mathcal{R}_2 p_o e^{i\omega(t-2L/c+x/c)}}{1 - \mathcal{R}_1\mathcal{R}_2 e^{i2\omega L/c}}. \tag{18}$$

This is a general result. Now, consider the special case of total reflection at the two ends, $\mathcal{R}_1 = \mathcal{R}_2 = 1$. Then

$$p(x,t) = p_o e^{i\omega t}\frac{e^{-i\omega x/c} + e^{-i2\omega L/c}e^{i\omega x/c}}{1 - e^{-i2\omega L/c}}. \tag{19}$$

If we multiply the numerator and denominator by $e^{i\omega L/c}$ and use Eqs. (11), we obtain

$$p(x,t) = p_o e^{i\omega t}\frac{\cos\omega(L-x)/c}{i\sin(\omega L/c)}, \tag{20}$$

which represents a "standing" wave. The wave is "standing" because it does not propagate.

The amplitude of the standing wave will tend to infinity if the denominator $\sin(\omega L/c)$ tends to zero; then the oscillation is said to "resonate." This occurs if

$$\frac{\omega L}{c} = n\pi \quad\text{or}\quad L = n\frac{\lambda}{2} \quad (n = 1, 2, \cdots), \tag{21}$$

that is, if the length of the segment equals an integral multiple of half-wavelength.

Problems

3.24 When the frequency tends to zero, show that the phase angle θ tends to zero and the modulus M tends to a constant. With suitable assumptions with regard to an arterial tree at the peripheral end (microcirculation), derive an expression for M as the frequency tends to zero.

Note: That the dynamics modulus of input impedance can be much smaller than the static impedance (resistance at zero frequency) is of great importance and interest. Compare this with some of our daily experiences. We can often shake a small tree if we do it at the right frequency, whereas the tree would not deflect very much if the same force is applied statically. In the circulatory system, this means that we can get blood to move with much smaller driving pressure when it is done dynamically.

3.25 Explain why can we extract information on input impedance from measurements such as those illustrated in Figure 3.11:2. One may use the Fourier analysis approach (see Sec. 3.8, Example 2). Express both the pressure and flow wave forms in Fourier series, and then compare them with the complex representation of waves discussed in Section 3.11.

3.26 In Section 3.8, Example 5, we extolled Chebyshev polynomials as the basis of generalized Fourier series. Can you develop a formal theory of input impedance in terms of Chebyshev polynomials? What difficulty is there?

3.13 Pressure and Velocity Waves in Large Arteries

The pressure and flow waves in arteries are generated by the heart. The conditions at the aortic valve and the capillary blood vessels are the end conditions of the arterial system. Major features along the length of the large arteries are explainable by the simple analysis presented in preceding sections, but the explanation of the major features of flow in regions close to the aortic valve must take the three-dimensional geometry of the left ventricle, the valve, and the aorta into consideration, together with three-dimensional fluid dynamics and the dynamics of the solid structures involved. Similarly, the analysis of flow in regions of vessel bifurcation, atherosclerosis, aneurysm, stenosis, or dilatation require extensive numerical calculation. Some of these problems are discussed in Sections 3.14 to 3.19. Here we present some features of flow in the aorta.

Figure 3.13:1 shows simultaneous recordings of the pressure in the left ventricle and in the ascending aorta immediately downstream from the aortic valve. When heart contracts, pressure rises rapidly in the ventricle at the beginning of systole and soon exceeds that in the aorta, so that the aortic valve opens, blood is ejected, and aortic pressure rises. During the early part of the ejection, ventricular pressure exceeds aortic pressure. About halfway through ejection, the two pressure traces cross, and the heart is faced with an adverse pressure gradient. The flow and pressure start to fall. Then a notch in the aortic pressure record (the *dicrotic notch*) marks the closure of the aortic valve. Thereafter the ventricular pressure falls very rapidly as the heart muscle relaxes: The aortic pressure falls more slowly, with the elastic vessel serving as a reservoir. The major feature of the pressure wave in the aorta is explained by the windkessel theory (see Sec. 2.1),

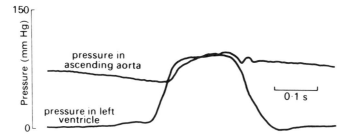

FIGURE 3.13:1. Pressure in the left ventricle and ascending aorta of the dog. From Noble, (1968) The contribution of blood momentum to left ventricular ejection in the dog. *Circ. Res.* **23**: 663–670. Reproduced by permission of the American Heart Association.

but the details can only be determined when all the waves are accounted for.

The change of pressure wave with distance from the aortic valve is shown in Figure 3.13:2. First we see a shift of the profile to the right, suggesting a wave propagation. We also see a steepening and increase in amplitude, while the sharp dicrotic notch is gradually lost. This increase of systolic pressure with distance from the heart in a tapered tube is a dynamic phenomenon in an elastic branching system. In a steady flow in a rigid tube of similar taper, the pressure must go down in the direction of flow unless there is deceleration. In the present case, however, the *mean* value of the pressure, averaged over the period of a heartbeat, still decreases with increasing distance from the aortic valve. It is difficult to see it in Figure 3.13:2 because the fall in mean pressure is only about 4 mm Hg (0.5 kPa) in the

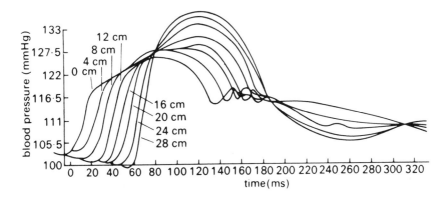

FIGURE 3.13:2. Simultaneous blood pressure records made at a series of sites along the aorta in the dog, with distance measured from the beginning of the descending aorta. From Olson, R.M. (1968) Aortic blood pressure and velocity as a function of time and position. *J. Appl. Physiol.* **24**: 563–569. Reproduced by permission.

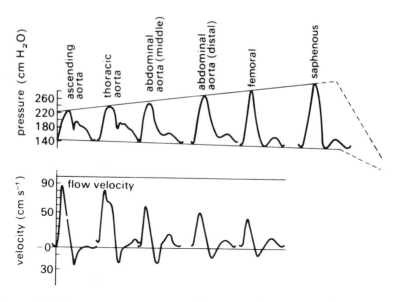

FIGURE 3.13:3. Pressure and velocity waves at different sites in the arteries of a dog. From McDonald (1974), by permission.

whole length of the aorta, while the amplitude of the pressure oscillation between systole and diastole nearly doubles.

This process of amplification of the pressure pulse continues into the branches of the aorta, as illustrated in Figure 3.13:3. In the dog, it continues to about the third generation of arterial branches. Thereafter both the oscillation and the mean pressure decrease gradually downstream along the arterial tree until it reaches the level of microcirculation.

Figure 3.13:3 also shows the variation of the flow velocity along the aorta. That the pressure and velocity waves are different is an indication of reflection of waves at junctions. The velocity waves do not steepen with distance, nor does the peak systolic velocity increase downstream.

3.14 Effect of Taper

One of the simplifying assumptions made in the preceeding sections is that the tube is circular and cylindrical in shape and is straight. Real blood vessels are often curved and of variable cross section. The nonuniform cross-sectional area is associated with branching (see Sec. 3.1) and elastic deformation of the vessel wall in response to a nonuniform pressure with a finite gradient (see Sec. 3.4). The taper is generally very mild, and it is possible to evaluate its effect approximately without extensive calculations.

FIGURE 3.14:1. Approximation of a stepwise tapered tube by a continuously tapered blood vessel.

Let a smoothly tapered tube be approximated by a stepwise tapered tube, as shown in Figure 3.14:1. Each site of step change may be regarded as a junction of two tubes, and the method of analysis presented in Section 3.11 can be applied. We have learned in Section 3.11 that the rate of energy transfer in a reflected wave is proportional to the square of $\mathcal{R}$, and therefore if $\mathcal{R} \ll 1$ it is quite negligible. $\mathcal{R}$, in this case, as is given in Eq. (3.11:8), is proportional to the difference in cross-sectional areas of the two segments divided by the sum of the cross-sectional areas, and is obviously very small if the taper is mild. $\mathcal{R}^2$ is another order of magnitude smaller. Therefore we conclude that very little energy is reflected as the wave travels along a slowly varying tube, and we may analyze the wave's development as if all the energy were transmitted.

The rate of transfer of energy in a progressive wave across any cross section of a vessel is shown in Eq. (3.11:10) to be equal to p^2/Z, the square of the oscillatory pressure divided by the characteristic impedance of the tube. If all the energy is transmitted, then p^2/Z is a constant and we have

$$p = \text{const.} \cdot Z^{1/2}. \tag{1}$$

Thus, in a gradually tapering artery, the amplitude of the pressure wave is proportional to the square root of the characteristic impedance. Since the characteristic impedance is pc/A, we see that Z increases as A decreases if pc were constant. Hence the amplitude of the pressure wave increases as the wave propagates down a tapering tube with decreasing cross section. The amplitude of the aortic *flow* pulse, proportional to p/Z [see Eq. (3.11:5)], will correspondingly decrease, being proportional to $Z^{-1/2}$.

These predicted features are evident in the records shown in Figures 3.13:2 and 3.13:3. However, a quantitative comparison of the predictions with the experimental results shows that the peaking is overestimated by the theory. One of the reasons for this is the neglecting of viscous effects of the blood and blood vessel; the other reason is the inaccuracy of the theory. The theory is more accurate if the taper is small. But how small is small? To answer this question one should turn to mathematics. We can reduce the general equations of motion and boundary conditions to a dimensionless form. Then we recognize two characteristic lengths, the tube radius and the wave length. For the taper, the proper dimensionless parameter is the rate of change of tube radius per unit wavelength. If this rate is not very small, the theory is not very accurate. Let $\xi = x/\lambda$, where x is the axial coordinate and λ is the wavelength, and let $D_r = D/D_0$, the ratio of the blood vessel at x to that at the entry section $x = 0$. Then the preceding description means that the effect of taper is to be judged by the derivative $dD_r/d\xi$. Now

$$\frac{dD_r}{d\xi} = \frac{dD_r}{dx}\frac{dx}{d\xi} = \lambda\frac{dD_r}{dx} = \frac{\lambda}{D_0}\frac{dD}{dx}. \tag{2}$$

Hence the effective taper is the product of the wavelength λ, expressed in tube diameter at the entry, and the rate of change of the diameter dD/dx.

Now, in the case of the aorta, the wavelength of the first few harmonics is longer than the length of the aorta, and the effective taper $dD_r/d\xi$ may be fairly large by virtue of large values of λ/D_0. For these harmonics the inaccuracy of the simple theory may have led to the overestimation of the peaking mentioned earlier.

Problem

3.27 The cross-sectional area of the abdominal aorta is about 40% of the area of the thoracic aorta. The Young's modulus is about twice that of the thoracic aorta. Show that the ratio of the characteristic impedances in these aorta is 3.5 and that the pressure pulse amplitude in the abdominal aorta is expected to be 85% higher than that in the thoracic aorta. How much additional increase in pressure pulse amplitude is expected because of the reflection of the wave at the iliac junction?

3.15 Effects of Viscosity of the Fluid and Viscoelasticity of the Wall

At the beginning of this chapter we analyzed steady flow of a viscous fluid in tube. In Sections 3.8 to 3.12, however, we treated pulsatile flow of blood as if it has no viscosity. The major justification for this has been suggested in Section 3.9 namely, that in human arterial blood flow the Reynolds and Womersley numbers are much larger than 1 in the large arteries, so that the boundary layers are very thin campared with the vessel radius. The boundary layers mediate the ideal fluid solution to the real fluid no-slip condition on the solid wall. The Reynolds and Womersley numbers decrease toward the periphery, become smaller than 1 in arterioles, capillaries, and venules. Hence the influence of viscosity is felt more and more as blood flows toward the peripheral vessels. In the microcirculation the entire flow field is dominated by viscous stresses.

Even in large arteries, where the Reynolds and Womersley numbers are large, the viscosity of the fluid still has a profound influence. Viscous stresses play a dominant role in determining stability and turbulence in the arteries, and in determining whether the streamline will separate (diverge) from the wall of the vessel at branching points or at segments where a sudden change in cross section occurs, such as in stenosis or aneurysm.

Physically, because viscosity is a dissipation mechanism, one expects that it would reveal itself in the attenuation of velocity and pressure in the direction of propagation. Associated with attenuation will be phase changes. As far as their effect on the wave propagation is concerned, the effect of viscoelastic dissipation in the vessel wall is found to be more significant than the viscous dissipation of the blood.

To include viscosity of blood and viscoelasticity of the vessel wall in the pulsatile flow analysis, consider large arteries and make the following simplifying hypotheses:

a. The fluid is homogeneous and Newtonian.
b. The wall material is isotropic and linearly viscoelastic.
c. The fluid motion is laminar.
d. The motion is so small that squares and higher order products of displacements and velocities, and their derivatives, are negligible.

Then the field equations are the linearized Navier–Stokes equations for the blood, the linearized Navier's equation for the wall, and the equation of continuity. The boundary conditions are the continuity of shear and normal stresses and velocities at the fluid–solid interface, and appropriate conditions on the external surface of the tube and at the ends of the tube. The governing equations are given in Section 2.6 in rectangular cartesian coordinates. Changing to polar coordinates to describe an axisymmetric traveling wave in a tube of incompressible viscoelastic material, we have, for the fluid,

$$\frac{\partial v_r}{\partial t} = -\frac{1}{\rho}\frac{\partial p}{\partial r} + v\left(\frac{\partial^2 v_r}{\partial r^2} + \frac{1}{r}\frac{\partial v_r}{\partial r} - \frac{v_r}{r^2} + \frac{\partial^2 v_r}{\partial x^2}\right). \tag{1}$$

$$\frac{\partial v_x}{\partial t} = -\frac{1}{\rho}\frac{\partial p}{\partial x} + v\left(\frac{\partial^2 v_x}{\partial r^2} + \frac{1}{r}\frac{\partial v_x}{\partial r} + \frac{\partial^2 v_x}{\partial x^2}\right), \tag{2}$$

$$\frac{\partial v_x}{\partial x} + \frac{\partial v_r}{\partial r} + \frac{v_r}{r} = 0, \tag{3}$$

and, for the tube wall,

$$\frac{\rho_w}{\mu^*}\frac{\partial^2 u_r}{\partial t^2} = \frac{\partial^2 u_r}{\partial r^2} + \frac{1}{r}\frac{\partial u_r}{\partial r} - \frac{u_r}{r^2} + \frac{\partial^2 u_r}{\partial x^2} - \frac{1}{\mu^*}\frac{\partial \Omega}{\partial r}, \tag{4}$$

$$\frac{\rho_w}{\mu^*}\frac{\partial^2 u_x}{\partial t^2} = \frac{\partial^2 u_x}{\partial r^2} + \frac{1}{r}\frac{\partial u_x}{\partial r} + \frac{\partial^2 u_x}{\partial x^2} - \frac{1}{\mu^*}\frac{\partial \Omega}{\partial x}, \tag{5}$$

$$\frac{\partial u_x}{\partial x} + \frac{\partial u_r}{\partial r} + \frac{u_r}{r} = 0, \tag{6}$$

where v_x, v_r, v_θ are the velocity components of the fluid; u_x, u_r, u_θ are the displacement components of the wall; μ^* is the dynamic modulus of rigidity of the wall; and Ω is a pressure, which must be introduced since we have assumed the material to be incompressible. v is the *kinematic viscosity* of the blood, μ is the coefficient of viscosity, ρ is the density of the blood, ρ_w is the density of the wall material, and $v = \mu/\rho$.

If the external surface of the tube is stressfree, and if the inner and outer radii of the tube are a and b, respectively, then the boundary conditions are

$$v_r = 0 \qquad\qquad \text{at } r = 0, \qquad (7)$$

$$\partial v_x/\partial r = 0 \qquad\qquad \text{at } r = 0, \qquad (8)$$

$$v_r = \partial u_r/\partial t \qquad\qquad \text{at } r = a, \qquad (9)$$

$$v_x = \partial u_x/\partial t \qquad\qquad \text{at } r = a, \qquad (10)$$

$$\mu\left(\partial v_r/\partial x + \partial v_x/\partial r\right) = \mu^*\left(\partial u_r/\partial x + \partial u_x/\partial r\right) \quad \text{at } r = a, \qquad (11)$$

$$-p + 2\mu\left(\partial v_r/\partial r\right) = -\Omega + 2\mu^*\left(\partial u_r/\partial r\right) \qquad \text{at } r = a, \qquad (12)$$

$$\mu^*\left(\partial u_r/\partial x + \partial u_x/\partial r\right) = 0 \qquad\qquad \text{at } r = b, \qquad (13a)$$

$$-\Omega + 2\mu^*\left(\partial u_r/\partial r\right) = 0 \qquad\qquad \text{at } r = b. \qquad (13b)$$

The solution that satisfies the boundary conditions (7) and (8) may be posed in the following form*:

$$v_x = -\sum_{n=0}^{N} i\left\{A_1\gamma_n J_0\left(i\gamma_n r\right) + A_2\kappa_n J_0\left(i\kappa_n r\right)\right\}\exp i\left(n\omega t - \gamma_n x\right), \qquad (14)$$

$$v_r = -\sum_{n=0}^{N} i\gamma_n\left\{A_1 J_1\left(i\gamma_n r\right) + A_2 J_1\left(i\kappa_n r\right)\right\}\exp i\left(n\omega t - \gamma_n x\right), \qquad (15)$$

$$p = \sum_{n=0}^{N} A_3 J_0\left(i\gamma_n r\right)\exp i\left(n\omega t - \gamma_n x\right), \qquad (16)$$

$$u_r = \sum_{n=0}^{N} -i\gamma_n\left\{A_4 J_1\left(k_n r\right) + B_4 Y_1\left(k_n r\right) + A_5 J_1\left(i\gamma_n r\right)\right.$$
$$\left. + B_5 Y_1\left(i\gamma_n r\right)\right\}\cdot\exp i\left(n\omega t - \gamma_n x\right), \qquad (17)$$

$$u_x = \sum_{n=0}^{N} -\left\{k_n A_4 J_0\left(k_n r\right) + k_n B_4 Y_0\left(k_n r\right) + i\gamma_n A_5 J_0\left(i\gamma_n r\right)\right.$$
$$\left. + i\gamma_n B_5 Y_0\left(i\gamma_n r\right)\right\}\cdot\exp i\left(n\omega t - \gamma_n x\right), \qquad (18)$$

$$\Omega = \sum_{n=0}^{N}\left\{A_6 J_0\left(i\gamma_n r\right) + B_6 Y_0\left(i\gamma_n r\right)\right\}\exp i\left(n\omega t - \gamma_n x\right), \qquad (19)$$

---

* In a strict notation the A's and B's should have a subscript n because their values may be different for each n.

where ω is the angular frequency, n is the harmonic number, γ_n is a constant named the *propagation constant* of the nth harmonic, N is a constant, the A's and B's are complex constants, J_0 and J_1 are Bessel functions of the first kind, Y_0 and Y_1 are Bessel functions of the second kind, and

$$\kappa_n^2 = (in\omega/v) + \gamma_n^2, \quad k_n^2 = n^2\omega^2\rho_w/\mu^* - \gamma_n^2. \tag{20}$$

$$A_1 = (i/n\omega\rho)A_3, \quad A_6 = n^2\omega^2\rho_w A_5, \quad B_6 = n^2\omega^2\rho_w B_5. \tag{21}$$

When these solutions are substituted into the boundary conditions (9) to (13), six linear, homogeneous, and simultaneous equations in six unknown coefficients, $A_1, A_2, \ldots, B_5$, are obtained. For a nontrivial solution the determinant of the coefficients of $A_1, A_2, \ldots, B_5$ must vanish. This determinantal equation,

$$\Delta(\gamma_n, \kappa_n, k_n, \mu, \mu^*, a, b) = 0, \tag{22}$$

is the frequency equation for the pulse wave. If $\gamma_n = \beta + i\alpha$ is solved with other parameters assigned, then β is the *wave number*, α is the *attenuation coefficient*, and $C_p = \omega/\beta$ is the *phase velocity*.

The dynamic elastic modulus of the vessel wall, μ^*, is a complex number if the wave motion is represented by complex exponential functions listed in Eqs. (14) to (19). Since μ^* is complex, the solution γ_n is almost certain to be complex also, thus yielding the attenuation coefficient in association with each characteristic wave number. It is evident that extensive numerical calculations are necessary to obtain detailed information.

One thing that becomes evident from the general equations is that Eq. (22) has many solutions. One of them is akin to the *flexural mode* discussed in Section 3.8 and is an improvement of that solution. All the others are new modes not considered before in this book. These include *longitudinal modes*, in which the principal motion consists of motion of the vessel wall in the longitudinal direction; *torsional modes*, in which the principal motion is the torsional oscillation of the vessel wall; and higher modes of these three types with higher frequencies, shorter wavelengths, and different attenuation. Many of these theoretical wave types have been found in in vivo measurements.

A general reader would probably have little interest in the numerical details and is thus referred to the original papers. Furthermore, it is possible to relax some or all of the simplifying assumptions listed earlier and to study mathematically the arterial blood flow problem in greater depth. A great deal has been published. Historically, Euler (1775) was the first to write down the governing equations of arterial blood flow. He suggested that the relationship between blood pressure p and cross-sectional area A be represented by $A = A_0p(c + p)^{-1}$ or $A = A_0(1 - e^{-p/c})$. Euler's equations were solved later by Lambert (1958). Young (1808, 1809) was the first to derive the wave speed formula given in Section 3.8, Eq. (15). Lamb (1897–1898) derived phase velocities of two types of long waves in arterial wall. Joukowsky (1900) used Riemann's method of characteristics to solve Euler's equation. Witzig (1914) was the first to investigate the effect of fluid viscosity. Hamilton and Dow (1939) studied the relationship between arterial pulse waves and cardiac ejection and stroke volume of the heart. Jacobs (1953) was the first to investigate the effect of arterial wall mass and nonlinear elasticity as well as fluid viscosity. King (1947), Morgan and Kiely (1954) and Morgan and Ferrante (1955) were the first ones to investigate the effect of viscoelasticity. Womersley (1955a,b) compared theory with experiments, considered tethering, branching, and longitudinal variation of the cross-sectional area, and gave exhaustive attention to computational details. Landowne (1958) experimentally induced waves in human brachial and radial arteries. Taylor (1959, 1966a,b) used impedance methods and electric analogs. McDonald (1960) made thorough examinations of theories and experiments from a physiological point of view. Klip (1958, 1962, 1967) presented extensive studies on measuring techniques, effects of fluid viscosity, wall thickness, elasticity, viscoelasticity, and three modes of axisymmetric and asymmetric waves. Rubinow and Keller (1972) paid attention to the effect of external tissues, and found significant effects. Anliker and Raman (1966), Anliker and Maxwell (1966), Jones et al (1971), and Maxwell and Anliker (1968) studied Korotkof sound, dispersion, initial strain, and found axisymmetric waves mildly dispersive, asymmetric waves highly dispersive. Van der Werff (1973) developed a periodic method of characteristics. Pedley (1980) has given a thorough review of recent advances in the theory of blood flow in large arteries. Extensive review and bibliography are given in Attinger (1964), Bergel (1972), Patel and Vaishnav (1980), Sramek et al. (1995), Valenta (1993), and Wetterer and Kenner (1968).

3.16 Influence of Nonlinearities

Of all the effects, the most difficult one to evaluate is the effect of nonlinearities. In Euler's equation of motion [Eq. (13) of Section 2.6], the connective acceleration term, Dv_i/Dt, is nonlinear, and it is the principal difficulty of hydrodynamics. The viscous force term, $\partial\sigma_{ij}/\partial x_j$, becomes nonlinear if the constitutive equation of the fluid is non-Newtonian. Blood is non-Newtonian, and the effect of nonlinear blood viscosity is especially important with regard to flow separation at points of bifurcation in pulsatile flow. In the equation governing the blood vessel wall, the most significant nonlinearity comes from the finite strain and nonlinear viscoelasticity.

To check the effects of nonlinearity means to compare the solutions of the linearized equations and boundary conditions with those of nonlinear ones. This is usually impossible because the available solutions of nonlinear problems are limited. Sometimes, however, we can discuss the effects of nonlinearity on the basis of dimensional analysis and comparison with experimental evidence. Several examples follow.

Convective Acceleration in Wave Propagation

In the wave analysis of Section 3.8, the local convective acceleration $u_j(\partial u_i/\partial x_j)$ is neglected against the transient acceleration $\partial u_i/\partial t$. Now, let $\tilde{u}$ represent a characteristic velocity of the disturbed flow due to wave motion, ω the circular frequency of the wave, and c the wave speed relative to the mean flow. Then the period of oscillation is $2\pi/\omega$, the wavelength is $2\pi c/\omega$, and the orders of magnitude of the two accelerations are

$$\text{Transient,} \quad \frac{\partial u_i}{\partial t}: \quad \frac{\tilde{u}}{(2\pi/\omega)}, \tag{1}$$

$$\text{Convective,} \quad u_j\frac{\partial u_i}{\partial x_j}: \quad \tilde{u}\frac{\tilde{u}}{(2\pi c/\omega)}. \tag{2}$$

Hence the condition that the convective acceleration is negligible compared with the transient acceleration is that $(2) \ll (1)$, that is, if

$$\frac{\tilde{u}}{c} \ll 1. \tag{3}$$

In large arteries, the maximum value of $\tilde{u}/c$ is about 0.25, which is large enough to suggest nonlinear effects. In smaller peripheral arteries, the linearity condition is better justified. In certain rare disorders the arterial wall becomes floppy and c is so low that $\tilde{u}/c$ approaches 1. In aortic valve incompetence, the upstroke of the pulse wave becomes very steep, $\tilde{u}$ becomes quite large, and a nonlinear effect is expected. The effect is to increase acceleration and pressure drop. The pressure wave form is therefore steepened at the peak of the velocity wave and flattened at the valley.

Effect of Nonlinear Elasticity of Vessel Wall on Wave Propagation

The incremental Young's modulus of the arterial wall increases with the tensile stress in the vessel wall. An increase in Young's modulus of the vessel wall increases the wave speed [see Eq. (3.8:15)]. Experimental evidence for this is shown in Figure 3.16:1, which was obtained by Anliker (1972) using short trains of high-frequency pressure oscillations generated in the dog aorta and superposed on pulse waves. The wavelength of such high-frequency oscillations is short, so that several cycles can be recorded at a downstream observation site before the reflected wave from the iliac junction returns to that site and distorts the recording. The results show that the wave speed is higher in systole than in diastole. This is due partly to the increase in the speed of the progressive waves, c, and partly to the higher mean flow velocity, $U(t)$, at systole. According to Section 3.10, pressure perturbations superposed on a steady flow have a velocity of propagation equal to the velocity c plus the steady flow velocity.

Apply the same principle to a single harmonic pressure wave. At a peak (high pressure) the velocity of propagation is higher than the mean veloc-

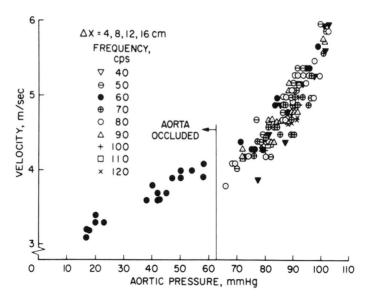

FIGURE 3.16:1. Relationship between the speed of sinusoidal pressure waves of various frequencies in the aorta of the dog and the instantaneous aortic pressure. In this case the diastolic pressure varied between 65 and 85 mm Hg, while the systolic pressure was between 85 and 100 mm Hg. Data points corresponding to pressure below the diastolic were obtained by occluding the aorta for about 10 s. The steeper rise of the wave speed with pressure between systole and diastole is attributed to a stiffening of the aorta with pressure and an increase in mean flow. From Anliker, M. (1972) by permission of Prentice-Hall, Inc. Englewood Cliffs, N.J.

ity. At a valley (low pressure) the velocity is lower. The wave form becomes distorted with the peak trying to overtake the mean, while the valley tries to lag behind. A similar phenomenon in the sea causes the water to tumble over at the crest and form "breakers," and in airstream over the wings of a high-speed airplane to form shock waves. Thus, one is led to expect shock waves in arteries and veins—especially in veins, which operate at a lower pressure, and hence are softer and have a lower wave speed. Countering this nonlinear effect are the shortness of the blood vessel, which does not make enough running distance to develop the steepening wave front, the existence of numerous side branches, which would "bleed" the pressure buildup, and the viscous dissipation in the vessel wall.

3.17 Flow Separation from the Wall

It is a well-known observation by designers of diffusers, wind tunnels, and airplanes that if the angle between the wall and the main flow direction is too large, the streamlines of the flow may detach from the wall and create

FIGURE 3.17:1. Flow separation from a curved wall. The laminar boundary layer has a Reynolds number of 20,000 based on distance from the leading edge (not shown). Because it is free of bubbles, the boundary layer appears as a thin dark line at the left. It separates tangentially near the start of the convex surface, remaining laminar for the distance to which the dark line persists, and then becomes unstable and turbulent. From Werlé (1974), an ONERA photograph, by permission.

a so-called separated region. An example is shown in Figure 3.17:1, which shows a photograph of a water stream containing minute air bubbles that make the streamlines visible. As the channel widens in the direction of flow, the velocity slows down and the pressure increases. The fluid in the boundary layer at the wall has to move against a pressure gradient, and after a certain distance the boundary layer becomes unstable and turbulent, and the stream leaves the wall and forms a jet. The "separated" region is a "dead water" region in which eddying motion occurs. A similar explanation can be given to the separated flow of a "stenosis" in a circular cylindrical tube, as shown in Figure 3.17:2.

In a pulsatile flow, the point of separation and the size of the separated region may vary with time. Figure 3.17:3 shows photographs of flow visualization in a mold of a human atherosclerotic abdominal aorta. Separation distal to an atherosclerotic plaque (arrow) is seen in the pulsatile flow at a mean Reynolds number of 500 and a Womersley number of 8.1. The zone of separation is seen to be greater during end disastole (top) than during peak flow (bottom).

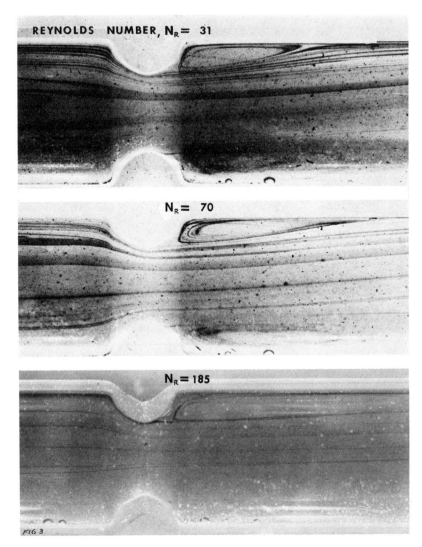

FIGURE 3.17:2. Flow separation in a model of stenosis in a circular cylindrical tube at Reynolds numbers of 31, 70, and 185. From Lee, J.S., and Fung, Y.C. (1971) "Flow in nonuniform small blood vessels." *Microvasc Res.* **3**: 272–287, by permission.

In the separated region, the flow is unsteady and the shear stress is lower than that in the unseparated regions. Atherosclerotic plaques in animals seem to correlate with the sites of flow separation.

3.18 Flow in the Entrance Region

In the simple solutions presented in Sections 3.2, 3.8, and 3.15, we do not specify the velocity and pressure distribution at the ends of the tube. We found solutions that satisfy the differential equations and boundary condi-

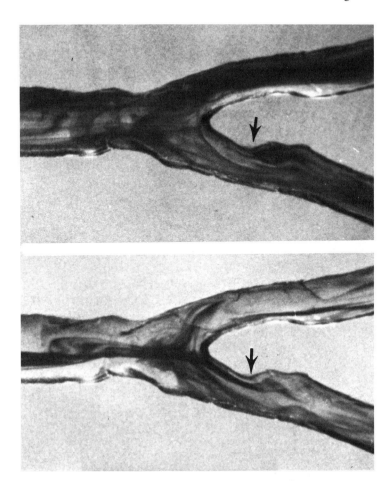

FIGURE 3.17:3. Flow separation distal to an atherosclerotic plaque during pulsatile flow in a mold of a human abdominal aorta. Flow was from left to right and was made visible with a dye. Separation is shown on the medial surface of the right common iliac artery (arrow). From F.J. Walburn, H.N. Sabbah, and P.D. Stein, "Flow visualization in a model of an atherosclerotic human abdominal aorta," *J. Biomechanical Engineering*, **103**: 168–170, 1981. Reproduced by permission.

tions on the wall of the blood vessel. These solutions are valid if the distribution of pressure and velocity over the cross sections at the ends of the tube was exactly as specified by the mathematical expressions. Any deviation from these conditions will call for a modification of the solution. In reality, the aorta is joined to the heart, small arteries are joined to larger arteries, and the end conditions practically never agree exactly with those specified by these simple solutions. Therefore, a modification is almost always required.

In steady flow, the Navier–Stokes equations belong to the so-called elliptical type of equations in the theory of partial differential equations. In equations of this type all of the partial derivatives of the highest order have the same sign. Elliptic equations have the property that any "self-equilibrating" local changes in boundary conditions (no net flow, no net force) over a small area will influence the solution only in the immediate neighborhood of the changed area, in the sense that the influence will die out as the distance from the changed area increases (St. Venant's principle). Applying this general property to the specific problem at hand, we conclude that if the net inflow and net pressure force remain the same, any changed boundary conditions at the ends of the tube will cause changes in the solution only in the immediate neighborhood of the ends. Therefore, any redistribution of velocity and pressure at the entry section without changing the resultant flow and force will affect only the "entrance region."

Consider the following example. A circular cylindrical tube is attached to a large reservoir that contains a Newtonian fluid. In the presence of a steady pressure gradient, the fluid flows into the tube. Far away from the entry section, the velocity profile is parabolic. At the entry section, a good approximation is to consider the velocity as uniformly distributed. Now we ask, how fast does the velocity profile change from the uniform distribution at the entry section to the parabolic (Poiseuillean) profile downstream?

The answer depends on the Reynolds number of flow. Results of a detailed analysis are presented in Figure 3.18:1. The Reynolds number, N_R, is denoted by Re in this figure, and is defined as

$$N_R = \frac{\rho a U}{\mu},\tag{1}$$

where ρ is the fluid density, a is the tube radius, U is the averaged speed of flow over the cross section of the tube, and μ is the viscosity of the fluid. The velocity in the axial direction is denoted by u, and that in the radial direction is denoted by v. In the figure, the dimensionless velocities u/U and v/U are plotted as functions of radial distance, r/a, and axial distance, x/a. It is seen that the effect of the entry flow is well limited to the region $0 \le x/a < 30$ for the Reynolds numbers considered. It is interesting to note that the maximum value of the radial component of velocity takes place at about $x/a = 0.2$ and $r/a = 0.7$. The maximum value of the radial component of velocity is about 30% of the mean speed of flow for Reynolds numbers less than 1. This ratio becomes about 0.1 for the case of Reynolds number equals to 50. Therefore, the radial component of the velocity is not negligibly small compared with the axial component of velocity in the intermediate range of Reynolds numbers.

Boundary Layer Growth in the Entry Region When the Reynolds Number Is Much Larger Than 1

When blood flows from a vessel into a branch (see Fig. 3.11:1), the new branch introduces new wall surfaces on which frictional force acts through

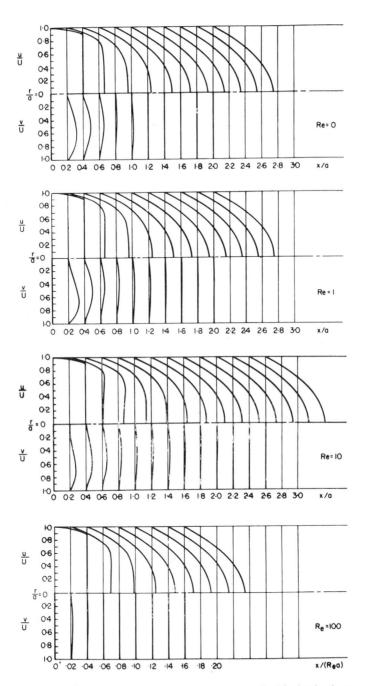

FIGURE 3.18:1. Examples of entry flow into a circular cylindrical tube from a reservoir. The flow in the entry region is seen to depend on the Reynolds number. u is axial velocity, v is radial velocity, and U is the mean flow velocity in the tube. Distance downstream from the entry section is expressed in terms of the tube radius a. From Lew and Fung (1970), by permission.

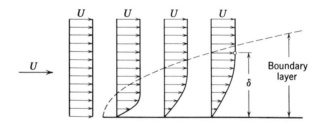

FIGURE 3.18:2. Velocity distribution in the boundary layer in the neighborhood of the leading edge of a flat plate.

viscosity, dissipates energy, and slows down the fluid in the neighborhood of the wall. If the Reynolds number, N_R, is much greater than 1, then the principal influence of the wall friction is limited to a thin boundary layer next to the wall, see Section 3.5. The thickness of the boundary layer increases as the distance from the leading edge of the new wall increases. A simple derivation of a law of the growth of the boundary layer thickness when the Reynolds number is large follows.

Consider a semi-infinite flat plate located on $y = 0, x \geq 0$ in a uniform flow of speed U in x-direction (Fig. 3.18:2). At various stations from the leading edge, the velocity profiles are illustrated in the figure. Consider the balance of forces acting on a small rectangular element of fluid in the boundary layer. The shear stress on the surface $dxdz$ is $\mu(\partial u/\partial y)dx\,dz$. This shear is variable with the depth y. In an element of thickness dy the net viscous force is

$$\mu \frac{\partial}{\partial y}\left(\frac{\partial u}{\partial y}\right)dy\,dx\,dz. \tag{2}$$

This is balanced by the inertial force, which is equal to the mass $\rho\,dx\,dy\,dz$ times the convective acceleration $u(\partial u/\partial x)$:

$$\rho u \frac{\partial u}{\partial x} dx\,dy\,dz. \tag{3}$$

Hence, on equating Eqs. (2) and (3), we have

$$\rho u \frac{\partial u}{\partial x} = \mu \frac{\partial}{\partial y}\left(\frac{\partial u}{\partial y}\right). \tag{4}$$

To get an estimate of the magnitude of these terms when the Reynolds number is large ($\gg 1$) so that the boundary layer thickness δ is small, note that u varies from 0 to U when y changes from 0 to δ. Hence $\partial u/\partial y$ is of the order of magnitude of U/δ, and the right-hand side of Eq. (4) is proportional to $\mu U/\delta^2$. Similarly, the left-hand side is proportional to $\rho U^2/x$ because u varies from U to a small fraction of U when x is varied from 0 to x. Hence we have

$$\rho \frac{U^2}{x} = K^2 \mu \frac{U}{\delta^2}, \tag{5}$$

where K is some constant. Solving for δ, we obtain the important result that

$$\delta = K \sqrt{\frac{\mu x}{\rho U}}. \tag{6}$$

Thus the boundary layer thickness δ increases with $\sqrt{x}$. The boundary layer is shaped like a parabola. Theory and experiments show that the constant K is equal to 4.0 when the boundary layer thickness is defined as the height at which the velocity is 95% of the free stream value.

This result is very useful. It tells us in a general way how the boundary layer grows, and therefore, of course, how the shear stress is distributed. The shear stress on the wall is given by

$$\mu \frac{\partial u}{\partial x} \sim \mu \frac{U}{\delta} \sim \sqrt{\frac{\mu \rho U^3}{x}}. \tag{7}$$

Thus, the shear stress is large at the leading edge and decreases as $x^{-1/2}$ when x increases. Hence, in a bifurcating artery, we expect the shear stress to be high at the tip of the new branches. If high shear stress is conducive to atherosclerosis, then this identifies the regions to watch.

This analysis is applicable to large arteries and to gas flow in trachea and large bronchi. For small vessels, whose Reynolds number approaches 1 or less, a more accurate analysis is needed. Results of such an analysis by Lew and Fung (1970) for the entry flow in Reynolds number range 0 to 100 have been presented in Figure 3.18:1 and have been discussed earlier. Other references are given in the paragraphs to follow.

Inlet Length

For an entry flow into a circular cylindrical tube with a uniform axial velocity at the entry section, it is common to define the *inlet length* as the distance through which the velocity has redistributed itself approximately into a parabolic profile. Since the approach to the parabolic profile is asymptotic, there is no unique definition of inlet length. Boussinesq (1891) defined the *inlet length* as the distance from the entry section to a point where the deviation from the parabolic profile is less than 1%. He obtained the result that the inlet length, L, is equal to $0.26aN_R$. With the same definition, Schiller (1922) obtained an inlet length of $0.115aN_R$, whereas Targ (1951) obtained the result

$$\frac{L}{a} = 0.16 N_R. \tag{8}$$

Lew and Fung (1970) showed that Targ's result is quite good for Reynolds numbers greater than 50. For smaller Reynolds numbers, however, Eq. (8)

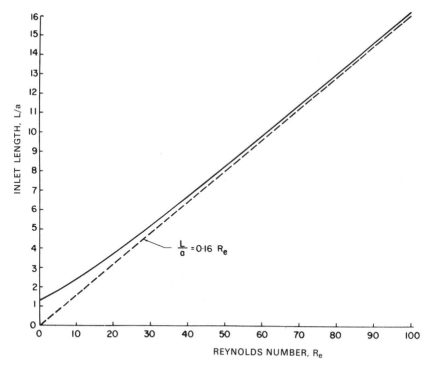

FIGURE 3.18:3. Change of inlet length versus Reynolds number in entry flow into a circular cylindrical tube from a reservoir. From Lew and Fung (1970), by permission.

does not apply. When the Reynolds number tends to zero, the inlet length tends to a constant $1.3a$. This is shown in Figure 3.18:3. The low Reynolds number case is applicable to the microcirculation. In the capillary blood vessels the Reynolds number is in the range 10^{-4} to 10^{-2}, and the inlet length is about 1.3 times the tube radius.

Entry Region in Oscillatory Flow

If there is an oscillation superposed on a steady flow, the entry region is also transient. Let us consider the case of large Reynolds number. The transient oscillation generates a boundary layer whose thickness is estimated in Section 3.5 to be proportional to $(v/\omega)^{1/2}$, where v is the kinematic viscosity and ω is the angular frequency. More detailed analysis yields a thickness equal to

$$\delta_1 = 6.5 \left(\frac{v}{\omega}\right)^{1/2}. \tag{9}$$

This boundary layer is created by the balance of the viscous stresses in the boundary layer against the inertial force associated with the transient acceleration $\partial u/\partial t$ in the free stream. On the other hand, Eq. (6) shows that the thickness of the convective boundary layer grows with increasing distance from the leading edge, x, according to the formula

$$\delta_2 = 4(vx/U)^{1/2},\tag{10}$$

where U is the velocity in the free stream immediately outside the boundary layer. For an unsteady flow entering a tube, both transient and convective accelerations exist, and the two boundary layers named above merge into one. Since in the resulting boundary layer the viscous stresses have to balance both of these accelerations, the boundary layer thickness must be at least as thick as either δ_1 or δ_2. Now, $\delta_1 = \delta_2$ when x is equal to

$$L \doteq 2.64 \frac{U}{\omega}.\tag{11}$$

For $x < L$ and $\delta_1 > \delta_2$, the transient acceleration tends to dominate. For $x > L$ and $\delta_2 > \delta_1$, the convective acceleration tends to dominate. Thus L is said to be an *unsteady entry length* (Caro et al., 1978, p. 321). Detailed analysis is given by Atabek (1962, 1980).

Problem

3.28 In the aorta of a dog, with a diameter of 1.5 cm, a mean velocity of 20 cm/s, a heart rate of 2 Hz ($\omega = 4\pi$), and an amplitude of the largest unsteady component of about 40 cm/s, show that the steady inlet length is about 36 cm and the unsteady entry length is about 10 cm. Thus the entire aorta is an entrance region.

  In the dog femoral artery, $a = 0.2$ cm and $U = 10$ cm/s; show that the entry length is about 1.2 cm.

3.19 Curved Vessel

Curvature of a vessel has a profound effect on flow. We may consider the physical factors as follows. Assume, as is sketched in Figure 3.19:1, that the velocity profile is flat at the entry section. Let the Reynolds number be much larger than 1. Then in the entry region the boundary layer is very thin and the flow in the core is like that of a nonviscous fluid. In this case the velocity is higher near the inner wall, in a manner similar to vortex flow. Further downstream the boundary layer grows. Eventually the boundary layer reaches the center of the tube. The flow is then said to be *fully developed*. In the boundary layer the centrifugal force will force the fluid toward the outer wall, and the velocity profile will be distorted, with a peak lying

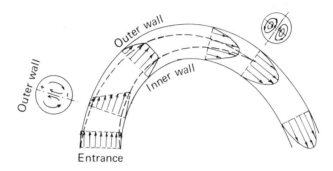

FIGURE 3.19:1. A sketch of entry flow into a curved pipe, showing the development of the boundary layer, velocity profile, and secondary flow.

beyond the centerline. The pressure fields corresponding to these velocity fields are not uniform in the cross sections, and *secondary flow* develops, as sketched in the figure. This flow is called *secondary* because it is super-imposed on the "primary" axial flow.

Secondary flow in curved tubes has been utilized by heart–lung machine builders to promote oxygenation of blood. In these machines the tubing is silastic, and thus permeable to oxygen. When the tube is curved and blood flows in the tube and oxygen flows outside the tube, the secondary flow stirs up the blood in the tube and results in faster oxygenation.

Extensive mathematical analyses of flow in curved tubes and branched tubes have been made by Pedley and are summarized in his book (Pedley, 1980). Entrance flow to curved tubes is presented by Yao and Berger (1975).

Other Nonuniformities

Uniformity is special; nonuniformity is the rule of nature. Uniform geometry is unique; nonuniform geometry has infinite variety. In this chapter we have used an infinitely long, straight, circular cylinder as the idealized uniform geometry of a blood vessel and have considered tapering, curvature, and the entrance region as nonuniformities. There are many other important nonuniformities in arteries. For example, stenosis, or local narrowing of the vessel, has great importance in pathology. Dilatation, or local enlargement of the vessel, is important in the study of diseases. Arteries branch off like branches on a tree. The detailed flow condition at each branching point is of interest, because at such a site there is a stagnation point where the velocity and velocity gradient are zero, and not far away is a region with a high velocity gradient. The shear stress acting on the wall is nonuniform.

3.20 Messages Carried in the Arterial Pulse Waves and Clinical Applications

Clinical applications of pulse wave studies are generally aimed at the following:

a. Discovering and explaining diseases of the arteries such as atherosclerosis, stenosis, and aneurysm. Locating sites that need surgical treatment.
b. Inferring the condition of the heart.
c. Diagnosing diseases anywhere in the body.

The approach to any of these objectives is to extract information from the characteristics of the waves. The most ancient method is to use fingers as probes. Any abnormality in the condition of the body affects pulse waves, which carry the message from distant sites.

The idea of extracting information from the arterial pulse waves to gain information about the heart and other organs of the body, however, remains an ideal. If we were able to read all the messages in the arterial pulse waves, then all we need for noninvasive diagnosis is to observe these waves in some conveniently located arteries (such as the radial artery on the wrist). If the messages are clear and unequivocal, then the art of noninvasive diagnosis would have been moved ahead a big step.

The idea of using pulse waves for diagnosis has been with us for a long time. In China, the oldest classic on arterial pulse waves is the *Nei Jing* (內經) mentioned in Chapter 1, Section 1.10. It was followed by *Nan Jing* (難經, first or second century B.C., authorship attributed to Qin Yue-Ren, 秦越人). *Jing* means classic, *Nan* means difficult as an adjective, or question as a verb). The book of Nan Jing sought to answer difficult questions, including those concerned with pulse waves. In the Eastern Han Dynasty, Chang Chi (張機, 字仲景) (probably 150 to 219 AD) wrote the books *The Influence of External Factors* (傷寒論) and the *Abstracts of the Golden Chest* (金匱要略), which systematically organized the Han Dynasty's 300 years' clinical experience in using pulse waves in diagnosis. Then in Tzin Dynasty (晉代), Wang Shu-He (王叔和) (201 to 285 AD) wrote *Mai Jing*, the *Book on Pulse Waves* (脈經). These became the classics of Chinese medicine, and their ideas and methods have been continuously developed and are being used in the Orient to this day.

The presentations given in these classics are descriptive and speculative, using similes and words to describe and classify the pulse waves. Empirically, abnormal waves were related to disease states. Clearly, the tasks of good recording, clear analysis, physiological experimentation, and rational explanation are left to modern researchers! The author's beginning studies are very limited in scope [Dai et al. (1985), Xue and Fung (1989)]. A large literature exists in Chinese.

3.21 Biofluid and Biosolid Mechanics of Arterial Disease

The belief that fluid and solid mechanics of the blood vessel are important to humanity is based on the fact that approximately 60% of all human deaths are caused by disorders of the cardiovascular system. One serious disease is atherosclerosis, which seems to be uniquely associated with abnormal stress and strain. *Atherosclerosis* is a form of arteriosclerosis (*artery* + *skleros*, hard) in which deposits of yellowish plaques containing cholesterol, lipoid material, and lipophages are formed within the intima and inner media of arteries. The atherosclerotic lesion begins with the accumulation of neutrophils and smooth muscle cells in the intima. As the lesion progresses, these cells become filled with lipids (called *foam cells* when this happens), a raised area of intima (called a *fatty streak*) appears, and gradually a thick, fibrous plaque is formed. The *plaque* has a cap composed of foam cells embedded in a matrix of elastic fibers and other connective tissue components overlying extracellular lipids and cell debris. Eventually the plaques become calcified.

The most intriguing feature of atheroscelerosis is that it is very nonuniformly distributed in the body. It occurs only at certain points on the arterial tree, as illustrated in Figure 3.21:1 published by DeBakey et al. (1985). In this figure the prevalent locations for atherosclerosis are marked by ink blots. It is seen that these spots are very limited in length. Data by Lei et al. (1995) say the same thing. What is so special about these locations? It is known that many factors influence atherogenesis, including food, high cholesterol diet, cigarette smoking, diabetes, high blood pressure, exercise, lifestyle, etc. These factors affect the whole body. Histological studies of the blood vessel found no specific markers for any protein of the blood vessel destined for atherogenesis. This leaves one to ask, why is it that of the many miles of blood vessels in a human only a few feet are destined to have atherosclerosis? Could it be that the stress and strain in these few feet of blood vessels are special? The answers given by fluid and solid mechanics theorists are the following.

Fluid Mechanical Theories

Fluid mechanics theorists point out that the places where atherogenesis is likely to occur are places that have a relatively complex geometry, a fairly large Reynolds number, and a lower than average wall shear stress. The flow patterns in these locations are complex, unsteady, and sometimes turbulent, as illustrated in Figure 3.21:2, which was obtained by Motomiya and Karino (1984) from a glass model of the human carotid sinus. The complexity of the flow pattern implies a complex, spatially nonuniform shear stress acting on the blood vessel wall. The unsteadiness and turbulence imply an inter-

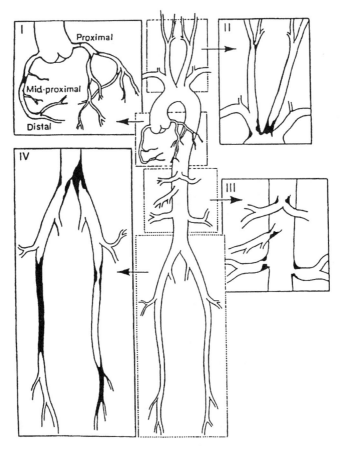

FIGURE 3.21:1. Distribution of atherosclerotic occlusive disease in humans. (Reproduced from the *Annals of Surgery* 201:116, 1985, with permission from the J.B. Lippincott Company, Philadelphia, PA, and Dr. Michael E. DeBakey.) From Thubriker and Robicsek, Ann Thorac Surg 59: 1594–1603, 1995.

mittent loading acting on the vessel wall. If turbulence is a marker for atherogenesis, then the rate of change of stress and strain must be important in the atherogenesis process.

The reaction of a blood vessel wall to pressure and shear of the flowing blood has been studied for nearly a hundred years in connection with the autoregulation of local blood flow. Autoregulation refers to a phenomenon in which a local increase of blood pressure in many organs will cause an immediate increase of vessel diameter and flow due to elasticity. This is followed by a contraction of the vascular smooth muscle in the vessel wall, reducing the vessel diameter and returning the flow to the normal level even though the local blood pressure remained high. This phenomenon is described in greater detail in Section 5.12, and recent advances in the

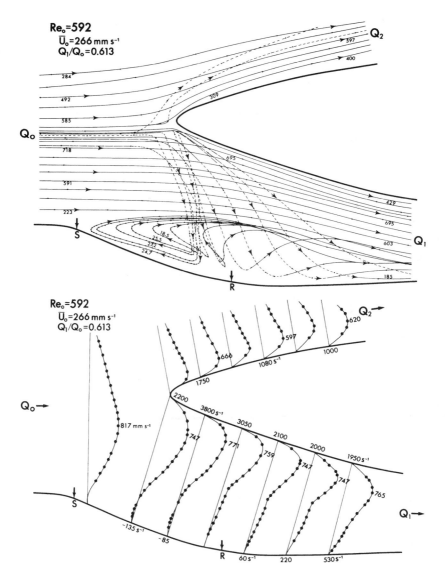

FIGURE 3.21:2. *Top:* Detailed flow patterns in the human carotid artery bifurcation in steady flow, showing the formation of a recirculation zone (paired spiral secondary flows) and a counter-rotating double helicoidal flow (both located symmetrically on both sides of the common median plane) in the internal carotid artery. The solid lines are the paths of particles in or close to the common median plane, and the dashed lines are the paths that are far away from the common median plane (projection of the particle paths on the common median plane). The arrows at R and S denote the respective locations of the separation and stagnation points. The numbers on the streamlines (particle paths) indicate the particle translational velocities in millimeters per second. *Bottom:* Distribution of fluid (particle) velocity and wall shear rate in the common median plane of a human carotid artery bifurcation in steady flow. The measured maximum velocity at each cross section is given in millimeters per second. From Motomiya and Karino (1984), reproduced by permission.

mechanics of smooth muscles are discussed in Section 7.11. The smooth muscle contraction is influenced significantly by the shear stress of the flowing blood. The phenomenon appears to be physiological and reversible. See Bohr et al. (1980), a handbook on vascular smooth muscles.

In Section 3.3, it was mentioned that the assumption of optimal design has lead to the conclusion that the average shear stress imposed by blood on the arterial wall is the same in all arteries, large or small, of a tree. This uniform shear concept has been tested and found to be quite good; and it has been suggested that the sites of atherogenesis are the sites of reduced blood shear. These sites of reduced shear are often associated with flow separation and turbulence. Flow separation and turbulence may be the factors that explain the difference between arteries and veins with regard to atherogenesis. Atherosclerosis is not a problem in veins, although blood shear acting on veins is smaller than that acting on arteries because at similar flow the cross-sectional area of the veins is considerably larger.

With this background, one would expect atherogenesis to be associated with disturbed wall shear stress. Fry (1968, 1977) initiated studies on the effect of increased shear stress on the endothelium. He showed that endothelial cells remodel their shape when the direction of flow is changed. The work of Barbee, Chien, Davies, Dewey, Frangos, Gimbrone, Ingber, Nerem, Resnick, Sato, Wang, and many others on the effect of shear stress on the endothelium followed, opening up a most fruitful field of research.

Caro et al. (1969, 1971) identified the locations of atherogenesis with regions of reduced wall shear stress. They felt that transport of lipoprotein within the arterial wall and across the endothelium is a major factor in atherogenesis. Research in this area by Carew, Chien, Nerem, Tarbell, Weinbaum, and others also opened up a very useful field of research.

The possibility that the shear stress acting on the endothelium due to blood flow may be closely related to atherogenesis has caused many researchers to focus attention on the effect of shear stress on the molecular biology of endothelial cells, stress-sensitive gene expression, signal transduction, mechanosensitive ion channels, etc.

Solid Mechanical Views

To persons well versed in the analysis of stresses in elastic shells, the sites of prevalent atherosclerosis shown in Figure 3.21:1 suggest that they are sites of high stress. Most blood vessels are circular cylindrical tubes. At the sites of bifurcation, constriction, or dilation, the vessels become three-dimensional curved shells of rather complex geometry. Analysis is difficult without using a finite-element program. Generally speaking, for a vessel subjected to internal pressure and longitudinal stretch, stress concentration occurs under the following conditions:

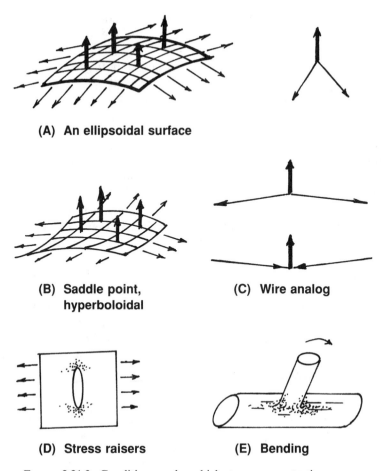

(A) An ellipsoidal surface

(B) Saddle point,
    hyperboloidal

(C) Wire analog

(D) Stress raisers

(E) Bending

FIGURE 3.21:3. Conditions under which stress concentrations occur.

1. Increased radius of curvature of the shell
2. Saddle shape of the shell, with one principal positive curvature and the other principal with negative curvature
3. In the neighborhood of a hole (a small side branch appears as a hole in the wall of the main vessel)
4. Bending of the wall

One can get a feeling about these conditions by considering the simple cases shown in Figure 3.21:3. For item (1), the effect of the radius of curvature, we note that the equation of equilibrium of a membrane, which has two principal curvatures, $1/R_1$ and $1/R_2$ (m^{-1}), and two corresponding principal membrane stress resultants, T_1 and T_2 (Newton/m), subjected to a pressure loading of p (N/m^2), is (Fig. 3.21:3A)

$$p = \frac{T_1}{R_1} + \frac{T_2}{R_2}. \tag{6}$$

Hence, if R_1 or R_2 or both are increased while p remains constant, then T_1, T_2 have to be increased. For item (2), if the shell is locally saddle shaped, then one of the curvatures, say, $1/R_2$, becomes negative. Then (Fig. 3.21:3B)

$$p = \frac{T_1}{R_1} - \frac{T_2}{R_2}. \tag{7}$$

To balance p, T_1 has to be increased. You may get a basic feeling for this situation by looking at the nearly straight wire shown in Figure 3.21:3C. If the wire is subjected to a lateral force p, then to balance p the tension T_1 would have to be very large if the wire is nearly straight. A curved membrane of saddle shape is a three-dimensional analog of a nearly straight wire.

For item (3), we recall that a hole in a flat plate is a stress raiser. See Figure 3.21:3D. The analog of a side branch in a large vessel is shown in Figure 3.21:3E. Finally, it is clear that bending a shell introduces bending stress in the wall, which is superimposed on the membrane stresses T_1, T_2 described in Eqs. (6) and (7). Bending is inevitable if the membrane stress resultants alone cannot keep the shell in equilibrium. Then shear and bending moments have to be introduced. In the example of a tube with an elliptic cross section shown in Figure 3.21:3E, internal pressure will inflate the tube to make the cross section more round. So, bending stress is introduced in the wall.

Now, look at the carotid sinus shown in Figure 3.21:4. In the sinus the cross section is locally bulged. At B and points below A and C, the vessel wall became saddle shaped; these are markers of high membrane stress. In panel II of Figure 3.21:1, the darkened prevalent sites are at bifurcation points. At each of these sites, before atherosclerotic plaque was formed, there is a local increase in the radius of curvature, some parts are saddle shaped, and in some parts bending is necessary under internal blood pressure. In panel III, all the sites are in the neighborhood of bifurcations where there are increases in stress. In panel I of Figure 3.21:1, some of the sites of coronary arteries are near bifurcation points, some are in the saddle side of the curved blood vessel, and some have elliptical cross sections. In panel IV, some bifurcation areas can be recognized. However, some sites in panels I and IV cannot be so easily explained; some sites have no obvious stress raisers.

The stress distribution in the neighborhood of a bifurcation point of a bovine circumflex coronary artery has been analyzed by Thubrikar and Robicsec (1995). The isostress contours of the maximum principal stress in the vessel wall are shown in Figure 3.21:5. The stress increases from one contour to the next toward the distal and proximal lips of the ostium. The stress was determined for a pressure increase from 80 to 120 mm Hg using the finite-element method.

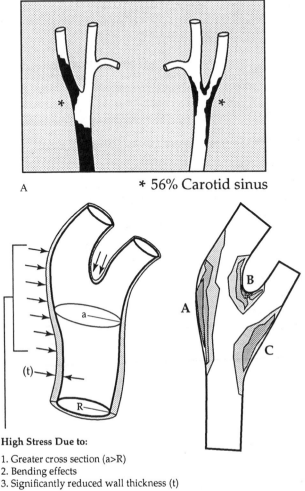

High Stress Due to:

1. Greater cross section (a>R)
2. Bending effects
3. Significantly reduced wall thickness (t)

B

FIGURE 3.21:4. (A) Locations of atherosclerotic disease at the carotid bifurcation. 56% indicates the relative frequency of occlusive lesions in the brachiocephalic vascular system. (B) (Left) Schematic presentation of the carotid bifurcation indicating that the stresses are high at the crotch and in the sinus bulb region. Also, the parameters that result in the high stresses are shown. (Right) Maximum principal isostress contours at the human carotid artery bifurcation. The stresses are high at all three locations B, A, and C, being the highest at B. The stress contours were obtained from the finite element analysis for a pressure (pulse) loading of 40 mm Hg. From Thubrikar and Robicsec (1995). Reprinted with permission from the Society of Thoracic Surgeons (The Annals of Thoracic Surgery, 1995, vol. 59, p. 1598).

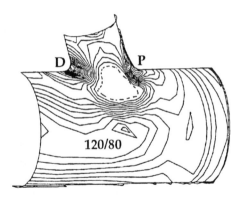

FIGURE 3.21:5. Distribution of stress contours in the arterial branch indicating that at both the distal lip (*D*) and the proximal lip (*P*) of the ostium the stresses are high and localized. Fatigue failure is expected to occur at either of these two locations, mainly as a result of the pulsatile pressure experienced by the artery. From Thubrikar and Robicsec (1995). Reprinted with permission from the Society of Thoracic Surgeons (The Annals of Thoracic Surgery, 1995, vol. 59, p. 1601).

The most interesting case reported by Thubrikar and Robicsec (1995) and Thubrikar et al (1988, 1990) refers to a human coronary artery that had a severely occlusive atherosclerotic plaque in the epicardial segment but was completely normal in the intramyocardial segment. The atherosclerotic lesion stopped abruptly at the entry of the artery into the myocardium. The reason suggested for this is that the pressure acting on the outside of the epicardial artery facing the pericardial fluid must not be far from atmospheric pressure, whereas the pressure outside the intramyocardial arterial segment is the muscle pressure, which must be much higher than atmospheric pressure. Thus, the transmural pressure of the artery was abruptly reduced as the vessel turned into the myocardium; hence, the stress was reduced.

Stress in Cells and Cell's Response to Stress

The fluid and solid mechanics views are not in conflict, but differ in emphasis. Both theories must explain why in a person who reacts to diet, exercise, and environmental factors, only a limited number of sites on the arterial tree are at risk for atherogenesis. Both have to explain how the foam cells are formed. Atherogenesis must be a part of the general subject of tissue remodeling. Other aspects of tissue remodeling are discussed in Fung (1990) and Taber (1995), and by other authors mentioned in previous sections.

Mechanics of Diseased Blood Vessels

Definitive studies of atherosclerotic plaque, thrombosis, stenosis, collapse of diseased arteries, flow in diseased vessels, medical treatment, and surgery of diseased arteries and veins, etc. are, of course, improtant. We refer to papers by Aoki and Ku (1993), Ku et al. (1985), McCord and Ku (1993), and Zarins et al. (1987).

Other Biological Studies and Hypotheses of Atherogenesis

McGill et al. (1957) identified the prelesion sites as areas of in vivo Evans blue dye uptake in the dog. Fry (1973) confirmed McGill's observations in the pig, noting that the local pattern of dye uptake is "virtually identical to the pattern of sudanophilia in early experimental atherosclerosis." In other words, the endothelial cells at these sites are chemically different from the rest of the vasculature. For a long time, biologists have searched for the biochemical mechanism of the formation of foam cells, fatty streak, lesion progression, and mature atherosclerotic plaque. Virchow (1856) believed that a form of low-grade injury to the artery wall resulted in a type of inflammatory insudate, which in turn caused increased passage and accumulation of plasma constituents in the intima. Rokitansky (1852) believed that an encrustation of small mural thrombi existed at sites of arterial injury and that these thrombi went on to become plaque. Ross and Glomset (1973), Ross (1988) combined these two notions and proposed his *response-to-injury* hypothesis. Caro et al. (1971) proposed the mass transport mechanism hypothesis. E.P. Benditt and J.M. Benditt (1973) proposed the monoclonal hypothesis. Friedman et al. (1983) proposed geometric shape to be a risk factor. Schwartz et al. (1991) proposed a unifying hypothesis, emphasizing the importance of focal arterial lesion–prone sites. It is clear that no theory is complete that does not tie all aspects together. The trend lies in clarifying the mechanics of molecular and cellular events in such a way that the microscopic and macroscopic levels can be unified.

References

Anliker, M. (1972). Toward a nontraumatic study of the circulatory system. In *Biomechanics: Its Foundations and Objectives* (Y.C. Fung, N. Perrone, and M. Anliker, eds.), Prentice-Hall, Englewood Cliffs, NJ, pp. 337–379.

Anliker, M., and Maxwell, J.A. (1966). The dispersion of waves in blood vessels. In *Biomechanics* (Y.C. Fung, ed.), American Society of Mechanical Engineers, New York, pp. 47–67.

Anliker, M., and Raman, K.R. (1966). Korotkoff sounds at diastole—a phenomenon and dynamic instability of fluid-filled shells. *Int. J. Solids Structures.* **2**: 467–492.

Anliker, M., Histand, M.B., and Ogden, E. (1968). Dispersion and attenuation of small artificial pressure waves in canine aorta. *Circ. Res.* **23**: 539–551.

Aoki, T., and Ku, D.N. (1993). Collapse of diseased arteries with eccentric cross-section. *J. Biomech.* **26**: 133–142.

Atabek, H.B. (1962). Development of flow in the inlet length of a circular tube starting from rest. *Z. Angew. Math. Phys.* **13**: 417–430.

Atabek, H.B. (1980). Blood flow and pulse propagation in arteries. In *Basic Hemodynamics and its Role in Disease Processes* (D.J. Patel and R.N. Vaishnav, eds.), University Park Press, Baltimore, MD, pp. 253–361.

Attinger, E.O. (ed.) (1964). *Pulsatile Blood Flow*. McGraw-Hill, New York.

Benditt, E.P., and Benditt, J. M. (1973). Evidence for a monoclonal origin of human atherosclerotic plaques. *Proc. Natl. Acad. Sci. USA*, **70**: 1753–1756.

Bergel, D.H. (ed.) (1972). *Cardiovascular Fluid Dynamics*, Vols. 1 & 2. Academic Press, New York.

Bohr, D.F., Somlyo, A.P., and Sparks, H.V., Jr. (eds.) (1980). *Handbook of Physiology, Sec. 2. The Cardiovascular System*. Vol. 2, *Vascular Smooth Muscles*. American Physiological Society, Bethesda, MD.

Boussinesq, J. (1891). Maniere dont les vitesses, se distrib. depui l'entree—Moindre longueur d'un tube circulaire, pour qu'un regime uniforme s'y etablisse. *Comptes Rendus*, **113**: 9, 49.

Caro, C.G., Fitzgerald, J.M., and Schroter, R.C. (1969). Arterial wall shear and distribution of early atheroma in man. *Nature* **223**: 1159–1161.

Caro, C.G., Fitzgerald, J.M., and Schroter, R.C. (1971). Atheroma and arterial wall shear. Observation, correlation and proposal of a shear dependent mass transfer mechanism for atherogenesis. *Proc. Roy. Soc. Lond. B.* **177**: 109–159.

Caro, C.G., Pedley, T.J., and Seed, W.A. (1974). Mechanics of the circulation. In *Cadiovascular Physiology* (A.C. Guyton, ed.), Chapter 1, Medical and Technical Publishers, London.

Caro, C.G., Pedley, T.J., Schroter, R.C., and Seed, W.A. (1978). *The Mechanics of the Circulation*. Oxford University Press, Oxford.

Dai, K., Xue, H., Dou, R., and Fung, Y.C. (1985). On the detection of messages carried in arterial pulse waves. *J. Biomech. Eng.* **107**: 268–273.

DeBakey, M.E., Lawrie, G.M., and Glaeser, D.H. (1985). Patterns of atherosclerosis and their surgical significance. *Ann. Surg.* **201**: 115–131, Lippincott Co., Philadelphia, PA.

Deshpande, M.D., and Giddens, D.P. (1980). Turbulence measurements in a constricted tube. *J. Fluid Mech.* **97**: 65–90.

Euler, L. (1775). Principia pro motu sanguins per arterias determinado. *Opera posthuma mathematica et physica* anno 1844 detecta, ediderunt P.H. Fuss et N. Fuss. Petropoli, Apud Eggers et socios, Vol. 2, pp. 814–823.

Friedman, M.H., and Deters, O.J. (1987). Correlation among shear rate measures in vascular flows. *J. Biomech. Eng.* **109**: 25–26.

Friedman, M.H., Deters, O.J., Mark, F. F., Bargeron, C.B., and Hutchins, G.M. (1983). Arterial geometry affects hemodynamics. A potential risk factor for atherosclerosis. *Atherosclerosis* **46**: 225–231.

Fry, D.L. (1968). Acute vascalar endothelial changes associated with increased blood velocity gradients. *Circ. Res.* **22**: 165–197.

Fry, D.L. (1973). Responses of the arterial wall to certain factors. In *Atherosclerosis: Initiating Factors*. Ciba Foundation Symp. Elsevier, Amsterdam, p. 93.

Fry, D.L. (1977). Aortic Evans blue dye accumulation: Its measurement and interpretation. *Am. J. Physiol.* **232**: H204–H222.

Fry, D.L., Griggs, Jr., D.M., and Greenfield, Jr., J.C. (1964). In vivo studies of pulsatile blood flow: the relationship of the pressure gradient to the blood velocity.

In *Pulsatile Blood Flow* (E.O. Attinger, ed.), McGraw-Hill, New York, Chap. 5, pp. 101–114.

Fung, Y.C. (1990). *Biomechanics: Motion, Flow, Stress, and Growth.* Springer-Verlag, New York.

Fung, Y.C. (1993a). *A First Course in Continuum Mechanics.* 3rd ed. Prentice-Hall, Englewood Cliffs, N.J.

Fung, Y.C. (1993b). *Biomechanics: Mechanical Properties of Living Tissues.* 2nd ed. Springer-Verlag, New York.

Fung, Y.C., and Liu, S.Q. (1993). Elementary mechanics of the endothelium of blood vessels. *J. Biomech. Eng.* **115**: 1–12.

Fung, Y.C., Fronek, K., and Patitucci, P. (1979). On pseudo-elasticity of arteries and the choice of its mathematical expression. *Am. J. Physiol.* **237**: H620–H631.

Giddens, D.P., Zarins, C.K., and Glagov, S. (1990). Response of arteries to near-wall fluid dynamic behavior. *Appl. Mech. Rev.* **43**: S98–S102.

Hamilton, W.F., and Dow, P. (1939). An experimental study of the standing waves in the pulse propagated through the aorta. *Am. J. Physiol.* **125**: 48–59.

Jacobs, R.B. (1953). On the propagation of a disturbance through a viscous liquid flowing in a distensible tube of appreciable mass. *Bull. Math. Biophys.* **5**: 395–409.

Jones, E., Anliker, M., and Chang, I.D. (1971). Effects of viscosity and constraints on the dispersion and dissipation of waves in large blood vessels. I & II. *Biophys. J.* **11**: 1085–1120, 1121–1134.

Joukowsky, N.W. (1900). Ueber den hydraulischen Stoss in Wasserheizungsrohren. *Memoires de l'Academie Imperiale des Science de St. Petersburg*, 8 series, Vol. 9, No. 5.

Kamiya, A., and Togawa, T. (1972). Optimal branching of the vascular tree (minimum volume theory). *Bull. Math. Biophys.* **34**: 431–438.

Kamiya, A., and Togawa, T. (1980). Adaptive regulation of wall shear stress to flow change in the canine carotid artery. *Am. J. Physiol.* **239**: H14–H21.

Kamiya, A., Bukhari, R., and Togawa, T. (1984). Adaptive regulation of wall shear stress optimizing vascular tree function. *Bull. Math. Biol.* **46**: 127–137.

Kassab, G.S., and Fung, Y.C. (1995). The pattern of coronary arteriolar bifurcations and the uniform shear hypothesis. *Ann. Biomed. Eng.*, **23**: 13–20.

Kassab, G.S., Rider, C.A., Tang, N.J., and Fung, Y.C. (1993). Morphometry of the pig coronary arterial trees. *Am. J. Physiol.*, **266**: H350–H365.

King, A.L. (1947). Waves in elastic tubes: velocity of the pulse wave in large arteries. *J. Appl. Phys.* **18**: 595–600.

Klip, W. (1958). Difficulties in the measurement of pulse wave velocity. *Am. Heart J.* **56**: 806–813.

Klip, W. (1962). *Velocity and Damping of the Pulse Wave.* Martinus Nijhoff, The Hague.

Klip, W. (1967). Formulas for phase velocity and damping of longitudinal waves in thick-walled viscoelastic tubes. *J. Appl. Phys.* **38**: 3745–3755.

Korteweg, D.J. (1878). Ueber die Fortpflanzungesgeschwindigkeit des Schalles in elastischen Rohren. *Ann. Physik. Chemie* **5**: 525–542.

Ku, D.N., Giddens, D.P., Zarins, C.K., and Glagov, S. (1985). Pulsatile flow and atherosclerosis in the human carotid bifurcation. *Arteriosclerosis* **5**: 293–302.

Lamb, H. (1897–1898). On the velocity of sound in a tube, as affected by the elasticity of the walls. *Phil. Soc. Manchester Memoirs Proc.*, lit. A, **42**: 1–16.

Lambert, J.W. (1958). On the nonlinearities of fluid flow in nonrigid tubes. *J. Franklin Inst.* **266**: 83–102.

Lanczos, C. (1952). Introduction. In *Tables of Chebyshev Polynomials.* National Bureau of Standards, Appl Math, Ser. 9, U.S. Govt. Printing Office, Washington, D.C., pp. 7–9.

Landowne, M. (1958). Characteristics of impact and pulse wave propagation in brachial and radial arteries. *J. Appl. Physiol.* **12**: 91–97.

Lei, M., Kleinstreuer, C., and Truskey, G.A. (1995). Numerical investigation and prediction of atherogenic sites in branching arteries. *J. Biomech. Eng.* **17**: 350–357.

Lew, H.S., and Fung, Y.C. (1970). Entry flow into blood vessels at arbitrary Reynolds number. *J. Biomech.* **3**: 23–38.

Liebow, A.A. (1963). Situations which lead to changes in vascular patterns. In: *Handbook of Physiology, Section 2: Circulation*, Vol. 2. Washington, D.C.: American Physiological Society, pp. 1251–1276.

Lighthill, M.J. (1978). *Waves in Fluids.* Cambridge University Press, London, UK.

Ling, S.C., and Atabek, H.B. (1972). A nonlinear analysis of pulsatile flow in arteries. *J. Fluid Mech.* **55**: 493–511.

Liu, S.Q., Yen, M., and Fung, Y.C. (1994). On measuring the third dimension of cultured endothelial cells in shear flow. *Proc. Natl. Acad. Sci. U.S.A.* **91**: 8782–8786.

Maxwell, J.A., and Anliker, M. (1968). The dissipation and dispersion of small waves in arteries and veins with viscoelastic wall properties. *Biophys. J.* **8**: 920–950.

McCord, B.N., and Ku, D.N. (1993). Mechanical rupture of the atherosclerotic plaque fibrous cap. *Bioengineering Conf.* ASME, BED Vol. 24, pp. 324–326.

McDonald, D.A. (1960, 1974). *Blood Flow in Arteries.* 1st ed., 2nd ed. Williams & Wilkins, Baltimore, MD.

McGill, H.C. Jr., Geer, J.C., and Holman, R.L. (1957). Sites of vascular vulnerability in dogs demonstrated by Evans Blue. *AMA Arch. Pathol.* 64: 303–311.

Moens, A.I. (1878). *Die Pulskwive.* Brill, Leiden, The Netherlands.

Morgan, G.W., and Ferrante, W.R. (1955). Wave propagation in elastic tubes filled with streaming liquid. *J. Acoust. Soc. Amer.* **27**: 715–725.

Morgan, G.W., and Kiely, J.P. (1954). Wave propagation in a viscous liquid contained in a flexible tube. *J. Acoust. Soc. Amer.* **26**: 323–328.

Motomiya, M., and Karino, T. (1984). Flow patterns in the human carotid artery bifurcation. *Stroke* **15**: 50–56.

Murray, C.D. (1926). The physiological principle of minimum work. I. The vascular system and the cost of blood volume. *Proc. Natl. Acad. Sci. U.S.A.* **12**: 207–214; *J. Gen. Physiol.* **9**: 835–841.

Nerem, R.M., Seed, W.A., and Wood, N.B. (1972). An experimental study of the velocity distribution and transition to turbulence in the aorta. *J. Fluid Mech.* **52**: 137–160.

Oka, S. (1974). *Rheology-Biorheology.* Syokabo, Tokyo. (In Japanese).

Patel, D.J., and Vaishnav, R.N. (eds.) (1980). *Basic Hemodynamics and its Role in Disease Process.* University Park Press, Baltimore, MD.

Pedley, T.J. (1980). *The Fluid Mechanics of Large Blood Vessels.* Cambridge University Press, London.

Poiseuille, J.L. (1841). Recherches expérimentales sur le mouvement des liquides dans les tubes de très petits diamètres. *Compte Rendus, Académie des Sciences.* Paris.

Prandtl, L. (1904). Über Flüssigkeitsbewegung bei sehr Kleiner Reibung. *Proc. 3rd Intern. Math. Congress.* Heidelberg.

Rodbard, S. (1975). Vascular caliber. *Cardiology* **60**: 4–49.

Rokitansky, C. von (1852). *A Manual of Pathological Anatomy.* Translated by G.E. Day. Vol. 4. London, The Sydenham Society.

Rosen, R. (1967). *Optimality Principles in Biology.* Butterworth, London.

Ross, R. (1988). The pathogenesis of atherosclerosis. In *Heart Disease* (E. Braunwald, ed.) Saunders, Philadelphia, Chapter 35, pp. 1135–1152.

Ross, R., and Glomset, J. (1973). Atherosclerosis and the arterial smooth muscle cell. *Science* **180**: 1332.

Rubinow, S.I., and Keller, J.B. (1972). Flow of a viscous fluid through an elastic tube with applications to blood flow. *J. Theo. Biol.* **35**(2): 299–313.

Schiller, L. (1922). Die Entwicklung der laminaren Geschwindigkeitsverteilung und ihre Bedeutung für Zahigkeitsmessungen. *Z. Angew. Math. Mech.* **2**: 96–106.

Schlichting, H. (1968). *Boundary Layer Theory,* 6th ed. McGraw-Hill, New York.

Schwartz, C.J., Valente, A.J., Sprague, E.A., Kelley, J.L., and Nerem, R.M. (1991). The pathogenesis of atherosclerosis: An overview. *Clin. Cardiol.* **14**: 1–16.

Skalak, R. (1966). Wave propagation in blood flow. In *Biomechanics Symposium* (Y.C. Fung, ed.). American Society of Mechanical Engineers, New York, pp. 20–40.

Skalak, R. (1972). Synthesis of a complete circulation. In *Cardiovascular Fluid Dynamics* (D.H. Bergel, ed.), Vol. 2. Academic Press, New York, pp. 341–376.

Sramek, B.B., Valenta, J., and Klimes, F. (eds.) (1995). *Biomechanics of the Cardiovascular System,* Czech Technical Univ. Press, Prague.

Szegö, G. (1939). *Orthogonal Polynomials,* 4th ed. American Math. Soc. Colloquium Vol. 23.

Taber, L.A. (1995). Biomechanics of growth, remodeling, and morphogenesis. *Appl. Mech. Rev.* **48**: 487–545.

Targ, S.M. (1951). *Basic Problems of the Theory of Laminar Flows.* Moscow (In Russian).

Taylor, M.G. (1959). An experimental determination of the propagation of fluid oscillations in a tube with a viscoelastic wall. *Phys. Med. Biol.* **4**: 63–82.

Taylor, M.G. (1966a). Use of random excitation and spectral analysis in the study of frequency-dependent parameters of the cardiovascular system. *Circ. Res.* **18**: 585–595.

Taylor, M.G. (1966b). Input impedance of an assembly of randomly branching elastic tubes. *Biophys. J.* **6**: 29–51, **6**: 697–716.

Thoma, R. (1893). *Untersuchungen über die Histogenese und Histomechanik des Gefassystemes.* Stuttgart: Enke.

Thubrikar, M.J., and Robicsec, F. (1995). Pressure-induced arterial wall stress and atherosclerosis. *Ann. Thorac. Surg.* **59**: 1594–1603.

Thubrikar, M.J., Baker, J.W., and Nolan, S.P. (1988). Inhibition of atherosolerosis associated with reduction of arterial intramural stress in rabbits. *Arteriosclerosis* **8**: 410–420.

Thubrikar, M.J., Roskelley, S.K., and Eppink, R.T. (1990). Study of stress concentration in the walls of the bovine coronary arterial branch. *J. Biomech.* **23**: 15–26.

Valenta, J. (ed.) (1993). *Biomechanics*, Academia, Prague, Elsevier, Amsterdam.

Van der Werff, T.J. (1973). Periodic method of characteristics. *J. Comp. Phys.* **11**: 296–305.

Virchow, R. (1856). Phlogose und thrombose im gefassystem. *Gessamette Abhandlungen zur Wissenschaftlichen Medicin.* Frankfurt-am-Main, Meidinger Sohn u.Co. p. 458.

Werlé, H. (1974). *Le Tunnel Hydrodynamique au Service de la Recherche Aérospatiale.* Publication No. 156, ONERA, France. Paris.

Wetterer, E., and Kenner, T. (1968). *Grundlagen der Dynamik des Arterienpulses.* Springer-Verlag, Berlin.

Witzig, K. (1914). *Über erzwungene Wellenbewegungen zaher, inkompressibler Flüssigkeiten in elastischen Rohren.* Inaugural Dissertation, Universitat Bern, K.J. Wyss, Bern.

Womersley, J.R. (1955a). Method for the calculation of velocity, rate of flow, and viscous drag in arteries when the pressure gradient is known. *J. Physiol.* **127**: 553–563.

Womersley, J.R. (1955b). Oscillatory motion of a viscous liquid in a thin-walled elastic tube-I: The linear approximation for long waves. *Phil. Mag.* **46**(Ser. 7): 199–221.

Xue, H., and Fung, Y.C. (1989). Persistence of asymmetry in nonaxisymmetric entry flow in a circular cylindrical tube and its relevance to arterial pulse wave diagnosis. *J. Biomech. Eng.* **111**: 37–41.

Yao, L.S., and Berger, S.A. (1975). Entry flow in a curved pipe. *J. Fluid Mech.* **67**: 177–196.

Yih, C.S. (1977). *Fluid Mechancs.* West River Press, Ann Arbor, MI.

Young, T. (1808). Hydraulic investigations, subservient to an intended Croonian lecture on the motion of the blood. *Phil. Trans. Roy. Soc. London* **98**: 164–186.

Young, T. (1809). On the functions of the heart and arteries. *Phil. Trans. Roy. Soc. London* **99**: 1–31.

Zamir, M. (1976). The role of shear forces in arterial branching. *J. Gene. Biolo.* **67**: 213–222.

Zamir, M. (1977). Shear forces and blood vessel radii in the cardiovascular system. *J. Gen. Physiol.* **69**: 449–461.

Zarins, C.K., Zatina, M.A., Giddens, D.P., Ku, D.N., and Glagov, S. (1987). Shear stress regulation of artery lumen diameter in experimental atherogenesis. *J. Vasc. Surg.* **5**: 413–420.

4
The Veins

4.1 Introduction

Veins normally contain about 80% of the total volume of blood in the systemic vascular system. Any change in the blood volume in the veins will affect blood flow through the heart. The most important feature of the systemic veins is, therefore, their compliance. A compliant structure runs the danger of collapsing; the problem with the vein is that it is often collapsed.

Both systemic and pulmonary veins resemble arteries in histological structure. Their anatomy and material properties have been discussed in Chapter 8 of *Biomechanics: The Mechanical Properties of Living Tissues* (Fung, 1993b). Compared with the arteries, there are several important differences: (1) The pressure in a vein is normally much lower than that in an artery at the same general location. (2) The veins have thin walls and some of them may be collapsed in normal function. (3) Blood flows in veins from the periphery toward the heart. (4) Many veins have valves that prevent blackflow. Veins more distal to the heart have more valves. The vena cava and the pulmonary veins have none. The smallest venules have no valves.

Normally, the internal pressure in the vein is higher than the external pressure, and the flow in the vein is similar to that in the artery. In certain situations, however, the external pressure may be raised or the internal pressure decreased in some parts of the vein to such a degree that the external pressure exceeds the internal pressure. In this situation, some vessels or parts of vessels may be collapsed. The controlling factor is the transmural pressure, Δp, which is equal to the internal pressure, p_i, minus the external pressure, p_e. Since, in general, the internal pressure in a blood vessel falls along the length because of frictional loss, Δp falls continuously from the entry to the exit section if the external pressure remains constant. In a dynamic condition, three eventualities may occur in a vein:

a. $\Delta p > 0$ at entry, $\quad$ $\Delta p > 0$ at exit,
b. $\Delta p > 0$ at entry, $\quad$ $\Delta p < 0$ at exit,
c. $\Delta p < 0$ at entry, $\quad$ $\Delta p < 0$ at exit.

If a vein collapses whenever $\Delta p < 0$, then in case (a) the vein functions normally and the principles discussed in the preceding chapter apply. In case (c), the vein will be collapsed and the flow will be greatly reduced. In case (b), the entry section is open and patent, but the exit section may be collapsed. Case (b) is the really interesting one. How effective is the choking at the exit section? A possible scenario is the following: The choking is effective, the flow stops, the pressure drop becomes zero, the whole tube has a Δp equal to that of the entry section, and the conditions of (a) then prevail, and flow starts again. But if flow starts, the pressure drops along the tube, and the exit section is choked again. This may lead to the dynamic phenomenon of "flutter," or a limiting steady-state flow, as described later.

It will be shown (Sec. 4.5) that in some instances of case (b), flow is controlled by the choking section, but the actual value of Δp at the exit section is quite immaterial as long as it is negative. An analogy may be drawn between this and a waterfall, or sluicing in industry or flood control: The volume flow rate in a waterfall depends on the conditions at the top of the fall and is independent of how high the drop is. Thus, the phenomenon of flow in case (b) is described as the *waterfall* phenomenon, or as *sluicing*.

The waterfall phenomenon occurs in a number of important organs: the airways and capillaries of the lung, vena cava, the heart, etc. It occurs in thoracic arteries during resuscitation maneuvers, and in brachial arteries while measuring blood pressure by cuff and Korotkoff sound. The same phenomenon also occurs in male and female urethra in micturition, and in human-made instruments such as the blood pump and the heart–lung machine. It thus pays to understand it well. Accordingly, in this chapter we consider the concept of elastic stability (Sec. 4.2), study the pressure–area relationship of veins in partially collapsed states (Secs. 4.3, 4.4), clarify steady and unsteady flows in veins (Secs. 4.5 to 4.7), and investigate some aspects of the dynamics of flow in collapsible tubes, such as flutter and the Korotkoff sound.

If the vascular waterfall phenomenon is important, we should know where the waterfalls are on the venous tree. This seemingly simple question does not have an easy answer. In Section 4.9 we show that pulmonary veins will not collapse, that is, they remain patent when the blood pressure is smaller than the alveolar gas pressure (i.e., Δp is negative) in the physiological range. Thus, one should not assume that every vein must collapse when Δp is negative. There are other factors to be considered, for example, tethering, or the relationship to the neighboring structures. Pulmonary capillary blood vessels, however, will collapse to zero thickness under negative transmural pressure. Hence (Sec. 4.10), we can show that pulmonary waterfalls, or sluicing gates, are located at the exit ends of capillaries, where the blood enters the venules. Once this is known, blood flow analysis is relatively easy.

The ultimate fate of sluicing may be finite flow, or may be no flow, depending on conditions at the sluicing gate. The factors that decide the outcome are the following:

1. Size of the vessel. Is its diameter much larger than the diameter of the red blood cells? Or is it comparable to the red cell diameter?
2. Shape of the gate. Is it like a circular tube? Or is it like a slit? A red cell can go sidewise through a slit. The thickness of a red cell is an order of magnitude smaller than its diameter.
3. Structure of and stress in the wall of the gate. Is the vessel beyond the gate patent? Or is it also collapsed? Is the gate narrow? Or is it a long tunnel?
4. The external condition. Outside the sluicing gate, is there a large fluid chamber? Or is it only a small interstitial space, with soft tissue surrounding it?

These factors indicate that there are differences between the microcirculation and macrocirculation, there are differences between a tube in a box in the laboratory and blood vessels in organs, and each organ has its own characteristics. Waterfall phenomena are discussed in Chapters 5 to 8. In this chapter, we deal mainly with large vessels whose diameter is much larger than that of red cells.

4.2 Concept of Elastic Instability

The collapsing of a vein under negative transmural pressure is a phenomenon due to elastic instability, similar to the buckling of thin-walled structures, such as airplane wings and fuselages, submarine hulls, and underwater pipelines. This topic is dealt with in the theory of elastic stability, an outline of which is given here.

Adjacent Equilibrium

A structure carries a certain load that causes a certain deformation. Let this system be disturbed slightly by a small additional load. If the small additional load causes a small additional deformation, and the structure will return to the basic configuration when the disturbance is removed, then the structure is said to be *stable*. If the small disturbance causes a large deformation, and the structure will not return to the basic configuration when the disturbance is removed, then the structure is said to be *unstable*.

Let us illustrate this concept with the classical theory of *columns*. (A column is a beam used to support an axial load.) Leonhard Euler (1707–1783) considered a column made of a material that obeys Hooke's law of elasticity, simply supported at both ends, and subjected to an end thrust, as illustrated in Figure 4.2:1(a). Under the load P, there is an axial

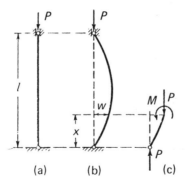

FIGURE 4.2:1. An Euler column used to support an axial load P. (a) Straight equilibrium configuration. (b) Deflected equilibrium configuration. (c) A free-body diagram of a portion of the column showing external and internal forces and moment acting on the column.

compressive stress $-P/A$ in all sections of the column, where A is the cross-sectional area.

Let us disturb the equilibrium by imposing some infinitesimal lateral loads on the column (loads acting in a direction perpendicular to the column axis). If the column is stable, this will cause only an infinitesimal lateral deformation. If, however, the load P is so large as to cause the column to be unstable, then under the infinitesimal disturbance a finite (i.e., "large") deformation of the column will be possible. Let us consider an unstable column, and assume that under an infinitesimal lateral disturbance a finite deflection w, as illustrated in Figure 4.2:1(b), occurs. A free-body diagram of a segment of the column is shown in Figure 4.2:1(c), in which P and M represent the thrust and bending moment in the column. Taking moment about the lower end yields a condition of equilibrium,

$$M - Pw = 0. \tag{1}$$

The bending moment M is assumed to be directly proportional to the curvature of the column (see Fig. 4.2:2, and Fung, 1993a, Section 1.11, p. 28, Eq. (1.11–32)), so that

$$M = -EI\frac{d^2w}{dx^2}, \tag{2}$$

where E is the Young's modulus of elasticity of the material and I is the moment of inertia of the cross-sectional area. Combining Eqs. (1) and (2) yields the differential equation,

$$EI\frac{d^2w}{dx^2} + Pw = 0. \tag{3}$$

The boundary conditions for a column simply supported at both ends are

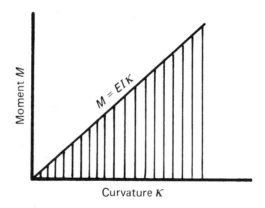

FIGURE 4.2:2. The relationship between the bending moment and curvature of a beam (or column). This relationship is assumed to be linear in the classical theory of beams that are made of material obeying Hooke's law. It is shown as the straight line $M = EI\kappa$ in the figure. The area under the curve (shaded) represents the work done by gradually bending the beam to the curvature κ. The curvature is equal to $-d^2w/dx^2$ if $dw/dx \ll 1$, and if the signs of w and x are chosen, as illustrated in Figure 4.2:1(b). Then Eq. (2) applies. If dw/dx is not negligibly small compared with 1, then the exact expression for curvature should be used: $\kappa = (d^2w/dx^2)[1 + (dw/dx)^2]^{-3/2}$.

$$w = 0 \quad \text{at} \quad x = 0 \quad \text{and} \quad x = L. \tag{4}$$

Equations (3) and (4) have an obvious solution, $w = 0$, which is the answer for a stable column. To look for a nontrivial solution for an unstable column, we note that Eq. (3) is satisfied by

$$w = c_1 \sin\sqrt{\frac{P}{EI}}x + c_2 \cos\sqrt{\frac{P}{EI}}x, \tag{5}$$

where c_1, c_2 are integration constants. The boundary conditions of Eq. (4) are satisfied if

$$c_2 = 0 \quad \text{and} \quad 0 = c_1 \sin\sqrt{\frac{P}{EI}}L. \tag{6}$$

The last equation is satisfied if

$$\sqrt{\frac{P}{EI}}L = n\pi, \qquad (n = 1, 2, \cdots), \tag{7}$$

that is, if

$$P = \frac{n^2\pi^2 EI}{L^2}, \qquad (n = 1, 2, \cdots). \tag{8}$$

Thus, if the load P equals the value specified by Eq. (8), the column can deform into a buckled form with

$$w = c_1 \sin \frac{n \pi x}{L} \tag{9}$$

without any lateral load. The amplitude c_1 is arbitrary. The load P given by Eq. (8) is the *critical load for neutral stability*. As the axial load P is increased gradually from zero, instability is encountered for the first time when $n = 1$ and $P = P_{cr}$:

$$P_{cr} = \frac{\pi^2 EI}{L^2}, \tag{10}$$

which is known as the *Euler load*.

This method of derivation yields the critical condition, but it says nothing about what happens when P is slightly greater than P_{cr}. What happens to the column when $P \geq P_{cr}$ becomes clear if the column was slightly crooked initially: It will collapse suddenly! (See Prob. 4.2.) When $P = P_{cr}$, the deflection given by Eq. (9) will have a finite amplitude. The column cannot carry a load larger than P_{cr} unless lateral deformation is prevented by some special means.

Energy Method

An alternative method to derive the results presented above uses the concept of work and energy. Consider again the Euler column. When the column is deflected [Fig. 4.2:1(b)], there is some strain energy stored in it. To evaluate the strain energy, note that if the material obeys Hooke's law of elasticity, the bending moment is proportional to the curvature with a constant of proportionality EI [see Eq. (2)]. When a segment of straight column of unit length is bent gradually to a curvature κ, the work done on the column is stored as strain energy, and is equal to the area under the moment–curvature curve, as shown in Figure 4.2:2. Thus, the strain energy per unit length of the column is the area of the triangle, $\frac{1}{2} EI \kappa^2$. For small deflection, the curvature $\kappa = d^2w/dx^2$. Integrating throughout the column, we obtain the strain energy,

$$V = \frac{EI}{2} \int_0^L \left(\frac{d^2 w}{dx^2} \right)^2 dx. \tag{11}$$

In the meantime, the ends of the column are moved closer because the column is bent. The axial load P does the work Pu, where u is the distance that the ends approached each other. The column itself, opposing P, does the work $-Pu$. To compute u, note that the vertical projection of an element of length dx in the deformed state is (Fig. 4.2:3)

$$dx \cos \frac{dw}{dx} = dx \left[1 - \frac{1}{2} \left(\frac{dw}{dx} \right)^2 + \cdots \right].$$

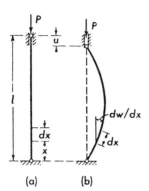

FIGURE 4.2:3. Euler column. (a) Straight and in equilibrium. (b) Deflected and in equilibrium.

The total vertical projection of the deformed column is, therefore,

$$L - u \doteq \int_0^L \left[1 - \frac{1}{2}\left(\frac{dw}{dx}\right)^2 \right] dx \doteq L - \frac{1}{2}\int_0^L \left(\frac{dw}{dx}\right)^2 dx.$$

Hence, when the slope dw/dx is $\ll 1$,

$$u = \frac{1}{2}\int_0^L \left(\frac{dw}{dx}\right)^2 dx. \tag{12}$$

The total energy needed to produce the deflection w is

$$V - Pu = \frac{EI}{2}\int_0^L \left(\frac{d^2w}{dx^2}\right)^2 dx - \frac{P}{2}\int_0^L \left(\frac{dw}{dx}\right)^2 dx. \tag{13}$$

It is zero if

$$P = \frac{EI\int_0^L \left(\dfrac{d^2w}{dx^2}\right)^2 dx}{\int_0^L \left(\dfrac{dw}{dx}\right)^2 dx}. \tag{14}$$

This is the value of P at which the column can deform by itself into a curve w without an external supply of energy. It signals an instability. We look for a function $w = w(x)$ that yields the smallest value of P, which is P_{cr}. Since both the numerator and denominator are quadratic in derivatives of w, the quotient does not depend on the absolute magnitude of the deflection but only on its shape.

There are several ways of finding this smallest value. We may use the calculus of variations, finding and solving the differential equation and boundary conditions for w as in the method of adjacent equilibrium. It will be found that these equations are exactly Eqs. (3) and (4). We may also assume a general expression for the unknown function $w(x)$, such as a Fourier series,

$$w(x) = \sum_{n=1}^{\infty} c_n \sin \frac{n\pi x}{L}, \tag{15}$$

which satisfies the boundary conditions, and then determine the coefficient c_n so that P from Eq. (14) becomes a minimum. When this is done, it is found that $c_n = 0$ for $n > 1$ and that c_1 is arbitrary while P becomes equal to the Euler load, Eq. (10).

Recapitulation

Both the adjacent equilibrium method and the energy method examine a disturbed basic system. In the adjacent equilibrium method, the differential equation and boundary conditions governing the disturbed system are studied, while looking for the eigenvalues at which a non-trivial solution exists. In the energy method, the problem is converted into a variational principle. Physically, a perturbed deformation that can occur in the system without additional external energy input is sought.

It is a special merit of the energy method that an approximate value of P_{cr} can be found by introducing into Eq. (14) some plausible function $w(x)$. The approximate function $w(x)$ can often be obtained by experimental observations. Since the assumed, plausible function $w(x)$ is not necessarily the one that makes P a minimum, the approximate value of P found by this method is always higher than the correct one. The approximation can be very good if the function $w(x)$ is skillfully chosen.

Problems

4.1 In Section 4.1 it was said that in general the internal pressure in a blood vessel falls along the length because of frictional loss. Under what conditions can exceptions to the general rule be found?

4.2 Consider a slightly crooked column with an initial deflection,

$$w_0 = A \sin \frac{\pi x}{L}.$$

When the column is loaded by P (see Fig. 4.2:1), an additional deflection w occurs. The bending moment in a cross section at x is then equal to $P(w_0 + w)$. The second term in Eq. (3) must be changed to $P(w_0 + w)$. Equation (4) remains valid. Find the solution. Show that when $P = P_{cr}$, the ratio c_1/A tends to ∞.

4.3 Show that the column can be strengthened if a lateral support is introduced to make $w = 0$ at some intermediate station x between 0 and L.

4.3 Instability of a Circular Cylindrical Tube Subjected to External Pressure

Now let us turn to a problem of more direct physiological interest: the collapse of a circular cylindrical tube under external pressure. The general theory of cylindrical shells is very complicated, so we shall deal explicitly with only one of the simplest cases: the deformation of a cylinder into another cylinder with a cross section of different shape. After the solution to this simple case is obtained, we shall then survey the general cases.

Consider a cylindrical tube made of a material that obeys Hooke's law of elasticity, as shown in Figure 4.3:1(a). Let the wall thickness be uniform, and equal to h. Assume that h is very small compared with the radius of the cylinder R. The cylinder is subjected to an internal pressure p and an external pressure p_e, and there is no force acting on the cylinder in the direction of the generating lines.

To visualize the forces and moments acting in the walls of the deformed tube, let us cut a small rectangular element out of the strained cylinder by two planes parallel to the tube axis and perpendicular to the middle surface of the wall, and two planes perpendicular to the cylinder axis at a unit distance apart. The isolated element is shown in Figure 4.3:1(b) as a free body,

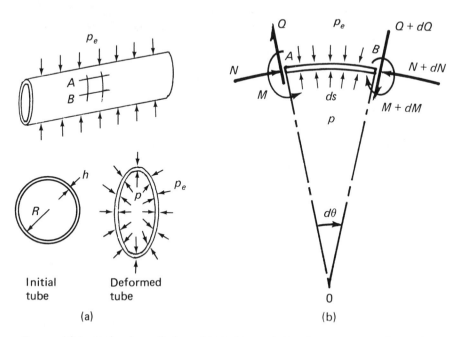

FIGURE 4.3:1. A circular cylinder subjected to an internal pressure p and an external pressure p_e. (a) Dimensions and loads. (b) Forces and moments.

where the points A and B are on the middle surface of the tube wall, and on planes containing generating lines. The plane of the figure is perpendicular to the tube axis. The location of the point A or B will be described by the arc length s measured in the circumferential direction from an arbitrary origin C on the middle surface of the tube wall. Let the length of the arc AB be infinitesimal and equal to ds. Let N denote the mean compressive stress resultant acting in the section at A. Let M denote the bending moment at A, which is a couple in a plane parallel to the plane of the figure. Q is the shear force per unit length across the wall, in the direction perpendicular to the middle surface of the wall. A pressure p acts on the concave side of the cylinder, and a pressure p_e acts on the convex side of the cylinder. The lines OA, OB are normal to the strained middle surface and contain an angle $d\theta$. The forces and couple at B are $Q + dQ, N + dN$, and $M + dM$, respectively.

The equations of equilibrium can be obtained by considering the balance of forces and moments acting on this element. By taking moments about B and neglecting small quantities of the second order, we get

$$dM + Q\,ds = 0,$$

from which we obtain

$$\frac{dM}{ds} = -Q. \tag{1}$$

By summing all forces in the direction BO, we get

$$Q + dQ - Q\cos d\theta - N\sin d\theta + \left(p_e - p\right)ds = 0.$$

Neglecting small quantities of the second and higher orders, we obtain

$$dQ - N\,d\theta + \left(p_e - p\right)ds = 0,$$

or

$$\frac{dQ}{ds} - N\frac{d\theta}{ds} + p_e - p = 0. \tag{2}$$

The ratio $d\theta/ds$ is, of course, the curvature of the wall.

Finally, by summing all forces in the direction of $N + dN$, we obtain

$$N + dN - N\cos d\theta + Q\sin d\theta = 0,$$

or

$$\frac{dN}{ds} + Q\frac{d\theta}{ds} = 0. \tag{3}$$

Combining Eqs. (1) and (2) we obtain

$$\frac{d^2 M}{ds^2} + N\frac{d\theta}{ds} = p_e - p. \tag{4}$$

Now we need an expression connecting the bending moment and wall deformation. In simple beams, the moment M is equal to EI times the change of curvature [see Eq. (4.2:2)]. In the theory of plates and shells, it can be shown that a similar result holds if the wall is thin,

$$M = E'I(\kappa - \kappa_0), \tag{5}$$

where $E'I$ is the flexural rigidity,

$$E'I = \frac{Eh^3}{12(1 - v^2)}, \tag{6}$$

and κ and κ_0 are the curvatures of the wall in the deformed and initial states. E is the Young's modulus of the material, v is the Poisson's ratio, and I is the cross-sectional area moment of inertia per unit length, $h^3/12$. On introducing ξ for the change of curvature,

$$\xi = \kappa - \kappa_0, \tag{7}$$

Eqs. (4) and (3) can be written in the form

$$E'I \frac{d^2\xi}{ds^2} + N(\xi + \kappa_0) = p_e - p, \tag{8}$$

$$\frac{dN}{ds} + Q(\xi + \kappa_0) = 0. \tag{9}$$

These are differential equations governing the deformation of a cylinder into another cylinder. Let us apply them to the case in which the cylinder is initially circular with a radius of R. Then

$$\kappa_0 = \frac{1}{R} = \text{constant.} \tag{10}$$

In the basic state, the cylinder remains circular and the hoop stress is

$$N_0 = (p_e - p)R. \tag{11}$$

If the pressure p_e is so large that the cylinder becomes neutrally unstable, a small deformation into a noncircular cylinder is possible. Let the change of curvature ξ be small compared with κ_0. Then Eq. (9) can be approximated by

$$\frac{dN}{ds} = -Q\kappa_0, \tag{12}$$

whereas Eqs. (1) and (5) yield

$$Q = -\frac{dM}{ds} = -E'I \frac{d\xi}{ds}. \tag{13}$$

Substituting Eq. (13) into Eq. (12) and integrating, we get

$$N = -\kappa_0 \int Q \, ds = E'I\kappa_0\xi + N_0, \tag{14}$$

where N_o is a constant given by Eq. (11), because when ξ vanishes, N must be reduced to N_0. Now substituting Eq. (14) into Eq. (4), noting that $d\theta/ds = \kappa = \xi + \kappa_0$, and neglecting the second-order term ξ^2, we obtain

$$E'I\frac{d^2\xi}{ds^2} + \left(N_0 + E'I\kappa_0^2\right)\xi = 0. \tag{15}$$

The solution of this equation is

$$\xi = c\cos\left(ms + k\right), \tag{16}$$

where

$$m^2 = \frac{N_0 + E'I\kappa_0^2}{E'I}, \tag{17}$$

and c and k are arbitrary constants. Now, since the displacement must be continuous, ξ must be such that its value is repeated when s is increased by $2\pi R$, the circumference of the cylinder. Thus,

$$m2\pi R = 2n\pi, \tag{18}$$

n being an integer; that is,

$$\frac{N_0 + E'I\kappa_0^2}{E'I}R^2 = n^2,$$

or, by Eq. (10),

$$N_0 = \frac{E'I}{R^2}\left(n^2 - 1\right); \tag{19}$$

and, by Eq. (11),

$$p_e - p = \frac{E'I}{R^3}\left(n^2 - 1\right). \tag{20}$$

These are the critical values of pressure at which the cylinder can deform into a shape described by Eq. (16). Here, n must be an integer. If $n = 1$, then $p_e - p = 0$ and we obtain the initial, no-load condition. As the transmural pressure $p_e - p$ is gradually increased, the cylinder becomes unstable for the first time when $n = 2$, at which the pressure equals the *critical pressure of buckling*:

$$\left(p_e - p\right)_{cr} = \frac{3EI}{R^3} = \frac{Eh^3}{4\left(1 - v^2\right)R^3}. \tag{21}$$

The corresponding shape of the buckled cross section is elliptical.

Problems

4.4 Let w be the radial displacement of the cross section of a circular cylinder as illustrated in Figure 4.3:2. Derive an expression for the curvature of the deformed cylinder, κ, in terms of w. It is simpler to use polar coordinates and to express the radius vector $\rho = R + w$ as a function of the polar angle θ. Simplify the expression when $w \ll R$. Then use Eqs. (7), (10), (16), (18), and $n = 2$ to show that the deformed cross section is elliptical.

4.5 Discuss the deformation patterns corresponding to $n = 3$.

4.6 What is the compressive stress in the tube wall at the critical buckling condition?

4.7 Equation (17) of Section 3.8, $h\langle\sigma_\theta\rangle = p_i r_i - p_0 r_0$, gives the average value of the circumferential stress σ_θ in terms of the external and internal pressures and radii, (p_0, p_i) and (r_0, r_i) respectively. Apply this formula to the femoral artery, arterioles, capillaries, and venules. Use the data for h, r_0, r_i from Table 3.1:1, p. 110, and p_i above atmospheric from Figure 5.3:4, p. 277. Assume the external pressure p_0 to be atmospheric (a compressive stress of about $1.013 \times 10^5 \, \mathrm{Nm^{-2}}$ or $1 \, \mathrm{kgf/cm^2}$). Show that the mean circumferential stress σ_θ is negative (compressive).

If both the external and intenal pressures acting on a circular cylindrical tube are atmospheric, then the circumferential stress σ_θ is, of course, compressive and equal to the atmospheric pressure. Would this compressive stress cause instability of the blood vessel?

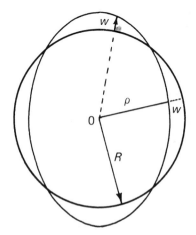

FIGURE 4.3:2. A circular cylinder deformed. The radial displacement w is shown.

Cells grown under atmospheric pressure are subjected to a compressive stress. Cells grown in deep sea are subjected to very high compressive stress. Is their stability affected?

These considerations show that a uniform hydrostatic pressure has no influence on the stability of a structure. Straightforward as it is, this point was subjected to considerable debate in the literature of the 1960s.

Stability of Thin Ring Under External Pressure

The foregoing reasoning applies equally well to a thin ring if E' were replaced by E, and if I were regarded as the moment of inertia of the cross section of the ring and p_e the external thrust per unit length applied to the ring. The critical value of the thrust that would cause collapse of the ring is then

$$p_e = \frac{3EI}{R^3}. \tag{22}$$

Postbuckling Behavior of Uniform Circular Tube

So far we have considered the critical conditions of buckling. If the external pressure exceeds the critical value given by Eq. (21), large deformations of the tube will occur, which are referred to as *postbuckling modes*. To determine these modes, Eqs. (1), (5), and (8) to (10) must be integrated with the boundary conditions such that the moment, shear, stress resultant, and curvature must be single valued and continuous, so that when the arc length s becomes $2\pi R$, the values of M, Q, N, ξ at $s = 2\pi R$ must be equal to their values at $s = 0$. A condition of symmetry may be imposed so that $d\xi/ds$ and Q vanish at $s = 0$ and $2\pi R$.

It can be proved that these equations have a nontrivial solution (i.e., one with $\xi \neq 0$) for every value of $p_e - p$ greater than the critical value given by Eq. (21), and for some $n \geq 2$. The solution obtained by Flaherty et al. (1972), by numerical integration, is shown by the dashed curve in Figure 4.3:3. Of great interest is the relationship between the cross-sectional area of the tube in the postbuckling modes and the presure difference $p - p_e$. Figure 4.3:3 shows this relationship expressed in the nondimensional variables

$$\tilde{p} = \frac{p - p_e}{\left(E'I/R^3\right)} \quad \text{and} \quad \alpha = \frac{A}{\pi R^2}. \tag{23}$$

The denominators in $\tilde{p}$ and α are also written as K_p and A_0, respectively. The long–dashed curve in Fig. 4.3:3 refers to initially perfect circular cylinders. These cylinders will not buckle for $\tilde{p}$ above -3. Buckling starts at $\tilde{p} = -3$. For $\tilde{p} < -3$, postbuckling, the cross-sectional area decreases rapidly with decreasing $\tilde{p}$ until the opposite sides of the tube touch. The buckling patterns are illustrated in Figure 4.3:4. Upon further increase in external pressure, the contact area increases and the open portion of the cross section is reduced in size but remains similar in shape. For this "self-similar" type of deformation, Flaherty et al. obtained the relationship

$$-\tilde{p} = \alpha^{-3/2}. \tag{24}$$

The dashed curve shown in Figure 4.3:3 is the relationship obtained for the mode $n = 2$. The mode $n = 3$ can occur when $\tilde{p} \leq -8$. At a sufficently large

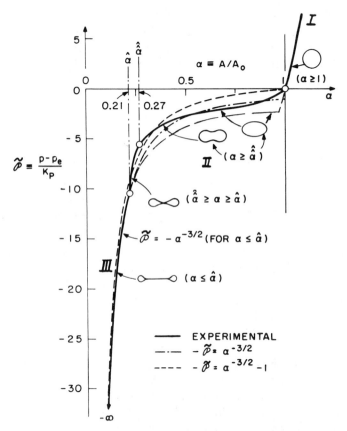

FIGURE 4.3:3. Behavior of a collapsible tube. Dimensionless transmural pressure difference, $\tilde{p}$, versus dimensionless area ratio, α. The solid curve shows a typical experimental curve for thin-walled latex tube and, adjacent to it, typical cross-sectional shapes for the different ranges of α. The dot–dash curve represents Eq. (24), which coincides with the solid curve for $\alpha < \hat{\alpha}$. The short–dashed curve represents the equation $-\tilde{p} = \alpha^{-3/2} - 1$. The curve with long dashes represents the theoretical result given by Flaherty et al. for cylinders whose cross sections are perfectly circular when $\tilde{p} = 0$. Point contact occurs at $\alpha = \hat{\alpha}$ and line contact occurs at $\alpha = \hat{\alpha}$. From Shapiro (1977), by permission.

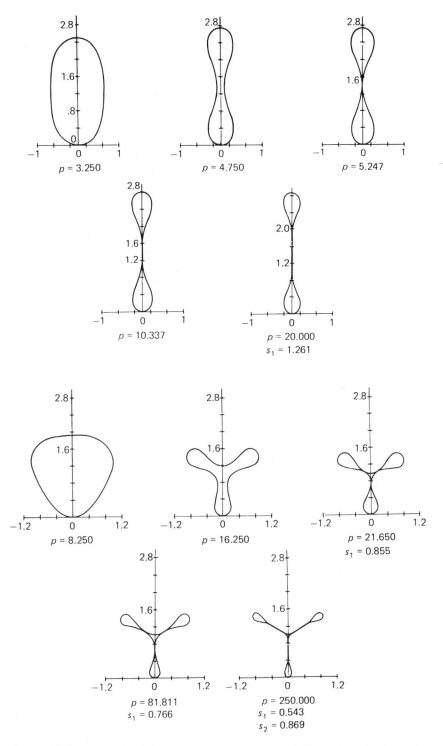

FIGURE 4.3:4. The postbuckling modes of an unsupported thin-walled circular cylinder subjected to uniform external pressure. The curves are the deformed cross-sectional shapes of the cylinder. From Flaherty et al. (1972), with permission of the author and the SIAM.

negative value of $\bar{p}$, the opposite sides of the tube also touch in the $n = 3$ mode, and the same self-similar solution described by Eq. (24) holds. Without taking special precautions to prevent the $n = 2$ mode from occurring, however, the modes $n \geq 3$ cannot be realized.

The solid curve in Figure 4.3:3 represents experimental results reported by Shapiro (1977). The other two curves in the figure represent two empirical formulas proposed by Shapiro [see Eqs. (8) and (10) of Sec. 4.4]. These curves apply to cylinders whose cross sections are elliptical in the unloaded condition.

Effect of Lateral Support

In the analysis presented thus far, the tube remains cylindrical (though not circular after buckling) because it is assumed that there is no lateral support. For a tube of finite length, the effect of lateral support on the stability can be very great. The critical pressure can be increased greatly by introducing reinforcing rings. Consider a tube *simply supported* at the ends, where the deflection and bending moments vanish. The analytical results in this case can be expressed in terms of two parameters, ϕ and λ:

$$\phi = \frac{Rp_{cr}\left(1 - v^2\right)}{Eh}, \quad \lambda = \frac{h^2}{12R^2}, \tag{25}$$

as described in Flügge (1960) and Timoshenko and Gere (1961). Figure 4.3:5 shows the relationship of ϕ to the ratio of the length L between supports and the tube radius R, for two values of λ. The value of n in this figure is the number of circumferential waves similar to n in Eq. (18), resulting in the deformation described by Eq. (16).

Effects of Axial Thrust, Initial Imperfections, and Material Nonlinearity

Thin cylinder stability is also very sensitive to axial thrust and initial imperfections. These effects have been studied quite thoroughly because of their importance to engineering. For a critical review see Fung and Sechler (1960, 1974).

The application of these engineering studies to physiology may be questioned on the ground that biological tissues do not obey Hooke's law. For example, in Chapter 8 of *Biomechanics: Mechanical Properties of Living Tissue* (Fung, 1993b), we have shown that the Young's modulus E of an artery increases with increasing stress level. A full analysis of cylindrical shell instability for a nonlinear pseudoelastic material is very complicated. For a rough estimation, it is reasonable to use the incremental modulus E at the prevailing state of stress in formulas such as Eq. (2) to obtain the critical pressure for buckling.

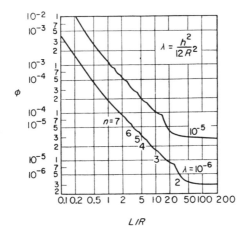

FIGURE 4.3:5. The critical buckling condition of a thin-walled circular cylinder of length L simply supported at both ends, and subjected to uniform external pressure. The critical pressure p_{cr} is contained in the parameter ϕ, which is the ordinate. The abscissa is the ratio L/R. R is the cylinder radius, n is the number of waves around the circumference. From Timoshenko, S. (1936) *Theory of Elastic Stability*. McGraw-Hill, New York, by permission.

4.4 Vessels of Naturally Elliptic Cross Section

If the cross section of the tube is elliptical in the stress-free state, then under increasing external pressure the tube will be compressed and distorted in a sequence of shapes similar to those sketched in Figure 4.4:1. At first, the eccentricity will be increased. Then the middle portion will be bent inward. Eventually, the opposite walls will touch and compress flat in the middle, leaving only the corners open.

The basic equations governing the bending of noncircular tubes subjected to external pressure are similar to those of the preceding section. Equations (1) through (9) of Section 4.3 remain valid, but Eq. (10) of Section 4.3 is no longer true. The initial curvature, κ_0, is not a constant, but varies around the circumference. An example is

$$\kappa_0 = \frac{1}{R}\left(1 + \varepsilon \cos \frac{s}{R}\right), \tag{1}$$

where s is the arc length measured on the middle surface of the vessel wall in the circumferential direction, as in Figure 4.3:1; R is the radius of an equivalent circle that has the same circumference as the noncircular tube cross section; and ε ($\ll 1$) is a small parameter. If this is used in Eqs. (8) and (9) of Section 4.3, we obtain the basic equations

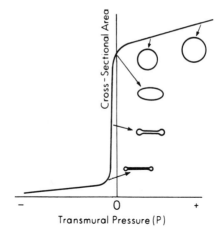

FIGURE 4.4:1. A sequence of cross-sectional shapes that occur as a naturally elliptic cylinder is subjected to gradually increasing external pressure. At first the eccentricity is increased; then it buckles. The successive postbuckling modes become very
similar to those of Figure 4.3:4.

$$E'I\frac{d^2\xi}{ds^2} + N\left[\xi + \frac{1}{R}\left(1 + \varepsilon\cos\frac{s}{R}\right)\right] = p_e - p, \tag{2}$$

$$\frac{dN}{ds} + Q\left[\xi + \frac{1}{R}\left(1 + \varepsilon\cos\frac{s}{R}\right)\right] = 0, \tag{3}$$

whereas Eqs. (1) and (5) of Section 4.3 remain valid:

$$Q = -\frac{dM}{ds} = -E'I\frac{d\xi}{ds}. \tag{4}$$

Combining Eqs. (3) and (4) yields

$$\frac{dN}{ds} - E'I\left[\xi + \frac{1}{R}\left(1 + \varepsilon\cos\frac{s}{R}\right)\right]\frac{d\xi}{ds} = 0. \tag{5}$$

Equations (2) to (4), or (2) and (5), are sets of ordinary differential equations to be solved together with the boundary conditions that ξ, N, and Q
are periodic,

$$\xi(2\pi R) = \xi(0), \quad N(2\pi R) = N(0), \quad Q(2\pi R) = Q(0); \tag{6}$$

and a condition of symmetry,

$$\frac{d\xi}{ds} = 0, \quad Q = 0 \quad \text{at } s = 0. \tag{7}$$

The analysis is more complex than that of the preceding section because ξ may not be considered infinitesimal in the present problem. There is no simple way to find analytic solutions, but a numerical solution is feasible.

Experimental results on latex rubber tubing and a vein are shown in Figures 4.3:3 and 4.4:2, respectively. In neither of these two examples was the geometry of the tubing in its natural, unstressed state accurately recorded. Hence, the data were not sufficient to check the theory. The effective Young's modulus in compression and the bending rigidity $E'I$ of the vein are also unknown. The general trend, however, can be understood. When the internal pressure exceeds the external pressure, $p - p_e > 0$, the tube cross-sectional area increases with increasing $p - p_e$. When the external pressure exceeds the internal pressure, $p - p_e < 0$, the cross-sectional area decreases rapidly with increasing external pressure.

In Figure 4.3:3, a chain–dot curve,

$$-\tilde{p} = \alpha^{-n}, \quad n = \tfrac{3}{2}, \tag{8}$$

is shown. This is an extension of Eq. (24) of Section 4.3 beyond the range ($\alpha \le \hat{a}$) in which the equation was originally derived. Here

$$\tilde{p} = \left(p - p_e\right)\frac{R^3}{E'I}, \quad \alpha = \frac{A}{A_0}, \tag{9}$$

where A_0 is the cross-sectional area in the unstressed state. Another curve indicated by short dashes is Shapiro's (1977) approximate formula,

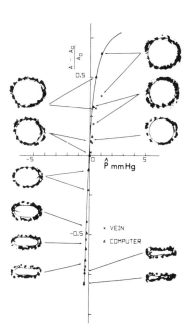

FIGURE 4.4:2. Relationship between transmural pressure and cross-sectional area of dog vena cava. Experimental results from Moreno et al. (1970). Reproduced by permission.

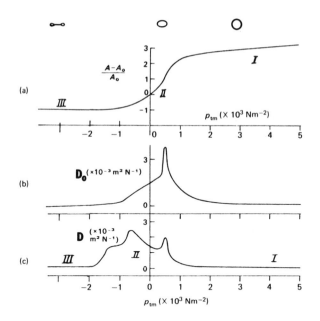

FIGURE 4.4:3. Elastic properties of a segment of a dog vein. Here the abscissa is the transmural pressure measured in units of kPa (each kPa $\doteq$ 10 cm H$_2$O). The ordinate of (a) is $\alpha - 1$, that is, the change of area $A - A_0$ divided by the area at zero transmural pressure. The ordinate of (b) is $D_0 = (1/A_0)(dA/d(p - p_e))$, in units of kPa^{-1}. The ordinate of (c) is the distensibility D defined in Eq. (11). The difference between (b) and (c) shows that D_0 is not a good approximation of D. Data from Moreno et al. (1970) and Attinger (1969). Adapted from Caro et al. (1978), by permission.

$$-\tilde{p} = \alpha^{-n} - 1; \quad n = \tfrac{3}{2}. \tag{10}$$

It has the merit that it passes through the point $\tilde{p} = 0$, $\alpha = 1$.

The most important feature of these curves relevant to the analysis of flow is the *distensibility* of the elastic tube defined by

$$D = \frac{1}{A} \frac{dA}{d(p - p_e)}. \tag{11}$$

Here, A is the cross-sectional area and $p - p_e$ is the transmural pressure. The variation of D with pressure $p - p_e$ for a vein is plotted in Figure 4.4:3(c). The curve can be divided into three zones: In zones I and III, the distensibility is small; in zone II it is larger. It is expected that the interaction between a flowing fluid and the wall deformation will be most vigorous in zone II. That this is indeed the case will be seen in Sections 4.6 and 4.8.

We could also mark zones I, II, and III on Figures 4.4:2 and 4.3:3. With regard to the points of demarcation between the zones, there is a difference between those vessels that are naturally circular and those that are naturally elliptic or otherwise nonaxisymmetric. In all cases, there is no clearly definable point of demarcation between zone II and zone III (in other words, the division is arbitrary). For the naturally elliptic or otherwise nonaxisymmetric vessels (as in Fig. 4.4:2, and also for the solid curve in Fig. 4.3:3), neither is there a definable point of demarcation between zone I and zone II. But, for a naturally circular cylindrical tube, as is shown by the long-dashed curve in Figure 4.3:3, there is a well-defined critical buckling point, and that point marks a sharp change of distensibility between zone I and zone II. This is clearly seen from the sudden change of the slope of the long-dashed curve for a circular cylinder at the critical buckling point in Figure 4.3:3. For circular cylindrical tubes with very thin walls (with a very small wall thickness to radius ratio, say, <0.01), the buckling load is so small that it is not too far wrong to take the point of demarcation between zone I and zone II at $p - p_e = 0$.

For arteries and arterioles, the wall thickness/radius ratio is much larger than that of the veins, and the critical buckling load is much larger. For some smaller arteries, severe reduction of cross-sectional area does not occur until the transmural pressure reaches -10 to -20 mm Hg.

4.5 Steady Flow in Collapsible Tubes

Blood flow in veins is similar to that in the arteries except that (a) veins often have valves that prevent reverse flow, and (b) sometimes the transmural pressure in the vein is negative, that is, the blood pressure in the vein is smaller than the pressure external to the vein. Transmural pressure in the arteries can be negative too, for example, in arteries in an arm under a pressure cuff in blood pressure measurement, but normally the arterial blood pressure is so high that this does not occur. In the vein, the blood pressure is low and the transmural pressure can become negative by hydraulic gradient, muscle action, or for other reasons. For example, the pressure in the vena cava is nearly atmospheric. If we raise our hands above our heads, the weight of blood in the vein will reduce the hydrostatic pressure in the veins of the hands by 60 or 70 cm H_2O below that of the vena cava. If the tissue pressure outside the veins remains nearly atmospheric, then the transmural pressure would be negative 60 or 70 cm H_2O.

The feature introduced by the negative transmural pressure is the great distensibility of the vessel in a certain range of pressure. This is evident from Figure 4.4:3 in region II. It is also seen Figures 4.4:2 and 4.3:3. Now the phase velocity of propagation of waves of small amplitude [see Sec. 3.8, Eq. (4)] is

$$c = \left(\frac{A}{\rho} \frac{dp}{dA} \right)^{1/2} = \left[\frac{A}{\rho} \frac{d(p - p_e)}{dA} \right]^{1/2}. \tag{1}$$

If $Ad(p - p_e)/dA$ is low, c is small and the blood flow velocity u may be comparable with, or even large than, c. This reminds us of transsonic and supersonic flow in gas dynamics. Indeed, the analogy with gas dynamics and with channel flow of a liquid with a free surface is very close. To see this consider a one-dimensional, unsteady, frictionless flow in a collapsible tube, a gas flow in a rigid tube, and a liquid flow in a uniform, horizontal open channel. The equation of motion is identical for each of the three cases:

$$\frac{\partial u}{\partial t} + u\frac{\partial u}{\partial x} = -\frac{1}{\rho}\frac{\partial p}{\partial x}, \tag{2}$$

where ρ is the mass density of the fluid, p is the pressure in the flowing fluid, u is the velocity, t is time, and x is longitudinal distance. The equations of continuity are

$$\text{For a collapsible tube}: \quad \frac{\partial A}{\partial t} + \frac{\partial}{\partial x}\left(Au\right) = 0, \tag{3a}$$

$$\text{For gas flow}: \quad \frac{\partial \rho}{\partial t} + \frac{\partial}{\partial x}\left(\rho u\right) = 0, \tag{3b}$$

$$\text{For channel flow}: \quad \frac{\partial h}{\partial t} + \frac{\partial}{\partial x}\left(hu\right) = 0. \tag{3c}$$

Here A is the cross-sectional area in the first case, ρ is the mass density in the second case, and h is the height of the free surface above the bottom in the third case. The phase velocity of propagation of small perturbations, c, is in these three cases

$$c^2 = \frac{A}{\rho}\frac{d(p - p_e)}{dA}, \tag{4a}$$

$$c^2 = \left(\frac{dp}{d\rho}\right) \text{ at constant entropy,} \tag{4b}$$

$$c^2 = \frac{h}{\rho}\frac{dp}{dh} = gh, \tag{4c}$$

where g is gravitational acceleration. Thus the analogy is seen. Those readers who are familiar with gas dynamics may recall the shock waves, supersonic wind tunnel, Laval nozzle for the steam turbine, convergent section to accelerate fluid in the subsonic regime, sonic throat, and divergent section to accelerate fluid in the supersonic regime. Those familiar with open channel flow may recall flow over a dam and hydraulic jump. One could anticipate the existence of analogous phenomena in blood flow in collapsible vessels. One anticipates also, of course, that similar phenomena occur in air flow in the airways, the Korotkoff sound in arteries, urine flow in the urethra, etc.

Flow Limitation

Consider a simple example of flow of a nonviscous liquid in a collapsible tube connected to two reservoirs (Fig. 4.5:1). The tube is mounted by rigid connections to the reservoirs, and is enclosed in a chamber containing air or water at an adjustable pressure, p_e. Such a tube is known as a *Starling resistor*, first used by the English physiologist Starling (1866–1927) in his heart–lung machine (Knowlton and Starling, 1912).

Consider flow in such a tube. The flow depends on the inlet pressure, p_1, the outlet pressure, p_2, and the external pressure, p_e, or, more precisely, on the pressure differences, $p_1 - p_e$, $p_2 - p_e$, because the tube cross section depends only on the transmural pressure. If $p_1 - p_e$ were fixed and $p_2 - p_e$ were gradually decreased, the velocity of flow would increase, but in the meantime the tube cross-sectional area would decrease. The flow, being a product of velocity and cross-sectional area, would increase at first, but eventually may be limited.

What happens in such a tube may be illustrated by an example. Figure 4.5:2 shows the result of Conrad's (1969) experiment with the flow of water in a Starling resistor whose outlet was connected to a flow resistor and then exposed to the atmosphere. The chamber pressure, p_e, was fixed at 3.3 kPa ($\sim$33 cm H_2O). The inlet pressure, p_1, was varied. The pressure–flow relationship is shown in Figure 4.5:2(a). If self-excited oscillations (flutter) occurred, the flow was averaged over time. The numbers indicated on the curve of Figure 4.5:2(a) correspond to the photographs shown in Figure 4.5:2(b), which are a sequence of side views of the tube at successive stages in the experiment. For these photographs, the flow was from the right to the left.

The pressure–flow relationship may be separated into three regimes—I, II, III—as marked on Figure 4.5:2(a). (a) In regime I, both $p_1 - p_e$ and $p_2 - p_e$ are positive, there is no buckling of the tube, and the flow Q is essentially proportional to $p_1 - p_2$. (b) In regime II, $p_1 - p_e > 0$, $p_2 - p_e < 0$, buckling occurs towards the outlet, the tube cross section changes rapidly as the

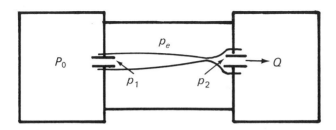

FIGURE 4.5:1. An experimental arrangement in which a collapsible tube is used as a Starling resistor. The tube connects the reservoir at the left, which has a pressure p_0, to the reservoir at the right at a lower pressure.

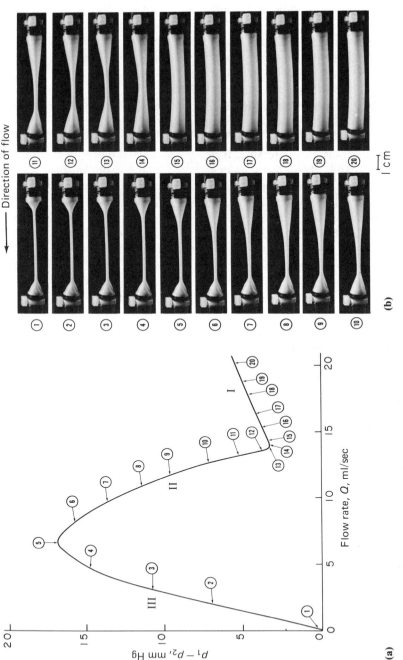

(a)

(b)

FIGURE 4.5:2. (a) Pressure–flow relationship in a Starling resistor. Ordinate: $p_1 - p_2$ or $(p_1 - p_e) - (p_2 - p_e)$. Abscissa: flow rate, Q. For this figure, $p_e = 3.3 \times 10^3$ N m^{-2} was kept fixed. The outlet of the tube was connected to a flow resistance and then was exposed to the atmosphere. The resistance R was varied. The inlet transmural pressure, $p_1 - p_e$, was kept constant. The exit pressure, p_2, is equal to $Q \cdot R$ plus atmospheric pressure. The numbers on the curve refer to the photographs in (b). (b) Side views of the collapsible segment of tubing at different stages of the experiment. Flow goes from the right to the left. Numbers correspond to the numbered positions on the graph in (a). From Conrad (1969), © 1969 IEEE. Reproduced with permission.

transmural pressure falls, and the tube may flutter (see Sec. 4.8). (c) In regime III, $p_1 - p_e < 0$, $p_2 - p_e < 0$, the tube is buckled, and the cross section becomes dumbbell-shaped, as described in Sections 4.3 and 4.4. The flow rate is small, but again it is roughly proportional to the pressure drop $p_1 - p_2$.

The most fascinating part is regime II, in which the great distensibility of a collapsed tube takes effect. The tube starts to buckle at the downstream end at the condition corresponding to panel 14 in Figure 4.5:2(b), or point 14 in Figure 4.5:2(a). As flow is further reduced, the pressure p_2 decreases because in Conrad's experimental setup, p_2 is equal to atmospheric pressure plus Q times the resistance, which is fixed, so that p_2 decreases if Q decreases. When $p_2 - p_e$ decreases, the buckle deepens, the resistance to flow increases, and the pressure drop required to maintain the rate of flow increases. The progressive collapse of the tube can be seen from panels 14 to 5 of Figure 4.5:2(b). At point 5, virtually the entire tube is collapsed and p_1 is approximatley equal to p_e.

The tube often flutters in regime II. This unsteady condition is discussed in Section 4.8.

In a different experiment by Holt (1969), p_1 and p_e ($p_1 > p_e$) were fixed while the downstream pressure p_2 was varied. Then, as can be seen from Figure 4.5:3, the flow increased as $p_2 - p_e$ was decreased, as long as $p_2 - p_e$ was positive. But, when $p_2 - p_e$ was negative, the flow was practically independent of the values of $p_2 - p_e$: No increase of flow could be obtained by lowering the pressure at the exit end. The condition of *flow limitation* was reached.

Figure 4.5:3 illustrates flow limitation in a collapsible tube. A simplified analysis of this situation is instructive. Consider laminar flow in an elastic tube at a large Reynolds number so that Bernoulli's equation holds:

$$p + \tfrac{1}{2}\rho u^2 = p_0, \tag{5}$$

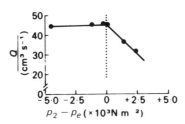

FIGURE 4.5:3. Illustration of flow limitation in a thin-walled rubber tube used as a Starling resistor. The upstream transmural pressure, $p_1 - p_e$, was fixed, whereas the downstream transmural pressure, $p_2 - p_e$, was continuously decreased. The mean flow rate was measured. In case of flutter, the flow was averaged over time. When $p_2 - p_e$ is negative, the flow rate is independent of $p_2 - p_e$. From Holt (1969), © 1969 IEEE. Reproduced by permission.

where p_0 is the stagnation pressure, p is the static pressure, ρ is fluid mass density, and u is velocity. The volume flow rate, Q, is

$$Q = Au = A\sqrt{\frac{2}{\rho}(p_0 - p)} = A\sqrt{\frac{2}{\rho}\left[(p_0 - p_e) - (p - p_e)\right]}. \tag{6}$$

A is the cross-sectional area, which is a function of $p - p_e$. Although p varies with distance down the tube, Q remains constant along the tube, of course. Now if $p_0 - p_e$ is fixed and we change $p - p_e$, then the rate of change of Q is

$$\frac{dQ}{d(p - p_e)} = -\frac{A}{\rho}\left\{\frac{2}{\rho}\left[(p_0 - p_e) - (p - p_e)\right]\right\}^{-1/2}$$

$$+ \frac{dA}{d(p - p_e)}\left\{\frac{2}{\rho}\left[(p_0 - p_e) - (p - p_e)\right]\right\}^{1/2}$$

$$= -\frac{A}{\rho u} + \frac{dA}{d(p - p_e)}u = -\frac{A}{\rho u} + \left[\frac{\rho}{A}\frac{dA}{d(p - p_e)}\right]\frac{A}{\rho}u.$$

The factor in the bracket in the last term is $1/c^2$, according to Eq. (1). Hence we obtain

$$\frac{dQ}{d(p - p_e)} = \frac{A}{\rho u}\left(\frac{u^2}{c^2} - 1\right). \tag{7}$$

Thus the flow will increase with decreasing $p - p_e$ only if $u < c$. A maximum, that is, a limitation, is reached when $u = c$. If $u > c$, then a further decrease in $p - p_e$ decreases the flow! Thus $u = c$ signifies flow limitation, and the section where $u = c$ is analogous to the sonic section in the Laval nozzle or supersonic wind tunnel. It *chokes* the flow. The ratio

$$S = \frac{u}{c} \tag{8}$$

is called the *speed index* by Shapiro (1977), and it plays a central role in liquid flow through a collapsible tube, as the Mach number does in gas dynamics.

The maximum flow is given by Eq. (6) with $u = c$, and is $Q_{\max} = Ac$. If the pressure–area relationship follows Eq. (8) or (10) of Section 4.4, then we have

$$c = \left[\frac{A}{\rho}\frac{d(p - p_e)}{dA}\right]^{1/2} = \left[\frac{A}{\rho}\frac{E'I}{R^3}\frac{d\tilde{p}}{d\alpha}\frac{1}{A_0}\right]^{1/2}$$

$$= \left[\frac{1}{\rho}\frac{E'I}{R^3}n\alpha^{-n}\right]^{1/2} = c_0\alpha^{-n/2}, \qquad \left(c_0 = \sqrt{\frac{E'In}{\rho R^3}}\right), \tag{9}$$

where c_0 is the value of c when $\alpha = 1$, that is, when the tube is in the no-load state, with $p - p_e = 0$. A_0 is the cross-sectional area at the no-load state. Hence we have

$$\frac{Q_{max}}{A_0 c_0} = \frac{Ac}{A_0 c_0} = \alpha^{1-n/2}. \tag{10}$$

If we use Eq. (8) or Eq. (10) of Section 4.4 as an approximation, we obtain, respectively,

$$\frac{Q_{max}}{A_0 c_0} = \left(-\tilde{p}\right)^{(n-2)/2n} \quad \text{or} \quad \left(1 - \tilde{p}\right)^{(n-2)/2n}. \tag{11}$$

In either case it is seen that the maximum flow rate depends solely upon the local transmural pressure difference at the "sonic" section, where $S = 1$, irrespective of the upstream driving pressure.

This analysis is based on the assumption that the frictional loss is negligible and that the tube cross-sectional area depends solely on the transmural pressure. Thus the rheology of the fluid and longitudinal tension in the tube are neglected. If these assumptions do not apply, as in the lung, and in microcirculation, then the conclusions are not valid. Flow limitation in the lung is discussed in Sections 4.10 and 6.15 to 6.19.

Problem

4.8 If the tube has a uniform cross section in the no-load state, then $-\tilde{p}$ must be the largest at the outlet section. Show that if $n = 2$ in Eqs (8) or (10) in Section 4.4, the maximum possible flow rate does not depend on $-\tilde{p}$ at all. If $n = \frac{3}{2}$, determine $Q_{max}/A_0 c_0$ for $-\tilde{p} = 5$, 10, and 15.

General One-Dimensional Steady Flow

Shapiro (1977) presented a full analysis of one-dimensional steady flow through a collapsible tube, including considerations of friction, gravity, and variations of external pressure or of muscle tone. He considered the external pressure, p_e, the initial cross-sectional area of the tube, A_0, the radius R, and the bending rigidity, $E'I$, as functions of x, the distance along the tube. He demonstrated the crucial role of the speed index S. For example, in a naturally uniform tube, when $S < 1$ friction causes the area and pressure to decrease in the downstream direction, and the velocity to increase. In contrast, when $S > 1$ the area and pressure increase along the tube, while the velocity decreases—exactly opposite to the former case. In general, whatever the effect of certain changes of A_0, p_e, etc. in a "subcritical" flow ($S < 1$), the effect is of opposite sign in a "supercritical" flow ($S > 1$). As a consequence, a variety of unexpected and remarkable features can be found. Some of these may be counterintuitive. As an illustration, consider a small

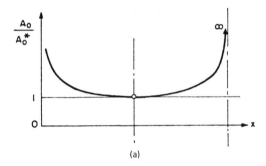

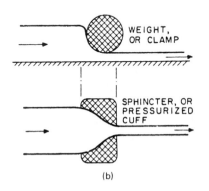

FIGURE 4.5:4. Several examples given by Shapiro (1977) in which a smooth transition through the critical condition $S = 1$ is possible. See the text for details. Reproduced from Shapiro (1977), by permission.

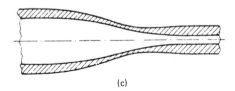

change of external pressure, p_e, while all other independent variables (A_0, friction, elasticity, etc.) are held constant. If p_e increases, both A and p decrease in subcritical flow, but they increase in supercritical flow!

The full analysis is intriguing, but quite complex. Let us quote here only Shapiro's examples (Fig. 4.5:4) of situations in which continuous passage of flow in regime $S < 1$ through $S = 1$ into $S > 1$ might occur. The pressure–area law given by Eq. (10) of Section 4.4 is assumed. Figure 4.5:4(a) shows the case with pure variation of initial cross-sectional area, A_0, along the tube axis. A_0^* denotes the area at the location where $S = 1$. Assuming that the transition is from subcritical to supercritical flow, then the pressure decreases continuously in the axial direction, and the area of the actual, deformed cross section, A, would decrease continuously in the axial direction also (not shown in the figure). Figure 4.5:4(b) shows the transition caused by a clamp, a cuff, or a sphincter, through their effect on changing the external pressure, p_e. Both the fluid pressure and the area decrease continuously in the axial direction. $S = 1$ occurs at a point in the region where a sharp constriction exists. Figure 4.5:4(c) shows the case with a pure change in the bending stiffness parameter $E'I/R^3$. This is physically realizable in a tube with variable wall thickness. Both the cross-sectional area and the pressure decrease continuously through the transition. Details can be found in Shapiro (1977).

In general it is not possible to have supercritical flow indefinitely along an originally uniform tube. Hence, one must consider the transition from supercritical flow back to subcritical flow. In the analogous case of a supersonic wind tunnel, this is done by a shock wave. Shapiro showed that in an elastic tube the shock wave may be very thick (several tube diameters), and that it is possible to have a transition from one supercritical flow to another supercritical flow through a shock wave (unlike a shock wave in gas dynamics). Later studies have shown that for the shock recovery problem, the one-dimensional flow approximation is inadequate. For a realistic solution it is necessary to consider three-dimensional flow, taking the distortion of cross sections and nonaxisymmetric flow into consideration.

4.6 Unsteady Flow in Veins

Flow in veins is unsteady for of a variety of reasons, including the pulsatile action of the heart and transient actions of muscles. The heart affects the end conditions, the muscle affects the boundary conditions along the vessel, and a variety of other factors may affect the internal and external pressures in the vessel. As illustrated in Figure 4.6:1, if the venous system is idealized as a system of pipes, then the right atrium controls the end condition at

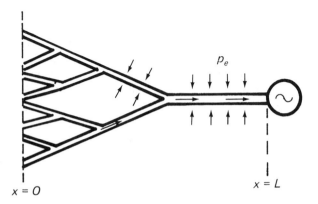

FIGURE 4.6:1. Schematic drawing of the venous system.

$x = L$, the capillaries control the pressure at $x = O$, while the muscles, body fluids, and neighboring organs control the pressure external to the veins. The muscle action may be indirect, as happens when one uses abdominal muscles and the diaphragm to control the pressure surrounding the vena cava, or it may be direct, as happens in venules and veins in the muscles of the arms and legs. Other causes of transient changes could arise from breathing, exercise, change of posture, flying, diving, etc. In all cases, the valves in the vein act to stop reversed flow but offer very little resistance to forward flow. These valves are very flexible, tenuous membranes. The hydrodynamic principles of valve action in the vein are the same as those in the heart and are discussed in Section 2.5.

Pulsatile Flow into the Right Heart

The venous tree begins at the venules and ends at the right atrium. The periodic fluctuation of pressure in the right atrium is shown in Figure 2.4:2 in Chapter 2 (the fifth curve from the top). In the right atrium of humans, the mean pressure (averaged over time) is on the order of 3 cm H_2O, and the amplitude of pressure fluctuations is on the order of 7 cm H_2O. In the venules, the mean pressure is on the order of 25 cm H_2O, and the amplitude of fluctuations is on the order of 4 cm H_2O. The fluctuations in the pressure and flow waves in the venules are not, however, in phase with the pressure and flow waves at the right atrium, owing to wave propagation, reflection at branching points, and attenuation due to the viscoelasticity of blood vessels. The situation is rather analogous to the flow of a river into an ocean. Just as the influence of the tide does not extend too far upstream, the influence of the pressure fluctuations in the right atrium dies out gradually in the veins. This is illustrated in Figure 4.6:2, which shows pressure waves at four sites along the venous tree of a human. The big disturbance in trace (c) is almost certainly due to the presence of a valve in the vein as the catheter was gradually withdrawn through it. The wave form in the superior vena cava is almost identical to that in the right atrium. The wave in the subclavian vein shows a phase lag from that in the vena cava, consistent with the propagation of disturbances away from the source.

The velocity waves in the venae cava of humans are presented in Figure 4.6:3, which shows that in each cardiac cycle there are two main oscillations of flow velocity, which are out of phase with the pressure oscillations in the right atrium. The pressure pulse in the jugular vein is presented in Figure 2.4:2. The pulsation in the jugular vein is visible to the naked eye and is palpable by the finger, and hence is used clinically to detect abnormalities.

Propagation of Progressive Waves in Veins

Studies of the propagation of progressive waves in veins are few. Some predictable features may be gathered from the principles discussed in

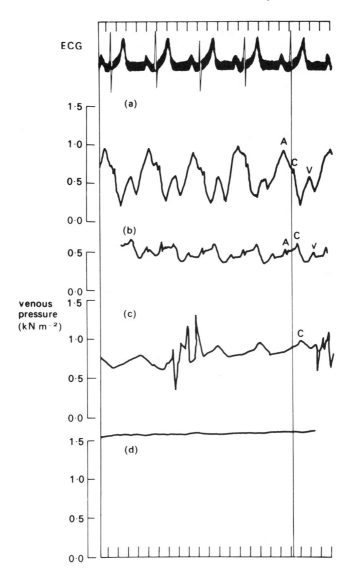

FIGURE 4.6:2. Pressure fluctuations in human venous system. From top to bottom: (a) the superior vena cava, (b) the subclavian vein, (c) the axillary vein, (d) the brachial vein. The pressures were measured sequentially with a single catheter. The values were relative to the atmosphere. The subject was supine so that all sites were at approximately the same level. The big disturbance in c, the axillary vein, is almost certainly due to withdrawing the catheter through a valve. From Caro et al. (1978), and by courtesy of Dr. G. Miller, Brompton Hospital, London. Reproduced from Caro et al. (1978), by permission.

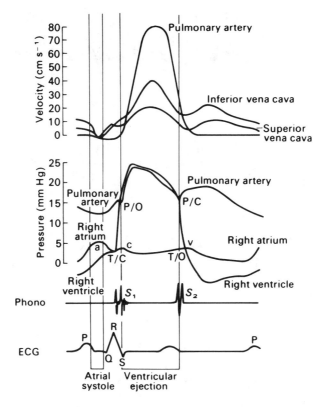

FIGURE 4.6:3. The velocity and pressure of flow into and out of the right atrium, coordinated with the phonocardiogram and ECG. T/C indicates the closure of tricuspid valve, P/O pulmonary valve open, P/C pulmonary valve closed, and T/O tricuspid valve open. From Wexler et al. (1968). Reproduced by permission of the American Heart Association.

Chapter 3; see also Problems 4.9 to 4.11 at the end of this section.

Experiments done by Anliker et al. (1969) on anesthetized dogs reveal a number of interesting features of venous blood flow. They used high-frequency sinusoidal pressure waves of small amplitude generated either by a small sinusoidal pump introduced into the abdominal vena cava or by an electromagnetic wave generator attached to the vein's outer wall. A short train of high-frequency pressure waves was generated each time. Its propagation was detected by two pressure sensors mounted on catheter tips and placed in the vein at a known distance apart (from 4 to 8 cm). The high frequency and short wave train were used to avoid distortions due to wave reflection. The small amplitude was used to ensure linearity of the system. Their results on the wave speed as a function of transmural pressure are shown in Figure 4.6:4. The different symbols represent different frequencies of the waves. It is seen that the wave speed varies from 2 m/s to 6 m/s over the applied pressure range, and that the wave speed increases with increasing transmural pressure, reflecting the rapid increase in the incremental Young's modulus of the vessel wall, as well as the change in the ratio of vessel wall thickness to diameter [see Eq. (2) in Problem 4.9]. The wave speed at any given transmural pressure is essentially independent of the frequency of oscillation; that is, the system is nondispersive in this frequency range.

The results of Anliker et al. on the attenuation of wave amplitude with distance propagated along the vein are shown in Figure 4.6:5. It is seen that the amplitude falls off exponentially with distance. The ratio x/λ is distance (x) measured in units of wavelength (λ). The figure shows that the a/a_0 versus x/λ relationship is almost unaffected by the frequency of the waves. In other words, the amount of attenuation *per wavelength* is almost independent of frequency in the frequency range tested. The experimental results may be expressed by the equation

$$a = a_0 e^{-kx/\lambda}, \tag{1}$$

in which a is the amplitude of the wave, a_0 is the amplitude at $x = 0$, x is the distance measured along the vessel, λ is the wavelength, and k is a constant. The value of k derived from the vena cava of the dog was found to lie in the range 1.0 to 2.5 (attenuation of 63% to 92% per wavelength), compared with the range 0.7 to 1.0 for the aorta.

There are at least three causes for this attenuation. One is the viscosity of the blood, but since both the Womersley number, N_W, and the Reynolds number, N_R, are quite large, the effect of blood viscosity is small. Another is radiation or transmission to tissues or fluids surrounding the vessel, but

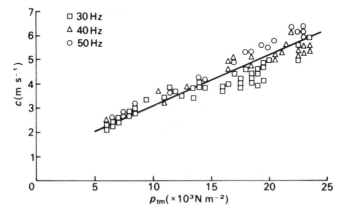

FIGURE 4.6:4. Speed of propagation c of short trains of sinusoidal waves of small amplitude imposed on the abdominal vena cava of a dog plotted against the transmural pressure, p_{tm}. From Anliker et al. (1969). Reproduced by permission.

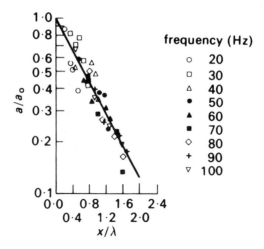

FIGURE 4.6:5. Attenuation of the amplitude of a short train of sinusoidal waves of small amplitude imposed on the abdominal vena cava of a dog with the distance of propagation. The ratio of the wave amplitude a to the amplitude a_0 at a fixed site is plotted against the ratio x/λ, the distance distal to the fixed site (x) divided by the wavelength of the wave (λ). From Anliker et al. (1969). Reproduced by permission.

since the experiment was performed on veins exposed to air (open chest and abdomen), this dissipation is minimal. The third cause, viscoelasticity of the vessel wall, must be predominantly responsible for this attenuation.

The characteristic of the viscoelasticity of the vessel wall revealed by this experiment is that the attenuation per wavelength is almost independent of frequency. Translated into a statement about the energy dissipated by the vessel wall in executing sinusoidal motion, it says that the energy dissipated per cycle of oscillation is almost independent of frequency. In other words, the energy dissipation mechanism is almost independent of the strain rate. Although this was verified in this experiment only in the frequency range 20 to 100 Hz, it was also found to be true in wide frequency ranges in a variety of experiments reported in *Biomechanics: Mechanical Properties of Living Tissues* (Fung, 1993b). It is in keeping with a general observation that living tissues are *pseudoelastic* (Fung, 1993b, p. 301).

Problems

4.9 Assume that a vein is circular cylindrical in the unstressed state, that its pressure–area relationship in the unbuckled state is derivable from membrane theory, and that in the buckled state Eq. (8) or (10) of Section 4.4 holds. Derive an analytic expression for the distensibility of the vessel, D, as defined by Eq. (11) of Sec. 4.4. It is well known that the Young's modulus E of a blood vessel wall varies with the stress in the wall [see Chapter 8 of *Biomechanics: Mechanical Properties of Living Tissues* (Fung, 1993b)]. For states of the vein near buckling of postbuckling, one may use the value of E at zero stress, $p - p_e = 0$. For veins subjected to positive inflation, $p - p_e > 0$, show that the membrane theory gives the formula

$$\frac{1}{D} = \frac{Eh}{d},$$
(2)

where E is the incremental Young's modulus for circumferential stretch, h is the wall thickness, and d is the vessel diameter. Using the data given in Figure 4.4:3, obtain an estimation of the incremental Young's modulus both for $p - p_e > 0$ and $p - p_e \leq 0$.

4.10 For progressive harmonic waves of small amplitude propagating in a cylindrical elastic tube, the wave speed c is given by the equation $c = (\rho D)^{-1/2}$, where ρ is the density of blood and D is the distensibility of the vessel. Using the values of D given in Figure 4.4:3 and ρ about $1\,g/cm^3$, show that the wave speed in a vein varies from 5 m/s or more at high transmural pressures, to a minimum of about 0.6 m/s when the vein is collapsing.

4.11 If the average velocity of blood in the inferior vena cava of a dog is between 10 cm/s and 20 cm/s, and the diameter of the vessel is about 1 cm, what is the Reynolds number (N_R) of the flow? What is the frequency parameter or Womersley's number (N_W) at a heart rate of 2 Hz? What is the wavelength (λ) if the wave speed is 2 m/s and frequency is 2 Hz?
    Answer: N_R lies between 250 and 500. $N_W \doteq 8$. N_R and N_W are smaller in smaller veins. $\lambda \doteq 1$ m.

4.12 On the basis of N_R and N_W estimated in the preceding problem, would the flow be laminar or turbulent? How long is the entrance length for the mean flow? How thick is the oscillatory boundary layer on the vessel wall? How long is the entrance length for the unsteady flow?
    Answer: Flow is laminar. The entrance length of the inferior vena cava is about 7.5 to 15 cm. The oscillatory boundary layer is thin. The unsteady flow entry length is about 4.5 cm.

4.7 Effect of Muscle Action on Venous Flow

If a vein is embedded in skeletal muscle, then the contraction of the muscle may squeeze the vein; this, in conjunction with the action of the valves in the vein, creates interesting control of blood flow. If a man stands still, the veins in his leg are filled with blood, the hydrostatic pressure gradient must prevail, and the pressure in his ankle vein may reach, say,

85 mm Hg. As he starts to walk, his muscles begin pumping the veins, which, because of their one-way valves, empty toward the heart. Segments of veins become empty or only partially filled, and the pressure drops. Measurements of ankle vein pressure were made a long time ago. Figure 4.7:1 shows the results of Pollack and Wood (1949). At first, the subject was motionless and the pressure in the ankle was 85 mm Hg. When the first step was taken, the pressure in the vein rose at first, and then dropped as the muscle relaxed. With successive steps, the pressure continued to drop until it reached, asymptotically, a maximum of 30 mm Hg and a minimum of 15 mm Hg. When the walking stopped, the pressure in the vein gradually rose to its previous level.

These events can be explained easily by squeezing and valve actions. One may deduce that when pressure in the vein decreased, the venous blood volume in the leg decreased, and indeed this was verified by plethysmography measurements. The squeezing action is effective only if the leg motion is fast enough to keep the vein partially empty. The relevant time constant is the time it takes the microcirculation to refill the vein (about 50 s at rest and 5 s during exercise). If the valves are incompetent, the action is somewhat less effective. But since the characteristic impedance of the vein closer to the heart is smaller than that closer to the microcirculation bed, squeezing of even valveless veins

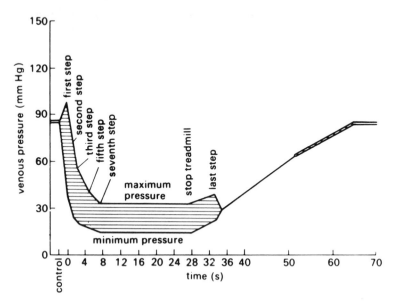

FIGURE 4.7:1. The change of pressure in an ankle vein of an erect subject as he stands and walks. The figure shows the maximum and minimum pressures. The subject is initially motionless, then walks, and finally again stands still. From Pollack and Wood (1949). Reproduced by permission.

can produce the effect just described, that is, pumping blood into the heart.

Respiratory Maneuvers

The indirect action of muscle on venous blood flow may be illustrated by respiratory maneuvers. Any movement that reduces thoracic pressure helps distend the thoracic vena cava. Movement that increases abdominal pressure squeezes blood in the abdominal veins into the thorax and heart. Thus, in quiet breathing enlargement of the thorax during inspiration reduces the intrathoracic pressure and causes blood to be drawn from the extrathoracic veins into the intrathoracic vessels and heart. Expiration has the opposite effect.

The Müller maneuver refers to a deep, sustained inspiration against a closed glottis (usually used by patients during x-ray photography to examine blood filling in the thorax). In such a maneuver, the alveolar and intrapleural pressures are reduced, as are the pressures in the right atrium and thoracic vena cava. The diaphragm muscle contracts and the diaphragm is lowered, so that the abdominal pressure and the pressure in the abdominal inferior vena cava are increased. Hence, upon initiation of Müller's maneuver, blood squeezed from the abdominal vena cava into the thorax, but the action is transient. Afterwards, a steady-state limiting flow of the type discussed in Section 4.5 (waterfall phenomena, sluicing flow) is established, and the flow is independent of the downstream (right atrium) pressure.

The Valsalva maneuver refers to a deep, sustained expiration against a closed glottis. In this case both the thoracic and abdominal pressures are increased, and blood flow in the thoracic and abdominal vena cava may be shut off. The flow is restored when the microcirculation into these venae cavae raises their pressures above the thoracic and abdominal pressures.

4.8 Self-Excited Oscillations

The best known self-excited oscillation is the flutter of flags in the wind. A very dangerous self-excited oscillation to avoid is the flutter of airplane wings in flight when a critical speed of flight is exceeded. Could veins, arteries, venous valves, heart valves, atria, and ventricles have self-excited oscillations due to blood flow? If they do, then the principle is the same: The elastic oscillation of the flag, the wing, or the blood vessel, induces fluid dynamical forces to act on the surface of the elastic body in such a way that energy can be imparted from the fluid to the elastic body, so that the kinetic energy of the body's oscillation can grow into a self-excited mode. The method of analysis of these systems is usually as follows: First, examine the structure, its mass and elasticity distribution, the flow, and the boundary conditions; see if the system can vibrate as an oscillator in the fluid flow; and determine its natural frequencies and the associated modes. Next, examine the fluid dynamic forces. The fluid forces (inertial, pressure, and shear) will be harmonic, but will be phase-shifted from the elastic deformation. Compute the work done by the fluid mechanical forces on the elastic body per cycle (integral of pressure × deflection × area, and shear stress × area × tangential movement). If the phase shift is such that this work done by the fluid is positive, then the oscillation will gain energy. If the work done is negative, then oscillation will subside. In the former case, self-excited oscillation will result

Natural heart valves and valves of the vein may flutter like a flag when the conditions are right. A type of artificial heart valve that operates as an airfoil may flutter like a wing at certain speeds of blood flow. A Starling resistor, as shown in Figure 4.5:1, may flutter in regime II, as illustrated in Figure 4.5:2(a). Further examples of the pressure–flow relationship in Starling resistors are given in Figure 4.8:1, which shows three curves derived from three different values of downstream resistance, with the chamber pressure, p_e, kept constant. Each curve is a continuous recording made while the flow rate was gradually increased. The self-excited oscillations that occur in some cases can be seen.

Ohba et al. (1984) examined the flow in a Starling resistor with respect to the effect of length and inner diameter of the downstream rigid pipe. Under certain downstream conditions, they found self-excited oscillations also in flow regimes in which the pressure drop increased with increasing flow rate. Further, Bertram et al. (1990, 1991, 1992) found a remarkable

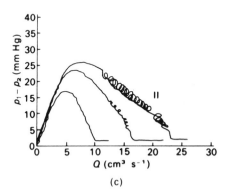

(c)

FIGURE 4.8:1. Flutter of a collapsible, thin-walled rubber tube in a certain range of flow. The tube was used as a Starling resistor, with the chamber pressure, p_e, kept fixed at 3.9 kPa. The outlet of the tube was connected to a flow resistance and then exposed to the atmosphere. The three curves were obtained by using three different resistors at the outlet. From Katz, Chen and Moreno (1969). Reproduced by permission.

variety of different types of oscillations over narrow regions of parameter space. Most oscillations observed were highly nonlinear, some were aperiodic, and often one type of oscillation changed back and forth to another type with some hysteresis. Cancelli and Pedley (1985) proposed a one-dimensional separated-flow model for collapsible tube flutter, taking into account the pressure loss due to flow separation, and hysteresis in the flow separation–recovery process, under the assumption that the tube elasticity can be represented by a "tube law" (see Sec. 4.3) and that *the longitudinal tension remains constant, unaffected by the radial and circumferential deformation.*

The last simplifying assumption just named is very severe from the point of view of plates and shells theory. In the conventional theory of elastic shells, it is always warned that this assumption of unchanging membrane stress is valid only if the deflection is very small compared with the thickenss of the tube wall. This restriction is more severe if the ends of the tube are clamped at fixed points in space, and less severe if the ends are free. In the Cancelli–Pedley theory, no limitation was placed on this restriction. The use of this assumption of constant membrane stress makes the mathematical problem linear. Removal of this assumption makes the problem nonlinear. Experience with "panel flutter" analysis in aeronautical engineering (Fung, 1954, 1958), however, has shown that the effect of nonlinearity is rich in consequences as far as flutter is concerned. The reason for this can be understood easily, as follows: With the ends of a tube fixed, a radial deflection increases the curved length of the wall and the longitudinal tension increases. The product of the tension and curvature is equivalent to a lateral load, which competes with the fluid pressure. In the meantime, because of the increase in tension, the elastic modulus of the blood vessel wall increases (Fung, 1993b, chapter 8). Thus, as the tube wall oscillates, the nonlinear elasticity and the product of tension and curvature interact with the fluid dynamic pressure with a variable phase shift. Hence, as we discussed at the beginning of this section, a mechanism of losing or gaining energy from the flow exists. If the vibration extracts energy from the blood stream, a self-excited motion ensues!

The mathematical problem of exploring the effect of nonlinearity due to the change in membrane stress caused by the lateral deflection is much simpler in the two-dimensional case of a channel than in the three-dimensional case of a circular cylindrical tube. Hence, a series of mathematical investigations was performed by Matsuzaki et al. in the case of a two-dimensional channel, as shown in Figure 4.8:2, which has a flexible segment sandwiched between two rigid segments. The entry section of length L_1 receives a flow from a large reservoir at a static pressure of P_r. The flexible section of fixed length L is subjected to an external pressure of P_e, and a set of nonlinear springs that makes the section obey the "tube law" discussed in Section 4.3 (relating the cross-sectional area of the channel to the transmural pressure). The longitudinal tension has a preset

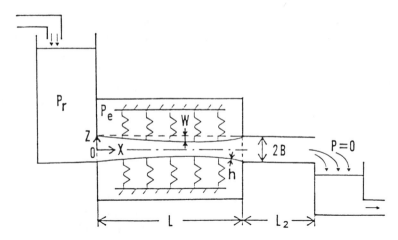

FIGURE 4.8:2. Two-dimensional channel model. From Matsuzaki (1995).

value when the deflection is zero and an elasticity for increased stretch, as discussed in the preceding paragraph. The exit section is rigid and of length L_2. The flow is mainly uniaxial, with a velocity component U in the x-axis direction. The velocity component v in the y-axis direction perpendicular to the x–z plane is assumed to be zero. In this case, if the deflection of the flexible wall is given by the equation $w = w(x, t)$, the length of the deflected surface in the longitudinal direction is

$$L + \Delta L = \int_0^L \left[1 + \left(\frac{\partial w}{\partial x}\right)^2\right]^{1/2} dx. \tag{1}$$

The membrane tension is

$$N_x = N_0 + Eh(\Delta L/L), \tag{2}$$

with E as Young's modulus, h as wall thickness, and N_0 as initial tension. The equation of equilibrium is

$$p_0 + p_w(x, t | w) - p_e = kw + N_x \frac{\partial^2 w/\partial x^2}{\left[1 + \left(\frac{\partial w}{\partial x}\right)^2\right]^{3/2}}, \tag{3}$$

where p_0 is the blood pressure in steady state without oscillation, p_w $(x, t | w)$ is the oscillatory fluid pressure that is a function of x and t and is induced by the deflection $w(x, t), p_e$ is the external pressure, and k is a spring constant of the elastic support sustaining the membrane. The nonlinearity due to longitudinal tension is given by the last term in Eq. (3). The fluid pressure, $p_w(x, t | w)$, could be nonlinear with respect to w, and k could be a function of w.

Matsuzaki and Fung (1977a,b, 1979) analyzed the stability of flow of a compressible or incompressible fluid in such a flexible channel with fixed ends with full aeroelastic equations, but calculated the fluid dynamical forces according to the potential flow theory. They found that the hydroelastic instability is in the nature of *divergence* (nonoscillatory) and not *flutter* (oscillatory). Thus, it is concluded that in order to explain flutter we cannot neglect the viscosity of the fluid and cannot use the linearized potential theory.

Matsuzaki and Matsumoto (1989a) then focused on flow separation and extended Cancelli and Pedley's (1985) work to the channel case, including the effect of nonlinear membrane tension induced by large deflection. They showed that the longitudinal tension is important in controlling flutter. They also derived stable deformed states of the membranes in certain conditions. In Matsuzaki (1995) these divergence and flutter modes are compared with Ohba et al.'s (1984) experimental results. Further results on the tube case are given by Jensen and Pedley (1989) and Jensen (1990).

In vivo evidence of self-excited flutter in veins is lacking. There are reports about seeing such oscillations in large veins during surgery. Gardner et al. (1977) reported the collapse of the human inferior vena cava near the diaphragm during deep inspiration. Perhaps an important contribution of these analyses lie in the light they throw on related problems, such as voice formation. Schoendorfer and Shapiro (1977) discussed the use of a collapsible tube as a prosthetic vocal source. Matsuzaki (1995) presented a theory of vocal chord and voice formation. These are illustrations of the value for a general understanding.

4.9 Forced Oscillation of Veins and Arteries Due to Unsteady Flow, Turbulence, Separation, or Reattachment

When a dynamic system is subjected to a transient external load, the system vibrates. Hence, when the blood flow is unsteady or turbulent, the fluid dynamic forces cause the blood vessel wall to vibrate. Unless the unsteadiness or turbulence is caused by the wall vibration itself, the mathematical structure of the problem is quite different from the self-excited oscillations discussed in the preceding section, which deals with unsteady flow due to wall motion itself. Flow separation from the wall at the bifurcation points of the blood vessel, or at a stenosis, is a major cause of turbulence or unsteadiness of blood flow in arteries and veins. The reattachment of flow to the wall may cause additional transient loading. Vibrations due to these causes may be considered as forced oscillations.

Possible separation of streamlines from a wall is illustrated in Figure 3.17:1 for a channel flow at a Reynolds number of 20,000, and in Figure 3.17:2 for flow in tubes at Reynolds numbers of 31, 70, and 185. Flow separation is usually caused by an adverse pressure gradient in the boundary layer in a divergent channel whose cross-sectional area increases in the direction of flow. That the pressure gradient is adverse in a divergent channel can be seen from Bernoulli's equation [Eq. (3) of Sec. 1.6], because in a divergent channel the velocity of flow decreases in the direction of flow, and therefore the pressure increases in the same direction. The adverse pressure gradient makes the velocity profile in the boundary layer flatter, and eventually causes the boundary layer flow to be unstable and the streamlines to separate from the wall.

Oscillatory Pressure Due to Boundary Layer Separation

Since flow separation is important not only to arterial and venous blood flow, but also to the generation of the Korotkoff sound in arteries, to post-stenotic dilatation, to forced expiratory flow in the airways, coughing, etc., we would like to know more about it. A review is given below.

One of the early experimental studies on a rigid pipe with a divergent boundary was made by Gibson (1910). Kline and his associates (1957, 1959) performed systematic experiments on two-dimensional divergent channels by means of a flow visualization technique in order to clarify the separation mechanism. They found four flow regimes: (1) No appreciable stall. (2) Large transitory stall. (3) Fully developed stall. (4) Jet flow. The term *stall* means backflow at a wall. For a fixed inflow velocity, these regimes are influenced by total divergence angle, the ratio of the wall length to throat width, and inflow turbulence. Their work, pertinent to diffuser designs, covers a Reynolds number range that is too large for physiological applications. For the moderate Reynolds number range of physiological interest, Young and Tsai (1973) measured the pressure drop of flow through tubes with constrictions, and Matsuzaki and Fung (1976, 1977a) studied the separation phenomenon in two-dimensional channels. The model used by Matsuzaki and Fung is sketched in Figure 4.9:1. It consists of a channel, one wall of which was movable to change the divergent angle. The inflow from the upper tank was very quiet (nonturbulent). The convergent segment of the channel was designed so that the flow had a uniform velocity profile at the throat. In each experiment the divergence angle θ was fixed, and the pressure differential between two points p_1, p_2 was measured. The length of the convergent segment, the throat, the divergent segment, and the distance between p_1 and p_2 were, respectively, 3.30, 2.54, 22.9, and 7.62 cm.

Figure 4.9:2 shows the experimental results on the pressure differential Δp (equal to $p_2 - p_1$ minus the hydrostatic head ρgz, where z is the vertical distance between p_1 and p_2) as a function of the speed of flow expressed as the Reynolds number at the throat:

$$N_R = \frac{\bar{u}W}{\nu},$$

FIGURE 4.9:1. Schematic diagram of a "two-dimensional" test apparatus for the study of flow separation in a divergent channel. From Matsuzaki and Fung (1976). Reproduced by permission.

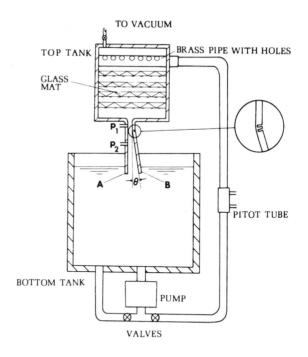

where $\bar{u}$, W, and v are, respectively, the mean velocity at the throat, the throat width, and the kinematic viscosity of the fluid, water. Note that when the divergence angle $\theta = 0$, Δp is negative according to Poiseuille's formula. For the divergence angle $\theta = 2.5°$, Δp is positive and increases parabolically with N_R, in accordance with the Bernoulli equation. For $\theta = 5°$, $7.5°$, and $10°$, Δp increases with N_R, first parabolically, then more or less linearly. For $\theta = 10°$, fluctuating flow was observed at higher flow rates. For $\theta = 12.5°$, there is a sudden drop of Δp at a point A if the flow speed is increasing, and a sudden increase at B if the flow is decreasing. At flow speeds greater than those at A and B, the amplitude of fluctuation is much greater. Beyond point C, the frequency of fluctuation decreases to a much lower value. If the free stream disturbances of the inflow are increased, points A, B, C would move to the left. This same type of hysteresis occurs at $\theta = 15°$. When $\theta = 17.5°$, the hysteresis loop becomes insignificant, but the Δp versus N_R relationship appears to be rather complex. Fluctuations in Δp were quite strong at all speeds when $\theta = 15°$ and $17.5°$.

The results of crossplotting the curves of Figure 4.9:2 for varying divergence angles but constant Reynolds numbers are shown in Figure 4.9:3. The divergence angle at which a discontinuous jump in Δp occurs is seen to vary with the Reynolds number.

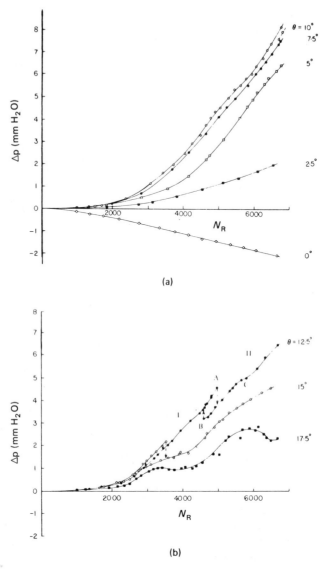

FIGURE 4.9:2. Variation of the pressure gradient ($\Delta p/\Delta x$) in a divergent channel with the speed of flow ($\bar{u}$) at several values of angle of divergence (θ). The speed of flow is expressed in terms of the Reynolds number $N_R = \bar{u}W/v$, where W is the width of the throat and v is the kinematic viscosity of fluid. The pressure difference (Δp) was measured at a distance $\Delta x = 7.62$ cm. From Matsuzaki and Fung (1976), by permission.

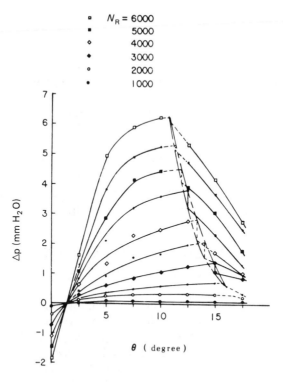

FIGURE 4.9:3. The variation of the pressure differential (Δp) with the divergence angle (θ) at several specific values of Reynolds number (N_R). The region in which a sharp change in pressure differential occurs is bounded by two chained curves. The interpolated portions of these curves are shown as dotted lines. From Matsuzaki and Fung (1976), by permission.

Thus, the complex character of flow in divergent channels is seen. However, this information is still insufficient for analysis of flutter because the boundary conditions are stationary in these experiments.

Korotkoff Sounds

Korotkoff sounds are used in measuring blood pressure by inflating a cuff on the arm. If a stethoscope bell is placed over the brachial artery, nothing can be heard at all in normal circumstances. However, when a wide cuff containing an inflatable bag is wrapped around the upper arm and the bag is inflated to pressures sufficient to collapse the artery and stop the flow, and then the pressure is allowed to fall slowly, characteristic sounds are heard from the brachial artery (the Korotkoff sounds). It has been shown by angiographic and ultrasound techniques that the first appearance of the sound coincides with the onset of blood flow through the collapsed artery segment. As the bag pressure falls, the sound becomes louder and more extended in time until it reaches a maximum intensity and begins to diminish, and then disappears. At a pressure just below that where the sound begins to diminish, there is a change in the character of the sound, known as *muffling*. The sound loses its ringing, staccato quality and becomes a thumping. It is not certain which particular sounds are associated with the cuff pressure being equal to the systolic and diastolic pressures in the heart, but by internationally agreed convention the cuff pressures at the appearance and the muffling of the Korotkoff sounds are taken as the systolic and diastolic pressures, respectively. McCutcheon and Rushmer (1967) and Ur and Gordon (1970) have provided basic experimental data on the Korotkoff sound. Anliker and Raman (1966), Pedley (1980), and Wild et al. (1977) have done extensive mathematical analyses on the subject.

4.10 Patency of Pulmonary Veins When the Blood Pressure Is Exceeded by Airway Pressure

Not all veins collapse when the internal pressure is smaller than the external pressure. Such is the case with pulmonary veins. Although the pulmonary venous vessel wall is thin and flexible, it is tethered by the interalveolar septa, which lend support and stabilize the vessel wall. This fact is very important to the understanding of pulmonary blood flow. Historically, it was in the lung that the "waterfall" phenomenon was first discovered. But until the stability of the venous vessels was proven by Fung et al. (1983), one could only speculate on where the "falls" are on the vascular tree. With the proof, we know that the falls are not on the venous tree. They are certainly not on the arterial tree. So if they exist, they must be located on the capillary blood vessels.

To demonstrate the patency of pulmonary veins, Fung et al. (1983) perfused the pulmonary blood vessels of the cat with a low-viscosity (20 cp) silicone elastomer (with 3% stenous ethylhexoate and 1.5% ethyl silicate hardening agent freshly added), After perfusing at a pressure of 25 mm Hg for 20 min, the flow was stopped and the perfusion pressure (p_v) was lowered to a desired level and held constant. In this selected static condition the silicone elastomer hardened. After hardening, the heart and lungs were removed and suspended in 10% KOH solution to corrode away the tissue. Then the arterial tree was gently separated from the venous tree, and casts of the two trees were obtained.

Patent Vessels

If a blood vessel is collapsed so that its internal cross-sectional area vanishes, then, after a process of tissue corrosion, that vessel will disappear from the tree. In lungs prepared in this manner, the capillary blood vessels dis-

FIGURE 4.10:1. A silicone elastomer cast of the venous tree of a cat lung whose veins were subjected to a pressure difference $p_v - p_A$ of -17 cm H_2O. Pleural pressure $= 0$ (atmospheric); airway pressure $= 10$ cm H_2O; blood vessel pressure $= -7$ cm H_2O.

appeared when the transmural pressure, $p_v - p_A$, was sufficiently negative. Here p_v stands for blood pressure in the vein and p_A stands for airway pressure. Figure 4.10:1 shows a venous tree cast of a cat lung prepared with a perfusion pressure (measured at the level of the left atrium) that was 17 cm H_2O lower than the alveolar gas pressure. Casts made at other negative values of $p_v - p_A$ look similar. We conclude, therefore, that the pulmonary veins do not collapse at these negative transmural pressures.

Smallest Open Veins, Branching Pattern of Venous Tree, and Elasticity and Compliance Constants of Venules

In our silicone rubber casts of the pulmonary venous tree, the smallest veins that remained open under negative transmural pressure are those smallest twigs that remained on the tree and did not fall off. Hence, by measuring the dimensions of these smallest twigs, we can determine the dimensions of the smallest vessels that did not collapse.

There is, however, the logical possibility that under increasing negative transmural pressure the larger vessels become smaller while the smallest vessels collapsed and disappeared. To make sure that this did not happen, we need to verify that the branching pattern of the entire tree did not change, that the same number of generations remained as the negative transmural pressure was increased.

There are two ways to describe vascular trees. One uses *generations*, analogous to a familial tree, and counts the number of offspring in each generation, (Weibel, 1963). The other uses the system of *Strahler* (1964), which is used in geography to describe rivers (Cumming et al., 1968). In the Strahler system, the smallest blood vessel is said to be of *order* 1. When two vessels of order 1 meet, the next larger vessel is called a vessel of *order* 2. Two order 2 vessels meet to form a larger vessel of order 3, etc. But if an order 1 vessel meets an order 2 vessel, the order number remains at 2. If a vessel of order 2 meets a vessel of order 3, the combined trunk's order remains at 3, and so on. If the branching pattern of the vascular tree were that of "symmetric" bifurcation, with every parent vessel yielding two equal offspring, then the ratio of the number of vessels of order n to that of $n + 1$, called the *branching ratio*, is 2. However, the pulmonary vascular tree does not bifurcate symmetrically, and the branching ratio is close to 3. In this situation the Strahler system gives a more accurate description of the branching pattern.

The results of our measurement are given in Tables 4.10:1 and 6.14:2 (p. 396). Table 4.10:1 (from Fung et al., 1983) presents the diameters and lengths of the small pulmonary veins. Table 6.14:2 (from Yen and Foppiano, 1981) presents the branching pattern of the entire venous tree of the right lung of the cat. It was found that there are 11 orders of pulmonary veins between pulmonary capillaries and the left atrium. The lungs of five cats were prepared at $p_v - p_A = -7$ cm H_2O, two cats were prepared at $p_v - p_A = -2$ cm H_2O, and one was prepared at $p_v - p_A = -17$ cm H_2O. All were found to have 11 orders of veins. Hence the structure of the venous tree was not changed when $p_v - p_A$ was varied from -2 to -17 cm H_2O; and the meaningfulness of identifying the smallest vessels in all casts as order 1 is assured.

The data in Table 4.10:1 show that the diameters of the smallest open venules (vessels of order 1) are in the range of 22 to 27 μm. If the diameter is assumed to be a function of the pressure $p_v - p_A$ and a linear regression is assumed, then by the least squares method we obtain the rate of decrease of the diameter (compliance constant) to be approximately 0.32 μm per cm H_2O of pressure, or 1.24% per cm H_2O based on the average diameter at $p_v - p_A = -2$ cm H_2O for the vessels of order 1.

Earlier, Sobin et al. (1978) measured the diameters of the *smallest* noncapillary blood vessels in each photomicrograph of the cat lung and found

TABLE 4.10:1. Diameters and Branching Ratios of the First Four Orders of Small Pulmonary Veins of the Cat and Lengths of Veins of Orders 3, 4, and 5*

		$p_v - p_A = -2$ cm H$_2$O		$p_v - p_A = -7$ cm H$_2$O	$p_v - p_A = -17$ cm H$_2$O
		Cat CFB (RLL)	Cat CFK (RLL)	Cat CET (LLL)	Cat CJG (RLL)
Diameter	Order 1	27.4 ± 10.1	24.2 ± 8.2	22.2 ± 4.1	23.6 ± 9.4
(μm)	Order 2	42.8 ± 10.0	41.0 ± 9.1	40.5 ± 7.3	37.0 ± 11.1
mean ± SD	Order 3	77.6 ± 19.8	73.9 ± 8.9	67.7 ± 11.5	71.0 ± 18.1
	Order 4	160.3 ± 39.5	142.0 ± 33.9	136.3 ± 44.7	126.1 ± 25.7
Diameter ratio		1.80	1.80	1.84	1.77
Length	Order 3	0.59 ± 0.27	0.56 ± 0.42	0.50 ± 0.21	0.58 ± 0.37
(mm)	Order 4	1.57 ± 0.60	1.47 ± 0.63	1.55 ± 1.11	1.48 ± 0.76
mean ± SD	Order 5	2.23 ± 1.58	2.44 ± 0.77	2.38 ± 1.74	2.69 ± 1.75
Length ratio		1.94	2.08	2.18	2.15
$\log \dfrac{N_i}{N_{i+1}} = B$		0.462	0.465	0.516	0.487
Average branching ratio		2.90	2.92	3.27	3.07

* $p_{PL} = 0$ (atmospheric); $p_A = 10$ cm H$_2$O; p_v = variable, measured at the level of the left atrium; N_i = the number of vessels of order i; p_v = blood pressure in vein; p_A = alveolar gas pressure; p_{PL} = pleural pressure.

that these smallest vessels have mean diameters 14.5, 16.6, 17.0, 18.2 and 21.0 μm when $p_v - p_A$ = (positive) + 0, 3, 10.25, 17, and 23.75 cm H_2O respectively. The compliance constant is 0.274 μm per cm H_2O, or 1.61% per cm H_2O, which is quite consistent with the mean value found earlier for the smallest venules. Hence, we conclude that the compliance constant of the venules does not change much as the transmural pressure $p_v - p_A$ changes from positive to negative values in the range −17 to −24 cm H_2O. Furthermore, since the smallest vessels must be smaller than the average order 1 vessels, we conclude that the smallest vessels listed in Table 4.10:1 are venules of order 1.

Table 4.10:1 lists also the *diameter ratio* (the ratio of the diameters of successive orders of vessels), the average length of the vessels of successive orders, the *length ratio* (the ratio of the average length of successive orders of vessels), and the *branching ratio*. The last quantity is obtained by plotting the number of branches in successive orders on semilog paper and finding the slope. The regression line yields

$$\log\left(N_i / N_{i+1}\right) = B,$$

where N_i is the number of vessels of order i and B is a constant. The *branching ratio* is then given by 10^B. It is seen that the branching ratio is about 3.0.

These results tell us that all the pulmonary veins remain open when the blood pressure is less than the airway pressure by as much as −17 cm H_2O. In contrast, the capillaries would collapse completely when $p_v - p_A$ is −1 cm H_2O (see Sec. 6.8). This difference in behavior between capillaries and venules is important for hemodynamics. It is the basis of our conclusion that the sluicing gates in the zone 2 condition are located at the junctions of the capillaries and venules (Fung and Sobin, 1972b, p. 473).

Compliance of Pulmonary Veins under Negative Transmural Pressure

To clarify the behavior of the pulmonary veins further, we measured the change of the diameters of these vessels with respect to blood pressure. The diameters were measured on x-ray films of isolated lungs after perfusing them first with saline, and then with $BaSO_4$ suspension. The method is presented in Yen et al. (1980).

Our results are shown in Figure 4.10:2, in which the distensibility of vessels in the diameter range of 100 to 200, 200 to 400, 400 to 800, and 800 to 1200 μm and with pleural pressures of −20, −15, −10, and −5 cm H_2O is presented. On the ordinate is shown the percentage change in diameter normalized with respect to D_{10}, the diameter at a $\Delta p = p_v - p_{PL}$ of 10 cm H_2O. This Δp is the difference between the venous perfusion pressure (p_v) and the pleural pressure (p_{PL}). The range 100 to 200, etc. indicates all vessels whose diameters fall in these ranges when $\Delta p = 10$ cm H_2O. At other Δp, the same vessels were followed. It appears from Figure 4.10:2 that a linear

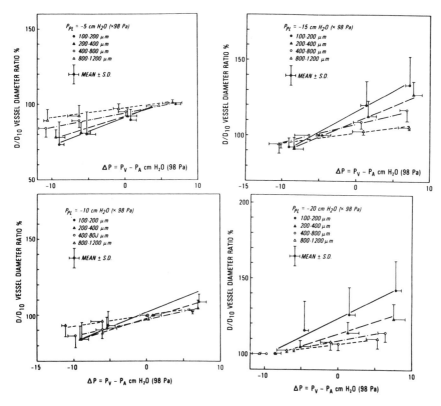

FIGURE 4.10:2. Percentage change in diameter of pulmonary veins of the cat as a function of blood pressure. The airway pressure, p_A, is zero. The values of the pleural pressure, p_{PL}, are noted in the figure. The vessel diameter is normalized against its value when $p_v - p_A$ is 10 cm H_2O, at which the vessel cross section is circular. The nominal size of a vessel is its diameter at $p_v - p_A = 10$ cm H_2O. From Yen and Foppiano (1981). Reproduced by permission.

relationship between pressure and diameter change exists for all these vessels in the ranges tested. In these figures, the vertical bars indicate the standard deviation (SD) of the D/D_{10} ratio of the vessels studied, whereas the horizontal bars indicate the SD of the pressure difference $p_v - p_A$ in these vessels. The scatter of $p_v - p_A$ was caused by the effect of gravity on the different heights of the branches, which resulted in different hydrostatic pressures in the vessels.

A linear regression line is assumed for each group of vessels (in a given diameter range), and slope and intercept of the regression line are determined by the method of least squares using the mean values of D/D_{10} and $p_v - p_A$. These regression lines are plotted in Figure 4.10:2. Their slopes are called the *compliance coefficients* and are listed in Table 4.10:2. The unit of the compliance coefficient is the percent change in diameter per cm H_2O

TABLE 4.10:2. The Compliance Coefficient (% Change of Diameter per cm H_2O Change in $p_v - p_A$) of Cat Pulmonary Veins when $p_v - p_A$ Is Negative*

Vessel diameter range (μm)	p_{PL} (cm H_2O)	Compliance coefficient
	−5	1.98
	−10	2.05
100–200	−15	2.79
	−20	2.44
	−5	1.83
	−10	1.44
200–400	−15	2.01
	−20	1.60
	−5	0.98
	−10	1.08
400–800	−15	1.16
	−20	0.93
	−5	0.79
	−10	0.71
800–1200	−15	0.57
	−20	0.58

*The airway pressure is 0 (atmospheric). The pleural pressure (p_{PL}) is listed. The compliance constant is the slope of the linear regression line over a range $-10 < p_v - p_A < 10$ cm H_2O determined for each group of vessels.

of pressure change. The table shows that smaller veins of the cat are more compliant than larger veins. Furthermore, for smaller veins (100 to 400 μm) inflation to a higher lung volume results in a higher compliance coefficient initially until a pleural pressure of −15 cm H_2O is reached, whereas a further increase in lung volume decreases compliance. For larger veins (400 to 1,200 μm), the compliance coefficients are smaller and their variation with the degree of inflation is not as significant.

The difference between the mechanical properties of pulmonary veins and the peripheral veins becomes evident if one compares the curves of Figure 4.10:2 with those of Figures 4.3:3, 4.4:2, and 4.4:3. Whereas the compliance of the peripheral veins increases greatly when the transmural pressure becomes negative, the compliance of the pulmonary veins changes hardly at all when the corresponding transmural pressure, $p_v - p_A$, changes from positive to negative.

Why Are the Pulmonary Veins So Stable?

The stability of the pulmonary veins against negative transmural pressure $p_v - p_A$ may be explained by the pull provided by the tension in the interalveolar septa on the veins. The relationship between the interalveolar septa and pulmonary venules is illustrated by the typical example shown in Figure 4.10:3. Larger vessels may be pulled by four or more interalveolar septa. A few appear to be tethered by only two septa. We have never seen an isolated, untethered vessel in the lung.

For each blood vessel, there are tethering interalveolar septa either parallel to or intersecting it at various angles. Tensile stresses prevail in the interalveolar septa as long as the lung is inflated and the alveoli are open, and they tend to distend the blood vessel. Between the successive septa the vessel wall is subjected to the pressure of the blood, p_v, in the inside and alveolar gas pressure, p_A, on the outside; the net pressure acting outward is $p_v - p_A$. When $p_v - p_A$ is negative, the transmural pressure acts inward.

As we have seen in Section 4.3, the stability of a circular cylindrical shell can be greatly improved by adding lateral and longitudinal supports. Figure 4.3:5 shows that the shorter the length of the cylinder L, and the larger the wave number n, the higher is the critical buckling pressure. Now, for a pulmonary vein, those interalveolar septa intersecting it perpendicular to its axis determine the length L (being the distance between successive septa, or approximately the alveolar diameter), whereas those tethering it parallel to its axis determine n (e.g., $n = 3$ for the vessel shown in Figure 4.10:3). If there were no tethering interalveolar septa, the buckling mode of the veins would correspond to $n = 2$, and L/R = vessel length/vessel radius, which is very large. The great stabilizing influence of tethering is then very clear from Figure 4.3:5.

Another factor that improves the stability of pulmonary veins by tethering arises from the nonlinear mechanical behavior of the blood vessel wall material. As is well known, the vessel wall has an exponential stress–strain relationship (see Fung, 1993b, chapter 8), and the incremental Young's modulus increases with increasing stress. The tethering interalveolar septa stresses the blood vessel wall and increases its elastic modulus. As we have seen in Sections 4.2 to 4.4, the critical buckling load is directly proportional to the elastic modulus; hence, an increase in the elastic modulus improves the stability.

Conclusions

We conclude that in the normal range of airway, pleural, and blood pressures in the zone II condition, the pulmonary veins, including venules, will not collapse. The elasticity of the blood vessels is such that the slope of the diameter versus $p_v - p_A$ curve remains almost constant for $p_v - p_A$ in the range of +10 to −17 cm H_2O. There is no sudden change of the pressure–diameter relationship as the transmural pressure changes from positive to negative values. This is in strong contrast to the curves shown in Figure 4.4:2. The cause for this patency is the tension exerted on the vessels by the interalveolar septa that are attached to the outer walls of the vessels.

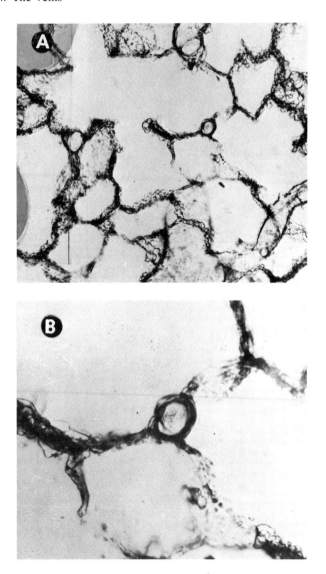

FIGURE 4.10:3. (A) Montage of photomicrographs of silicone elastomer–filled cat lung showing two venules in the central portion of the figure, each tethered by three interalveolar septa. Another venule, which was cut lengthwise, is seen at the lower right corner of the photograph. All noncapillary vessels in this figure are venous. Gelatin-embedded preparation, cresyl violet stain, 90-μm thick sections. (B) Enlargement of a portion of (a), showing tethering of a venule by three interalveolar septa. These figures are enlargements of a small portion of the montage presented in Figure 6.13:1, located at the upper right corner of that figure, near the letter "a." The dark shadows at the left border of (a) are marks of the arterial regions, as explained in Section 6.13. From Sobin et al. (1980). Reproduced by permission.

Although this conclusion is based on the cat lung, it should be applicable to other mammalian lungs because their basic structures are similar, as can be seen from the many illustrations in Krahl (1964), Miller (1947), von Hayek (1960), and Weibel (1963) listed at the end of Chapter 6.

Pulmonary veins that remain patent under the condition $p_v - p_A < 0$ have been reported earlier by Glazier et al. (1969), Macklin (1946), Howell et al. (1961), and Permutt and Riley (1963). For large veins the patency has been explained by Mead and Whittenberger (1964), and Lai-Fook (1979). The new data given in Tables 4.10:1 and 4.10:2 and Figures 4.10:2 and 4.10:3 show that the smallest venules follow the same trend.

4.11 Waterfall Condition in the Lung

In the preceding section we have shown that the pulmonary venules and veins remain patent when the blood pressure (p) is smaller than the gas pressure in the alveoli (p_A), provided that the lung is inflated. On the other hand, we know (see Chap. 6, Sec. 6.8) that the pulmonary capillary blood vessels can and will be collapsed if the transmural pressure $p - p_A$ is negative, see Sobin et al, 1972. It follows that the "waterfall" condition (see Sec. 4.1), if it occurs in the lung, will occur in the capillary blood vessels. The important fact about the capillaries in this regard is that they are very small. They are so small that the Reynolds number of flow and the Womersley number at the frequency of cardiac pulsation are much less than one (usually $<10^{-3}$). Hence, the inertial force in the blood vessel due to transient and convective accelerations is very small compared with the gradients of viscous stresses and pressure. In this condition, the phenomena of collapsing, choking, flow limitation, etc. discussed in Sections 4.2 to 4.7 can occur, but the vessel will not flutter (Fung and Sobin, 1972b). Flutter (see Sec. 4.8) is a phenomenon caused by a competition between the inertial, pressure, and elastic forces. For flow in capillary blood vessels the inertial force is almost absent, and the balance of the elastic and pressure forces results in a static deformation. Thus waterfall occurs, but statically.

In the lung, the waterfall phenomenon is a common occurrence, and the discussion presented here is necessary for its understanding. The phenomenon was first described by Permutt et al. (1962) and Permutt and Reiley (1963). Further clarification is given in Chapter 6.

Problem

4.13 Show that if a circle is deformed into an ellipse of the same circumferential length, then the ratio of the mean diameter of the ellipse (defined as the average of the major and minor axes) to the diameter of the circle is 0.998 when $a/b = 1.25$, 0.976 when $a/b = 1.75$, and 0.965 when $a/b = 2.0$, a being the major diameter and b being the minor diameter. In other words, if an inextensible circle is deformed into an ellipse with the major axis twice as large as the minor axes, the mean diameter of the ellipse is only 3.5% smaller than that of the circle. Therefore, if the blood vessel cross sections were circular originally and were deformed into elliptical shapes under a negative transmural pressure, then a collection of random samples of the projected widths will yield a mean diameter quite close to the mean diameter of the original circular vessels. The standard deviation will reflect the degree of deformation as well as the variation in the original diameter.

References

Anliker, M., and Raman, K.R. (1966). Korotkoff sounds at diastole—a phenomenon of dynamic instability of fluid-filled shells. *Int. J. Solids Struct.* **2**: 467–491.

Anliker, M., Wells, M.K., and Ogden, E. (1969). The transmission characteristics of large and small pressure waves in the abdominal vena cava. *IEEE Trans. Biomed. Eng.* **BME-16**: 262–273.

Attinger, E.O. (1969). Wall properties of veins. *IEEE Trans. Biomedical Eng.* **BME-16**: 253–261.

Bertram, C.D., Raymond, C.J., and Pedley, T.J. (1990, 1991, 1992). Mapping of instabilities for flow through collapsed tubes of differing length; and Application of nonlinear dynamics concepts. *J. Fluids Struct.* **3**: 125–153; **4**: 3–36; **5**: 391–426.

Cancelli, C., and Pedley, T.J. (1985). A separated-flow model for collapsible-tube oscillations. *J. Fluid Mech.* **157**: 375–404.

Caro, C.G., Pedley, T.J., Schroter, R.C., and Seed, W.A. (1978). *The Mechanics of the Circulation*. Oxford University Press, Oxford.

Conrad, W.A. (1969). Pressure flow relationship in collapsible tubes. *IEEE Trans. Biomed. Eng.* **BME-16**: 284–295.

Cumming, G., Henderson, R., Horsfield, K., and Singhal, S.S. (1968). The functional morphology of the pulmonary circulation. In *The Pulmonary Circulation and Interstitial Space* (A. Fishman and H. Hecht, eds.). University of Chicago Press, Chicago, pp. 327–338.

Flaherty, J.E., Keller, J.B., and Rubinow, S.I. (1972). Post buckling behavior of elastic tubes and rings with opposite sides in contact. *SIAM J. Appl. Math.* **23**: 446–455.

Flügge, W. (1960). *Stresses in Shells*. Springer-Verlag, Heidelberg.

Fung, Y.C. (1954). The static stability of a two-dimensional curved panel in a supersonic flow, with applications to panel flutter. *J. Aeronaut. Sci.* **21**: 556–565.

Fung, Y.C. (1958). On two-dimensional panel flutter. *J. Aeronaut. Sci.* **25**: 145–160.

Fung, Y.C. (1993a). *A First Course in Continuum Mechanics*, 3rd ed. Prentice-Hall, Englewood Cliffs, NJ.

Fung, Y.C. (1993b). *Biomechanics: Mechanical Properties of Living Tissues*, 2nd ed. Springer-Verlag. New York.

Fung, Y.C., and Sechler, E.E. (1960). Instability of thin elastic shells. In *Structural Mechanics*. Proc. of Symp. on Naval Structure Mechanics (J.N. Goodier and N. Hoff, eds.). Pergamon Press, New York.

Fung, Y.C., and Sechler, E.E. (eds.) (1974). *Thin Shell Structures: Theory, Experiment and Design*. Prentice-Hall, Englewood Cliffs, NJ.

Fung, Y.C., and Sobin, S.S. (1972a). Elasticity of the pulmonary alveolar sheet. *Circ. Res.* **30**: 451–469.

Fung, Y.C., and Sobin, S.S. (1972b). Pulmonary alveolar blood flow. *Circ. Res.* **30**: 470–490.

Fung, Y.C., Perrone, N., and Anliker, M. (1972). *Biomechanics: Its Foundations and Objectives.* Prentice-Hall, Englewood Cliffs, NJ.

Fung, Y.C., Sobin, S.S., Tremer, H., Yen, M.R.T., and Ho, H.H. (1983). Patency and compliance of pulmonary veins when airway pressure exceeds blood pressure. *J. Appl. Physiol: Respir., Exerc. Environ. Physiol.* **54**: 1538–1549.

Gardner, A.M.N., Turner, M.J., Wilmshurst, C.C., and Griffiths, D.J. (1977). Hydrodynamics of blood flow through the inferior vena cava. *Med. Biol. Eng. Comp.* **15**: 248–253.

Gibson, A.H. (1910). On the flow of water through pipes and passages having converging or diverging boundaries. *Proc. Roy. Soc. London* A, **83**: 366–378.

Glazier, J.B., Hughes, J.M.B., Maloney, J.E., and West, J.B. (1969). Measurements of capillary dimensions and blood volume in rapidly frozen lungs. *J. Appl. Physiol.* **26**: 65–76.

Holt, J.P. (1969). Flow through collapsible tubes and through in situ veins. *IEEE Trans. Biomedical Eng.* **BME-16**: 274–283.

Howell, J.B.L., Permutt, S., Proctor, D.F., and Riley, R.L. (1961). Effect of inflation of the lung on different parts of pulmonary vascular bed. *J. Appl. Physiol.* **16**: 71–76.

Jensen, O.E. (1990). Instabilities of flow in a collapsed tube. *J. Fluid Mech.* **220**: 623–659.

Jensen, O.E., and Pedley, T.J. (1989). The existence of steady flow in a collapsible tube. *J. Fluid Mech.* **206**: 339–374.

Kamm, R.D., and Pedley, T.J. (1989). Flow in collapsible tube: a brief review. *J. Biomech. Eng.* **111**: 177–179.

Katz, A.I., Chen, Y., and Moreno, A.H. (1969). Flow through a collapsible tube: Experimental analysis and mathematical model. *Biophys. J.* **9**: 1261–1279.

Kline, S.J. (1959). On the nature of stall. *J. Basic Eng., Trans. ASME* **81**, Ser. D: 305–320. See also Kline et al., *J. Basic Eng., Trans. ASME* **81**: 321–331.

Kline, S.J., Moore, C.A., and Cochran, D.L. (1957). Wide-angle diffusers of high performance and diffuser flow mechanisms. *J. Aeronaut. Sci.* **24**: 469–471.

Knowlton, F.P., and Starling, E.H. (1912). The influence of variations in temperature and blood pressure on the performance of the isolated mammalian heart. *J. Physiol.* (London) **44**: 206–219.

Lai-Fook, S.J. (1979). A continuum mechanics analysis of pulmonary vascular interdependence in isolated dog lobes. *J. Appl. Physiol.* **45**: 419–429.

Macklin, C.C. (1946). Evidences of increase in the capacity of the pulmonary arteries and veins of dogs, cats and rabbits during inflation of the freshly excised lung. *Rev. Canad. Biol.* **5**: 199–232.

Matsuzaki, Y. (1995). Unsteady flow in a collapsible tube: analysis and experiment. *Proc. Fourth China, Japan, USA, Singapore Conf. Biomechanics.*

Matsuzaki, Y., and Fung, Y.C. (1976). On separation of a divergent flow at moderate Reynolds numbers. *J. Appl. Mech.* **43**: 227–231.

Matsuzaki, Y., and Fung, Y.C. (1977a). Unsteady fluid dynamic forces on a simply-supported circular cylinder of finite length conveying a flow, with applications to stability analysis. *J. Sound Vibr.* **54**: 317–330.

Matsuzaki, Y., and Fung, Y.C. (1977b). Stability analysis of straight and buckled two-dimensional channels conveying an incompressible flow. *J. Appl. Mech.* **44**: 548–552.

Matsuzaki, Y., and Fung, Y.C. (1979). Nonlinear stability analysis of a compressible flow. *J. Appl. Mech.* **46**: 31–36.

Matsuzaki, Y., and Matsumoto, T. (1989). Flow in a two-dimensional collapsible channel with rigid inlet and outlet. *J. Biomech. Eng.* **111**: 180–184, 1989.

McCutcheon, E.P., and Rushmer, R.F. (1967). Korotkoff sounds: An experimental critique. *Circ. Res.* **20**: 149–169.

Mead, J., and Whittenberger, J.L. (1964). Lung inflation and hemodynamics. In *Handbook of Physiological* Sec. 3, *Respiration*, Vol. 1. (W.O. Fenn and H. Rahn, eds.). American Physiology Society, Washington, DC., pp. 477–486.

Moreno, A.H., Katz, A.I., Gold, L.D., and Reddy, R.V. (1970). Mechanics of distension of dog veins and other very thin-walled tubular structures. *Circ. Res.* **27**: 1069–1079.

Ohba, K., Yoneyama, N., Shimanaka, Y., and Maeda, H. (1984). Self-excited oscillations of flow in collapsible tube, I. *Technol. Rep. Kansai Univ.* No. 25, pp. 1–13.

Pedley, T.J. (1980). *The Fluid Mechanics of Large Blood Vessels*, Cambridge University Press, Cambridge.

Permutt, S., and Riley, R.L. (1963). Hemodynamics of collapsible vessels with tone: Vascular waterfall, *J. Appl. Physiol.* **18**: 924–932.

Permutt, S., Bromberger-Barnea, B., and Bane, H.N. (1962). Alveolar pressure, pulmonary venous pressure, and the vascular waterfall. *Med. Thorac.* **19**: 239–260.

Pollack, A.A., and Wood, E.H. (1949). Venous pressure in the saphenous vein at the ankle in man during exercise and changes in posture. *J. Appl. Physiol.* **1**: 649–662.

Schoendorfer, D.W., and Shapiro, A.H. (1977). The collapsible tube as a prosthetic vocal source. *Proc. San Diego Biomed. Symp.* **16**: 349–356.

Shapiro, A.H. (1977). Steady flow in collapsible tubes, *J. Biomech. Eng.* **99**: 126–147.

Sobin, S.S., Fung, Y.C., Tremer, H., and Rosenquist, T.H. (1972). Elasticity of the pulmonary interalveolar microvascular sheet in the cat. *Circ. Res.* **30**: 440–450.

Sobin, S.S., Lindal, R.G., Fung, Y.C., and Tremer, H.M. (1978). Elasticity of the smallest noncapillary pulmonary blood vessels in the cat. *Microvas. Res.* **15**: 57–68.

Sobin, S.S., Fung, Y.C., Lindal, R.G., Tremer, H.M., and Clark, L. (1980). Topology of pulmonary arterioles, capillaries and venules in the cat. *Microvac. Res.* **19**: 217–233.

Strahler, A.N. (1964). Quantitative geomorphology of drainage basin and channel networks. In *Handbook of Applied Hydrology: Compedium of Water Resources Technology* (V.T. Chow, ed.). McGraw-Hill, New York.

Timoshenko, S., and Gere, J.M. (1961). *Theory of Elastic Stability*, 2nd ed. McGraw-Hill, New York.

Ur, A., and Gordon, M. (1970). Origin of Korotkoff sounds. *Am. J. Physiol.* **218**: 524–529.

Weibel, E.R. (1963). *Morphometry of the Human Lung*. Springer-Verlag, Berlin.

Wexler, L., Bergel, D.H., Gabe, I.T., Makin, G.S., and Mills, C.J. (1968). Velocity of blood flow in normal human venae cavae. *Circ. Res.* **23**: 349–359.

Wild, R., Pedley, T.J., and Riley, D.S. (1977). Viscous flow in collapsible tubes of slowly-varying elliptical cross-section. *J. Fluid Mech.* **81**: 273–294.

Yen, R.T., and Foppiano, L. (1981). Elasticity of small pulmonary veins in the cat. *J. Biomech. Eng. Trans. ASME* **103**: 38–42.

Yen, R.T., Fung, Y.C., and Bingham, N. (1980). Elasticity of small pulmonary arteries in the cat. *J. Biomech. Eng. Trans. ASME* **102**: 170–177.

Yen, R.T., Zhuang, F.Y., Fung, Y.C., Ho, H.H., Tremer, H., and Sobin, S.S. (1983). Morphometry of cat's pulmonary venous tree. *J. Appl. Physiol.* In press.

Young, D.F., and Tsai, F.Y. (1973). Flow characteristics in models of arterial stenosis. I. Steady flow. *J. Biomech.* **6**: 395–410.

5
Microcirculation

5.1 Introduction

In the preceding chapters we studied the flow of blood in large blood vessels in which the main feature is a balance between the pressure forces, inertia forces, and forces of tissue and muscle. Only in the boundary layer are the viscous friction forces important. The boundary layer thickness grows with increasing distance from the entry section, and in a long tube the boundary layer on the wall eventually becomes so thick as to fill the entire tube. However, arteries divide and divide again. The vessel diameter decreases with each division, and soon the Reynolds and Womersley numbers become quite small, the entry length becomes only a small multiple of the vessel diameter, and the flow becomes fully developed even in relatively short vessels, and the analysis given in Section 3.2 becomes applicable.

The Reynolds and Womersley numbers tend to 1 at the level of the terminal arteries. Further downstream, in the arterioles, capillaries, and venules, both the Reynolds number and the Womersley number become less than 1. In these vessels, the inertia force becomes less important, and the flow is determined by the balance of viscous stresses and the pressure gradient. This is the realm of the microcirculation.

Small Reynolds numbers ($\ll 1$) are not the only characteristics of the microcirculation. At least three other features are unique:

a. The individuality of blood cells must be recognized.
b. Exchange of fluid and other matters between blood and tissue surrounding the blood vessel occurs.
c. The smooth muscle of the microvasculature operates to regulate the flow locally.

We have already discussed the role of red blood cells in the rheology of blood in microvessels in Chapter 5 of *Biomechanics: Mechanical Properties of Living Tissues* (Fung, 1993), and more will be said in the present chapter. Fluid exchange is discussed in Chapter 9 of *Biomechanics: Motion, Flow, Stress, and Growth* (Fung, 1990). The regulation of blood flow locally by arterioles is discussed in Section 5.13.

266

A striking characteristic of the capillary circulation is the continuous variation in flow of the red blood cells. Changes are seen in the velocity, direction of movement, hematocrit, and number of capillaries with blood cells. During the ebb and tide of flow, a continuous circulation of blood cells persists in certain channels, whereas in the majority of capillaries the circulation of blood cells is intermittent. The blood stream courses from an arteriole to a venule through two or three pathways at one time. At a subsequent time, the blood cells spill over into numerous side branches so that as many as 15 to 20 capillary vessels contain flowing red cells originating from a single arteriole. These features can be explained on the basis of mechanics, with the active contraction of smooth muscle on the one hand, and the basic laws of continuum mechanics on the other. As usual, we shall start with anatomy, then proceed to the mechanical properties of tissues, and finally, to system dynamics. The importance of the subject is unquestionable because the whole purpose of the heart and arteries is to carry blood to the capillaries to nourish the cells of the body.

Each organ has a unique microvascular bed. For example, the lung has arterial and venous systems that branch like trees, but the pulmonary capillary blood vessels have a topology that is entirely different from a tree. The heart has treelike coronary arterial and venous systems, but the topology of coronary capillaries is not treelike. To understand the flow in any organ requires information on its vasculature. We illustrate mechanical analysis of the lung, heart, and skeletal muscle in Chapters 6 to 8. In this chapter, we focus on some general features of microcirculatory mechanics.

5.2 Anatomy of Microvascular Beds

There are infinite variations in the detailed geometry of microvascular beds, just as there are infinite varieties of irrigation fields in agriculture. Figure 5.2:1 shows three major models. In (a) an artery supplies a number of parallel microvessels that drain into a vein. In (b) an arterial network and a venous network are nearly parallel. Many microvessels connect various points of the arterial network to points of the venous network. In (c), arteries supply a region that is drained by veins and lymph vessels. Three different things are sketched in (c). One is a *glomerulus*, whose capillaries have a large caliber and many local dilatations. The second is a *sinusoid*, which has a dense network and dilates easily. The third is a *sinus*, which is a wide capillary that does not form any meshes or networks. To help visualize the differences between these models, analogous agricultural irrigation channels are sketched on the side. The analog to blood pressure is elevation of the irrigation channels.

With the topology illustrated in these sketches, it is not surprising that the flow in all microvessels is not the same: Some carry stronger currents, some less. The channels in which flow is more robust than their neighbors

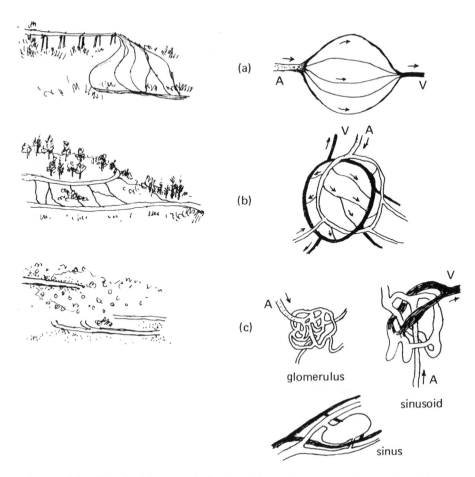

FIGURE 5.2:1. Sketch of four topological models of microvascular beds and their irrigation analogs.

are called *thoroughfares*. Some microvessels, especially in cases (a) and (c), are so short that they appear to connect an artery to a vein directly, to form shortcuts that bypass the capillary network: These are called *anastomoses*. Thoroughfares and anastomoses are important flow control gates. The draining of fluid is assisted by lymph vessels, which are analogous to subterranean drainage in agriculture.

These conceptual models can help us understand the structure of various organs. Figure 5.2:2 is a photomicrograph of rat cremaster muscle after intravenous injection of carbon to visualize the vascular pattern. The small arteries and veins run parallel to each other, the arterioles branch off at right angles, and the capillary networks are arranged so that the capillaries run parallel to the muscle fibers. The photograph shows two layers of muscle fibers that lie approximately at right angles to each other. Figure 5.2:3 is a schematic abstraction of the vascular structure.

Readers interested in seeing pictures of many different kinds of microvascular beds in various organs may consult Kaley and Altura (1977), Johnson (1978), and Wiedeman et al. (1981). The commonly used terminology is explained next.

Arterioles and Venules

Most authors base the terminology of arterioles and venules on histological characteristics. Thus a *terminal arteriole* is a final arterial ramification of

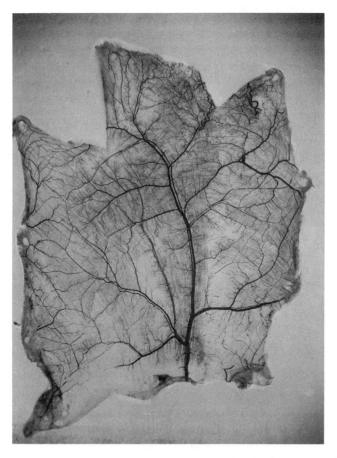

FIGURE 5.2:2. Photomicrograph of rat cremaster muscle after intravenous injection of carbon to visualize the vascular pattern. Note the parallel course of small arteries and veins, and the interarcading pattern of the terminal arterioles and venules (×20). From Smaje et al. (1970). Reproduced by permission.

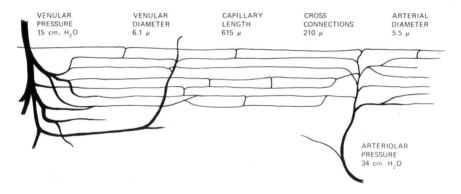

FIGURE 5.2:3. Abstraction of the cremaster muscle capillary blood vessel network, showing the type of arrangement of the vessels and mean values obtained in the experiments, Capillary density = 1,300/mm^2; distance between capillaries = 34 μm; capillary surface area = 244 cm^2/cm^3 muscle; red cell velocity = 700 μm/s; capillary filtration coefficient = 0.001 $\mu^3/\mu^2 \cdot s \cdot cm \cdot H_2O$ difference. From Smaje et al. (1970). Reproduced by permission.

10 to 50 μm diameter endowed with a continuous single layer of smooth muscle cells and scant supporting connective tissue. A *metarteriole* is one with discontinuous smooth muscle cells in the wall. The *precapillary sphincter* is the last smooth muscle cell along any branch of a terminal arteriole. An *arteriole* is a small artery ranging approximately between 10 and 125 μm in diameter, having more than one smooth muscle layer, and having a nerve association in the outermost muscle layer. The inner wall of these vessels is lined with endothelial cells that are flat in shape and abut a basement membrane and an elastic layer. Fine elastic fibers are interspersed among the muscle cells.

Analogously, a *postcapillary venule* is a vessel with a diameter of 8 to 30 μm formed by, and as a continuation of, two to four confluent venous capillaries, with an increasing number of pericytes as the lumen increases. A *collecting venule* is 10 to 50 μm in diameter, with one complete layer of pericytes and a complete layer of veil cells, and occasional smooth muscle cells. A *muscular venule* is 50 to 100 μm in diameter, with a thick wall of smooth muscle cells that sometimes overlap to form two layers. The confluence of muscular venules forms larger, *small collecting veins*, 100 to 300 μm diameter, with a prominent media of continuous layers of smooth muscle cells.

These descriptive definitions lack precision, and their use is not universally adhered to by all authors. Furthermore, not all organs have all these vessels. For example, neither the metarteriole, nor precapillary sphincter, have been found in skeletal muscle. Also, the precapillary arteries (i.e., arterioles) in the lung do not have layers of smooth muscles, and thus do not fit the definition of an arteriole given earlier. In these cases, counting generations from capillaries is the simplest way to identify the hierarchy of the vascular tree, as illustrated in Section 4.10 of Chapter 4, and Section 6.2 of Chapter 6.

Capillaries

The wall of a capillary blood vessel consists of a single layer of endothelial cells surrounded by a basement membrane, which splits to enclose occasional cells called *pericytes*. The pericytes are thought to have the potential to become smooth muscle cells.

The endothelial cells of the capillary wall have their edges closely apposed to one another. In electron microscopy, apposing cell membranes appear very close, with a gap of 10 to 20 nm. At certain points, the intercellular clefts are sealed by tight junctions or *maculae occludens*, which are formed by close apposition or fusion of the external leaflets of the plasmalemma. In certain areas (e.g., in the brain) these junctions form an uninterrupted seal, that is, *zonulae occludens*, preventing the passage of molecules with a radius of 2.5 nm or greater. They play an important role in determining the permeability of the endothelium to water and other molecules. A schematic drawing is shown in Figure 5.2:4.

The appearance of the endothelial cell lining of arteries and veins may be different in different organs, as illustrated in Figure 5.2:5. In the first, *continuous* type, the endothelial cells are joined tightly together. In the vessels of striated muscle, the cells may be quite flat and thin. In postcapillary venules, they may be cuboidal and form a thick layer. In the second, *fenestrated type* the endothelial cells are so thin that the opposite surfaces of their membrane are very close together, and form small circular areas known as *diaphragms of fenestrate*, approximately 25 nm thick, and of the

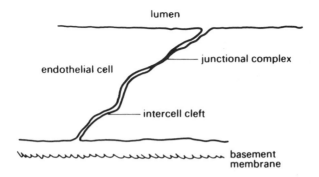

FIGURE 5.2:4. Schematic view of the endothelial layer of the capillaries showing the intercell cleft and junctional complex between adjacent cells viewed in cross section. From Caro et al. (1978, p. 369). Reproduced by permission.

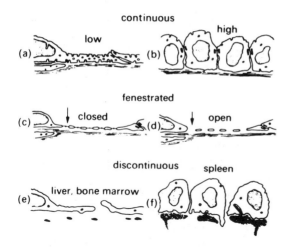

FIGURE 5.2:5. Schematic view of different types of endothelial lining, classified according to the degree of continuity. From Majno (1965). Reproduced by permission.

order of 100 nm across. Adjacent endothelial cells are still tightly joined. This type of vessel has been described in three groups of organs: (1) endocrine gland; (2) structures engaged in the production or absorption of fluids (e.g., renal glomerulus, choroid plexus of the brain, ciliary body of the eye, intestinal villus); and (3) retia mirabilia (e.g., renal medulla, fish swim bladder).

The third type of endothelium is the *discontinuous* type, in which there are distinct intercellular gaps and discontinuous basement membrane. These occur in those vessels commonly called *sinusoids*. They are common in organs whose primary functions are to add or to extract from the blood whole cells as well as large molecules and extraneous particles, for example, liver, spleen, and bone marrow.

Lymphatics

The *initial lymphatic capillary* has a single layer of endothelium surrounded by a basement membrane, and lacks smooth muscle in its walls. The lymphatic endothelial cells contain more microfilaments than the systemic capillary endothelial cells. The basement membrane of the initial lymphatic capillaries is incomplete, and there are filaments running through it to the endothelial cell membrane from the collagen and elastin fibers of the interstitial space. The intercellular junction of adjoining lymphatic endothelial cells is similar to that of blood capillaries, except that open junctions are seen in the initial lymphatic capillaries. The gap of the open junction may be some micrometers across, but is usually about 0.1 μm. These open gaps play an important role in the uptake of material and its retention in the lymphatics.

The initial lymphatic capillaries are blind sacs. They merge to form *collecting lymphatics*, which transport material to the vein. Collecting lymphatics have an endothelium, a basement membrane, and a layer of smooth muscle cells. They have valves that are either truncated cones or paired leaflets originating from opposite sides of the lymphatic wall. These valves ensure unidirectional movement of lymph.

How is the initial lymphatic sac filled? What holds the sacs open? What drives the fluid into the sacs and flow through the first valve? The answers to these questions are discussed in Section 8.11 of Chapter 8.

Interstitial Space

The complement to the space occupied by the blood vessels, lymphatics, and cells of a tissue is called the *interstitial space* of that tissue. It is mainly connective tissue, containing collagen, elastin, hyaluronic acid, and other substances, either bathed in some kind of fluid or embedded in a gel (see Chapter 9 of the companion volume, *Biomechanics: Motion, Flow, Stress and Growth*, Fung, 1990).

Nerves

Sympathetic fibers invest the aorta, large and small arteries and veins, and to a variable degree the networks of the arteriolar vessels and muscular venules in each organ. There appears to be no direct innervation of the capillary blood vessels and collecting venules, although nerve fibers may be found in the capillary region. Sympathetic fibers are usually superimposed on the smooth muscle layer of the blood vessel wall, but do not make direct synaptic contact with the vascular smooth muscle cells (Norberg and Hamberger, 1964).

Stimulation of sympathetic nerves generally constricts the blood vessel. The blood vessels in most organs maintain constriction during tonic sympathetic stimulation. However, the contribution of sympathetic nerves to resting or basal vascular tone varies greatly among different organs. Acute denervation has no effect on cerebral blood flow, whereas denervation in skin increases flow 5- to 10-fold. In intestine, liver, spleen, and lymph nodal tissue the response to a continued stimulus is transient and the vascular bed "escapes" from the constrictor influence within a few minutes.

If the effect of the sympathetic transmitter norepinephrine is blocked by an α-adrenergic blocking agent, a vasodilator influence acting through β-receptors is unmasked. When the β-receptor is also blocked, the catecholamine is without effect. Sympathetic activity may also cause vasodilation through effects on tissue metabolism, as seen in cardiac muscle and liver.

The parasympathetic system produces vasodilation in those tissues that are innervated by it. The vasodilation action involves either an alteration in metabolism or (in the salivary and sweat glands) release of an enzyme that diffuses into the tissue spaces.

The central nervous system has various sites where neural control mechanisms of vasomotor tone are located (Folkow and Neil, 1971; Sagawa et al., 1974). There are baroreceptors on the aortic arch and carotid sinus that inhibit medullary vasomotor activity when stimulated. The effect of the reflex mechanisms differs considerably from organ to organ. Reduction of pressure in the carotid sinus causes strong constriction in muscle and has a lesser effect on the kidney, liver, and intestine. Brain apparently is not significantly affected. Distention of pulmonary veins and the atrial junction also inhibit sympathetic tone arising from the medullary neurons. There are similar receptors in the atria, ventricles, and the lung.

The chemoreceptors in carotid sinus and the aortic arch affect the peripheral blood vessels in a complex manner. If other factors are held constant, stimulation of either the aortic or carotid bodies causes reflex vasoconstriction. However, the normal increase in ventilation that accompanies chemoreceptor activity overrides this reflex response and causes vasodilation.

In addition to control by the central nervous system, the vascular smooth muscle cells respond to changes in tension, stretch, and metabolities. These responses are responsible for moment-to-moment regulation of vascular tone in individual organs. This local control mechanism is discussed in Section 5.12.

5.3 Pressure Distribution in Microvessels

Systematic measurement of pressure distribution in small blood vessels is usually done by a probe originally designed by Wiederhielm et al. (1964). Lipowsky and Zweifach (1977, 1978), Nellis and Zweifach (1977), Zweifach (1974) and Zweifach and Lipowsky (1977) have made extensive measurements of pressure distribution in microvessels. The arterial to venous distribution of intravascular pressure and velocity is shown in a typical microvascular bed of cat mesentery in Figure 5.3:1. The pressure decreases rapidly in arterioles of diameters in the range of 10 to 35 μm. The decline in pressure within the true capillaries and postcapillaries is much more gradual.

The pressure gradient, dp/dL, can be measured by using two pressure probes inserted into two side branches of a given vessel and occluding flow in these side branches. The two pressure readings divided by the distance between the stations gives the pressure gradient. As is shown in Figure 5.3:2, the pressure gradient increases as vessel size decreases.

Figures 5.3:1 and 5.3:2 show the data obtained from cats whose central arterial pressures were in the normal range of 101 to 142 mm Hg. Data were also obtained in hypertensive cats with arterial pressures in the range of 142 to 194 mm Hg, and hypotensive cats with pressures of 60 to 100 mm Hg. These are shown in Figure 5.3:3. The pressure drop in arterioles in hypertensive cats was larger than in normals, whereas that in hypotensive cats

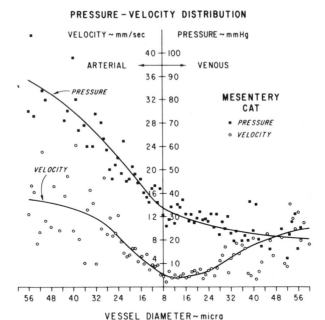

FIGURE 5.3:1. Arterial to venous distribution of intravascular pressure and velocity in the mesentery of the cat. Vessel diameter (abscissa) is taken to be representative of the functional position of each vessel in the microvascular network. Each data point represents the average value of three to five individual measurements at the abscissa (diameter) value. The solid curves are piecewise cubic spline fits of the data and are statistically representative of the arterial to venous trends. From Zweifach and Lipowsky (1977, p. 386). Reproduced by permission of the American Heart Association, Inc.

was smaller. Pressures in the capillaries and postcapillary vessels were similar in the two groups, with the difference between the mean pressures being only 3 to 5 mm Hg. Thus it appears that arterioles control the blood pressure in such a way that capillary pressure is maintained in the normal range while central arterial pressure may fluctuate. In the skeletal muscle of the cat, the pressure drop in the arterioles is even stronger (Fig. 5.3:4).

Shunts

It was observed that in the skin blood pressures in about 10% of the venules were sometimes higher than expected (approximately 70 mm Hg compared with the usual value of 30 mm Hg). This is taken to be evidence of the existence of shunts (*thoroughfare* channels) in the microvascular bed. These shunts are more direct low-resistance pathways connecting arterioles to venules.

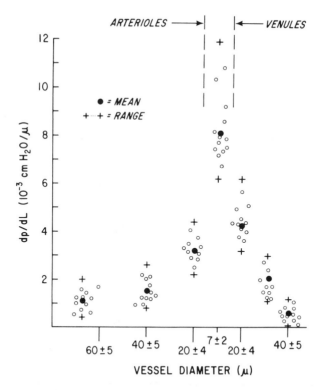

FIGURE 5.3:2. Reduction in pressure measured by two microprobes separated by the maximum distance possible between branches of cat mesentery. In arterioles, dp/dL was measured along a 1,500 to 2,500 μm interval; in venules, longer segments were usually available, 2,000 to 3,000 μm. In the capillary region, the interprobe separation was usually 200 to 350 μm. Values indicated cover a diameter range of $\pm 5 \mu$m for each of the categories listed, except for capillaries, in which the range is $\pm 2 \mu$m. From Zweifach (1974, p. 850). Reproduced by permission of the American Heart Association, Inc.

Temporal Variation

The pressure wave forms in the microvessels are illustrated in Figure 5.3:5. Note that cardiac oscillations have an amplitude of 1 to 2 mm Hg normally, and 2 to 4 mm Hg if the precapillary sphincter is dilated. Also note the random fluctuation of a period on the order of 15 to 20 s, with an amplitude of 3 to 5 mm Hg. A third type of pressure variation is not shown in the figure, it is more substantial and lasts longer, on the order of 10 mm Hg and 5 to 8 min, followed by a return to the steady-state condition in about 2 to 3 min.

That cardiac pulses must leave ripples in the capillaries is not surprising. The records shown in Figure 5.3:5 show them clearly. These waves are attenuated in the direction of propagation, reflecting the fact that in the capil-

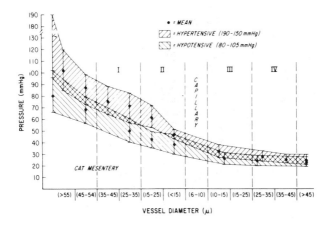

FIGURE 5.3:3. Pressure profiles for cats with extreme pressures (hypertension and hypotension). Note the trend for pressure to converge in the precapillary (15 to 20 μm) region. Postcapillary pressures were essentially the same for all groups (12 hypertensive and 8 hypotensive cats). From Zweifach (1974, p. 849). Reproduced by permission of the American Heart Association, Inc.

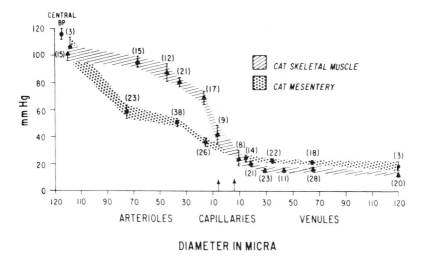

FIGURE 5.3:4. Micropressure distribution in two different tissues in relation to the average diameter of given vessels. Vertical bars express ±SEM, and the number of measurement is given in parentheses. Central BP indicates central arterial pressure and is an average from all experiments. From Fronek and Zweifach (1974). Reproduced by permission of the American Heart Association, Inc.

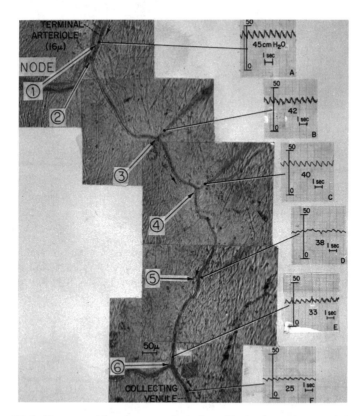

FIGURE 5.3:5. Photographic reconstruction of capillaries from the arteriole to collecting venule with direct recordings of pressure taken at the points indicated. The capillaries ranged from 7.5 to 9 μm in width. Note the persistence of the pulse pressure throughout. Each reading was taken at a side branch so as not to interrupt the flow through the feeding capillary. From Zweifach (1974, p. 858). Reproduced by permission.

laries viscous stress dissipates pressure fluctuations. By measuring the phase shift of a wave from the arterial end to the venular end of a capillary, the wave speed is estimated to be about 7.2 cm/s.

5.4 Pressure in the Interstitial Space

The pressure in the interstitial space is of great importance in the study of fluid movement in tissue. If we treat the capillary blood vessel wall as a semipermeable membrane, the rate of movement of water across the wall (from the blood into the interstitium) may be assumed to obey Starling's hypothesis:

$$\dot{m} = K\left(p_b - p_t - \pi_b + \pi_t\right), \tag{1}$$

where p stands for hydrostatic pressure, π stands for osmotic pressure, the subscript b stands for blood, t stands for tissue, $\dot{m}$ represents the volume flow rate per unit area of the membrane, and K is the *permeability constant*. In order to calculate the fluid transfer rate $\dot{m}$, we must know all five quantities, p_b, p_t, π_b, π_t, and K; they are equally important. Since fluid movement in the tissue space is very important to health (too much or too little fluid in the tissue means edema or dehydration), the measurement of p_t, and π_t has engaged attention of physiologists for a long time.

Methods and results of measuring the interstitial pressure in the tissues of various organs are discussed in Chapter 9 of *Biomechanics: Motion, Flow, Stress and Growth* (Fung, 1990).

5.5 Velocity Distribution in Microvessels

In large blood vessels the velocity of red blood cells can be measured by an electromagnetic flow meter because red cells are charged particles. For microvessels, a small electromagnetic flow meter is not available. For capillary blood vessels in a suitable preparation, it is possible to measure red cell velocity by optical means because red cells are optically dense in mercury arc emission light. By observing a red cell in a capillary blood vessel at two stations of known distance apart, and by measuring the time of travel between the stations, the velocity of the red cell can be calculated. This is called the *two-slit method*. In practice, the transmitted light at any station fluctuates with time because red cells are nonuniformly distributed. By recording the transmitted light intensity at two stations, one can calculate the cross-correlation of the two signals. Now, if the signal of the first station is delayed by a time period τ and then correlated with the signal from the second station, we can obtain the cross-correlation function as a function of τ. The value of delay time τ at which the correlation is the maximum is then taken to be the time required for red cells to move from the first station to the second. A division of the known distance between stations by the lapse time τ yields the average red cell velocity.

This method can be used with confidence in narrow capillaries in which the red cells move in single file. For larger vessels, the velocity of red cells in any cross section varies from a maximum at the center to near zero at the wall. Hence, if we measure red cell velocity, we have to explain carefully where it is measured and whether it is the centerline velocity or is an average over the entire cross section. Since the diameters of the red cells of most mammals are in the range of 6 to 9 μm, it is possible to select a microscope objective so that the red cells in a focal plane are clearly seen, while others not in focus are blurred. Thus, if the vessel is several cell diameters large, you can focus up and down, and get an impression of the velocity profile. Now, for quantitative work you will probably use a photoelectric multiplier to record the optical signals. The question then arises as to what is seen by the photoelectric tube. Wayland (1982) emphasized the fact that if a photomultiplier is used to "observe" the cells, the tube integrates the signals both from those cells sharply in focus and those cells not so sharply in focus. If you then focus up and down, the photomultiplier will give you a velocity profile that is *blunter* than the real one. This is an important factor to remember. For example, if laminar flow of blood in a glass tube is observed, Baker and Wayland (1974) found that the photoelectric tube yields a maximum velocity of red cells on the centerline of the tube equal to 1.6 times the mean velocity of flow, instead of the theoretical factor of 2.0. Lipowsky and Zweifach (1978) confirmed the result for tubes of diameter from 80 to 17 μm, and erythrocytes of diameter about 6 μm, but the scatter of the data was large. Lee et al. (1983) have shown that this ratio depends on light intensity and tube diameter.

Usually we wish to know the flow rate of whole blood. Thus, if red cell velocity is measured, we should know the ratio of the average red cell velocity to the mean velocity of whole blood. In a large blood vessel this ratio approaches 1, since the cells, which are small compared with the tube, will be convected with the plasma. In a microvessel, the ratio will be larger than 1 because the cells tend to concentrate in the center of the tube, where the velocity is higher than the average. If the vessel is so narrow that the red cells effectively plug up the tube, then the particle velocity and the whole blood velocity tend to be equal again, because the "leak back" (see *Biomechanics: Mechanical Properties of Living Tissues* (Fung, 1993, Chapter 5, Sec. 5.8, p. 195) of the plasma in the narrow gap between the red and the tube wall is small.

To clarify this ratio, a model experiment was done by Yen and Fung (1978) with gelatin pellets simulating red cells, and silicone fluid simulating plasma (see Sec. 5.6). The results are shown in Figure 5.5:1 and Table 5.5:1. The velocity ratio $v/\overline{V}$ (average particle velocity v divided by the mean velocity of whole blood $\overline{V}$) is seen to be independent of flow velocity but depends strongly on the ratio of the diameter of the cell, D_c, to that of the tube, D_t. When the hematocrit is 10% the velocity ratio is about 1.21 when the diameter of the tube is equal to that of the cell, ($D_c/D_t = 1.0$). It increases to 1.47 when D_c/D_t is 0.67, and to 1.49 when D_c/D_t is 0.5. Each point in Figure 5.5:1 represents an experimental measurement. The scatter of the data is seen to be very large, reflecting the fact that the exact value of the velocity ratio $v/\overline{V}$ depends on the incidental factor of particle configuration relative to the tube. The red cells deform severely in such a flow, and can assume all kinds of configuration—with their axes parallel to the cylinder axis, or perpendicular to it, or at some angles in between. For larger tubes Lee et al. (1983) have obtained a good empirical formula for $v/\overline{V}$.

From measurements on red cell velocity, one can compute the mean flow velocity using the velocity ratio described earlier. Multiplication by the

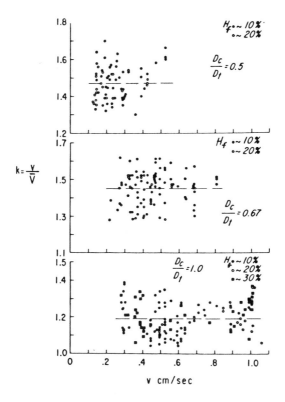

FIGURE 5.5:1. Experimental results of test model for $D_c/D_t = 0.5, 0.67$, and 1.0, where D_c is the diameter of the simulated red cells and D_t is the diameter of the tube. The ratio k between the average particle velocity v and the mean flow velocity $\overline{V}$ of the whole blood (simulated plasma and cells) is plotted against v for different feed-tube hematocrit H_f values. Dotted lines are mean values of k computed for combined values of H_f for each value of D_c/D_t. From Yen and Fung (1978). Reproduced by permission.

TABLE 5.5:1. Ratio of Particle Velocity to Mean Flow Velocity*

	Hematocrit			
D_c/D_t	10%	20%	30%	Combined data
1.0	1.21 ± 0.012	1.17 ± 0.012	1.20 ± 0.016	1.19 ± 0.008
0.67	1.47 ± 0.010	1.42 ± 0.014		1.45 ± 0.008
0.5	1.49 ± 0.014	1.45 ± 0.014		1.48 ± 0.011

*Experimental results of test model for $D_c/D_t = 1.0$, 0.67, and 0.5. Means ± SE of the ratio between particle velocity and mean flow velocity are listed, corresponding to feed-tube hematocrit, and are combined for each value of D_c/D_t. D_c is the diameter of the simulated red blood cell. D_t is the diameter of the tube. Data from Yen and Fung (1978).

TABLE 5.5:2. Rheological Parameter* for Several Microvessels of the Cat Mesentery

Vessel type	Diameter (μm)	Length (μm)	$\langle V_{RBC}\rangle^\dagger$ (mm/sec)	$\langle \Delta P\rangle^\ddagger$ (cm H$_2$O)	$\overline{V}$ (mm/sec)	μ (cP)	τ_w (dyne/cm^2)
Arteriole	45.0	632.0	20.8	4.3	13.0	3.25	75.0
Arteriole	23.0	429.0	23.2	6.7	14.5	1.75	88.0
Capillary	7.0	481.0	0.67	11.1	—†	5.17	39.6
Venule	17.0	667.0	2.00	2.10	1.25	2.23	13.1
Venule	54.0	403.0	7.50	0.71	4.69	3.36	23.3

* Based on time-averaged red cell velocities and pressure drops for the first 2 s of each record.
† Baker and Wayland correction factor not applied. Flow is based on $\langle V_{RBC}\rangle$.
‡ ΔP is the pressure difference measured in a blood vessel at two points separated by a distance equal to the "length" listed in the table. $\overline{V}$ is equal to $\langle V_{RBC}\rangle/1.6$ according to Baker and Wayland (1974), μ is the apparent viscosity of blood. τ_w is the shear stress acting on the vessel wall calculated according to Eq. (19) of Section 3.2.
From Lipowsky and Zweifach (1977). Reproduced by permission.

measured cross-sectional area of the vessel then yields the flow. Zweifach and Lipowsky's (1977) results for the average values of the velocity of flow in the cat mesentery microvessels are shown in Figure 5.3:1.

Using two pressure probes of the type discussed in Section 5.3 to measure the difference in pressure Δp at a distance ΔL apart, a dimensional analyzer to measure vessel diameter, and the aforementioned velocity diode, Zweifach and Lipowsky (1977) obtained data on the volume–flow rate, $\dot{Q}$, and the pressure gradient, $\Delta p/\Delta L$, in microvessels. From these data one can calculate the specific resistance R according to the formula,

$$-\frac{\Delta p}{\Delta L} = \dot{Q}R. \tag{1}$$

According to Poiseuille's formula [Eq. (17) of Sec. 3.2], we have

$$R = \frac{8\mu}{\pi a^4}. \tag{2}$$

Hence, one can calculate the apparent viscosity of the blood, μ, in these vessels. Table 5.5:2 shows the rheological parameters culculated from data from several microvessels of the cat mesentery.

The hematocrit distribution in microvasculature can be measured by optical means (Lipowsky et al., 1980). Figure 5.5:2 shows the hematocrit in mesenteric microvasculature of the cat. The very significant decrease in hematocrit in microvessels is noteworthy.

5.6 Velocity–Hematocrit Relationship

By observing capillary blood flow in vivo, it is noted that the flow is non-homogeneous and unsteady. The major part of the unsteadiness is not due to the heartbeat, with which it is nonsynchronous. It changes much more

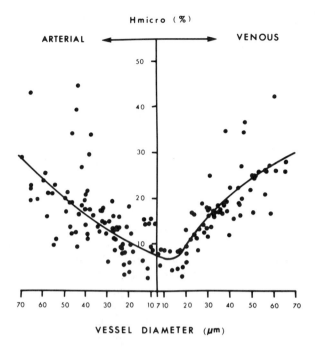

FIGURE 5.5:2. Arteriovenous distribution of hematocrit in the normal flow state of the mesenteric microvasculature of the cat. Each point represents hematocrit in a single unbranched micovessel of the indicated luminal diameter (abscissa). Hematocrit was obtained by in situ measurement of optical density and in vitro correlation with hematocrit following the technique of Jendrucko and Lee (1973). Hematocrits of microvessels smaller than $20\,\mu m$ diameter were obtained by rapid occlusion of the vessel and cell counting. The solid curve is a piece-wise cubic spline smoothing of the 150 individual measurements taken in 15 animals. Arteriolar hematocrits averaged 15.9%, and venular hematocrits averaged 17.3%. In contrast, systemic hematocrit averaged 33.8% ± 8.4% (SD). From Lipowsky et al. (1980). Reproduced by permission.

slowly, and without a definite period. In one period, red cells are seen rushing by; in another period, no red cells are seen at all. Since the volume fraction of red cells in whole blood is called *hematocrit*, the hematocrit in capillary blood vessels is thus said to be unsteady. One of the reasons for the unsteadiness of hematocrit in capillary blood vessels is the unsteadiness of velocity in these vessels. In *Biomechanics: Mechanical Properties of Living Tissues* (Fung, 1993, chapter 5, p. 186), it was explained that if a narrow capillary bifurcates into two equal daughters, the faster branch will get more red cells, thus having a higher hematocrit. Since any microcirculation circuit is derived from bifurcating blood vessels, there must exist a relationship between the distributions of blood flow velocity and hematocrit. Too many factors influence this relationship, however, and only in the simplest case can a definitive mathematical relationship be stated. This is the case with one vessel bifurcating into two equal daughters. This case has been studied by Fung (1973) theoretically, by Yen and Fung (1978) using a simulated model, by Gaehtgens (1980) using blood in micropipette and in vivo observations, and by Schmid-Schönbein et al. (1980a) in rabbit ear-chamber, with particular attention to leukocytes. In vivo observations are necessary, but accurate measurements are difficult, especially because a method to determine the capillary blood vessel diameter to a desired level of accuracy is still unavailable. Model experiments are simpler. By fixing a number of parameters, they yield precise mathematical relations more readily, and can be understood and analyzed more easily. In the following, we shall discuss a model approach.

In Yen and Fung's (1978) model, the plasma is simulated by a silicone fluid, the red blood cells are simulated by gelatin pellets, and the blood vessel is simulated by lucite tubes. The model was designed according to the principle of kinematic and dynamic similarity, at a Reynolds number in the range of 10^{-2} to 10^{-3}. At such a low Reynolds number, the angle of branching is unimportant, and therefore 90° was used. A schematic diagram of the apparatus is shown in Figure 5.6:1. It consists of a closed reservoir of simulated blood with an inverted **T** tube of lucite attached as shown. All branches are circular cylinders and of the same diameter.

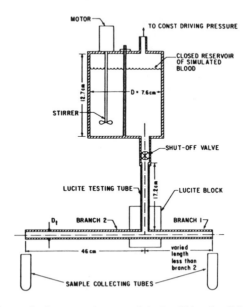

FIGURE 5.6:1. Schematic diagram of test model of red blood cell flow in a bifurcating capillary blood vessel. Three cylindrical tubes meet to form an inverted **T** joint. Flow in branch 1 is faster than that in branch 2. Inner diameters of test tubes, D_t, are 0.32, 0.48, and 0.64 cm. From Yen and Fung (1978). Reproduced by permission.

The hematocrit and velocity relationship in bifurcation branches was determined by allowing a gelatin-pellet suspension to flow through the **T** tube. The experiment consisted of collecting quantities of fluid from the two branches in a steady-state condition. Flow can be stopped by closing the shut-off valve after each test. Velocities in the two branches were varied by changing the lengths of the branches by cutting off a segment of one branch at a time. The pellet velocities were obtained by measuring the time required for a pellet to travel a known distance. The mean velocity of flow in each tube was computed from the volume discharge rate. The discharge hematocrit was measured by centrifuging the collected sample, and obtaining the ratio of the volume of the packed particles to the total volume. The tube hematocrit is computed according to the following consideration: In each tube the volume of the pellets that crossed a given cross section in unit time is equal to the product of (mean speed of tube flow) · (tube cross section) · (discharge hematocrit). It is also equal to the product (mean speed of pellets) · (tube cross section) · (tube hematocrit). Hence,

$$\frac{\text{Discharge hematocrit}}{\text{Tube hematocrit}} = \frac{\text{mean speed of pellets}}{\text{mean speed of tube flow}}. \tag{1}$$

Hematocrit of the feeding tube (H_f) was calculated as follows: Let $\overline{V}_f$ be the mean volume-flow rate in the feeding tube, $\overline{V}_1$ be that in branch 1, $\overline{V}_2$ be that in branch 2; and let v_f be the average particle velocity in the feeding tube, v_1 be that in branch 1, v_2 be that in branch 2. Without loss of generality, we may let $v_1 \geq v_2$ and $\overline{V}_1 \geq \overline{V}_2$. Then, by the principle of conservation of mass of the mixture and of the pellets, we have, since the cross-sectional area of all branches is the same,

$$\overline{V}_f = \overline{V}_1 + \overline{V}_2, \tag{2}$$

$$H_f v_f = H_1 v_1 + H_2 v_2. \tag{3}$$

The mean flow velocities $\overline{V}_f$, $\overline{V}_1$, $\overline{V}_2$ are related to the particle velocities v_f, v_1, v_2, as discussed in Section 5.5 and illustrated in Figure 5.5:1 and Table 5.5:1. It is seen that the ratio of particle velocity to mean flow velocity depends strongly on the ratio of cell to tube diameters, D_c/D_t, but does not vary significantly with hematocrit; nor does it vary with flow rate. Hence, for each value of D_c/D_t, we can replace Eq. (2) by

$$v_f = v_1 + v_2. \tag{4}$$

On substituting Eq. (4) into Eq. (3), we obtain the feed tube hematocrit, H_f,

$$H_f = \frac{H_1 v_1 + H_2 v_2}{v_1 + v_2}. \tag{5}$$

Figures 5.6:2 to 5.6:4 show the experimental results. In these figures, the ratio H_1/H_2 is plotted against the ratio v_1/v_2 for different feed-tube hematocrit H_f. The symbols v_1, v_2, H_1, H_2 are the particle velocities and tube hema-

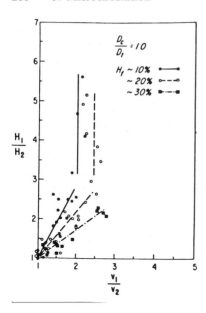

FIGURE 5.6:2. Experimental results of test model for $D_c/D_t = 1$. H_1/H_2 is plotted against v_1/v_2 for three different feed-tube hematocrits, H_f. From Yen and Fung (1978). Reproduced by permission.

tocrits in branches 1 and 2, respectively. It is seen that for narrow capillaries (with the cell diameter/tube diameter = $D_c/D_t = 1.0, 0.67,$ and 0.5), the branch with faster flow (branch 1) will have more cells. In general, for velocity ratios sufficiently smaller than a critical value, the hematocrit ratio can be expressed by the linear relationship given by

$$\frac{H_1}{H_2} - 1 = a\left(\frac{v_1}{v_2} - 1\right). \tag{6}$$

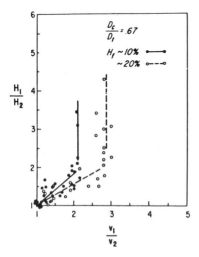

FIGURE 5.6:3. Experimental results of test model for $D_c/D_t = 0.67$. H_1/H_2 is plotted against v_1/v_2 for two different feed-tube hematocrits, H_f. From Yen and Fung (1978). Reproduced by permission.

FIGURE 5.6:4. Experimental results of test model for $D_c/D_t = 0.5$. H_1/H_2 is plotted against v_1/v_2 for two different feed-tube hematocrits, H_f. From Yen and Fung (1978). Reproduced by permission.

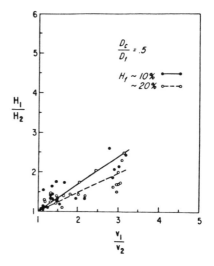

In Eq. (6), the dimensionless constant, a, depends on a number of factors, the most important of which are (a) the ratio of cell diameter to tube diameter, (b) the shape and rigidity of the pellets, and (c) the hematocrit in the feeding tube.

Figures 5.6:2 and 5.6:3 also show that for velocity ratios beyond a critical value nearly all the cells flow into the faster branch. The smaller the feeding tube hematocrit is, the smaller is the critical velocity ratio at which this phenomenon occurred. The critical velocity ratio lies in the range of 2 to 3.0 when $D_c/D_t = 1.0$, with the exact value depending on the feed hematocrit. The critical velocity ratio becomes higher when D_c/D_t decreases.

Theoretical Analysis of the Experiment

An approximate analysis of the forces that act on a particle at a bifurcation point will clarify the theme of this section. We shall show that if the mean velocities of flow in the two daughter branches are $\overline{V}_1$ and $\overline{V}_2$, with $\overline{V}_1 > \overline{V}_2$, then the resultant forces due to pressure and shear stress are both proportional to $\overline{V}_1 - \overline{V}_2$, and act in the direction of the daughter tube with higher velocity, $\overline{V}_1$.

First, consider disk-shaped flexible pellets. Such a pellet can assume all kinds of orientation in the tube. Let us consider two configurations, one in which the plane of the disk is perpendicular to the axis of the tube and another in which the plane of the disk is parallel to the axis of the tube. For conciseness, we say that the first case is plugging and the second is edge-on.

Consider the plugging case. Since the local velocity of flow must vanish on the wall of the tube and is maximal at the center, the flexible pellet must deform. Consistent with the principle of minimum potential energy, the deformation of a thin disk will be planar, that is, its midplane will deform into a developable cylindrical surface, which requires no membrane strain in the midplane of the disk. Most of the energy of deformation will be used in bending the midplane, and very little in stretching. (If h represents the thickness of the disk, then the bending rigidity is proportional to h^3, whereas the stretching rigidity is proportional to h. When $h \to 0$, it becomes much easier to bend than to stretch.) Let us sketch such a deformed pellet in a branching tube as shown in Figure 5.6:5(a) or 5.6:5(b). Let us consider three sections of the tubes as indicated by dotted lines in the figure: one in the mother tube and two in the daughter tubes, all at a distance L from the point of bifurcation. In a flow of a homogeneous viscous fluid before the arrival of the pellet, let the pressures at these sections be P_0, P_1, P_2,

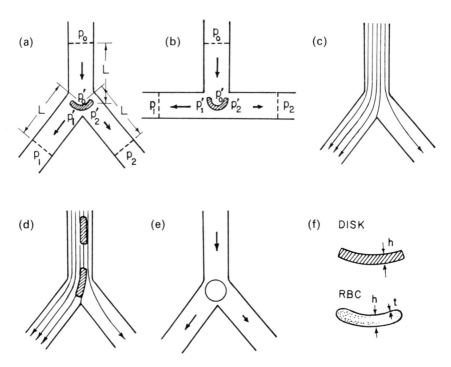

FIGURE 5.6:5. Illustration of the features of modeling. Flow is directed downward. Velocity in left branch (tube 1) is faster than that in right branch (tube 2). (a) A plugging pellet at the fork. (b) Similar to (a), but with the branching angle equal to 90°. (c) Streamline pattern in the case $\overline{V}_1 = 2\overline{V}_2$, as seen in the middle section. (d) Pellets flowing edge-on, and astride of the dividing stream surface. (e) Flow of a rigid spherical particle at the fork. (f) Cross sections of a pellet in the model and in a red blood cell. The pellet is a solid disk of homogeneous elastic material. A red blood cell is considered to be a thin shell filled with a liquid. From Yen and Fung (1978). Reproduced by permission.

respectively. Obviously $P_0 > P_2 > P_1$ on account of the assumption that $\overline{V}_1 > \overline{V}_2$. When the pellet arrives at the junction, let the pressures acting on the pellet surface facing the three branches be denoted by P_0', P_1', P_2', respectively, as indicated in the figure. Now, the values of P_0', P_1', P_2' depend on how tightly the pellet plugs up the flow. If the flow is momentarily plugged completely, then we have

$$P_0' = P_0, \quad P_1' = P_1, \quad P_2' = P_2. \tag{7}$$

If the flow can leak past the pellet, then $P_0' < P_0$, $P_1' > P_1$, $P_2' > P_2$. But they are related such that $P_1' < P_2' < P_0'$; and $P_0' = P_1' = P_2'$ only in the limiting case of no pellet. The force that acts on the pellet that pulls the pellet into branch 1 is equal to the difference of pressures acting on the two sides of the pellet multiplied by the projected area in branch 1; that is, $(P_0' - P_1')\pi D_c^2/8$. The pressure difference that pulls the pellet into the branch 2 is $(P_0' - P_2')\pi D_c^2/8$. Here D_c is the diameter of the pellet. The resultant force due to the pressure difference is F_p. In the case shown in Figure 5.6:5(b),

$$F_p = \left(P_0' - P_1'\right)\frac{\pi D_c^2}{8} - \left(P_0' - P_2'\right)\frac{\pi D_c^2}{8} = \left(P_2' - P_1'\right)\frac{\pi D_c^2}{8}, \tag{8}$$

and is directed into branch 1 because $P_1' < P_2'$. We can convert this expression into the mean velocities of flow as follows. Assuming Poiseuille flow, we have, from Eq. (18) of Section 3.2,

$$\overline{V} = \frac{D_t^2}{32\mu}\frac{\left(P_0' - P_0\right)}{L}, \quad \overline{V}_1 = \frac{D_t^2}{32\mu}\frac{\left(P_0' - P_1\right)}{L}, \quad \overline{V}_2 = \frac{D_t^2}{32\mu}\frac{\left(P_0' - P_2\right)}{L}, \tag{9}$$

where μ is the apparent viscosity of the fluid. If the flow is momentarily plugged so that Eq. (7) applies, then by using Eq. (7) in (9) and (8) we obtain

$$F_p = 4\mu L \pi \frac{D_c^2}{D_t^2}\left(\overline{V}_1 - \overline{V}_2\right). \tag{10}$$

This shows that the resultant force is proportional to the velocity difference of the two branches, and to the square of the particle-to-tube diameter ratio.

Next, consider the edge-on case. Here it is necessary to consider the shear stress and velocity gradient, and hence the velocity distribution. The easiest way to understand the velocity distribution in a branching flow is to look at the streamline pattern. A sketch of the streamline of a flow of a homogeneous fluid in a branching tube is shown in Figure 5.6:5(c). A streamline is a curve whose tangent is parallel to the velocity of a particle of the fluid lying at the point of tangency. A stream tube is a tube whose wall is composed of streamlines. Fluid contained in a stream tube will not leave the tube in a steady flow. Hence, the rate of volume flow is a constant in each stream tube. This consideration tells us how to quantitize the flow field by streamlines. Let the cross section of the mother tube be divided in to N equal areas. Let N stream tubes enclose these areas. The walls of these tubes are uniformly spaced far away from the bifurcation point. As the fluid velocity becomes nonuniform, the spacing of the stream tubes will vary. The higher the velocity, the more crowded the tubes will be. Figure 5.6:5(c) shows the distribution of the stream tubes of a flow of a homogeneous fluid in the case in which $\overline{V}_1 = 2\overline{V}_2$. Note that there is a dividing stream surface: All fluid on its left goes into tube 1, and all on its right goes into tube 2.

Now, let there be an edge-on pellet in the tube. If it lies entirely to the left of the dividing stream surface, it will flow into tube 1. If it lies to the right, it flows into tube 2. The borderline cases, in which the pellet lies astride the dividing stream surface, as shown in Figure 5.6:5(d), is the one that needs attention. In this case the streamlines are crowded on the left-hand side of the pellet, indicating that the flow velocity is high there. On the right the streamlines are rarified. This feature is accentuated especially in the neighborhood of the bifurcation point. The velocity gradients on the two sides can be assumed to be proportional to $\overline{V}_1/D_t$ and $\overline{V}_2/D_t$. The resultant shear force is therefore

$$F_s = \text{const} \cdot \frac{D_c^2}{D_t}\left(\overline{V}_1 - \overline{V}_2\right), \tag{11}$$

and is directed toward tube 1.

Pellets in other configurations and of other shapes will be acted on by stress resultants similar to those given by Eqs. (10) and (11), though different in magnitude. Thus, it is seen that the net pressure and shear forces tend to pull pellets to the faster side.

This analysis is simple because many details are left in a qualitative state. A more refined analysis must be based on the solution of the Stokes equation with appropriate boundary conditions. The question of simulating the flexibility of the red blood cells by pellets is discussed in Yen and Fung (1978). A more compact method of presentation with additional results is given by Schmid-Shönbein et al. (1980a,b).

Why Is the Hematocrit So Low in Capillaries?

Very careful measurements have been done on the hematocrit in capillary blood vessels in the mesentery and striated muscles of the rat (see Klitzman and Duling, 1979, and the literature reviewed therein). The results show that the hematocrit in microvessels is low compared with the hematocrit in large vessels. Typically, with an arterial hematocrit of 50% and a discharge hematocrit of $49 \pm 5\%$, the hematocrit in the capillary blood vessels varies from 10% to 26%, with an average of 18%. It is easy to explain the heterogeneity from fluid dynamics point of view, and it is clear that the capillary hematocrit should be lower than the arterial hematocrit, because the red cells move in the central region of the capillaries, whereas the plasma must contact the vessel wall where the velocity is zero. One could compute the required thickness of the retarded plasma layer at the vessel wall in order to explain the low capillary hematocrit, but one finds the result improbably large. Desjardins and Duling (1990) hypothesized that the glycocalyx on the endothelial cell membrane plays a role in regulating the capillary hematocrit. They showed that perfusing the vessel with a solution containing heparinase raised the tube hematocrit at least twofold without a significant change in red cell velocity. Other evidence is awaited.

5.7 Mechanics of Flow at Very Low Reynolds Numbers

As we have said earlier, blood flow in micro–blood vessels is characterized by small Reynolds and Womersley numbers. If we assume a velocity of flow of 1 mm/s, a vessel diameter of 10 μm, a viscosity of blood of 0.02 poise, a density of 1 g/cm^3, and a heart rate of 2 Hz, then the Reynolds number is 0.005 and the Womersley number is 0.0126, both much smaller than 1. This smallness is typical in microcirculation. At such small Reynolds and Womersley numbers, the viscosity effect becomes predominant. By comparison, the inertial forces due to transient and convective acceleration become negligible. Thus the Navier–Stokes equations [see Eq. (18) of Section 2.6] are simplified into

$$\mu\nabla^2 u = \frac{\partial p}{\partial x}, \quad \mu\nabla^2 v = \frac{\partial p}{\partial y}, \quad \mu\nabla^2 w = \frac{\partial p}{\partial z}, \tag{1}$$

where u, v, w are the velocity components in Cartesian coordinates x, y, z, and

$$\nabla^2 = \frac{\partial^2}{\partial x^2} + \frac{\partial^2}{\partial y^2} + \frac{\partial^2}{\partial z^2}.$$

If we differentiate the first of Eq. (1) with respect to y, the second of Eq. (1) with respect to x, and subtract, we obtain

$$\mu\nabla^2\left(\frac{\partial v}{\partial x} - \frac{\partial u}{\partial y}\right) = 0. \tag{2}$$

The quantities

$$\xi = \frac{\partial w}{\partial y} - \frac{\partial v}{\partial z}, \quad \eta = \frac{\partial u}{\partial z} - \frac{\partial w}{\partial x}, \quad \zeta = \frac{\partial v}{\partial x} - \frac{\partial u}{\partial y}, \tag{3}$$

are the Cartesian components of the vorticity. Hence,

$$\nabla^2\xi = 0, \quad \nabla^2\eta = 0, \quad \nabla^2\zeta = 0. \tag{4}$$

The equation of continuity of an incompressible fluid is

$$\frac{\partial u}{\partial x} + \frac{\partial v}{\partial y} + \frac{\partial w}{\partial z} = 0. \tag{5}$$

Differentiating the three Eqs. (1) with respect to x, y, z, adding, and using Eq. (5), we obtain

$$\nabla^2 p = 0. \tag{6}$$

Applying the Laplacian operator ∇^2 to Eqs. (1) and using Eq. (6), we have

$$\nabla^4 u = 0, \quad \nabla^4 v = 0, \quad \nabla^4 w = 0. \tag{7}$$

Thus, when the Reynolds and Womersley numbers are very small, the pressure p satisfies the Laplace Eq. (6), and the components of velocity satisfy the biharmonic Eq. (7). Equation (1) is known as the *Stokes equation*, and a flow obeying this equation is called a *Stokes flow*.

Stokes' Solution for a Falling Sphere

Consider a sphere of radius a falling in a fluid of viscosity μ. It is convenient to consider the sphere to be stationary, and the velocity of the flow relative to the sphere to be U at infinite distance from the sphere. Assume that the flow field is axisymmetric. If cylindrical coordinates (r, θ, z) are used, with the z axis vertical (Fig. 5.7:1), then

$$u = c\cos\theta, \quad v = c\sin\theta, \quad w = w, \tag{8}$$

in which c is the velocity component in the r direction. c and w are functions of r and z only. It follows from Eqs. (8) and (3) that

$$\xi = -2\omega\sin\theta, \quad \eta = 2\omega\cos\theta, \quad \zeta = 0, \tag{9}$$

in which 2ω is the *resultant vorticity* given by

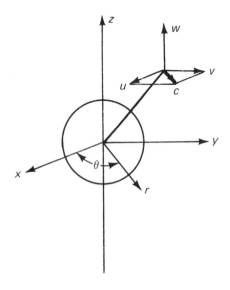

FIGURE 5.7:1. Use of cylindrical–polar coordinates for the falling sphere problem.

$$2\omega = \frac{\partial c}{\partial z} - \frac{\partial w}{\partial r}. \tag{10}$$

From Eq. (9) one obtains

$$\frac{1}{2}\nabla^2 \xi = -\left(\frac{\partial^2}{\partial z^2} + \frac{\partial^2}{\partial r^2} + \frac{1}{r}\frac{\partial}{\partial r} + \frac{1}{r^2}\frac{\partial^2}{\partial \theta^2} \right) \omega \sin\theta$$

$$= -\sin\theta \left(\frac{\partial^2}{\partial z^2} + \frac{\partial^2}{\partial r^2} + \frac{1}{r}\frac{\partial}{\partial r} - \frac{1}{r^2} \right) \omega$$

$$= -\frac{\sin\theta}{r}\left(\frac{\partial^2}{\partial z^2} + \frac{\partial^2}{\partial r^2} - \frac{1}{r}\frac{\partial}{\partial r} \right) r\omega = -\frac{\sin\theta}{r}L(r\omega) \tag{11}$$

if

$$L = \frac{\partial^2}{\partial z^2} + \frac{\partial^2}{\partial r^2} - \frac{1}{r}\frac{\partial}{\partial r}. \tag{12}$$

Similarly,

$$\frac{1}{2}\nabla^2 \eta = \frac{\cos\theta}{r}L(r\omega). \tag{13}$$

Thus, the three Eqs. (4) can be represented by the single equation (since $\zeta = 0$),

$$L(r\omega) = 0. \tag{14}$$

If we now introduce the *stream function* ψ so that

$$w = \frac{1}{r}\frac{\partial\psi}{\partial r}, \quad c = -\frac{1}{r}\frac{\partial\psi}{\partial z}, \tag{15}$$

we have

$$-\omega = \frac{1}{2}\frac{1}{r}\left(\frac{\partial^2}{\partial z^2} + \frac{\partial^2}{\partial r^2} - \frac{1}{r}\frac{\partial}{\partial r} \right)\psi = \frac{1}{2r}L\psi. \tag{16}$$

Thus Eq. (14) becomes

$$L^2\psi = 0. \tag{17}$$

Now, let us introduce spherical coordinates (R, ϕ, θ) with the polar axis pointing to the direction opposite to the direction of motion of the sphere (Fig. 5.7:2). Then

$$z = R\cos\phi, \quad r = R\sin\phi, \tag{18}$$

and Eq. (17) becomes

$$\left[\frac{\partial^2}{\partial R^2} + \frac{\sin\phi}{R^2}\frac{\partial}{\partial \phi}\left(\frac{1}{\sin\phi}\frac{\partial}{\partial \phi} \right) \right]^2 \psi = 0. \tag{19}$$

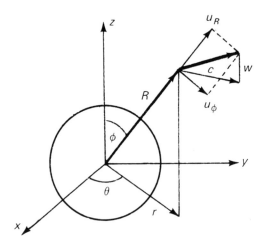

FIGURE   5.7:2. Spherical–polar coordinates.

This is satisfied by

$$\psi = \sin^2 \phi f(R), \tag{20}$$

if

$$\left(\frac{d^2}{dR^2} - \frac{2}{R^2}\right)^2 f(R) = 0. \tag{21}$$

The solution of Eq. (21) is

$$f(R) = \frac{A}{R} + BR + CR^2 + DR^4. \tag{22}$$

The boundary conditions are undisturbed flow at infinity and no-slip on the surface of the sphere, that is,

at $R = \infty$: $w = U$, $c = 0$, or $v_R = U\cos\phi$, $v_\phi = U\sin\phi$,

at $R = a$: $w = 0$, $c = 0$, or $v_R = v_\phi = 0$, $\tag{23}$

where v_R, v_ϕ are velocity components in the directions of increasing R and ϕ, respectively. It can be shown that

$$v_R = \frac{1}{R^2 \sin\phi}\frac{\partial\psi}{\partial\phi}, \quad v_\phi = -\frac{1}{R\sin\phi}\frac{\partial\psi}{\partial R}. \tag{24}$$

From Eqs. (20), (22), and (24), it is seen that the boundary conditions are satisfied by choosing

$$C = \tfrac{1}{2}U, \quad D = 0,$$

$$A = \tfrac{1}{4}Ua^3, \quad B = -\tfrac{3}{4}Ua. \tag{25}$$

Thus

$$v_R = U\cos\theta - 2\left(\frac{A}{R^3} + \frac{B}{R}\right)\cos\theta,$$

$$v_\phi = -U\sin\theta - \left(\frac{A}{R^3} - \frac{B}{R}\right)\sin\theta. \tag{26}$$

The problem is now solved. The streamlines $\psi =$ constant are plotted for a series of equidistant values of ψ in Figure 5.7:3.

A quantity of interest is the total hydrodynamic resistance encountered by the sphere. To evaluate this force, we must compute the stress tensor τ_{ij} from the constitutive equation

$$\tau_{ij} = -p\delta_{ij} + 2\mu\dot{e}_{ij} = -p\delta_{ij} + \mu\left(\frac{\partial v_i}{\partial x_j} + \frac{\partial v_j}{\partial x_i}\right)$$

and the stress vector acting on the surface of the sphere at radius $R = a$,

$$\overset{v}{T_i} = v_j\tau_{ij},$$

where $v = (v_1, v_2, v_3)$ unit vector normal to the surface. Clearly, v is the radius vector

$$\left(v_1, v_2, v_3\right) = \left(\frac{x}{R}, \frac{y}{R}, \frac{z}{R}\right).$$

By consideration of the axisymmetry of the flow field, it is evident that the resultant forces in the x and y directions are zero, and that the resultant force in the z direction is given by

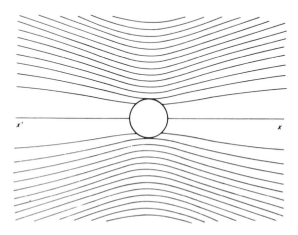

FIGURE 5.7:3. Streamlines in the flow caused by a sphere moving with a small constant velocity in the horizontal direction: Stoke's solution. From Lamb (1932), p. 599. By permission of Cambridge Univ. Press.

$$F = \oint \overset{v}{T_z} \, dS = 2\pi a^2 \int_0^\pi \overset{v}{T_z} \sin\phi \, d\phi. \tag{27}$$

It turns out that on the surface of the sphere, $R = a$,

$$\overset{v}{T_z} = -\frac{z}{a}p_o + \frac{3}{2}\mu\frac{U}{a}, \quad \overset{v}{T_x} = -\frac{x}{a}p_o, \quad \overset{v}{T_y} = -\frac{y}{a}p_o, \tag{28}$$

and the resultant force is

$$F = 6\pi\mu a U. \tag{29}$$

This solution, due to Stokes (1851), has been well verified experimentally, as shown in Figure 5.7:4, where the ordinate is the *drag coefficient*, C_D, defined by

$$C_D = \frac{F}{\frac{1}{2}\rho U^2 A}, \tag{30}$$

with A being the projected area πa^2 in this case. From Eq. (29) we have

$$C_D = \frac{12}{N_R}, \tag{31}$$

where N_R is the Reynolds number, $aU\rho/\mu$. Equation (31) is plotted as the solid line in Figure 5.7:4.

A note may be added with regard to the actual calculations leading to Eqs. (28) and (29). Instead of finding the full tensor τ_{ij} and then proceeding

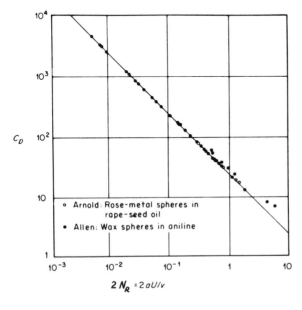

FIGURE 5.7:4. Experimental verification of Stoke's law, $C_D = 12/N_R$, which is represented by the straight line. By permission from Rouse (1959, p. 240).

as outlined earlier, the following steps will greatly lessen the labor. We note that

$$
\begin{aligned}
\overset{v}{T}_z &= \frac{z}{R}\tau_{zz} + \frac{y}{R}\tau_{zy} + \frac{x}{R}\tau_{zx} \\
&= \cos\phi\left(-p + 2\mu\frac{\partial w}{\partial z}\right) + \mu\frac{y}{R}\left(\frac{\partial w}{\partial y} + \frac{\partial v}{\partial z}\right) + \mu\frac{x}{R}\left(\frac{\partial u}{\partial x} + \frac{\partial u}{\partial z}\right) \\
&= -p\cos\phi + \mu\frac{\partial w}{\partial R} - \frac{\mu w}{R} + \frac{\mu}{R}\frac{\partial}{\partial z}(xu + yv + zw) \\
&= -p\cos\phi + \mu\frac{\partial w}{\partial R} - \frac{\mu w}{R} + \frac{\mu}{R}\frac{\partial}{\partial z}Rv_R.
\end{aligned}
$$

But

$$
w = v_R\cos\phi - v_\phi\sin\phi, \quad \frac{\partial v_R}{\partial z} = \frac{\partial v_R}{\partial R}\cos\phi - \frac{\partial v_R}{\partial\phi}\frac{\sin\phi}{R}.
$$

Hence

$$
\overset{v}{T}_z = -p\cos\phi + 2\mu\cos\phi\frac{\partial v_R}{\partial R} - \mu\sin\phi\frac{\partial v_\phi}{\partial R} + \frac{1}{R}\frac{\partial v_R}{\partial\phi} - \frac{v_\phi}{R}.
$$

Equation (28) is then obtained by substituting $R = a$ into this last equation.

A further alternative is to note that in spherical coordinates,

$$
\overset{v}{T}_R = \tau_{RR}\cos\phi - \tau_{R\phi}\sin\phi = \left(-p + 2\mu\dot{e}_{RR}\right)\cos\phi - \mu\dot{e}_{R\phi}\sin\phi.
$$

The strain rate components, $\dot{e}_{RR}$ and $\dot{e}_{R\phi}$, are given by

$$
\dot{e}_{RR} = \frac{\partial v_R}{\partial R}, \quad \dot{e}_{R\phi} = R\frac{\partial}{\partial R}\frac{v_\phi}{R} + \frac{1}{R}\frac{\partial v_R}{\partial\phi}.
$$

This classical work illustrates the method of formulation and analysis of a flow of very low Reynolds number. It is presented here not only for its beauty but also for its importance. The slow motion of small particles in a fluid is a common occurrence in the biological world. Equation (29) is used very frequently in connection with the flow of fluid containing particles; sedimentation of particles; centrifugation or ultracentrifugation of suspensions, colloids, and blood; isolation of tumor antigens; etc. The fluid does not have to be liquid, and the particle does not have to be solid. The formula is useful in the analysis of natural fog or smog formation, the atomization of liquids, etc., with many medical and industrial applications.

An analogous problem of a long circular cylinder slowly falling in a liquid in the direction perpendicular to the cylinder axis, that is, the two-dimensional version of Eqs. (1) and (23), has no solution. The lack of a solution of Eq. (1) for two-dimensional flow is called the *Whitehead paradox*. It illustrates the delicate nature of mathematical idealization of physical problems.

Problems

5.1 Consider a very slow falling of a sphere in a viscous fluid. Let $\Delta\gamma$ be the difference in specific weight of the sphere and the fluid. Using dimensional analysis, show that the only dimensionless parameter is another Stokes number, $S = U\mu/(a^2\Delta\gamma)$, if inertial forces are neglected. What other parameters intervene if the falling velocity is not small?

5.2 Apply Stokes' solution to the Problem 5.1. If the sphere is not accelerating, the hydrodynamic drag force is balanced by the weight of the sphere minus the buoyancy. Show that when the Reynolds number tends to zero, the Stokes number is

$$S = \frac{U\mu}{a^2\Delta\gamma} = \frac{2}{9}.$$

5.8 Oseen's Approximation and Other Developments

In Stokes' solution for the moving sphere, the inertial terms are entirely neglected. It is a limiting solution when the Reynolds number tends to zero. For small but finite Reynolds number, say, with N_R of the order of 0.1, some correction of the inertial terms is needed. Oseen (1910) proposed the following method to treat problems in which the flow field consists of a small perturbation of a constant mean flow, U. He made the substitution

$$v_1 = U + v_1', \quad v_2 = v_2', \quad v_3 = v_3', \tag{1}$$

in the Navier–Stokes equations and neglected the quadratic terms in the primed quantities. This leads to the following equation:

$$U\frac{\partial v_i'}{\partial x_1} = -\frac{1}{\rho}\frac{\partial p}{\partial x_i} + \nu\nabla^2 v_i', \quad (i = 1, 2, 3) \tag{2}$$

which is known as *Oseen's equation*. The approximation that Eq. (2) stands for is called *Oseen's approximation*.

When the moving sphere problem of the preceeding section is solved on the basis of Eq. (2), Oseen found that the resultant hydrodynamic force (drag) is

$$F = 6\pi\mu a U\left(1 + \tfrac{3}{8}N_R\right), \tag{3}$$

where N_R is the Reynolds number Ua/ν. The streamline pattern in this case is shown in Figure 5.8:1.

The difference between Eq. (3) of this section and Eq. (29) of Section 5.7 lies in the factor $(3/8)N_R$. One may question, however, whether the correction term is not fortuitous, because in a frame of reference moving with the sphere, the fluid near the sphere is almost at rest, and in that region inertial force is negligible and Stokes equation is well justified. Far away from

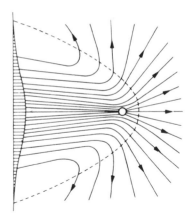

FIGURE 5.8:1. Streamlines in flow caused by a sphere moving with a small constant velocity to the right: Oseen's solution. The shaded profile at left is the horizontal velocity profile of the fluid at that location. From Schlichting (1962). Reproduced by permission.

the sphere, the flow velocity approaches U and Oseen's approximation is more accurate. But Eq. (3) was obtained by using Oseen's equation for the entire flow field. This question was answered by Proudman and Pearson (1957), who solved the Navier–Stokes equations and gave an improved Stokes solution in the neighborhood of the sphere and an improved Oseen solution at infinity, and matched the two solutions in a supposed common region of their validity. They obtained

$$F = 6\pi\mu a U\left[1 + \tfrac{3}{8} N_R + \tfrac{9}{40} N_R^2 \ln N_R + O\left(N_R^2\right)\right]. \tag{4}$$

To the order N_R, Eqs. (3) and (4) agree. Equation (4) is valid for Reynolds numbers that are so small that the square of N_R is negligible compared with 1.

The details of the solution of the falling sphere problem based on Oseen's equation, and other developments concerning forces acting on an oscillating sphere, forces acting on a sphere moving with arbitrary speed in a straight line, motion of a sphere released from rest, etc., can be found in the excellent textbook by C.S. Yih, *Fluid Mechanics* (1977).

It is instructive to compare the solutions to the problem of a moving sphere in an incompressible fluid by neglecting one term or another in the Navier–Stokes equation:

$$\rho\left(\frac{\partial u_i}{\partial t} + u_j \frac{\partial u_i}{\partial x_j}\right) = -\frac{\partial p}{\partial x_i} + \mu \nabla^2 u_i.$$

When the inertial force terms on the left-hand side are neglected, the streamlines are shown in Figure 5.7:3. When the fluid is nonviscous, $\mu = 0$, then the last term $\mu \nabla^2 u_i$ can be dropped; the solution is well known from potential flow theory, and the streamline pattern is shown in Figure 5.8:2. Comparison of Figures 5.7:3, 5.8:1, and 5.8:2 shows the tremendous differences implied by these assumptions.

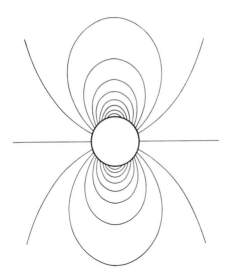

FIGURE 5.8:2. Streamlines in the flow caused by a sphere moving in a non-viscous fluid. Potential flow solution. From Lamb (1932, p. 128). Reproduced by permission.

The streamline pattern of the solution based on the Oseen equation already shows the development of a wake behind the sphere. The wake becomes more pronounced as the Reynolds number of flow increases. Separation of streamlines from the solid surface of the sphere seems to begin in the neighborhood of the rear stagnation point at a Reynolds number of about 5. As the Reynolds number increases, the separation region increases and standing vortices develop. The point of separation depends on the flow condition in the boundary layer—whether it is laminar or tubulent. These fascinating and important features of flow and drag variation are discussed in several books, for example, in Yih (1977).

5.9 Entry Flow, Bolus Flow, and Other Examples

Features of entry flow have already been discussed in Section 3.18. With the equations presented in Section 5.7, we can now discuss the mathematical side of the problem.

Entry Flow into a Circular Cylindrical Tube at a Very Low Reynolds Number

For a steady flow of an incompressible flow at a low Reynolds number, the Stokes approximation prevails and the equation of momentum becomes the Stokes equation [Eq. (1) of Sec. 5.7]:

$$-\nabla p + \mu \nabla^2 \mathbf{v} = 0. \tag{1}$$

Here, and in the rest of this section, letters printed in boldface denote vectors. The equation of continuity of an incompressible fluid is

$$\nabla \cdot \mathbf{v} = 0. \tag{2}$$

The boundary conditions are as follows:

a. The adherence of the fluid to the wall of the tube,

$$\mathbf{v} = 0 \quad \text{on the wall of the tube.} \tag{3}$$

b. $\mathbf{v} = $ a specified distribution at the inlet cross section of the tube.
c. $\mathbf{v}$ and p are given by the Poiseuille formulas far downstream of the tube.

The mathematical problem is fully specified by these equations. To solve them, Lew and Fung (1969a) split $\mathbf{v}$ and p into two parts,

$$\mathbf{v} = \mathbf{v}' + \mathbf{v}_\infty, \quad p = p' + p_\infty, \tag{4}$$

where $\mathbf{v}_\infty$ and p_∞ correspond to Poiseuille flow. $\mathbf{v}'$ and p' satisfy Eqs. (1) and (2) because $\mathbf{v}_\infty$ and p_∞ satisfy them. Equation (2) is satisfied if $\mathbf{v}'$ is derived from an arbitrary function $f(x, r)$ as follows:

$$\mathbf{v}' = \nabla \times \nabla \times \left[\hat{x} f(x, r) \right], \tag{5}$$

where $\hat{x}$ designates the unit base vector of the x-axis and r, θ, x are a system of cylindrical polar coordinates with the x-axis coinciding with the axis of the tube, and the origin of the coordinate system located at the entrance section. Since $f(x, r)$ is independent of θ, Eq. (5) can be expanded as

$$\mathbf{v}' = \hat{r} \frac{\partial^2 f}{\partial r \partial x} - \hat{x} \frac{1}{r} \frac{\partial}{\partial r} \left(r \frac{\partial f}{\partial r} \right). \tag{6}$$

Substitution of Eq. (5) into Eq. (1) yields the equation

$$-\nabla \left(p' - \mu \nabla^2 \frac{\partial f}{\partial x} \right) - \hat{x} \mu \nabla^4 f = 0, \tag{7}$$

which is satisfied if

$$p' = \mu \nabla^2 \frac{\partial f}{\partial x}, \tag{8}$$

$$\nabla^4 f = 0. \tag{9}$$

A general solution of Eq. (9), which has bounded values in the whole region of consideration, is

$$f(x, r) = \left(A + B\lambda \frac{x}{a} \right) e^{-\lambda x/a} J_0 \left(\lambda \frac{r}{a} \right)$$

$$+ \left[C \cos \left(\eta \frac{x}{a} \right) + D \sin \left(\eta \frac{x}{a} \right) \right] \left[I_0 \left(\eta \frac{r}{a} \right) + E \frac{r}{a} I_1 \left(\eta \frac{r}{a} \right) \right], \tag{10}$$

where A, B, C, D, E, λ, and η are arbitrary constants; a is the radius of the tube, J_0 is the Bessel function of the first kind of order zero, and I_0, I_1 are the modified Bessel functions of the first kind of order zero and one, respectively.

Since the governing differential equations and boundary conditions are linear, we can superpose solutions of the Eq. (10) type to construct a general solution. In particular, we construct terms such as

$$\sum_{n=1}^{\infty} A_n \left\{ \left(1 + k_n \frac{x}{a} \right) e^{-k_n x/a} J_0 \left(k_n \frac{r}{a} \right) \right.$$

$$\left. + \int_0^{\infty} C_n(\eta) \cos\left(\eta \frac{x}{a} \right) \left[I_0 \left(\eta \frac{r}{a} \right) - \frac{I_0(\eta)}{\frac{2}{\eta} I_0(\eta) + I_1(\eta)} \frac{r}{a} I_1 \left(\eta \frac{r}{a} \right) \right] d\eta \right\} \quad (11)$$

as part of the general solution, in which k_n is the nth zero of the Bessel function, $J_0(k_n) = 0$. With such superposition, one obtains a solution that is sufficiently general to satisfy all the boundary conditions if the velocity distribution at the inlet is axisymmetric. The details are given in Lew and Fung (1969a). The velocity distribution is shown on p. 185, Figure 3.14:2.

Entry Flow into Circular Cylinder at an Arbitrary Reynolds Number

Entry flow at a high Reynolds number (say, $N_R > 100$, with Reynolds number $N_R = \rho a U/\mu$ defined on the basis of the mean flow velocity U and the tube radius a) can be analyzed by the boundary layer theory, from which Targ (1951) obtained the result that the entry length L is

$$L = 0.16 a N_R, \quad (12)$$

where a is the tube radius, and L is defined as the distance of transition from a uniform inflow at the inlet to a section where the velocity profile differs from the Poiseuille profile by less than 1%. Entry flow at zero Reynolds number is discussed in the preceding paragraph, and we obtained the entry length

$$L = 1.3a. \quad (13)$$

For flow with Reynolds number between 0 and 100, Lew and Fung (1970b) used the Oseen approximation (Sec. 5.8),

$$\rho U \frac{\partial \mathbf{v}}{\partial x} = -\nabla p + \mu \nabla^2 \mathbf{v}, \quad (14)$$

where U is the average speed of flow over the cross section of the tube and $\mathbf{v}$ is the velocity. The equations of continuity and boundary conditions are the same as Eqs. (2) and (3). To solve these equations, Eqs. (4) to (6) remain valid, while Eq. (7) becomes

$$-\nabla\left[p'-\mu\left(\nabla^2-\frac{N_R}{a}\frac{\partial}{\partial x}\right)\frac{\partial f}{\partial x}\right]-\hat{x}\mu\left(\nabla^2-\frac{N_R}{a}\frac{\partial}{\partial x}\right)\nabla^2 f=0, \qquad (15)$$

where N_R is the Reynolds number $\rho a U/\mu$. This equation is satisfied if

$$p'=\mu\left(\nabla^2-\frac{N_R}{a}\frac{\partial}{\partial x}\right)\frac{\partial f}{\partial x}, \qquad (16)$$

$$\left(\nabla^2-\frac{N_R}{a}\frac{\partial}{\partial x}\right)\nabla^2 f=0. \qquad (17)$$

A solution of Eq. (17), which is bounded in the region of the tube, is

$$f(x,r)=\left[Ae^{-\lambda x/a}+Be^{-\left(\sqrt{4\lambda^2+N_R^2}-N_R\right)x/2a}\right]J_0\left(\lambda\frac{r}{a}\right)$$

$$+\left[C\cos\left(\eta\frac{x}{a}\right)+D\sin\left(\eta\frac{x}{a}\right)\right]I_0\left(\eta\frac{r}{a}\right)$$

$$+E\left\{\cos\left(\eta\frac{x}{a}\right)I_0\left[e^{i\alpha/2}\left(1+\frac{N_R^2}{\eta^2}\right)^{1/4}\eta\frac{r}{a}\right]\Bigg|_R\right.$$

$$\left.-\sin\left(\eta\frac{x}{a}\right)I_0\left[e^{i\alpha/2}\left(1+\frac{N_R^2}{\eta^2}\right)^{1/4}\eta\frac{r}{a}\right]\Bigg|_I\right\}$$

$$+F\left\{\cos\left(\eta\frac{x}{a}\right)I_0\left[e^{i\alpha/2}\left(1+\frac{N_R^2}{\eta^2}\right)^{1/4}\eta\frac{r}{a}\right]\Bigg|_I\right.$$

$$\left.+\sin\left(\eta\frac{x}{a}\right)I_0\left[e^{i\alpha/2}\left(1+\frac{N_R^2}{\eta^2}\right)^{1/4}\eta\frac{r}{a}\right]\Bigg|_R\right\}, \qquad (18)$$

where A, B, C, D, E, F, λ, and η are constants, J_0 is a Bessel function, and I_0 is a modified Bessel function, both of the first kind and zeroth order. $I_0|_R$ and $I_0|_I$ denote the real and imaginary part of I_0, respectively, and

$$\alpha=\tan^{-1}\frac{N_R}{\eta}. \qquad (19)$$

Equation (18) generalizes Eq. (10) of the case $N_R = 0$. By superposition of terms of the type given in Eq. (18), we can construct a full solution that generalizes Eq. (11) of the zero-N_R case. The details are given by Lew and Fung (1970b), and the results are sketched in Figure 3.14:2, on p. 185.

Motion of the Plasma Between
Red Cells in a Bolus Flow

A photograph of blood flow in a capillary blood vessel in the mesentery of a dog is given on p. 111 of *Biomechanics: Mechanical Properties of Living Tissues* (Fung, 1993). The red blood cells are seen to be severely deformed. In such a flow, the red cells seem to plug the capillary blood vessel. The motion of the plasma in the capillary between successive red cells is called *bolus flow*. The significance of bolus flow was pointed out by Prothero and Burton (1961). If streamlines are drawn relative to a frame of reference moving with the red cells, the streamlines will be seen to form eddies, thus creating a stirring mechanism that brings material on the centerline to the tube wall by convection. One might suspect that this action will help mass transfer. It was looked into and concluded that the influence of this on oxygenation and CO_2 removal is minor, but its effect on macromolecules may be significant (Aroesty and Gross, 1970).

The mathematical theory can be based on Stokes Eq. (1) and the continuity Eq. (2). Consider plug flow and an idealized geometry shown in Figure 5.9:1. We have the boundary conditions:

$$u(x,r) = -U \quad \text{for} \quad r = a \quad \text{and} \quad -L \le x \le L,$$
$$v(x,r) = 0 \quad \text{for} \quad r = a \quad \text{and} \quad -L \le x \le L,$$
$$u(x,r) = 0 \quad \text{for} \quad x = \pm L \quad \text{and} \quad 0 \le r \le a,$$
$$v(x,r) = 0 \quad \text{for} \quad x = \pm L \quad \text{and} \quad 0 \le r \le 0,$$

where a is the radius of the capillary, U is the velocity of the red cells, and $2L$ is the distance between the two consecutive red blood cells.

A solution given by Lew and Fung (1969b) is quite similar to that of the entry flow problem mentioned earlier. In fact, the results can be easily understood in the light of the entry flow solution. We have seen that at a very low Reynolds number of the entry length is about $1.3a$. Hence, if $L = 1.3a$, the velocity profile at midway between the cells will be almost para-

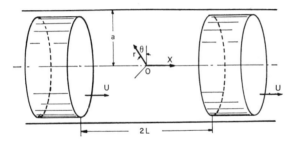

FIGURE 5.9:1. Idealized geometry of bolus flow. From Lew and Fung (1969b).

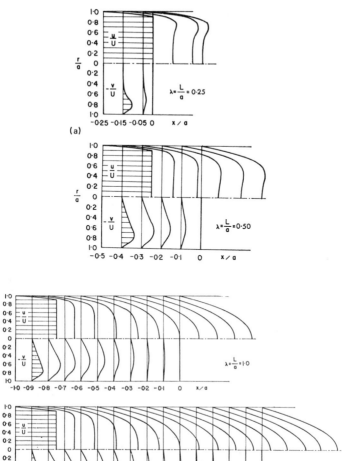

FIGURE 5.9:2. Velocity distribution of the plasma in a bolus flow induced by two adjacent red cells moving in a fixed tube. Figure shows the axial component of velocity u above the centerline and the radial component of velocity v below the centerline in each cross section. The velocity specified at the red cell is $u = U$ in $0 \leq r/a \leq 0.9$, where U is a constant. In the gap between the red cell and the tube wall, u decreases from U to 0 in $0.9 \leq r/a \leq 1$. On the red cell, $x = \pm L$, the vertical velocity $v = 0$ in $0 \leq r/a \leq 1$. The parameter λ is L/a, that is, (distance between cells)/(tube diameter). In the figures from the top downward, $\lambda = 0.25, 0.50, 1.0,$ and 1.5. From Lew and Fung (1969b). Reproduced by permission.

bolic, differing from it by less than 1%. For smaller L, the velocity profile would not have enough space to be readjusted to the parabolic profile; the u, v distributions are shown in Figure 5.9:2, and are quite similar to those shown in Figure 3.14:2, on p. 185.

5.10 Interaction Between Particles and Tube Wall

Mathematical models of particles moving in a circular cylindrical tube can be based on Eqs. (1) and (2) of Section 5.9. The boundary conditions are (a) zero velocity at the tube wall, (b) matching particle velocity and traction on the surface of the suspended particles, (c) a specification of either the total discharge or the pressure drop over some appropriate length of the capillary, and (d) on neutrally bouyant particles the net force on any particle and the net moment must be equal to zero. The condition (d) is called the *zero-drag condition*. For deformable particles, the equations of motion and appropriate constitutive equations of the particles must be satisfied, and the internal velocity and stress fields must match the fluid stresses on the boundary of the particles. The following are solved cases.

Rigid Particles

In a steady axisymmetric flow, solutions have been developed for a line of spheres (Bungay and Brenner, 1973; Hochmuth and Sutera, 1970; Wang and Skalak, 1969), spheroids (Chen and Skalak, 1970), disks (Aroesty and Gross, 1970; Lew and Fung, 1969b, 1970a; Tong and Fung, 1971), and biconcave rigid disk shapes (Skalak et al., 1972). The general features of the solutions are the same for all these shapes: (1) A particle on the axis of the capillary travels faster than the mean flow. (2) The pressure drop depends most strongly on the ratio of the maximum diameter of the particle to the tube diameter. (3) The pressure drop due to a given particle is independent of the presence of other particles if the spacing is greater than about one tube diameter. The last point is easily understood if one recalls the results discussed toward the end of Section 5.9. It follows that at low hematocrit the apparent viscosity of blood in a capillary vessel is proportional to the hematocrit.

The methods used to obtain these solutions are of four kinds. Skalak et al. (1972), Tong and Fung (1971), and Zarda et al. (1977) used the finite-element method. Chen and Skalak (1970), Hyman and Skalak (1972), Lew and Fung (1969b, 1970a), and Wang and Skalak (1969) used infinite series. Bungay and Brenner (1973) used asymptotic expansions. Skalak et al. (1972) used M.J. Lighthill's (1968) lubrication layer method (see Fung, 1993, p. 195). All authors treated axisymmetric flow, except Bungay and Brenner (1973), who treated an eccentrically located sphere, in which case the sphere rotates as it translates down the capillary along a line parallel to the axis of the tube.

Elastic Spheres

Tözeren and Skalak (1978, 1979) analyzed the steady flow of elastic spheres in a circular cylindrical tube, using a series expansion for the particle displacements and lubrication theory for the fluid motion. The results show that the apparent viscosity depends on the shear modulus of elasticity of the particle. In the later paper (1979), the authors extended the calculation to tapered tubes.

Flexible Red Cell Models

Zarda et al. (1977) analyzed the steady flow of a row of flexible red blood cells in a circular cylindrical tube in axisymmetric configuration. The membrane of the red cell is assumed to be elastic in shear and in bending, but its area is assumed to remain constant during any deformation. The tensile stress resultants, T_1, T_2 (dyne/cm), in principal directions are adopted from Skalak et al. (1973) in the form,

$$T_1 = B(\lambda_1^2 - 1)\lambda_1/2\lambda_2 + T_0,$$
$$T_2 = B(\lambda_2^2 - 1)\lambda_2/2\lambda_1 + T_0,$$

where λ_1 and λ_2 are the principal extension ratios and T_0 is an isotropic stress, analogous to the pressure in an incompressible fluid, introduced in response to the hypothesis that the cell membrance area is unchangeable. The coefficient B is an elastic modulus, which is taken to be 0.0005 dyne/cm in the computations.

The principal bending moments, M_1, M_2, are assumed to depend on the change of curvature, K_1, K_2, according to the formulas,

$$M_1 = D(K_1 + vK_2)/\lambda_2,$$
$$M_2 = D(K_2 + vK_1)/\lambda_1,$$

where D is a bending stiffness coefficient taken to be equal to 10^{-12} dyne/cm in the computations, and v is the Poisson's ratio. The interior of the red cell is assumed to be a fluid, so that in static condition the internal pressure is uniform. The exterior fluid is the blood plasma, a Newtonian fluid with a viscosity of 1.2 cp. The pressure drop over a typical length of the capillary containing one blood cell is assumed to be given, and the computations seek the shape and velocity of the red blood cells.

Some of their results for a line of cells in a tube with a hematocrit close to 26% are given in Figure 5.10:1. The assumed unstressed shape of the cell is the one given on pp. 114–115 of *Biomechanics: Mechanical Properties of Living Tissues* (Fung, 1993) at a tonicity of 300 mOsm. The figure shows the successive shapes of a cell for increasing dimensionless pressure drop, Δp. From the pressure drop and velocity of flow, the apparent viscosity of blood in the tube can be calculated. The results show that the apparent viscosity for a given initial cell/tube diameter ratio will generally decrease as the pressure drop increases. This is because at higher pressure drop the cells are deformed more and thereby pull away further from the tube wall.

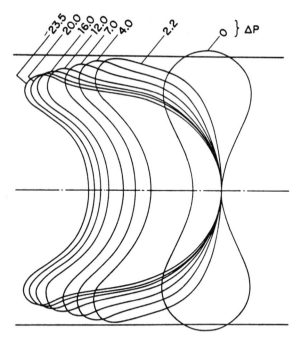

FIGURE 5.10:1. Shapes computed for red blood cells flowing axisymmetrically in a capillary. The unstressed radius of the red cell is 3.91 μm and is 5% larger than the radius of the tube. From Zarda et al. (1977). Reproduced by permission.

All the literature cited earlier is concerned with axisymmetric flow in circular cylinders. In reality, the non-axisymmetric cases are of great importance. As is discussed in Chapter 5 of Fung (1993), experiments have shown that the most common configuration for a red cell to enter a capillary blood vessel is edge on, with the axis of symmetry of the cell perpendicular to the axis of the cylinder. Furthermore, the "tank-treating" mechanism, first pointed out by Fischer et al. (1978) and Schmid-Schönbein and Wells (1969), and offers an elegant mechanism for reducing friction between a red cell and the blood vessel wall. As illustrated in Figure 5.10:2, the red cell is deformed into the shape of a slipper, the cell membrane tank-treads.

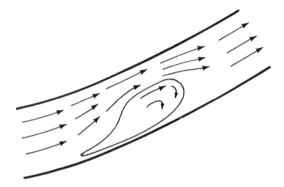

FIGURE 5.10:2. Illustration of tank treading of a red cell eccentrically placed in a tube reducing viscous drag.

The part of the cell membrane that is very close to the tube wall has small relative motion with respect to the wall. The part of the cell membrane closer to the center of the tube moves forward, thus reducing its effect on the motion of the plasma. If we compare this situation with the condition shown in Figure 5.10:1, we can expect that the resistance to the movement of the cell in the tube is much reduced in the asymmetric, tank-treading case.

5.11 Stokes Flow in Pulmonary Capillaries: Sheet Flow Around a Circular Post

Many capillary blood vessels do not look like long circular cylindrical tubes. Look at the photographs of pulmonary capillary blood vessels shown in Section 6.3, especially the plane views in Figures 6.3:1, 6.3:2, and 6.3:4. Compare them with the cross-sectional view shown in Figure 4.10:3 in Section 4.10. One sees that the pulmonary capillaries are organized into closely knit two-dimensional networks. Each network forms a sheet-like structure. An adult human has four billion sheets organized into a three-dimensional lung that consists of 300 million ventilation units called alveoli. Every alveolus is ventilated to the mouth. On each sheet, if one insisted on calling every segment of capillary blood vessel between bifurcation points a cylinder, then the length of each cylinder is equal or shorter than its diameter, and the axis is curved. The Poiseuille formula of Section 3.2 certainly will not apply. Actually, the geometry of the blood space in these sheets is much better simulated by a space between two parallel elastic membranes separated by a biaxial array of "posts," as sketched in Figure 6.3:3. For a red blood cell flowing in such a space, the situation may be compared with a car driving through an underground garage. The situation is depicted in Figure 6.9:1 in Section 6.9.

Whereas the anatomical details are described in Chapter 6, the fluid mechanics of the sheet flow is considered presently. The physical problem is Stokes flow of a viscous fluid containing blood cells between two elastic membranes separated by a biaxially periodic array of posts. As a first step to catch the essence, we study the Stokes flow of a homogeneous incompressible viscous fluid between two rigid walls separated by a single circular cylindrical post.

The geometry of the idealized problem is shown in Figure 5.11:1. The physiologically relevant range of the ratio of the sheet thickness, or plate spacing, $2h$, to the post diameter, $2a$, is an h/a ratio of about 2 or 3. If h/a is much smaller than 1, then the problem is reduced to that of Hele-Shaw (1898), who 100 years ago discovered that the streamlines for this kind of flow can be used to represent the lines of force around a metal cylinder in a dielectric medium in a magnetic field. A little later in the same year, Stokes proved that in Hele–Shaw flow, with $h/a \ll 1$, the stream function for the

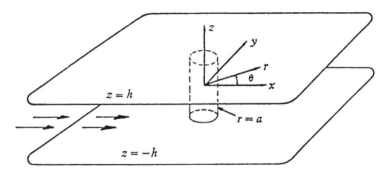

FIGURE 5.11:1. Stokes flow around a circular post as a prototype of pulmonary capillary sheet flow. Sketch showing coordinates and notations.

velocity averaged across the thickness is governed by a two-dimensional Laplace equation. But when h/a is of order 1 to 5, Stokes' solution does not apply. The following solution is given by Lee and Fung (1969).

The governing equations are Eqs. (1) to (7) of Section 5.7. The boundary conditions are no-slip on the walls of the plates and cylinder, and Poiseuille flow at infinity. In cylindrical polar coordinates $(\gamma,\ \theta,\ z)$ and rectangular cartesian coordinates (x, y, z) (see Fig. 5.11:1), the boundary conditions on the velocity vector $\mathbf{v}$ (v_x, v_y, v_z) are

$$\mathbf{v} = 0 \quad \text{at} \quad r = a, \tag{1}$$

$$\mathbf{v} = 0 \quad \text{at} \quad z = \pm h. \tag{2}$$

The flow at infinity is assumed to be a uniform Poiseuille flow. Therefore,

$$\mathbf{v} = U\left(1 - z^2/h^2\right)\mathbf{i} \quad \text{as} \quad r \to \infty, \tag{3}$$

where $\mathbf{i}$ is a unit vector in the x-direction and U is the velocity in the midplane $z = 0$ at large distance from the post.

To solve the biharmonic equation $\nabla^4\mathbf{v} = 0$ [Eq. (5.7:7)], we use Almansi's theorem (Fung, 1965, p. 207) that any biharmonic vector function $\mathbf{v}$ can be represented in the form

$$\mathbf{v} = z\mathbf{v}_1 + \mathbf{v}_2, \tag{4}$$

in which $\mathbf{v}_1$, and $\mathbf{v}_2$ are harmonic functions. Now, if $\phi_n(x, y)$ and $\bar{\phi}_n\,(x, y)$ are any two solutions of the differential equation

$$\frac{\partial^2 \phi_n}{\partial x^2} + \frac{\partial^2 \phi_n}{\partial y^2} - \alpha_n^2 \phi_n = 0, \tag{5}$$

where α_n is a constant, then the functions

$$\phi_n(x, y)\cos\alpha_n z, \quad \bar{\phi}_n(x, y)\sin\alpha_n z \tag{6}$$

must be harmonic. Hence Eq. (2) will certainly be satisfied in the z-component if v_z takes the form

$$v_z = \phi_n(x, y) q_n(z), \tag{7}$$

where

$$q_n(z) = \frac{\sin \alpha_n z}{\sin \alpha_n h} - \frac{z \cos \alpha_n z}{h \cos \alpha_n h}. \tag{8}$$

Substituting, Eq. (7) into Eqs. (1) and (5) of Section 5.7, on p. 291, and solving for v_x, v_y and p, one finds

$$v_x = -\frac{1}{\alpha_n^2} \frac{\partial \phi_n}{\partial x} \frac{dq_n}{dz},$$

$$v_y = -\frac{1}{\alpha_n^2} \frac{\partial \phi_n}{\partial y} \frac{dq_n}{dz},$$

$$p = -\frac{2\mu}{h} \phi_n \frac{\cos \alpha_n z}{\cos \alpha_n h}. \tag{9}$$

The no-slip conditions $v_x(\pm h) = v_y(\pm h) = 0$ are satisfied if α_n satisfies the following characteristic equation:

$$\sin 2\alpha_n h = 2\alpha_n h. \tag{10}$$

Approximate values of $2\alpha_n h$ are $(2n + \frac{1}{2})\pi \pm i \log(4n + 1)\pi, n = 1, 2, \ldots$

All roots of Eq. (10) are complex-valued, except $\alpha_0 = 0$, which corresponds to

$$v_x = \frac{\partial \phi_0}{\partial x} \left(1 - \frac{z^2}{h^2}\right),$$

$$v_y = \frac{\partial \phi_0}{\partial y} \left(1 - \frac{z^2}{h^2}\right),$$

$$v_z = 0, \quad p = -\frac{2\mu}{h^2} \phi_0, \tag{11}$$

where $\phi_0(x, y)$ is a harmonic function. This is Stokes' solution for Hele–Shaw flow.

To Eqs. (7) and (9), we can add a harmonic function of the following form that satisfies Eq. (2), and Eqs. (1) and (5) of Section 5.7:

$$v_x = -\frac{\partial \psi_n}{\partial y} \cos k_n z,$$

$$v_y = \frac{\partial \psi_n}{\partial x} \cos k_n z,$$

$$v_z = p = 0, \tag{12}$$

where $k_n = (2n + 1)\,\pi/2h$ with $n = 0, 1, 2, \ldots$ and $\psi_n(x, y)$ are functions satisfying Eq. (5), with α_n replaced by k_n.

A general solution is a series of the solutions given in Eqs. (7), (9), and (12). Upon transformation to cylindrical coordinates, and using Re to denote the real part of a complex number, we obtain

$$
v_r = \frac{\partial \phi_0(r,\theta)}{\partial r}\left(1 - \frac{z^2}{h^2}\right) - \mathrm{Re}\left[\sum_{n=1}^{\infty} \frac{1}{\alpha_n^2} \frac{\partial \phi_n(r,\theta)}{\partial r} \frac{dq_n(z)}{dz}\right]
$$

$$
- \sum_{n=0}^{\infty} \frac{1}{r} \frac{\partial \psi_n(r,\theta)}{\partial \theta} \cos k_n z, \tag{13}
$$

$$
v_\theta = \frac{1}{r}\frac{\partial \phi_0(r,\theta)}{\partial \theta}\left(1 - \frac{z^2}{h^2}\right) - \mathrm{Re}\left[\sum_{n=1}^{\infty} \frac{1}{\alpha_n^2 r} \frac{\partial \phi_n(r,\theta)}{\partial \theta} \frac{dq_n(z)}{dz}\right]
$$

$$
+ \sum_{n=0}^{\infty} \frac{\partial \psi_n(r,\theta)}{\partial r} \cos k_n z, \tag{14}
$$

$$
v_z = \mathrm{Re}\left[\sum_{n=1}^{\infty} \phi_n(r,\theta) q_n(z)\right], \tag{15}
$$

$$
p = -\frac{2\mu}{h}\left\{\frac{\phi_0(r,\theta)}{h} + \mathrm{Re}\left[\sum_{n=1}^{\infty} \phi_n(r,\theta)\frac{\cos \alpha_n z}{\cos \alpha_n h}\right]\right\}. \tag{16}
$$

The functions ϕ_n and ψ_n are solutions of Eq. (5), which is well known. The function ϕ_0 can be taken as any linear combination of the following:

$$
r^s \cos s\theta, \quad r^s \sin s\theta, \quad r^{-s} \cos s\theta, \quad r^{-s} \sin s\theta, \tag{17}
$$

Similarly, $\phi_n(r, \theta)$ can be taken as any linear combination of

$$
\begin{aligned}
I_s(\alpha_n r)\cos s\theta, \quad I_s(\alpha_n r)\sin s\theta, \\
K_s(\alpha_n r)\cos s\theta, \quad K_s(\alpha_n r)\sin s\theta,
\end{aligned} \tag{18}
$$

and $\psi_n(r, \theta)$ can be taken as any linear combination of

$$
\begin{aligned}
I_s(k_n r)\cos s\theta, \quad I_s(k_n r)\sin s\theta, \\
K_s(k_n r)\cos s\theta, \quad K_s(k_n r)\sin s\theta,
\end{aligned} \tag{19}
$$

where $s = 0, 1, 2, \ldots$ and I_s, K_s are modified Bessel functions of the first and second kind of order s, respectively. By virtue of the boundary condition Eq. (3), terms involving r^s ($s > 1$), $I_s(\alpha_n r)$, and $I_s(k_n r)$ for $s \geq 0$ must be excluded. For $K_s(\alpha_n r)$ to be admissible, Re $(\alpha_n) > 0$ is required.

Flow Around a Circular Post

For a circular post, it is sufficient to give an exact solution by taking $s = 1$. Hence, Eqs. (13) to (16) become

$$v_r = \left\{ \left(U + b\frac{a^2}{r^2} \right) \left(1 - \frac{z^2}{h^2} \right) - \mathrm{Re} \left[\sum_{n=1}^{\infty} \frac{A_n + iB_n}{\alpha_n} \frac{K_1'(\alpha_n r)}{K_1(\alpha_n a)} \frac{dq_n(z)}{dz} \right] \right.$$

$$\left. - \sum_{n=0}^{\infty} \frac{c_n}{k_n r} \frac{K_1(k_n r)}{K_1(k_n a)} \cos k_n z \right\} \cos \theta, \tag{20}$$

$$v_\theta = \left\{ \left(-U + b\frac{a^2}{r^2} \right) \left(1 - \frac{z^2}{h^2} \right) + \mathrm{Re} \left[\sum_{n=1}^{\infty} \frac{A_n + iB_n}{\alpha_n^2 r} \frac{K_1(\alpha_n r)}{K_1(\alpha_n a)} \frac{dq_n(z)}{dz} \right] \right.$$

$$\left. + \sum_{n=0}^{\infty} \frac{c_n K_1'(k_n r)}{K_1(k_n a)} \cos k_n z \right\} \sin \theta, \tag{21}$$

$$v_z = \mathrm{Re} \left\{ \sum_{n=1}^{\infty} (A_n + iB_n) \frac{K_1(\alpha_n r)}{K_1(\alpha_n a)} q_n(z) \right\} \cos \theta, \tag{22}$$

$$p = p_0 - \frac{2\mu}{h} \left\{ \left(\frac{Ur}{h} - b\frac{a^2}{rh} \right) \right.$$

$$\left. + \mathrm{Re} \left[\sum_{n=1}^{\infty} (A_n + iB_n) \frac{K_1(\alpha_n r)}{K_1(\alpha_n a)} \frac{\cos \alpha_n z}{\cos \alpha_n h} \right] \right\} \cos \theta. \tag{23}$$

Here $b, c_0, c_1, \ldots, A_1, B_1, \ldots, p_0$ are unknown real constants. Because all terms vanish at infinity, except the first terms in v_r and v_θ, Eqs. (20) to (22) satisfy the boundary condition Eq. (3). The unknown constants can be determined by Eq. (1).

As $\cos k_n z$ forms a complete set of orthogonal functions in the interval $(0, h)$, we can expand the velocities v_r and v_θ at $r = \alpha$ into a half-range Fourier series. Since

$$1 - \frac{z^2}{h^2} = \sum_{m=0}^{\infty} d_m \cos k_m z, \tag{24}$$

$$-\frac{1}{\alpha_n} \frac{dq_n}{dz} = \sum_{n=0}^{\infty} e_{nm} \cos k_m z, \tag{25}$$

where

$$d_m = \frac{32(-1)^m}{\pi^3 (2m+1)^3} \quad \text{and} \quad e_{nm} = \frac{4(-1)^m k_m \alpha_n}{h^2 (\alpha_n^2 - k_m^2)^2}, \tag{26}$$

v_r and v_θ from Eqs. (20) and (21) can be reduced into the following forms:

$$v_r = \sum_{n=0}^{\infty} C_n(r) \cos \theta \cos k_n z, \tag{27}$$

$$v_\theta = \sum_{n=0}^{\infty} D_n(r) \sin\theta \cos k_n z. \tag{28}$$

Then the no-slip condition on the surface of the post $r = \alpha$ requires that

$$C_n(a) = 0, \quad D_n(a) = 0 \quad (n = 0, 1, 2, \cdots). \tag{29}$$

As to v_z, the coefficients in the series in Eq. (22) are not easy to determine: The functions $q_1(z), q_2(z) \ldots$ are not orthogonal in the range $(-h \leq z \leq h)$. Lee and Fung (1969) renormalized the series in Eq. (22) to a series of orthonormal functions, then truncated the series and obtained the coefficients numerically. They found v_z to be much smaller than v_θ, v_r. Hence, alternatively, we can obtain the required number of equations by the method of collocation, applying Eq. (22) to the boundary condition Eq. (1) at a suitable number of points on the vertical cylinder.

Figure 5.11:2 shows the functions $v_r/\cos\theta$ and $v_\theta/\sin\theta$ at $z = 0$, and 10 times $v_z/\cos\theta$ at $z = 0.5\,h$ for the case $h/a = 1$. The horizontal axis $(r - a)/a$ is the distance to the cylinder expressed in units of cylinder radius. The v_θ depends very much on the thickness ratio, h/a, increasing from 0 to a maximum at $(r - a)$ about h, and the maximum value of v_θ decreases as h/a increases.

The velocity component v_z is seen to be very small compared with v_r and v_θ. Note that the scale for v_z is magnified 10 times in Figure. 5.11:2 as compared with v_r, v_θ. v_z is significant only within a distance about h from the surface of the cyclinder. For $h/a = 0.1$, $|v_z|$ is less than $0.002U$, and is too small to be plotted.

In Figure 5.11:3 velocity profiles at certain radii are plotted against z for $h/a = 1$. If h/a is much less than 1, then the velocity profiles for v_r and v_θ are nearly parabolic. The profile for v_z is described mainly by the eigenfunction $q_1(z)$ and is not parabolic.

With this solution, the resistance to flow can be computed from the frictional forces acting on the cylindrical post and the plates. If the drag of the post is denoted by D, we can define a dimensionless coefficient f_D,

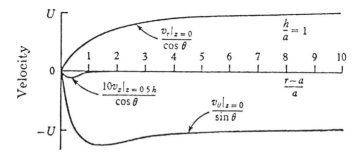

FIGURE 5.11:2. Velocity distributions versus $(r - a)/a$ for the cases $h/a = 1$. From Lee and Fung (1969). Reproduced by permission.

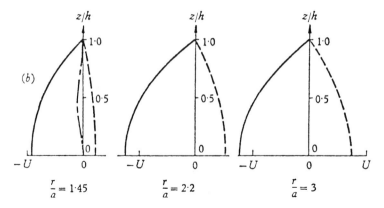

FIGURE 5.11:3. Velocity profiles in the z direction at several radial stations. $h/a = 1$. ——, $v_\theta/\sin\theta$; - - - -, $v_r/\cos\theta$; —·—·—, $10v_z/\cos\theta$. From Lee and Fung (1969). Reproduced by permission.

$$f_D = \frac{D}{4\pi\mu Uh}. \tag{30}$$

Due to the presence of the post, the stresses on the plates are modified by comparison with those for pure Poiseuille flow. The total additional resistance of the flow is not the drag acting on the post alone, but can be computed by summation of all the perturbation forces acting on a cylindrical surface at very large radius concentric with the post. At such a surface $r = r_0$ with $r_0 \to \infty$, perturbed strain rates and momentum fluxes are at most on the order of $1/r_0^3$, while perturbed pressure is on the order of $1/r_0$. Thus the resistance of the flow is solely due to the pressure over the surface $r = r_0$:

$$F_R = -\int_{-h}^{h}\int_{0}^{2\pi} p\Big|_{r=r_0} r_0 \cos\theta \, d\theta \, dz = \frac{4\pi\mu Ua^2}{h}\left(\frac{r_0^2}{a^2} - b\right). \tag{31}$$

The first term in parentheses is the resistance of pure Poiseuille flow. The second term is the additional resistance due to the presence of the post. We shall define a dimensionless coefficient f_R for the "additional" resistance as follows:

$$f_R = \frac{F_R}{4\pi\mu Uh} - \frac{r_0^2}{h^2} = -\frac{a^2}{h^2}b. \tag{32}$$

In Figure 5.11:4, f_D and f_R are plotted against h/a. As f_D and f_R represent, respectively, the drag on the post and the additional resistance on the post and the plates due to the presence of the post, the figure shows that for the same h the drag and resistance at first drop quickly as the radius of the post decreases and then gradually level off to small values. Figure 5.11:4 shows that the value of f_D is about twice f_R.

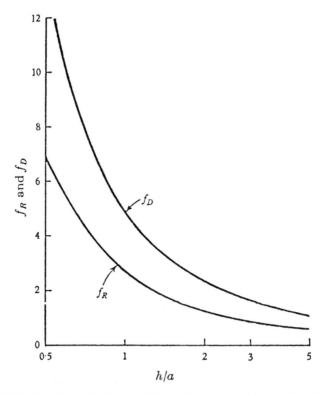

FIGURE 5.11:4. The dimensionless coefficients f_D, representing the drag force acting on the post, and f_R, representing the total resistance over and above the Poiseuillian value, as functions of the h/a ratio. From Lee and Fung (1969), by permission.

5.12 Force of Interaction of Leukocytes and Vascular Endothelium

As an example of using dimensional analysis and model testing to study the mechanics of the microcirculation, the first attempt at measuring the force of interaction of activated white blood cells and the blood vessel endothelium is recorded here. Red blood cells do not stick to the endothelium unless the latter is damaged. White blood cells, however, are often seen sticking to the endothelial wall or rolling slowly on it, while the plasma and the red cells whiz by around them. The sticking phenomenon is illustrated in Figure 5.12:1, which is a view taken from the omentum of an anesthetized rabbit. In similar views taken from a high-speed cinemicrograph, individual red blood cells can be identified. The motion of the red cells, plasma, and leukocytes was measured from successive frames of such a cinemicrograph by Schmid-Schönbein et al. (1975). From the kinematic data they determined the shear force acting on a leukocyte sticking to the wall of a venule by dimensional analysis and testing in dynamically similar models.

Dimensional Analysis

Two flow systems are said to be *kinematically similar* if their geometric shapes are similar, although their linear dimensions can be different. Two systems are said to be *dynamically similar* if the systems of differential equations describing their motion, written in dimensionless form, are the same. If two systems are both kinematically and dynamically similar, then

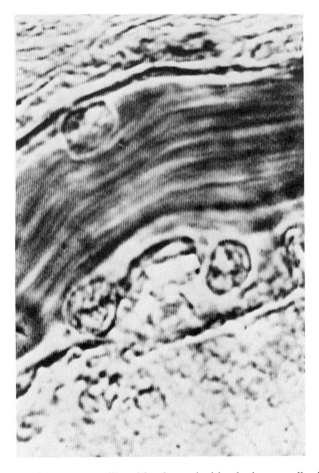

FIGURE 5.12:1. A 23 μm postcapillary blood vessel with a leukocyte adhering to the endothelium of rabbit omentum. The red cells move in a stream with clear streamlines, but individual cells are not recognizable. Photographed from video screen by Dr. Geert Schmid-Schönbein.

TABLE 5.12:1. List of Variables*

Symbol	Definition	Dimensions
d_c	Diameter of the leukocyte considered as a sphere	L
d_t	Diameter of the blood vessel	L
V_M	Maximum velocity of undisturbed flow in the blood vessel	LT^{-1}
V_c	Linear velocity of the centroid of the white blood cell	LT^{-1}
μ	Coefficient of viscosity of plasma	$ML^{-1}T^{-1}$
ρ	Density of plasma ($\bar{\rho}$ is that of the silicone oil)	ML^{-3}
H	Hematocrit of the blood, i.e., the volume fraction of the cellular content in the blood	—

* All symbols with an overbar refer to the model. For example, μ signifies the viscosity of plasma and $\bar{\mu}$ denotes the viscosity of the silicone oil in a model experiment.

the dimensionless variables (expressing stresses, velocities, displacements, etc. nondimensionalized by forming ratios to characteristic values of these variables) will have exactly the same values at corresponding points in the two systems. This is the basic *principle of similarity*.

In the blood flow problem, the principal variables describing the geometry and motion of our system are listed in Table 5.12:1. In this table, the hematocrit H is dimensionless. From the remaining six variables, three independent *dimensionless parameters* can be formed. The following is a convenient set:

$$d_c/d_t, \quad V_c/V_M, \quad V_M d_c \rho/\mu. \tag{1}$$

All other dimensionless parameters are functions of this set of parameters. For example, if we are interested in the resultant shear force, S (dynes), imparted to the white blood cell by the blood flow, we note that the ratio $S/(V_M \mu d_c)$ is dimensionless, and we can express this ratio as a function of the three dimensionless parameters listed in Eq. (1) and the hematocrit H:

$$\frac{S}{V_M \mu d_c} = f\left(\frac{d_c}{d_t}, \frac{V_c}{V_M}, \frac{V_M d_c \rho}{\mu}, H\right), \tag{2}$$

where $f(\ldots)$ denotes a relationship that can be determined either by solving the differential equations or by using a model experiment. We notice that the parameter $V_M d_c \rho/\mu$ is a *Reynolds number* (N_R). The other parameter $S/(V_M \mu d_c)$ will be called the *shear coefficient* and is denoted by C_S; thus,

$$C_S = \frac{S}{V_M \mu d_c}, \tag{3}$$

$$N_R = \frac{\rho V_M d_c}{\mu}, \tag{4}$$

and Eq. (2) can be written as

$$C_S = f\left(N_R, d_c/d_t, V_c/V_M, H\right). \tag{5}$$

If we use a bar over a variable to indicate that it belongs to a geometrically similar model, then

$$\overline{C_S} = f\left(\overline{N_R}, \overline{d_c}/\overline{d_t}, \overline{V_c}/\overline{V_M}, \overline{H}\right). \tag{6}$$

Our basic approach is to determine Eq. (6) by model experiments, then to identify $\overline{C_S}$ with C_S in Eq. (5), and to use Eq. (3) to compute the shear force S:

$$S = C_S \mu V_M d_c. \tag{7}$$

It is expected that the shear coefficient C_S is strongly dependent on the hematocrit. Since the hematocrit is stochastic in vivo, the shear force must fluctuate as time passes. The fluctuation is quasistatic, because the Reynolds number of flow is very small.

The Prototype and Its Dynamically Similar Model

The prototype of the leukocyte-endothelium interaction is shown by pictures, such as Figure 5.12:1. The essence is sketched in Figure 5.12:2, which shows a white blood cell adhering to a blood vessel wall. The cell is subjected to forces imparted on its surface by the plasma and the red blood cells, by pressure variations in the flowing blood, by a shear stress at the interface of the white blood cell and the endothelium, by a variable normal stress σ on the interface, and by a body force due to acceleration and gravity. The blood vessel is small. The leukocyte is either stationary or rolling slowly. The reynolds number of flow in the venule is small (of the order of 10^{-2}). The normal stress σ acting on the interface between the endothelium and the white blood cell must balance the forces and moments of the pressure and shear acting on the cell. Since the surface traction acting on the white blood cell has a resultant torque SL that does not vanish, where L respresents the moment arm, as shown in Figure 5.12:2, the normal stress σ on the interface cannot be uniform and must have a resultant moment equal to SL.

From the point of view of fluid dynamics the system is very complex for mathematical analysis. Hence it is proposed to gain a quantitative understanding through experimental measurements on a physical model.

The physical model is sketched in Figure 5.12:3. The plasma was modeled by silicone oil with a coefficient of viscosity of 47 poise. The red blood cells were modeled by gelatin pellets. The white blood cells were modeled by rigid spheres of certain diameters ($\overline{d_c}$ = 0.63 cm, 0.55 cm, and 0.4 cm) mounted on a lever and introduced into a tube through a boring 0.65 cm in diameter. The blood vessel was modeled by a circular cylindrical tube ($\overline{d_t}$ = 2.25 cm) connected to a container filled with a silicone oil–gelatin pellet suspension. The fluid was allowed to flow under gravity, using a valve to regu-

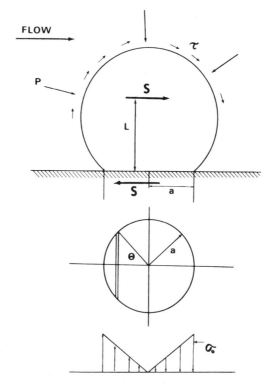

FIGURE 5.12:2. Cross-sectional view of an adhering white blood cell. P is the hydrostatic pressure, τ is the shear stress exerted on the surface of the white blood cell, and S is the resultant shear force. The two lower diagrams show the assumed contact area and the normal stress distribution on that contact area. A uniform normal stress acting on the contact area due to pressure over the cell is omitted from the lowest diagram: Only the part needed to resist rolling of the cell is shown. See text for other abbreviations. From Schmid-Schönbein et al. (1975). Reproduced by permission of the American Heart Association, Inc.

late the velocity. The force acting on the white blood cell was measured by a force transducer attached to the lever, as shown in Figure 5.12:3.

As a tracer for velocity measurement, tiny air bubbles were introduced into the silicone oil simply by pouring the oil into the container. A camera with a lens of short depth of field (3 mm) was focused on the plane of symmetry of the sphere and the tube. The velocity, V_M, was measured from 16-mm film recordings.

Results

At zero hematocrit, the model experiment was performed with spheres of several diameters. Figure 5.12:4 shows that $\overline{C_S}$ calculated according to Eq. (3) increases with increasing Reynolds number and increasing $\overline{d_c}/\overline{d_t}$:

$$\overline{d_c}\big/\overline{d_t} = 0.17\text{:}\quad \overline{C_S} = 6.2 + 80N_R,$$
$$\overline{d_c}\big/\overline{d_t} = 0.24\text{:}\quad \overline{C_S} = 8 + 80N_R,$$
$$\overline{d_c}\big/\overline{d_t} = 0.28\text{:}\quad \overline{C_S} = 8.8 + 80N_R.$$

When the hematocrit is not zero, the shear force on the white cell fluctuates with time (because the shear force is influenced by the red blood cells, whose concentration is never really uniform). The mean shear force, as well as the maximum deviations, increased with increasing velocity of flow. Figure 5.12:5 shows the data for a hematocrit of 30%. The solid circles indicate the mean values averaged over a period of about 25 s, and the vertical bars indicate maximum deviations from these mean values. For comparison, we also plotted the shear force at zero hematocrit as a straight line, showing the striking effect of the red cells.

On the other hand, the animal experiments yielded micrographs such as those shown in Figure 5.12:1. These micrographs can be analyzed to obtain data on the velocity of flow in the tube (V_M), the diameter of the white blood cell (d_c), and the diameter of the venule (d_t). From these data and the viscosity of the plasma, wa can calculate the Reynolds number (N_R). The values of $\overline{C_S}$ were then determined for the corresponding values of N_R and d_c/d_t by interpolating linearly from the experimental values such as those

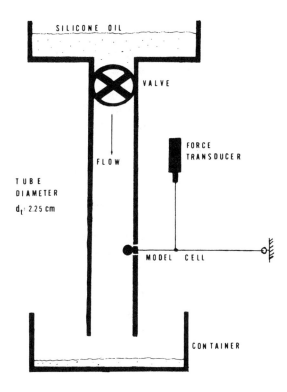

FIGURE 5.12:3. Cross-sectional view of the experimental model. The camera was focused precisely on the midplane of the tube in which the velocity of flow was determined. From Schmid-Schönbein et al. (1975). Reproduced by permission of the American Heart Association, Inc.

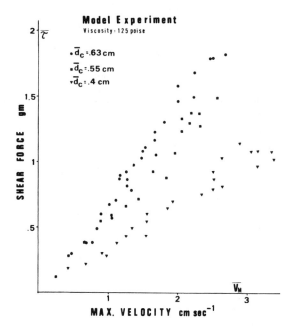

FIGURE 5.12:4. Shear force as a function of the maximum velocity of the tube flow for model white blood cells of different sizes at zero hematocrit. $\bar{d}_c$ = model white cell diameter. From Schmid-Schönbein et al. (1975). Reproduced by permission of the American Heart Association, Inc.

shown in Figures 5.12:4 and 5.12:5. The results show that the resultant shear force, S, acting on a white blood cell rolling on the endothelial wall lies in the range of 4 to 45×10^{-6} dyne at $H = 0$, and

$$8 \text{ to } 90 \times 10^{-6} \text{ dyne at } H = 20\%,$$
$$16 \text{ to } 180 \times 10^{-6} \text{ dyne at } H = 30\%,$$
$$21 \text{ to } 234 \times 10^{-6} \text{ dyne at } H = 40\%. \tag{8a}$$

H is the hematocrit. This is the average shear force (averaged over time); the instantaneous maximum values are higher.

To determine the *stress* of interaction between the white cell and the endothelium of the venule, we must know their area of contact. By estimating the contact length from the photographs and assuming the contact area to be circular, some rough estimation of the shear stress can be obtained. The results are

$$50 \text{ to } 200 \text{ dyne/cm}^2 \text{ at } H = 0,$$
$$100 \text{ to } 400 \text{ dyne/cm}^2 \text{ at } H = 20\%,$$
$$200 \text{ to } 800 \text{ dyne/cm}^2 \text{ at } H = 30\%,$$
$$265 \text{ to } 1060 \text{ dyne/cm}^2 \text{ at } H = 40\%. \tag{8b}$$

These are again estimated average values (over time). Instantaneous shear stress can be higher.

It would be interesting to estimate the maximum normal stress of interaction that is required to prevent rolling of the white blood cell over the endothelium. For this purpose, let us assume that the normal stress is a linear function of x, as shown in the lower sketch of Figure 5.12:2. Let the area of contact be a circle of radius a, and let the maximum normal stress at the outer edge be σ_0. Then the total moment is

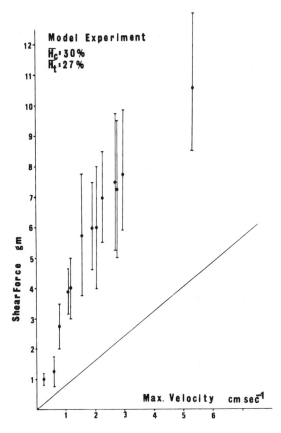

FIGURE 5.12:5. Shear force $\bar{S}$ acting on a model white cell with diameter $\bar{d}_c = 0.63$ cm plotted as a function of the average centerline velocity, $\bar{V}_M$, for a given hematocrit. The hematocrit in the container ($\bar{H}_c$) of 30% corresponds to a hematocrit in the tube ($\bar{H}_1$) of 27%. The solid circles correspond to a mean value integrated over a period of 10 to 15 s, and the vertical bars give the maximum deviations during this time period. For comparison, the shear force $\bar{S}$ at zero hematocrit for the same model cell is plotted in the lower part of the figure as a straight line through the origin. From Schmid-Schönbein et al. (1975). Reproduced by permission of the American Heart Association, Inc.

$$4 \int_0^{\pi/2} \sigma_0 a^3 \sin^2 \theta \cos^2 \theta \, d\theta = \frac{\sigma_0 a^3}{4} \pi, \tag{9}$$

which must be equal to the torque SL, as discussed earlier. Hence

$$\sigma_0 = \frac{4SL}{\pi a^3}. \tag{10}$$

For the assumed white blood cell considered previously, with $S = 10^{-5}$ dyne, $a = 2.5\,\mu m$, and $L = 3.5\,\mu m$, we obtain

$$\sigma_0 = 285 \ \text{dyne}/\text{cm}^2.$$

On the other hand, if we assume a uniform tension for $x < 0$ and a uniform compression for $x > 0$, we obtain a stress $\sigma = 3SL/(4a^3)$ 168 dyne/cm². The corresponding shear stress, assuming a constant stress distribution, is 50 dyne/cm².

5.13 Local Control of Blood Flow

The vascular smooth muscles respond to many substances as well as to mechanical and nervous stimulations. In the microvascular bed, changes in the chemical environment of the vascular smooth muscle cells occur due to metabolism of the tissue, release of catecholamines from the nerve fibers, or material transfer across the endothelium. These changes may affect the muscle tone, which in turn controls the vessel diameter and flow resistance. Exercise, drugs, metabolic conditions, and nutritional needs can thus affect local blood flow in an organ.

The vascular smooth muscle cells also respond to stress and strain. A step increase in tension may stimulate the muscle to contract and shorten. A step decrease in tension may cause the muscle to relax and lengthen. A step shortening of the muscle stimulates it to relax. A step stretch of the muscle causes it to contract. These active myogenic properties are called *Bayliss phenomena* (Bayliss, 1902). They are opposite to what one would expect from a passive material. We see them in some other muscles, for example, in the earthworm. If you press your finger on an earthworm, it will swell up when you take your finger off. If you stretch the earthworm it will contract when you release it.

In most organs the factors just mentioned act to control the arteriolar segment of the vascular tree in such a way that the capillary pressure remains relatively constant when blood pressure in the aorta changes. The following phenomena provide evidence of local control of blood flow.

Autoregulation of Blood Flow

Autoregulation is the tendency for blood flow to remain constant in the face of changes in local arterial pressure to the organ. It is seen in virtually all

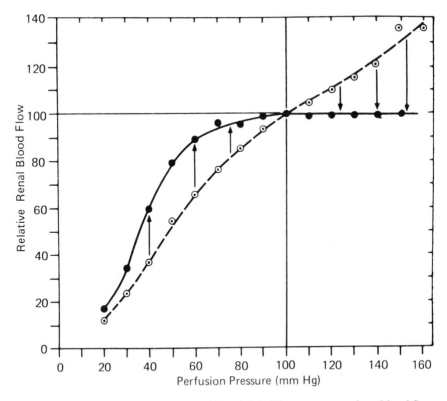

FIGURE 5.13:1. Autoregulation of renal blood flow. The two curves show blood flow in steady-state conditions (continuous line) and transient values following a sudden shift of perfusion pressure from the control level of 100 mm Hg (dashed line). From Rothe et al. (1971). Reproduced by permission.

organs of the body. It is most pronounced in brain and kidney. It is also evident in the myocardium, intestine, skeletal muscle, and liver. Figure 5.13:1 shows the pattern of steady-state flow in the kidney along with the transient values obtained immediately after pressure was altered. After a sudden change of arterial pressure in the range 90 to 160 mm Hg, the flow changed instantaneously but subsequently returned to its initial level. If the arterial pressure is reduced, there is often an initial decrease in diameter of the arterioles followed by a dilation, which causes the diameter to be increased above control levels, although the intravascular pressure is reduced. Periodic vasomotion of the arteriolar vessels often stops when arterial pressure is reduced.

Reactive Hyperemia

Reactive hyperemia is the period of elevated flow that follows a period of circulatory arrest. It exists in every vascular bed in the peripheral circulation. It is rapid on onset, with a notable flow elevation following even a few seconds of flow arrest in some vascular beds. The magnitude of the hyperemia is related to the duration of ischemia.

Venous–Arteriolar Response

The arteriolar beds of some organs (intestine, liver, and certain skeletal muscles) constrict when venous pressure is elevated. This is the myogenic response of the arterioles.

Functional Hyperemia

Functional hyperemia is the increase in blood flow that accompanies an increase in tissue activity. It occurs in skeletal and cardiac muscle, brain, intestine, stomach, salivary glands, kidney, and adipose tissue. There is evidence that endproducts of tissue metabolism play a role in functional hyperemia. PCO_2, H^+, lactate, pyruvate, adenosine, and sympathetic vasodilators (cholinergic and β-adrenergic) have vasodilator influences. Furthermore, potassium released from depolarized skeletal muscle cells at the initiation of exercise may diffuse to the vicinity of the vascular smooth muscle cells to cause vascular relaxation and a rapid increase in blood flow. Release of potassium may also contribute to the vasodilation in brain during increased cerebral activity.

Hormonal factors may also be important (Johnson, 1978). Physical factors may play a role too. For example, extravascular compression during muscular contraction may cause a relaxation of myogenic vascular tone in the arterioles of skeletal muscle. Emptying of the veins during muscle contraction increases the flow in limbs during exercise.

Myogenic Control

Figure 5.13:2 shows quantitative evidence of the Bayliss mechanism: elevation of static intravascular pressure in the arteriole during the no-flow condition leads to sustained contraction of the arteriole in mesentery (Johnson and Intaglietta, 1976). Further, a reduction of ambient pressure around an organ leads to sustained vasoconstriction (Greenfield, 1964). These are in line with reactive hyperemia discussed earlier. In these cases the flow is not increased so that the metabolite changes cannot be a significant factor. It has also been reported, however, that rapid pressure transients lead to powerful contraction, and that pulsatile perfusion leads to increased vascular tone. Hence the effect of blood flow, and the associated shear stress acting on the endothelium, is evident.

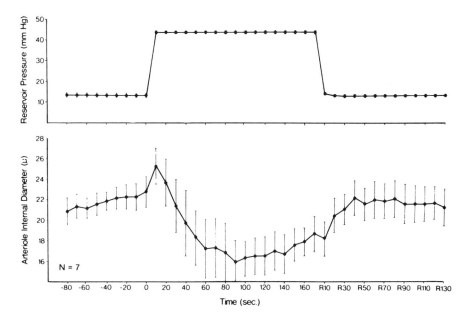

FIGURE 5.13:2. Effect of static pressure elevation in the vasculature of cat mesentery on arteriolar diameter. Data obtained during total flow arrest and equalization of arterial and venous pressure. From Johnson and Intaglietta (1976). Reproduced by permission.

What Is the Mechanism?

Many theories have been proposed to explain the phenomena described earlier but a resolution was difficult because most experiments were done in whole organs which allow multiple interpretations. As Paul Johnson (1980, 1991) said, the myogenic theory (Folkow, 1964) appears to be the winner. We shall return to this question after reviewing new results of experiments on isolated arteriole vessels in Section 7.10.

Problems

5.3 Give a complete mathematical formulation of the following problems. Write down all the required differential equations and boundary conditions. State all the assumptions clearly and explicitly.

A series of equally spaced rigid spheres suspended in an incompressible Newtonian viscous fluid flowing down a rigid circular cylindrical cylinder in a single file, in symmetric fashion (Wang and Skalak, 1969).

5.4 Replace the rigid spheres in the preceding problem by a series of droplets of oil that is immiscible with the carrying fluid. Surface tension should be considered.

5.5   Now consider flexible pellets resembling red blood cells. Revise the theory (Zarda et al., 1977).

5.6   Relax further. First, let the tube be elastic but impermeable to the fluid. Then let the wall of the tube be permeable so that fluid exchange between the fluid, the tube wall, and the medium outside the tube takes place. Make suitable assumptions with regard to the transport characteristics of the membrane and the movement of fluid outside the tube.

5.7   Further generalize the problems described earlier by letting the particles be red blood cells and the tube be capillary blood vessels. Formulate a mathematical theory.

5.8   Discuss conditions in capillary blood flow that are nonstationary in nature. Formulate these transient problems. One example is the famous Landis experiment in which the flow is suddenly stopped by compressing a capillary blood vessel with a microneedle. The movement of the red cell is then recorded and the result is used to study the permeability of the blood vessel wall (Lew and Fung, 1969c).

5.9   Again, think of the real capillary bed. Formulate a theory in which the blood vessels are not represented by cylindrical tubes. Discuss the practical significance of such an investigation.

Capillary blood vessels are usually not circular in cross section; nor are they cylindrical. The asymmetry and unevenness of the wall have a significant effect on the force of interaction between vessel walls and blood cells, especially the leukocytes. This is an important problem for future investigation.

5.10   Formulate a mathematical problem to analyze the tank-treading of a red cell flowing in a capillary blood vessel (see Sec. 5.10, and Secomb and Skalak, 1982).

5.11   Formulate a mathematical theory for the leucocyte problem discussed in Sec. 5.11.

References

Aroesty, J., and Gross, J.F. (1970). Convection and diffusion in the microcirculation. *Microvasc. Res.* **2**: 247–267.

Atherton, A., and Born, G.V.R. (1972). Quantitative investigations of the adhesiveness of circulating polymorphonuclear leucocytes to blood vessel walls. *J. Physiol.* **222**: 447–474.

Baker, M., and Wayland, H. (1974). On-line volumetric flow rate and velocity profile measurement for blood microvessels. *Microvasc. Res.* **7**: 131–143.

Bayliss, W.M. (1902). On the local reactions of the arterial wall to changes in internal pressure. *J. Physiol. (Lond.)* **28**: 220–231.

Bungay, P.M., and Brenner, H. (1973). The motion of a closely-fitting sphere in a fluid-filled tube. *Int. J. Multiphase Flow* **1**: 25–56.

Caro, C.G., Pedley, T.J., Schroter, R.C., and Seed, W.A. (1978). *The Mechanics of the Circulation.* Oxford University Press, Oxford.

Chen, T.C., and Skalak, R. (1970). Spheroidal particle flow in a cylindrical tube. *Appl. Sci. Res.* **22**: 403–441.

Desjardins, C., and Duling, B.R. (1987). Microvessel hematocrit: measurement and implications for capillary oxygen transport. *Am. J. Physiol.* **252**: H494–H503.

Desjardins, C., and Duling, B.R. (1990). Heparinase treatment suggests a role for the endothelial cell glycocalyx in regulation of capillary hematocrit. *Am. J. Physiol.* **258**: H647–H654.

Duling, B.R., and Desjardins, C. (1987). Capillary hematocrit—what does it mean? *News in Physiol. Sci.* **2**: 66–69.

Fischer, T.M., Stohr-Liesen, M., and Schmid–Schönbein, H. (1978). The red cell as a fluid droplet: Tank tread-like motion of the human erythrocyte membrane in shear flow. *Science* **202**: 894–896.

Folkow, B. (1964). Description of the myogenic hypothesis. *Circ. Res.* **15**: I.279–I.285.

Folkow, B., and Neil, E. (1971). *Circulation.* Oxford University Press, London.

Fung, Y.C. (1965). *Foundations of Solid Mechanics.* Prentice-Hall, Englewood Cliffs, NJ.

Fung, Y.C. (1973). Stochastic flow in capillary blood vessels. *Microvasc. Res.* **5**: 34–48.

Fung, Y.C. (1990). *Biomechanics: Motion, Flow, Stress, and Growth.* Springer Verlag, New York.

Fung, Y.C. (1993). *Biomechanics: Mechanical Properties of Living Tissues.* 2nd ed., Springer-Verlag, New York.

Gaehtgens, P. (1980). Flow of blood through narrow capillaries: Rheological mechanisms determining capillary hematocrit and apparent viscosity. *Biorheology J.* **17**: 183–189.

Greenfield, A.D.M. (1964). Blood flow through the human forearm and digits as influenced by subatmospheric pressure and venous pressure. *Circ. Res.* **14**:I.0–I.75.

Hele-Shaw, H.S. (1897, 1898). *Phil. Trans. Roy. Inst. Nav. Arch.* **41**: 21. **42**: 49.

Hele-Shaw, H.S., and Hag, A. (1898). *Royal Soc. Phil. Trans.* **A 195**: 303.

Hochmuth, R.M., and Sutera, S.P. (1970). Spherical caps in low Reynolds-number tube flow. *Chem. Eng. Sci.* **25**: 593–604.

Hyman, W.A., and Skalak, R. (1972). Non-Newtonian behavior of a suspension of liquid drops in fluid flow. *Am. Inst. Chem. Eng. J.* **18**: 149–154.

Jendrucko, R.J., and Lee, J.S. (1973). The measurement of hematocrit of blood flowing in glass capillaries by microphotometry. *Microvasc. Res.* **6**: 316–331.

Johnson, P.C. (1978). *Peripheral Circulation.* Wiley, New York.

Johnson, P.C. (1980). The myogenic response. In *Handbook of Physiology*, Sec. 2. *The Cardiovascular System*, Vol. 2. *Vascular Smooth Muscle* (D.F. Bohr, A.P. Somlyo, and H.V. Sparks, Jr., eds.). American Physiological Society, Bethesda, MD, pp. 409–442.

Johnson, P. C. (1991). The myogenic response. *News in Physiol. Sci.* **6**: 41–42.

Johnson, P.C., and Intaglietta, M. (1976). Contributions of pressure and flow sensitivity to autoregulation in mesenteric arterioles. *Am. J. Physiol.* **231**: 1686–1698.

Kaley, G., and Altura, B.M. (1977). *Microcirculation*, Vols. 1 & 2. University Park Press, Baltimore, MD.

Klitzman, B., and Duling, B.R. (1979). Microvascular hematocrit and red cell flow in resting and contracting striated muscle. *Am. J. Physiol.* **237**: H481–H490.

Lamb, H. (1932). *Hydrodynamics*, 6th ed. Cambridge University Press. Reprinted by Dover, New York.

Lee, J.S., and Fung, Y.C. (1969). Stokes flow around a circular cylindrical post confined between two parallel plates. *J. Fluid Mech.* **37**: 657–670.

Lee, T.Q., Schmid-Schönbein, G.W., and Zweifach, B.W. (1983). An application of an improved dual-slit photometric analyzer for volumetric flow rate measurements in microvessels. *Microvasc. Res.* **26**: 351–361.

Lew, H.S., and Fung, Y.C. (1969a). On the low-Reynolds-number entry flow into a circular cylindrical tube. *J. Biomech.* **2**: 105–119.

Lew, H.S., and Fung, Y.C. (1969b). The motion of the plasma between the red blood cells in the bolus flow. *J. Biorheology* **6**: 109–119.

Lew, H.S., and Fung, Y.C. (1969c). Flow in an occluded circular cylindrical tube with permeable wall. *Z. Angew. Math. Physik* **20**: 750–766.

Lew, H.S., and Fung, Y.C. (1979a). Plug effect of erythrocytes in capillary blood vessels. *Biophys. J.* **10**: 80–99.

Lew, H.S., and Fung, Y.C. (1970b). Entry flow into blood vessels at arbitrary Reynolds number. *J. Biomech.* **3**: 23–38.

Lighthill, M.J. (1968). Pressure-forcing of tightly fitting pellets along fluid-filled elastic tubes. *J. Fluid Mech.* **34**: 113–143.

Lipowsky, H.H., and Zweifach, B.W. (1977). Methods for the simultaneous measurement of pressure differentials and flow in single unbranched vessels of the microcirculation for rheological studies. *Microvasc. Res.* **14**: 345–361.

Lipowsky, H.H., and Zweifach, B.W. (1978). Application of the "two-slit" photometric technique to the measurement of microvascular volumetric flow rates. *Microvasc. Res.* **15**: 93–101.

Lipowsky, H.H., Usami, S., and Chien, S. (1980). In vivo measurements of "apparent viscosity" and microvessel hematocrit in the mesentery of the cat. *Microvasc. Res.* **19**: 297–319.

Majno, G. (1965). Ultrastructure of the vascular membrane. In *Handbook of Physiology*, Sec. 2 *Circulation*, Vol. 3. (W.F. Hamilton, and P. Dow, eds.). American Physiological Society, Washington, D.C. pp. 2293–2375.

Nellis, S.N., and Zweifach, B.W. (1977). A method for determining segmental resistances in the microcirculation from pressure-flow measurements. *Circ. Res.* **40**: 546–556.

Norberg, K.A., and Hamberger, B. (1964). The sympathetic adrenergic neuron. *Acta Physiol. Scand* . **63** (Suppl. 238).

Oseen, C.W. (1910). Über die Stokessche Formel und über die verwandte Aufgabe in der Hydrodynamik. *Arkiv Mat. Astron. Fysik* **6**(29).

Prothero, J., and Burton, A.C. (1961, 1962). The physics of blood flow in capillaries. I. The nature of the motion. *Biophys. J.*, **1**: 567–579. II. The capillary resistance to flow. *Biophys. J.* **2**: 199–212.

Proudman, I., and Pearson, J.R.A. (1957). Expansions at small Reynolds number for the flow past a sphere and a circular cylinder. *J. Fluid Mech.* **2**: 237–262.

Rothe, C.F., Nash, F.D., and Thompson, D.E. (1971). Patterns in autoregulation of renal blood flow in the dog. *Am. J. Physiol.* **220**: 1621–1626.

Rouse, H. (1959). *Advanced Fluid Mechanics*. Wiley, New York.

Sagawa, K., Kumoda, M., and Schramm, L.P. (1974). Nervous control of the circulation. In *Cardiovascular Physiology*, Vol. 1 (A.C. Guyton and C.E. Jones, eds). Butterworths, London, pp. 197–232.

Schlichting, H. (1962). *Boundary Layer Theory*. McGraw–Hill, New York.

Schmid-Schönbein, H., and Wells, R.E. (1969). Fluid drop-like transition of erythrocytes under shear. *Science* **165**: 288–291.

Schmid-Schönbein, G.W., Fung, Y.C., and Zweifach, B. (1975). Vascular endothelium-leucocyte interaction: Sticking shear force in venules. *Circ. Res.* **36**: 173–184.

Schmid-Schönbein, G.W., Skalak, R., Usami, S., and Chien, S. (1980a). Cell distribution in capillary networks. *Microvasc. Res.* **19**: 18–44.

Schmid-Schönbein, G.W., Usami, S., Skalak, R., and Chien, S. (1980b). The interaction of leukocytes and erythrocytes in capillary and postcapillary vessels. *Microvasc. Res.* **19**: 45–70.

Secomb, T.W., and Skalak, R. (1982). A two-dimensional model for capillary flow of an asymmetric cell. *Microvasc. Res.* **24**: 194–203.

Skalak, R., Chen, P.H., and Chien, S. (1972). Effect of hematocrit and rouleaux on apparent viscosity in capillaries. *Biorheology* **9**: 67–82.

Skalak, R., Tozeren, A., Zarda, P.R., and Chien, S. (1973). Strain energy function of red cell membranes. *Biophys. J.* **13**: 245–264.

Smaje, L., Zweifach, B.W., and Intaglietta, M. (1970). Micropressures and capillary filtration coefficients in single vessels of the cremaster muscle of the rat. *Microvasc. Res.* **2**: 96–110.

Stokes, G.G. (1851). On the effect of the internal friction of fluids on the motion of pendulums. *Trans. Cambridge Philosophical Soc.* **9**: 8; *Mathematical and Physical Papers*, Vol. 3, pp. 1–141.

Stokes, G. (1898). *Rep. British. Assoc.* **A 144** (also, papers, Vol. 3, p. 278).

Targ, S.M. (1951). *Basic Problems of the Theory of Laminar Flows* (in Russian). Moscow.

Tong, P., and Fung, Y.C. (1971). Slow viscous flow and its application to biomechanics. *J. Appl. Mech.* **38**: 721–728.

Tözeren, H., and Skalak, R. (1978). The steady flow of closely fitting incompressible elastic spheres in a tube. *J. Fluid Mech.* **87**: 1–16.

Tözeren, H., and Skalak, R. (1979). Flow of elastic compressible spheres in tubes. *J. Fluid Mech.* **95**: 743–760.

Wang, H., and Skalak, R. (1969). Viscous flow in a cylindrical tube containing a line of spherical particles. *J. Fluid Mech.* **38**: 75–96.

Wayland, H. (1982). A physicist looks at the microcirculation. *Microvasc. Res.* **23**: 139–170.

Wiedeman, M.P., Tuma, R.F., and Mayrovitz, H.N. (1981). *An Introduction to Microcirculation*. Academic Press, New York.

Wiederhielm, C.A., Woodbury, J.W., Kirk, S., and Rushmer, R.F. (1964). Pulsatile pressure in microcirculation of the frog's mesentery. *Am. J. Physiol.* **207**: 173–176.

Yen, R.T., and Fung, Y.C. (1978). Effect of velocity distribution on red cell distribution in capillary blood vessels. *Am. J. Physiol.* **235**: H251–H257.

Yih, C.S. (1977). *Fluid Mechanics*. Corrected edition, West River Press, Ann Arbor, MI.

Zarda, P.R., Chien, S., and Skalak, R. (1977). Interaction of a viscous incompressible fluid with an elastic body. In *Computational Methods for Fluid-Structure Inter-*

action Problems (T. Belytschko and T.L. Geers, eds.), American Society of Mechanical Engineers, New York, pp. 65–82.

Zweifach, B.W. (1974). Quantitative studies of microcirculatory structure and function. I. Analysis of pressure distribution in the terminal vascular bed. *Circ. Res.* **34**: 843–857. II. Direct measurement of capillary pressure in splanchmic mesenteries. *Circ. Res.* **34**: 858–868.

Zweifach, B.W., and Lipowsky, H.H. (1977). Quantitative studies of microcirculatory structure and function. III. Microvascular hemodynamics of cat mesentery and rabbit omentum. *Circ. Res.* **41**: 380–390.

6
Blood Flow in the Lung

6.1 Introduction

We shall now apply the general principles discussed in the preceding chapters to one organ, the lung. The purpose is to illustrate, in one concrete example, the use of physical principles to an organ with a specific anatomy, histology, and mechanical properties to explain and predict the function of that organ in quantitative terms.

A mathematical analysis of any system requires a set of hypotheses. A good theory uses as few ad hoc hypotheses as possible. The solution of any system of equations requires the specification of boundary conditions. The solution will be more valuable if the boundary conditions are realistic. The basic equations must contain descriptions of the geometry, structure, and material properties of the system. The more closely these descriptions are based on experimental observations and less idealized, the more truthful will be the solutions. This is the creed of theoreticians. This is our aim.

The function of the lung is to oxygenate the blood and to remove CO_2. Nature chooses to do this by diffusion and chemical reactions, and for this purpose blood is spread out into very thin layers or sheets so that the blood–gas interfacial area becomes very large. In an adult human lung with a pulmonary capillary blood volume on the order of 150 ml, the pulmonary capillary blood–gas exchange area is of the order of $70 \, m^2$, so that the average computed thickness of the sheets of blood in the pulmonary capillaries is only about $4 \, \mu m$. The thin membrane that separates the blood from the air is less than $1 \, \mu m$ thick; it consists of a layer of endothelial cells, an interstitium, and a layer of epithelial cells. Each sheet of blood, bounded by two membranes, forms an *interalveolar septum*. Several billion septa form a space structure, which may be compared with a honeycomb (Malpighi, 1661, see translation by Young, 1930), or a bowl of soap bubbles. The smallest unit of space bounded by *interalveolar septa* is called the *alveolus*. In an adult human there are about 300 million alveoli. Each alveolus is polyhedral. It is bounded by interalveolar walls, with one or more sides missing in order to ventilate to the atmosphere.

The blood vessels of the lung must provide an adequate transit time for the erythrocytes to go through the capillary sheets so that the gas exchange process can be completed. Oxygenation of hemoglobin requires the oxygen to move across the alveolar–capillary membrane, through the plasma, and across the erythrocyte membrane. About half of the total time for the diffusion of oxygen from the alveolus to the red cell is spent in traversing the alveolar–capillary membrane. Of the total of approximately 3 s required to move blood from the pulmonary valve to the left atrium in humans, about 1 s is spent in the alveolar capillary bed. To accomplish complete oxygenation in such a short period of time requires not only a fast oxygenation and CO_2 exchange process in the blood, but also a low resistance system of pulmonary blood vessels to handle the output of the heart. (The resting cardiac output is very close to one total blood volume per minute in nearly all mammals, or about 4.4 l/min in humans.) Furthermore, this low resistance system must be flexible, because cardiac output in strenuous exercise can increase fivefold! Nature provides a remarkable factor of safety for physiological stress loading.

The pulmonary vasculature is a low pressure system. This can be seen by comparing the pressures in the pulmonary artery and the aorta in Figure 2.4:2. The pressure in the largest pulmonary artery oscillates between 10 and 25 mm Hg; that in the largest pulmonary veins is only 4 or 5 mm Hg. A low-pressure system does not need a very strong container. Thus, the wall of the pulmonary artery is found to be thinner than that of the aorta. The structures of the smaller pulmonary arteries are similarly flimsier than those of the corresponding systemic arteries. The arterioles, which in the systemic circulation are thick walled and muscular, are thin walled in the lung. The capillary blood vessels, which in the systemic circulation are often embedded in gel-like tissues, are exposed to gas in the lung. Some capillary blood vessels in the systemic circulation are quite rigid and show negligible change in diameter with normal variation of blood pressure. In contrast, the capillary blood vessels in the lung are quite flexible: The thickness of the capillary sheet varies almost linearly with the blood pressure when the blood pressure is larger than the alveolar gas pressure, but collapses to zero when the blood pressure becomes smaller than the alveolar gas pressure. Pulmonary venules and veins are similarly thin walled. Thus, the container of the pulmonary blood is quite flexible, and the flow is significantly influenced by elastic deformation. The resistance of the pulmonary capillaries is strongly influenced by blood and air pressures. Pulmonary blood flow is not a linear function of the blood pressure gradient, because an increase in blood pressure distends the blood vessels, reduces the resistance, and increases the flow out of proportion.

The flexibility of the lung structure also reveals itself in other ways. One of the most dramatic effects is the influence of gravity on the distribution of blood flow in the lung. The flow per unit volume increases greatly in the direction of gravity. The reason is rather simple: Gravity affects the hydrostatic pressure in the blood. For humans in the upright position, the hydrostatic head at the apex of the lung may equal or exceed the pulmonary arterial pressure at the pulmonic valve. The blood pressure (the sum of the hydrostatic head and the pressure in the pulmonary artery minus the resistance loss to the point in question) at the apex of the lung may become smaller than the alveolar gas pressure, whereas that at the base may become more than twice that of the arterial pressure. As a result, the capillary blood vessel sheet thickness is reduced at the apex and increased at the base, and the flow at the apex may become choked off, whereas that at the base is much increased.

The major determinants of pulmonary blood flow are therefore (a) the topology of the blood vessels, (b) the pressures of the alveolar gas and blood, (c) the viscosity of blood as affected by the hematocrit and other factors, (d) the elasticity of the blood vessels, and (e) the intrapleural pressure that affects the size of the lung. These factors interact to produce a unique nonlinear pressure–flow relationship.

The lung, however, is a very complex organ. It has two systems of circulation: (a) the low-pressure pulmonary circulation discussed earlier and (b) the high-pressure bronchial circulation that perfuses the bronchi and blood vessel walls. It has an extensive lymphatic system. Although the pulmonary blood vessels are richly supplied with nerve fibers and have chemoreceptor areas, the vascular bed of the lung normally is largely free from neural and chemical control. However, it responds promptly to hypoxia, retains the features of potential neural and humoral control in disease, and responds to pharmacological doses of catecholamines, histamine, serotonin, and other agents. The pulmonary arterial system and the bronchial arterial system have some connections, and collateral circulation develops in abnormal states. Connections between the pulmonary arterial ramifications and pulmonary venous tributaries larger than the capillaries are not seen in a normal individual, but in some disease states arteriovenous shunts can be quite prominent.

The objective of the present chapter is limited to analysis of the main features of the mechanics of pulmonary circulation. It will be shown that after certain anatomical, histological, and rheological data are obtained, the problem can be formulated and solved analytically, and the results of theoretical analysis are in reasonable agreement with those of physiological experiments.

6.2 Pulmonary Blood Vessels

Figure 6.2:1 shows the relationship between the heart and lung. Blood flows from the right ventricle to the pulmonary artery, then to capillary blood vessels, the veins, and finally the left atrium. Figure 6.2:2 shows an enlarged view of capillary blood vessels in the interalveolar septa. Figure 6.2:3 shows

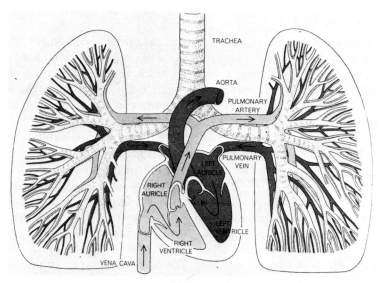

FIGURE 6.2:1. Relationship between the heart and lung in humans. Note that the pulmonary arteries lie close to the bronchi, whereas the pulmonary veins stand alone. Only the first few generations of the large pulmonary blood vessels and bronchi are shown in the figure. For smaller vessels, see Figures 6.2:2 and 6.2:3.

FIGURE 6.2:2. Scanning electron micrograph of cat lung illustrating how each inter-alveolar septum is shared by two alveoli. These septa are sheets of pulmonary capillary blood vessels. The wrinkly appearance of the septa is an artifact due to the cutting of the specimen and relieving of the stresses, thereby causing the contraction of elastin fibers, which cannot be fixed by commonly known fixing agents (see Sobin et al. (1982), and *Biomechanics: Mechanical Properties of Living Tissues*, Fung, 1993, p. 245).

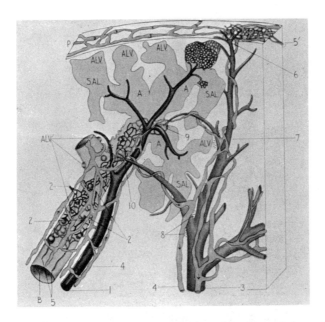

FIGURE 6.2:3. A schematic diagram of the pulmonary blood vessels, bronchial blood vessels, and lymphatic vessels in the lung and on the pleural surface. Drawing by William S. Miller (1947, p. 75). Reproduced by permission. B = *bronchiole*, leading to two *alveolar ducts*, one of which is shown here; A = atrium; ALV, ALV' = *alveoli*; S. AL. = *alveolar sac*; P = *pulmonary pleura*; 1 = *pulmonary artery*, dividing into capillaries; 2 = branches of pulmonary arteries distributed to bronchioles and ducts, and then broken up into capillaries, which unite with capillaries derived from the bronchial arteries; 3 = *pulmonary vein*; 4 = *lymphatics*; 5 = *bronchial artery* and capillaries; 5' = bronchial arterial supply in the pleura (in animals with a *thick* pleura); 6 = *pulmonary venule*; 7–10 are situations in which lymphoid tissue is found.

a schematic diagram of a lobule of the lung drawn by Miller (1947, p. 75) to illustrate the relation of the blood vessels to the air spaces. Note that arteries are adjacent to bronchi, whereas the veins stand alone. Bronchial vessels and lymphatics are also shown in the figure.

Figure 6.2:4 shows a polymer cast of the arteries of the rat left lung. Figure 4.10:1 on p. 253 is a similar photograph of a silicone cast of the veins of a cat lung. From these casts we see that the topology of the pulmonary arteries and veins is treelike. The branching patterns of these trees have been studied by Horsfield (1978), Singhal et al. (1973), Weibel (1963, 1973), and Yen et al. (1983, 1984). The most popular system is that of Strahler (see Sec. 4.9, p. 254). In this system the capillaries are counted as vessels of order 0. The smallest arterioles are called vessels of order 1. Two order 1 vessels meet to form a larger vessel of order 2, and so on. But if an order 2 vessel meets a vessel of order 1, the order number of the combined vessel remains 2. A similar counting system is used for the venous tree. The ratio of the number of vessels of

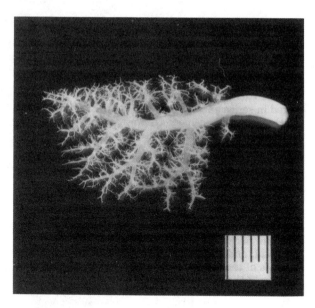

FIGURE 6.2:4. Polymer cast of the arterial tree of rat left lung. Length scale, 5 mm. From Jiang et al. (1994), reproduced by permission.

order n to that of order $n + 1$ is called the *branching ratio*. The ratio of the diameter of the vessels of order n to that of order $n + 1$ is called the *diameter ratio*. A length ratio is defined similarly. Yen et al. (1983) measured the branching pattern of the pulmonary arteries and veins of the cat, and their results are presented in Tables 6.14:1 and 6.14:2, p. 396. Singhal et al. (1973) measured the branching pattern of human pulmonary arteries; their results are given in Table 6.2:1. For the smallest arterioles (of order 1), Singhal et al. assume that each alveolus is supplied by one arteriole. For the cat we know this assumption is untrue. Each terminal arteriole of the cat supplies an average of 24 alveoli, and about 18 alveoli drain into one venule (Zhuang et al., 1985). Figure 6.2:5 shows the relationship between the pulmonary capillary bed and the arterioles of orders 1, 2, and 3 in the rat lung. This relationship is typical of other mammalian lungs as well.

Jiang et al. (1994) have pointed out that although the conventional Strahler's system correctly takes into account the asymmetric bifurcation of the arterial tree, it suffers from several difficulties: (a) The relationship between the vessel diameter and order number appears chaotic. Many larger vessels have smaller order numbers; many smaller vessels have larger order numbers. The standard deviations of diameters of all orders are very large, and the ranges of diameters of successive orders overlap widely. (b) All vessels of the same order are treated as parallel in the sense of electric circuits. Ignoring parallel and series connections may cause large errors in circuits. To improve accuracy, we define each vessel between two successive

TABLE 6.2:1. Integrated Data for the Total Pulmonary Arterial System of Humans

Order	Number of branches	Diameter (mm)	Length (mm)	End branches	Capillary bed* (%)
17	1.000	30.000	90.50	3.000×10^8	1.000×10^2
16	3.000	14.830	32.00	1.000×10^8	3.333×10
15	8.000	8.060	10.90	3.021×10^7	1.007×10
14	2.000×10	5.820	20.70	1.376×10^7	4.588
13	6.600×10	3.650	17.90	3.983×10^6	1.328
12	2.030×10^2	2.090	10.50	1.159×10^6	3.863×10^{-1}
11	6.750×10^2	1.330	6.60	3.470×10^5	1.157×10^{-1}
10	2.290×10^3	0.850	4.69	8.916×10^4	2.972×10^{-2}
9	5.861×10^3	0.525	3.16	4.805×10^4	1.602×10^{-2}
8	1.756×10^4	0.351	2.10	1.604×10^4	5.437×10^{-3}
7	5.255×10^4	0.224	1.38	5.358×10^3	1.786×10^{-3}
6	1.574×10^5	0.138	0.91	1.787×10^3	5.957×10^{-4}
5	4.713×10^5	0.086	0.65	5.975×10^2	1.992×10^{-4}
4	1.411×10^6	0.054	0.44	1.995×10^2	6.650×10^{-5}
3	4.226×10^6	0.034	0.29	6.664×10	2.221×10^{-5}
2	1.266×10^7	0.021	0.20	2.370×10	7.900×10^{-6}
1	3.000×10^8	0.013	0.13	1.000	3.333×10^{-7}

* Capillary bed (%) is the calculated percent of the total capillary bed supplied by one branch of a given order.
From Singhal et al. (1973).

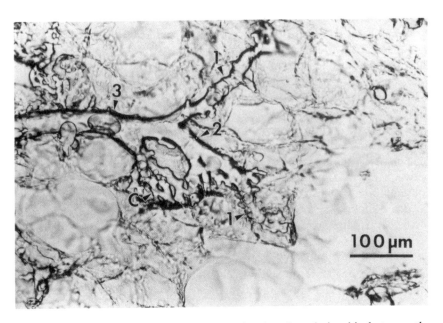

FIGURE 6.2:5. Photomicrograph of rat lung showing the relationship between the capillary bed and several orders of arteries. Arteries were perfused with catalyzed silicone elastomer containing 0.25% Cab-O-Sil. Section thickness, 40 μm. C = capillaries; 1 = arteries of *order 1*; 2 = arteries of *order 2*; 3 = artery of *order 3*. From Jiang et al. (1994), reproduced by permission.

TABLE 6.2:2. Elemental Data of Rat Left Pulmonary Arterial Tree

Order no.	Diameter of elements		No. of elements[†]	Length of elements		NS/NE*
	μm*	DR		mm*	LR	
11	1,639 ± 83	1.76	1	18.11 ± 1.81	6.86	8.00 ± 1.23
10	929 ± 97	1.54	5	2.64 ± 1.94	1.52	2.52 ± 1.62
9	602 ± 57	1.44	11 ± 1	1.74 ± 1.35	1.31	2.54 ± 1.88
8	417 ± 40	1.57	23 ± 1	1.33 ± 0.69	1.07	2.14 ± 1.20
7	266 ± 30	1.75	56 ± 2	1.24 ± 0.60	1.72	2.60 ± 1.33
6	152 ± 25	1.73	183 ± 18	0.72 ± 0.42	1.80	1.89 ± 1.04
5	88.1 ± 11	1.43	555 ± 38	0.40 ± 0.23	1.48	1.40 ± 0.66
4	61.5 ± 4.6	1.39	1,351 ± 71	0.27 ± 0.13	1.35	1.14 ± 0.37
3	44.4 ± 4.9	1.40	3,794 ± 319	0.20 ± 0.09	1.33	1.08 ± 0.45
2	31.7 ± 4.9	2.38	7,993 ± 852	0.15 ± 0.07	3.00	1.60 ± 1.14
1	13.3 ± 2.8		52,001 ± 4,378	0.05 ± 0.03		1.23 ± 0.56

Pleural pressure = atmospheric pressure; airway pressure = 10 cm H_2O; pressure in left atrium = 0 cm H_2O. DR (diam. ratio) = D_{n+1}/D_n; LR (length ratio) = L_{n+1}/L_n; NS/NE = ratio of no. of segments to no. of elements. Total number of elements in *orders 7–11* is the average of five casts; total number in *orders 1–6* is calculated from the connectivity matrix.
* Means ± SD; [†] means ± SE.
From Jiang et al. (1994).

points of bifurcation a *segment*, and segments connected in series as an *element*. (c) A vessel of order n may be connected to vessels of order $n + 1, n + 2, \ldots$, but this is not evident in the Strahler system. To correct these difficulties, the present author and Kassab et al. (1993) introduced a *diameter-defined Strahler system*, with the addition of the following rule: When two vessel segments of order n or less meet, the order number of their confluent vessel becomes $n + 1$ if and only if its diameter exceeds that of order n by a certain amount. Let D_n and SD_n be the mean and standard deviation of the diameters of the vessel segments of order n. Then the dividing line between the diameter of order n and that of order $n + 1$ lies halfway between $D_n + SD_n$ and $D_{n+1} - SD_{n+1}$. Jiang et al. (1994), using this rule, counted the number of elements and segments in each order, and measured the *connectivity matrix* whose component in row m and column n is the ratio of the total number of elements of order m springing from elements of order n divided by the total number of elements in order n. Jiang et al.'s (1994) results are given in Tables 6.2:2 and 6.2:3.

Corresponding data on dog and human lungs have been obtained by Michael Yen and his associates (Gan et al., 1993; Huang et al., 1996).

6.3 Pulmonary Capillaries

In the lung, the smallest unit of air space is the alveolus. The alveolus is bounded by networks of capillary blood vessels. The walls of each alveolus are shared by neighboring alveoli, and are called *interalveolar septa*. The

TABLE 6.2.3. Connectivity Matrix of Elements of Rat Left Pulmonary Arterial Tree

	1	2	3	4	5	6	7	8	9	10	11
1	0.19 ± 0.03	4.06 ± 0.15	2.48 ± 0.35	1.36 ± 0.40	0	0	0	0	0	0	0
2	0	0.12 ± 0.03	1.55 ± 0.09	0.93 ± 0.22	0	0	0	0	0	0	0
3	0	0	0.17 ± 0.07	2.26 ± 0.05	0.31 ± 0.03	0.05 ± 0.01	0.11 ± 0.05	0.13 ± 0.07	0	0	0
4	0	0	0	0.05 ± 0.02	2.00 ± 0.04	0.70 ± 0.06	0.62 ± 0.10	0.43 ± 0.12	0.39 ± 0.29	0	0
5	0	0	0	0	0.18 ± 0.02	1.92 ± 0.07	1.56 ± 0.13	0.85 ± 0.16	0.83 ± 0.26	0.50 ± 0.34	0
6	0	0	0	0	0	0.18 ± 0.02	2.05 ± 0.13	1.15 ± 0.15	1.00 ± 0.29	0.67 ± 0.33	0
7	0	0	0	0	0	0	0.05 ± 0.02	1.55 ± 0.14	1.06 ± 0.21	0.67 ± 0.33	0
8	0	0	0	0	0	0	0	0.08 ± 0.04	1.67 ± 0.23	0.83 ± 0.40	6
9	0	0	0	0	0	0	0	0	0.06 ± 0.06	1.67 ± 0.49	4
10	0	0	0	0	0	0	0	0	0	0.17 ± 0.17	2
11	0	0	0	0	0	0	0	0	0	0	0

Values are means ± SE. The component C_{mn} in *row m* and *column n* is the ratio of the total number of elements of *order m* springing from elements of *order n* divided by the total number of elements in *order n*.
From Jiang et al. (1994).

overriding fact that determines the topology of the capillary blood vessels is that all pulmonary alveolar septa in adult mammalian lungs are similar. Each septum contains one single sheet of capillary blood vessels and is exposed to air on both sides.

The topology of the network of the pulmonary capillary blood vessels is definitely not treelike; it is sheetlike. The dense network of the capillary blood vessels in an alveolar wall of the frog is shown in Figure 6.3:1 (Maloney and Castle, 1969). A similar picture of the cat lung is shown in Figure 6.3:2; cross-sectional views of the cat lung can be found in *Biomechanics: Mechanical Properties of Living Tissues* (Fung, 1993, Fig. 5.2:1 on p. 167, Fig. 5.2:2 on p. 169, and Figs. 8.10:1, 8.10:2 on pp. 361, 362). To characterize the geometry of such a network, we model the vascular space as a sheet of fluid flowing between two membranes held apart by a number of more or less equally spaced "posts" (Fig. 6.3:3). In plane view (a) this is a sheet with regularly arranged obstructions. The plane may be divided into

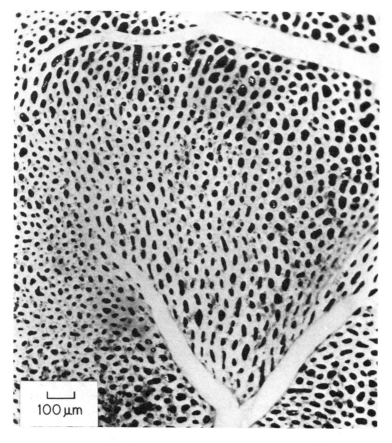

FIGURE 6.3:1. Photograph of a network of capillary blood vessels in the frog by Maloney and Castle (1969). Reprinted by permission.

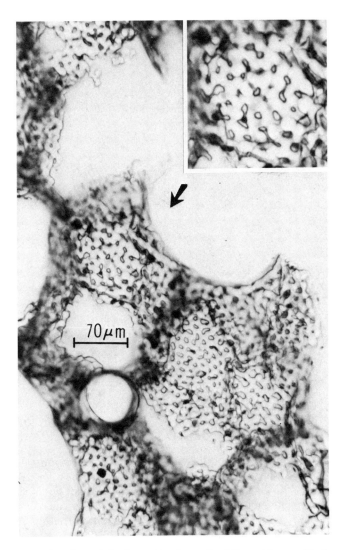

70 μm

FIGURE 6.3:2. Flat view of the interalveolar wall of the cat lung with the microvasculature filled with a silicone elastomer. This photomicrograph illustrates the tight mesh or network of the extensively filled capillary bed. The circular or elliptical enclosures are basement membrane stained with cresyl violet and are the nonvascular posts. Frozen section from gelatin-embedded tissue; glycerol-gelatin mount. The inset shows a detail from the region indicated by the arrow. From Fung and Sobin (1969), by permission.

a network of hexagons, with a circular post at the center of each hexagon. The *sheet-flow* model is therefore characterized by three parameters: L, the length of each side of the hexagon; h, the average height or thickness of the sheet; and ε, the diameter of the posts.

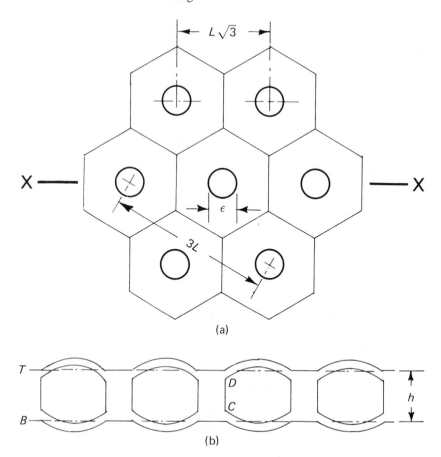

FIGURE 6.3:3. Sheet-flow model. (a) Plane view. (b) Cross section through *X—X* of (a). The cylindrical elements [circular in (a) and rectangular in (b)] are the "post." The space between the top and bottom walls is the flow channel. For bounding surfaces only, T, top, and B, bottom. Sheet thickness, h. C and D, contact of posts with endothelial surface at T and B. From Sobin et al. (1970). Reprinted by permission of the American Heart Association, Inc.

We shall define the *vascular-space tissue ratio* (VSTR) as the ratio of the vascular lumen volume to a certain circumscribing volume defined presently. As illustrated in Figure 6.3:3(b), the circumscribing volume is that enclosed between surfaces T and B. The tissues (epithelial, interstitial, and endothelial) external to the surfaces T and B are excluded. Thus VSTR does not represent the volumetric fraction of the total interalveolar septum occupied by blood; it represents only the fraction of the blood volume over a sum of the volumes of the vascular space and the posts. For the sheet model, the VSTR is independent of the height h and is equal to the percentage of the area occupied by blood in the plane cross section. From Figure 6.3:3(a), we have

$$\text{Area of a hexagon} = \frac{3\sqrt{3}}{2} L^2 .$$

$$\text{Area of a circle} = \pi \frac{\varepsilon^2}{4} . \tag{1}$$

Hence for sheet flow,

$$\text{VSTR} = 1 - \frac{\pi}{6\sqrt{3}} \frac{\varepsilon^2}{L^2} . \tag{2}$$

In actual practice, it is difficult to measure and evaluate L, ε, and h from microscopic preparations because of the random irregularities in the geometric appearance of the specimens. The VSTR, however, can be determined with greater confidence by planimetry and random sampling. The dimensions of the obstructing posts (ε) can be calculated from Eq. (2) by determining the VSTR and measuring $3L$, as indicated in Figure 6.3:3(a). Then

$$\frac{\varepsilon}{L} = \sqrt{\frac{6\sqrt{3}}{\pi}(1 - \text{VSTR})} . \tag{3}$$

This idea is illustrated in Figure 6.3:4. We choose an area that is large compared with the individual posts, and measure the area ratio of the vascular space and the circumscribing area, thus obtaining VSTR directly. ε/L is then

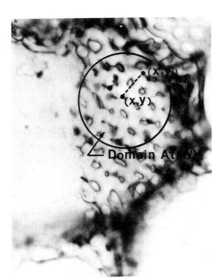

FIGURE 6.3:4. Domain of averaging around a point (x, y) in an alveolar sheet. From Fung and Sobin (1969). Reprinted by permission.

TABLE 6.3:1. Summary of Data for Elastomer-Filled Lungs of Vertically Positioned Cats at a Transpulmonary Pressure (= Alveolar − Pleural Pressures) of 10 cm H_2O*

Δp	VSTR (%)	Hexagon area (μm^2)	Post diameter (μm)	Interpost distance (μm)	No. fields analyzed
6.3	90.94 ± 1.94	239.51 ± 27.06	5.25 ± 0.83	10.21 ± 0.30	6
6.8	88.81 ± 2.10	275.34 ± 37.80	6.22 ± 0.66	10.34 ± 0.99	6
7.3	91.16 ± 1.50	256.53 ± 45.58	5.35 ± 0.80	10.60 ± 0.89	6
10.3	90.16 ± 1.84	202.08 ± 43.34	5.01 ± 0.82	9.15 ± 0.91	18
14.3	90.59 ± 2.12	193.25 ± 63.07	4.74 ± 1.02	8.99 ± 1.38	18
18.3	90.43 ± 2.64	203.54 ± 38.75	4.96 ± 1.11	9.24 ± 0.51	18

* Values are means ± SD. Δp is the transmural pressure difference, that is, vascular pressure minus alveolar air pressure. VSTR is the vascular space–tissue ratio expressed in percent. The hexagon area is the planimetered sheet area divided by the number of posts in the area. The interpost distance is the distance between post edges.
From Sobin et al. (1972).

calculated from Eq. (3). Let the measured average area of the posts be A_p and let that of the basic hexagon be A_S; then

$$L = \sqrt{\frac{2}{3\sqrt{3}} A_s} \qquad (4a)$$

$$\varepsilon = 2\sqrt{\frac{A_p}{\pi}}. \qquad (4b)$$

The interpost distance (the distance between post edges) is $L\sqrt{3} − \varepsilon$, which is the width of the channel for the passage of red cells in the plane of the alveolar wall.

The data for the cat lung are presented in Table 6.3:1. The VSTR for each individual lobe is not influenced by the blood pressure; the average value is 0.91. These data indicate that in the plane of the interalveolar wall, the capillary bed can occupy 91% of the area of the wall.

This high value of VSTR applies to the interalveolar septa. It does not apply to those alveoli that were noted by Miller (1947) to have capillary beds with a "coarse" mesh, such as pleura, peribronchial, and perivascular septa, and those abutting connective tissues.

To help visualize the vascular space in the pulmonary alveolar sheet, we present a composite drawing in Figure 6.3:5. The epithelium is lifted from the interstitium and pulled back. The elastin and collagen fibers in the interstitium are indicated but not drawn in detail. The cross section of an individual capillary blood vessel may appear like a tube in certain sections, but overall the vascular space is represented as a sheet. From the point of view of fluid mechanics, the sheet representation is particularly appropriate, because the streamlines will occupy the vascular space of a tube model as sketched in Figure 6.3:6, as if in a sheet. Each segment is so short that

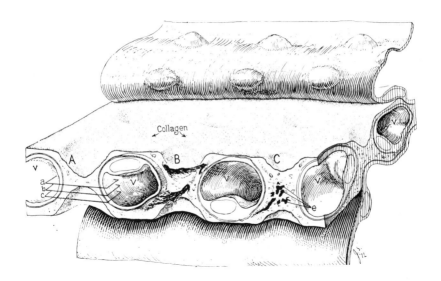

FIGURE 6.3:5. Composite drawing of the pulmonary interalveolar wall of the dog showing the interalveolar microvascular sheet, composed of a vascular compartment (V) (the capillary bed) and the avascular (intercapillary) posts (A, B, C). The alveolar epithelium has been pulled back to show the connective tissue matrix of the wall. Collagen converges on the post from the surrounding capillary wall. In post B collagen fiber bundles pass around within the post in a curving arrangement. From Rosenquist et al. (1973), by permission.

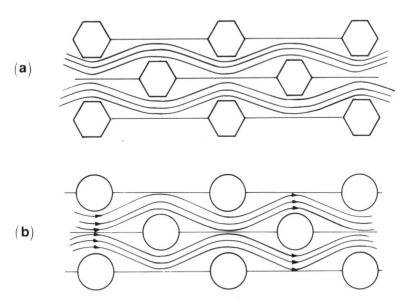

FIGURE 6.3:6. Streamlines of flow in an alveolar capillary network. (a) Flow in the middle plane of a tube model. Note that although the capillaries are considered to be tubes, the flow is entirely different from a Poiseuillean pattern. (b) Flow in sheet model. Note the similarity in the flow patterns between (a) and (b). From Fung and Sobin (1969), by permission.

347

the Poiseuillean velocity profile discussed in Section 3.2 does not have a chance to develop. The Poiseuillean formula, Eq. (17) of Section 3.2, certainly does not apply. Accurate theoretical analysis and experimental observations of the flow in a pulmonary alveolar sheet are presented in Sections 6.9 to 6.14.

In Section 6.8, the elasticity of the interalveolar septa is examined. It is shown that the deformation of the pulmonary capillary blood vessels in response to changes of pressures in the blood, alveolar gas, and pleura can be described and calculated according to the sheet model rigorously and without theoretical difficulty.

This discussion has referred to the pulmonary capillaries in the interalveolar septa that are responsible for oxygen and CO_2 exchange between blood and alveolar gas. There are other capillary blood vessels in the lung, of course. The large pulmonary blood vessels are themselves perfused by capillaries called *vasa vasorum*. The bronchi are perfused by systemic capillaries. At the junctional domain of the edges of neighboring alveoli there are "corner" vessels, which carry blood flow in the zone-1 condition (see Sec. 6.15). These capillaries are tubes.

6.4 Spatial Distribution of Pulmonary Capillaries: Shape of the Alveoli and Alveolar Ducts

Since the pulmonary capillary blood vessels lie at the core of the interalveolar septa, the three-dimensional distribution of the pulmonary capillaries is the same as that of the pulmonary alveoli. Experiments on lung morphology of animals in different posture and at zero gravity (in actual or simulated space flight) have convinced us that all alveoli of an animal are essentially the same. Every alveolus must have an opening for ventilation to the atmosphere. In mammalian lung, this is done by ventilating into an alveolar duct, which is connected to a bronchiole, then to the bronchi, trachea, and mouth and nose.

Many classic observations of the alveolar geometry were done by perfusing molten wax or low-melting point alloys or catalyzed polymers into the trachea, pushing the liquid into the periphery, letting the fluid solidify, corroding away the tissues, and then observing the cast. Naturally, the casts look like bunches of grapes (see detailed review by Miller, 1947, and a brief review by Fung (in *Biomechanics: Motion, Flow, Stress and Growth*, 1990, pp. 401–408). This led to the spherical shell model of the alveolus. The erroneous concept of lung instability under the influence of surface tension based on the spherical shell model was pointed out by Fung (1975a). The difficulties disappear when the pulmonary alveoli are correctly recognized as space–filling polyhedra (Fung, 1975b). The polyhedra should be approximately uniform (Fung, 1988). Modeling choice among regular polyhedra is limited, because there are only five regular convex polyhedra in three-

FIGURE 6.4:1. Several polyhedrons mentioned in the text. The three on the right-hand side are space filling. From Fung (1988), reproduced by permission.

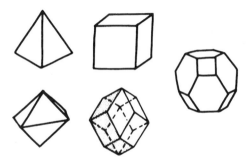

dimensional space: the tetrahedron, the cube, the octahedron (shown in Fig. 6.4:1), the pentagonal dodecahedron bounded by 12 regular pentagons, and the icosahedron bounded by 20 regular triangles (not shown). But the last two are not space filling, (although the dodecahedron needs only a slight distortion to fill the space), and the tetrahedron and octahedron are space filling only if they are used in combination. Morphometric observation of human and animal lungs rarely discern interalveolar septa as all triangles, or all rectangles, or all pentagons. Hence, these regular polyhedra are not good candidates. It is known, however, that the nonregular octahedron (with every face a triangle, not shown), the garnet-shaped rhombic dodecahedron (12 sided, with each side a rhombus, shown in Fig. 6.4:1), and the tetrakaidecahedron (14-sided polyhedron shown in Fig. 6.4:1, formed by cutting the six corners off a regular octahedron shown in lower left of Fig. 6.4:1 in such a way as to leave all edges equal in length, forming 6 square and 8 hexagonal faces) are space filling. Based on the morphometric observations described earlier, the 14-hedron is a reasonable candidate (Fung, 1988).

Since each alveolus must be ventilated by perforating at least a wall, Fung (1988) has proposed that the basic unit of respiration is a combination of the 14-hedron and the order-2 polyhedron shown in Figure 6.4:2, each of which consists of fourteen 14-hedra clustered around an empty 14-hedron at the center with all its walls perforated. All fourteen 14-hedra ventilate into the center. The second-order polyherdron is a basic unit of the alveolar duct. Several order-2 14-hedra can be assembled together to form ducts by perforating a few more membranes (see Fig. 11.6:6 of Fung, 1990). An assemblage of order-2 and order-1 14-hedra is space filling. Perforation of a membrane calls for reinforcement of the edges for structural integrity. With elastic deformation taken into account, Fung (1988) shows that this model fits the morphometric data of Hansen and Ampaya (1975) and Hansen et al. (1975) quite well, as well data by Oldmixon et al. (1988) concerning alveolar shapes, alveolar mouth geometry, lengths of alveolar sacs and ducts of successive generations, dihedral angles between interalveolar septa, patterns of alveolar walls in plane cross sections of lung parenchyma,

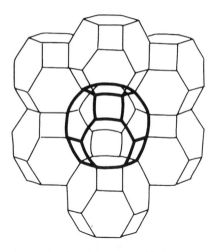

FIGURE 6.4:2. Sketch of an order-2 polyhedron with fourteen 14-hedra surrounding one central 14-hedron. Several 14-hedra in front are removed to reveal the central 14-hedron. To form a basic unit of lung structure, all faces of the central 14-hedron are removed, whereas all its edges are reinforced to take up load and to keep the structure in stable equilibrium. From Fung (1988), reproduced by permission.

the types of the vertices, and the ratio of the length of alveolar mouths to the surface area of the alveolar wall. Hence, for the analysis of circulation, our model is that the pulmonary capillaries are spatially distributed as order-1 and order-2 14-hedra.

This model is consistent with the observations of Miller (1947), Orsos (1936), von Hayek (1960), Weibel (1963), and Wright (1961). Dale et al. (1980), Stamenovic and Wilson (1985), Wilson and Bachofen (1982), have proposed other mathematical models.

6.5 Spatial Distribution of Pulmonary Arterioles and Venules

Since along a streamtube blood must flow out of an arteriole into the capillaries and then be collected into venules, we need to know the relative positions of the arterioles, capillaries, and venules. Are the arterioles and venules uniformly distributed separately in the lung and uniformly mixed together in space? Is each alveolus a unit of microcirculation? Is each alveolus supplied by one arteriole and drained by one venule? The answer to all of these questions turned out to be negative, according to Sobin et al. (1980). They perfused cat lung with a particulate polymer and developed four criteria to determine whether a blood vessel is arterial or venous. Then on each histological cross section of the lung parenchyma, they marked all arteries with white dots and all veins with red dots, and covered the area of white dots with one color and the area of red dots with another. A two-colored map emerged (Fig. 6.5:1). A two-color cartographic map can only show islands in an ocean. Our histological maps of the cat lung show that the arterial zones are islands and the venous zone is the ocean. From these maps, quantitative information about the arterial and venous zones in the lung is obtained by stereological methods. A morphological definition of

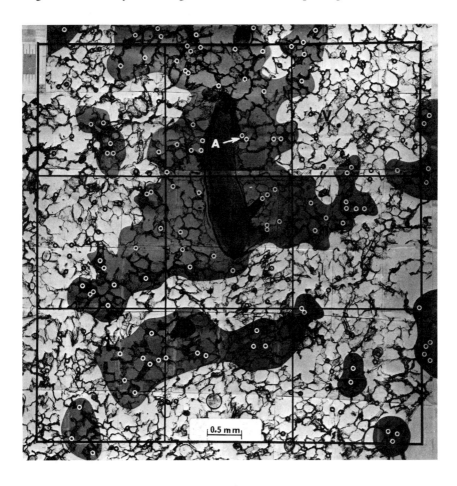

FIGURE 6.5:1. Map made from montage of individual photomicrographs of a histological section (50 μm thick) of cat lung. Vessels of 15 to 100 μm diameter are individually identified, arteries by white circles (arrow A) and veins by black circles (arrow V). A large artery lies in the middle of the section. The domains of the arteries are made darker by overlaying the areas with blue transparency film; they appear as isolated islands. The domain of the veins is continuous. Thus the arterial regions are like islands immersed in an ocean that contains only pulmonary veins. From Sobin et al. (1980), by permission.

the length of a capillary blood vessel is the average length of blood pathways connecting an arteriole and a venule. From maps like Figure 6.5:1, the average morphological length of capillary blood vessels in the cat lung, L, was found to be 556 ± 286 (SD) μm when the transpulmonary pressure was 10 cm H_2O. L is, of course, important, because the flow rate varies as L^{-1}, and the transit time varies with L^2 for given arterial and venous pressures [see Eq. (10) of Sec. 6.20, p. 432].

From maps like Figure 6.5:1, one can deduce also that, on average, each cat terminal pulmonary arteriole supplies 24.5 pulmonary alveoli and each terminal venule drains 17.8 alveoli (Zhuang et al., 1985). The average diameter of cat pulmonary alveoli is 163 μm; thus, blood leaving an arteriole will travel, on average, 3.4 alveoli before it enters a venule. The stereological principle of this calculation is given below.

In a sample containing randomly dispersed rods of uniform size, with radius r and length L ($L \gg r$), the number of rods per unit volume N_V is related to the number per unit area of the intersections of the rods with a test plane, N_A, by the equation (see Underwood, 1970, p. 91),

$$N_V = \frac{2N_A}{L}. \tag{1}$$

This formula applies to histology if the histological section is infinitely thin. For a specimen of finite thickness and containing vessels of various diameters, a modification is necessary (Zhuang et al., 1985). Consider the situation shown in Figure 6.5:2. The unit cube contains randomly dispersed rods. A plane P cuts the rods A and B. A parallel plane Q cuts rods A and C but not B. A slab bounded by planes P and Q cuts A, B, and C. If the slab represents a slice of the lung tissue, the rods represent the blood vessels, and a photograph is taken in the direction perpendicular to the slab, one will see in the photograph the cross sections of vessels A, B, and C. Now, all the randomly distributed vessels in the unit cube seen in the slab between P and Q can be classified into three types: those like A, cut by both

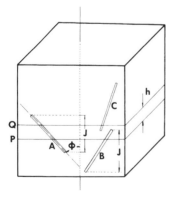

FIGURE 6.5:2. A sketch of the derivation of Eq. (5). See text for details. From Sobin et al. (1980), reproduced by permission.

planes P and Q; those like B, cut by P but not by Q; and those like C, cut by Q but not by P. Let the total number of vessels cut by the slab be counted and the result be denoted by N_S; then obviously N_S is larger than N_A, the intersection number in plane P. Let us write

$$N_S = N_A + N_C'. \tag{2}$$

Then obviously the number N_C' represents the number of all vessels of type C, located in a position that is cut by plane Q but not by P. But statistically, N_C' must be equal to N_B', the number of vessels of type B, cut by P but not by Q. For sufficiently thin slabs, N_B' must be a fraction of N_A. The fraction, N_B'/N_A, must be equal to $h/\bar{J}$, where h is the thickness of the slab and $\bar{J}$ is the average value of the vertical projections of all the vessels (i.e., the components of vessel lengths in the direction perpendicular to plane P). If all the vessels are of equal length L, we shall show presently that

$$\bar{J} = L/2. \tag{3}$$

Then

$$N_C' = N_B' = \frac{h}{\bar{J}} N_A = \frac{2h}{L} N_A; \tag{4}$$

and a substitution into Eq. (2) yields the desired formula (Zhuang et al., 1985):

$$N_A = \frac{N_S}{1 + \dfrac{2h}{L}}, \tag{5}$$

a formula giving the ideal value N_A to be used in Eq. (1) from the values of N_S, h and L measured from the tissue specimen. Equation (5) is valid for $h < L/2$, when $h \ll L$, $N_S \to N_A$, as it should be. Combining Eqs. (5) and (1), we obtain N_v, the number of vessels of length L in a unit volume:

$$N_v = \frac{2}{L} \left(1 + \frac{2h}{L}\right)^{-1} N_S. \tag{6}$$

It remains to prove Eq. (3). The vertical projection of a vessel of length L inclined at an angle ϕ to the normal of the plane P is $\bar{J} = L \cos \phi$. For randomly oriented vessels, the probability of finding a vessel inclined at an angle between ϕ and $\phi + d\phi$ to the vertical is $\sin \phi \, d\phi$. Hence

$$\bar{J} = \int_0^{\pi/2} L \cos \phi \sin \phi \, d\phi \bigg/ \int_0^{\pi/2} \sin \phi \, d\phi$$
$$= L/2.$$

Next, we need a stereological formula for the number of arterioles and venules per unit volume of tissue. The vessels marked by circles in Figure 6.5:1 include all arteries less than 100 μm in diameter; thus they consist of several orders of vessels of different lengths. Let vessels of order i be denoted by the subscript i. Then Eq. (6) can be written for vessels of order i as

$$N_{Si} = \frac{L_i}{2}\left(1 + \frac{2h}{L_i}\right)N_{vi} \qquad (i = 1, 2, 3 \ldots). \qquad (7)$$

Stereometry of pulmonary blood vessels by Yen et al. (1983, 1984) shows that the ratio of the numbers of vessels in successive order is approximately a constant. Thus, if N_{v1} denotes the *number of vessels of order 1 per unit volume* and B denotes the *branching ratio*, which is the average value of the ratio of the number of vessels of order i to that of order $i + 1$ in a given volume of the tissue, then

$$N_{v2} = \frac{N_{v1}}{B}, \quad N_{v3} = \frac{N_{v2}}{B} = \frac{N_{v1}}{B^2}, \quad \text{etc.} \qquad (8)$$

Using Eq. (8) in Eq. (7) and summing over i, we obtain

$$\sum_{i=1}^{n} N_{Si} = N_{v1}\sum_{i=1}^{n}\frac{L_i}{2}\left(1 + \frac{2h}{L_i}\right)\frac{1}{B^{i-1}}. \qquad (9)$$

According to the morphometric data on L_i, B, and the diameters presented in Section 6.2, we see that the upper limit of n is 3 for vessels of diameter $<100\,\mu m$. The term on the left-hand side, $\Sigma\,N_{Si}$, is the total number of vessels tagged by circles or bull's-eyes on a map such as that shown in Figure 6.5:1, divided by the area of the map expressed in terms of the area of the tissue specimen. With the measured values of $\Sigma\,N_{si}$, h, L_i, and B, we can compute the desired number N_{v1} from Eq. (9):

$$N_{v1} = \left[\sum_{i=1}^{n}\frac{L_i}{2}\left(1 + \frac{2h}{L_i}\right)\frac{1}{B^{i-1} - 1}\right]^{-1}\left(\sum_{i=1}^{n}N_{Si}\right). \qquad (10)$$

The results reported in the first two paragraphs of this section were obtained from these formulas by Zhuang et al. (1985).

6.6 Relative Positions of Pulmonary Arterioles and Venules, and Alveolar Ducts

One more piece of anatomical information is required: the manner in which pulmonary capillaries are connected to the terminal arterioles and venules. The capillary–venule connections are especially important because they are the "sluicing gates" which control the flow in a regime when the pulmonary venular pressure is lower than the alveolar gas pressure. The flow in this regime is discussed in Sections 6.16 to 6.19. Figure 6.6:1 shows a photograph of a venule draining capillary sheets perpendicular to it. Figure 6.6:2 shows a photograph of a venule draining a capillary sheet tangential to it, i.e., a

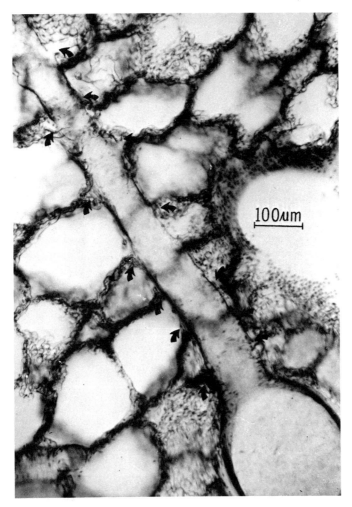

FIGURE 6.6:1. Connection between a venule and alveolar sheets in the cat lung. Courtesy of Dr. Sidney Sobin.

capillary sheet connected to the venule edgewise. A schematic drawing combining these features is shown in Figure 6.6:3(a). Here one interalveolar septum intersects a venule perpendicularly, whereas three others tether it longitudinally. Draining may occur along all the junctions according to the local pressure distribution and sluicing gate condition.

Finally, we can combine our information about the alveoli, alveolar ducts, arterioles, and venules, and present a model of the spatial relationship between them. The picture that emerges is shown in Figure 6.6:4. The alveoli are represented by small circles in the arterial regions and are left blank in the venous regions. The basic units of ducts, the order-2 14-hedra, are indi-

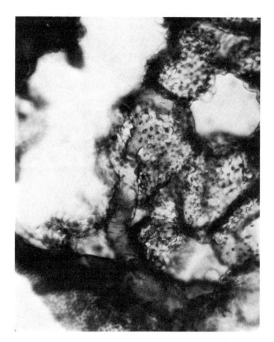

FIGURE 6.6:2. Another view of the junction of an alveolar sheet and a venule in the cat. From Sobin and Tremer (1966), by permission.

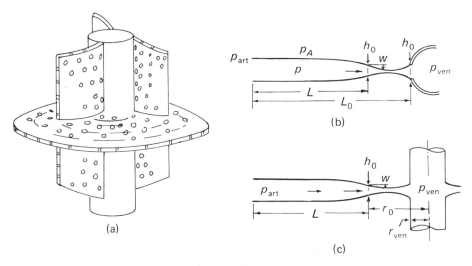

FIGURE 6.6:3. Schematic drawings of (a) relationship between a pulmonary venule and the neighboring capillary sheets; (b) a sluicing gate in the "one-dimensional" case, in which a sheet drains into a venule along an edge; (c) a sluicing gate in the case of a sheet draining axisymmetrically into a venule perpendicular to its plane. From Fung and Zhuang (1986). Reproduced by permission.

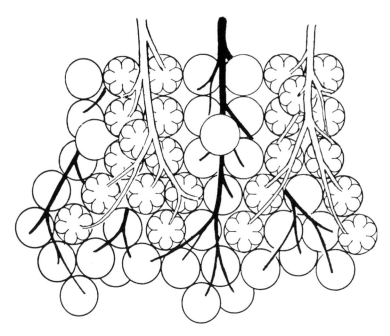

FIGURE 6.6:4. Conceptual illustration of how circulation and respiration systems are joined at the microvascular level. Arterioles are indicated by white color. Venules are indicated by black color. Alveoli are indicated by small loops. Order 2 14-hedra are indicated by larger circles surrounding those loops. Alveoli in venular regions are not drawn in. See text for details. From Fung (1989), reproduced by permission.

cated by larger circles. Alveolar ducts of generations 1 to 7 are composed of order-2 polyhedra, and ducts of generation 8 connect the isolated alveolus to longer ducts. For the cat lung, every order-2 polyhedron in an arterial region is perfused by 1.3 terminal arterioles (arteries of order 1); every order-2 polyhedron in the venous region is drained by 1.4 terminal venules (veins of order 1). The widths of the arterial regions and venous regions are both approximately twice the diameter of the order-2 polyhedron. The arterial regions are simply connected bodies; the venous region is a single multiconnected domain, a continuous phase that envelops all arterial regions.

Figure 6.6:4 illustrates schematically how the tops of the three groups of trees of airway, arteries, and veins of the lung are tied together.

6.7 Elasticity of Pulmonary Arteries and Veins

Three methods have been used in the author's laboratory to obtain the elasticity data on pulmonary arteries and veins. For larger specimens, the biaxial testing machine was used to obtain the stress-strain relationship of the vessel wall with the zero-stress state as reference (Debes and Fung, 1995).

For smaller intact pulmonary blood vessels with diameters larger than $100\,\mu\mathrm{m}$, the method of x-ray photography was used (Yen et al., 1980, 1981). For pulmonary vessels smaller than $100\,\mu\mathrm{m}$ in diameter, including the capillaries, the silicone elastomer method was used (Sobin and Tremer, 1966; Sobin et al., 1972, 1980). These methods were used also by Gan and Yen (1994), Huang et al. (1996), Yen (1988, 1989), and Yen and Sobin (1988). The testing of excised specimens in testing machine yields more detailed information. But vessels inside the lung cannot be excised as isolated tubes; hence we resort to the x-ray and elastomer methods. Some details and results follow.

Biaxial Test of Excised Specimens

The method of isolating a dog pulmonary artery specimen for testing is shown in Figure 6.7:1. Advantage is taken of the fact that at the zero-stress state (more precisely, the state of zero stress-resultants and zero stress-moments as a shell), the specimen is almost a flat plate. The flat shape makes the installation of the specimen into the TRIAX testing machine simple and without need for corrections. The TRIAX machine is described in Fung (1993, p. 296, Section 7.9) and Vawter et al. (1979). Typical experimental results are shown by the dotted curves in Figures 6.7:2 and 6.7:3. Here, since the deformation is large, stress is defined according to the definition of Kirchhoff, and strain is defined according to the definition of Green (see Eqs. (40) to (43) of Section 2.6, or, in greater depth, Chapter 10 of Fung (1990)). The stresses are denoted by S_{xx}, S_{yy}. The strains are denoted by E_{xx}, E_{yy}. The range of in vivo strain is unknown, but is estimated to be no larger than 0.4. It is seen from Figures 6.7:2 and 6.7:3 that the stress-strain relationship is linear in the physiological range. It was found also that the failure stress (at rupture) of the specimens is at least 10 times greater than the estimated physiological level. The hysteresis in cycle loading and unloading was found to be very small and insensitive to the frequency. Hence, the concept of pseudo elasticity (Fung et al., 1979, also described in Fung, 1993, p. 293) applies. The experimental results are fitted with the following *pseudo-elastic strain energy function*:

$$\rho_0 W^{(2)} = \frac{1}{2}\left(a_1 E_{xx}^2 + a_2 E_{yy}^2 + 2a_4 E_{xx} E_{yy}\right), \tag{1}$$

where a_1, a_2, and a_4 are material constants with units of stress. The superscript "2" indicates that this is a two-dimensional (plane stress) strain energy function. The subscripts x and y refer to the circumferential and longitudinal coordinates of the artery, respectively. By differention, the stress-strain relationship is:

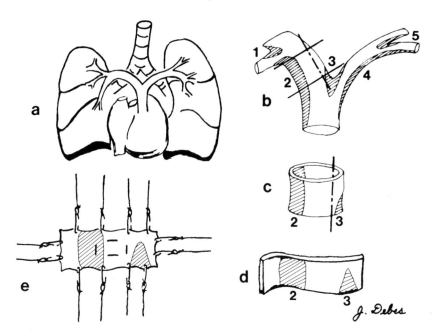

FIGURE 6.7:1. Specimen preparation and targeting method. **a**: chest is opened along mediastinum, exposing heart and lungs. **b**: first 3 generations of right pulmonary arterial tree were isolated. **c**: two transverse cuts at both ends near bifurcations of right pulmonary artery were made, yielding a ringlike cylinder. **d**: cylinder was cut radially, relieving residual stress, causing ring to spring open to approximate a flat rectangular slab with approximate dimensions $1.0 \times 3.0 \times 0.1$ cm. **e**: a rectangular target ($\sim 0.5 \times 1.0$ cm) was made in the center of each specimen by means of four wire markers made from 254-μm-diameter steel wire. Wires, each ~ 0.5 cm in length, were bent in the middle at a right angle, forming an "L" shape. One leg of "L" was pressed into specimen perpendicular to intimal surface, leaving other leg resting on intimal surface. Four markers defined the sides of a rectangle. Shaded numbered regions of the vessel wall in **b–e** are saddle shaped; nonshaded regions in **b–e** are ellipsoidal. Consequences of this difference are discussed in Debes and Fung (1995). From Debes and Fung (1995), reproduced by permission.

$$S_{xx} = a_1 E_{xx} + a_4 E_{yy},$$
$$S_{yy} = a_2 E_{yy} + a_4 E_{xx}. \tag{2}$$

The fitted coefficients a_1, a_2, a_4 are given in Debes and Fung (1995). The correlation is better than 97%. The theoretical curves, with the best fit constants are plotted by solid lines in Figure 6.7:2. For the high-stress case shown in Figure 6.7:3, Eq. (2) fits in the E_{xx} range from 0 to 1.3. The mean and standard deviations of the constants for all tested specimens are:

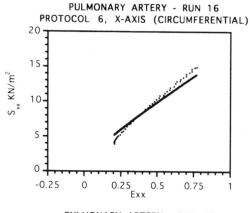

FIGURE 6.7:2. Loading curve test data of Green's strain vs. Kirchhoff stress in dog pulmonary artery. Points show experimental data, solid lines show theoretical fit. From Debes and Fung (1995), reproduced by permission.

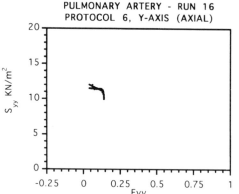

$$a_1 = 1.6185 \pm 0.3768$$
$$a_2 = 2.0210 \pm 1.8527$$
$$a_4 = 0.6429 \pm 0.5072 \tag{3}$$

The scatter of the elastic constants from one animal to another is large. We found no significant correlation between a_1, a_2, and a_4 and vessel size or body weight of the animals. Two possible reasons for the scatter may be suggested: The rapid pulmonary arterial tissue remodeling with blood pressure change (Fung and Liu, 1991, 1992, Liu and Fung, 1989) and the complex geometry of the pulmonary arterial tree, especially the local variation of the Gaussian curvature of the wall. Gaussian curvature is the product of the two principal curvatures of a shell: it is one of the most important quantities that determines the stress distribution in a shell subjected to lateral pressure. In the test specimen shown in Figure 6.7:1, the shaded regions are saddle-shaped while in vivo, with negative Gaussion curvature; the non-shaded regions are "ellipsoidal," with positive Gaussion curvature. Thus the in vivo stress due to blood pressure could be very nonuniform. Hence, according to the stress-dependent tissue remodeling mentioned earlier, the

FIGURE 6.7:3. Data of Green's strain vs. Kirchhoff stress at very high stress and strain in dog pulmonary artery. Nonlinearity sets in the range of stress far beyond the physiological range. From Debes and Fung (1995), reproduced by permission.

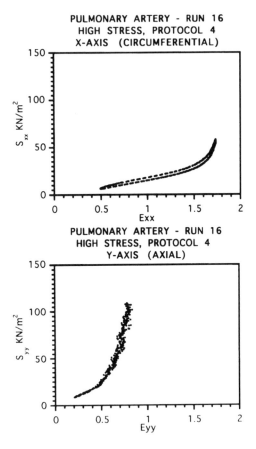

mechanical properties that remodel with the stress are expected to be nonuniform. Details are discussed in Debes and Fung (1995).

X-Ray Angiography of Small Vessels

To study pulmonary blood vessels larger than $100\,\mu$m in diameter, the lung was hung in a lucite box, and inflated by negative pleural pressure while the airway was exposed to atmosphere. The pulmonary vessels were first perfused with saline, then with a suspension of radio-opaque $BaSO_4$, either from the artery or from the vein. The $BaSO_4$ particles will not pass through the capillaries; hence, either the arterial tree or the venous tree can be photographed by x-ray. Photographs were taken while the perfusion pressure was varied step by step. Then the vessel diameters were measured and the pressure and diameter relationship was determined.

Figure 6.7:4 shows the results of Yen et al. (1980) for the pulmonary arteries of the cat. The diameter of each vessel, D, plotted on the ordinate, is

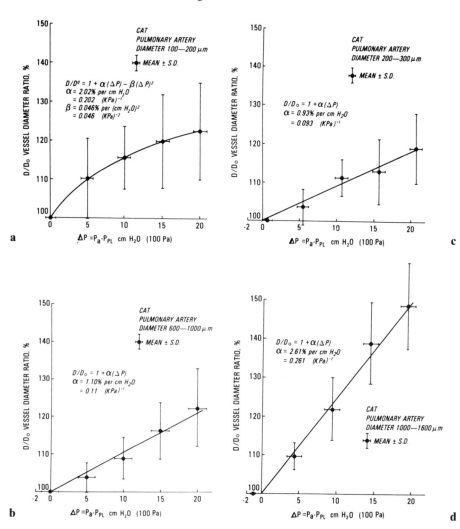

FIGURE 6.7:4. (a) Percentage change in diameter of the pulmonary arteries of the cat as functions of the pressure difference of the arterial blood pressure and the pleural pressure, $\Delta p = p_a - p_{PL}$, for vessels in the size range 100 to 200 μm. (b) Vessels in 200 to 300 μm range. (c) Vessels in 600 to 1,000 μm range. (d) Those in 1,000 to 1,600 μm range. The alveolar gas pressure, p_A, was zero (atmospheric). The pleural pressure, p_{PL}, was -10 cm H_2O. The vertical bars indicate the scatter (SD) of D/D_0 in different lungs studied. The horizontal bars indicate the SD of the pressures in different vessels. From Yen et al. (1980). Reproduced by permission.

normalized with respect to the diameter of that vessel when $p_a - p_{PL}$ is equal to zero, with p_a being the blood pressure and p_{PL} being the pleural pressure. Each vessel is identified by its diameter at $p - p_{PL}$ equal to zero and is classified into groups as shown in the figure. The abscissa is the pressure difference $p_a - p_{PL}$. The lung inflation pressure $p_A - p_{PL}$ was 10 cm H_2O.

Note that the pressure–diameter relationship for the pulmonary arteries is very different from that of the systemic arteries discussed in Chapter 8 of *Biomechanics: Mechanical Properties of Living Tissues* (Fung, 1993). Curves for the systemic arteries are nonlinear, with pressure an exponential function of diameter. Curves for pulmonary arteries are straight lines. The reason for this difference could be intrinsic, that the smaller pulmonary arteries have as wide a linear range of stress-strain relationship as exhibited by the large pulmonary arch in Figures 6.7:1 to 6.7:3. Or it could be due to tethering of the pulmonary arteries by the lung parenchyma. The pressure-diameter curves shown in Figure 6.7:4 are not that of the arteries alone, but of the artery–parenchyma combination. The lung parenchyma, consisting of the alveolar walls, are attached to the outside of the blood vessel. They are an integral part of the vessel. The vessels are tethered, embedded in an elastic medium. Together, they have a linear mechanical behavior in the strain range of concern.

The slopes of the regression lines of Figure 6.7:4 are called the *compliance constants*. They are determined and listed in Table 6.14:1, on p. 396, in Section 6.14.

The corresponding results for the pulmonary veins of the cat are shown in Figure 4.10:2, on p. 257, in which the ordinate is the vessel diameter (projected width) normalized with respect to D_{10}, the diameter of the vessel when $p - p_A = 10$ cm H_2O. This value of 10 cm H_2O was chosen because it was feared that the cross-sectional shape of the pulmonary veins may not be circular when $p - p_A$ is zero; but when $p - p_A$ is 10 cm H_2O, the cross section will not deviate far from being a circle; hence D_{10} serves better as a normalizing factor. A source of scatter of data is thus minimized.

A remarkable feature revealed by Figure 4.10:2 on p. 257 is the gentle change of the slope of the curves when $p - p_A$ is negative. The implications of this have been discussed at length in Section 4.10. It clearly indicates that the pulmonary veins do not collapse when the transmural pressure is negative. This stability is, again, due to tethering of the lung parenchyma, in which the veins are embedded (Fung et al., 1983).

If a linear relationship between blood pressure and vessel diameter is assumed for the data shown in Figure 4.10:2, then the slope of the regression line, the *compliance constant*, can be determined. Data for all orders of pulmonary veins are listed in Table 4.10:2 on p. 258 and Table 6.14:3 on p. 397 for several specified sets of values of p_A and p_{PL}. When the pleural pressure changes, the degree of lung inflation changes, and the compliance constant will change. Figure 6.7:5 shows the variation of the compliance constant with changing pleural pressure when p_A is zero (atmospheric).

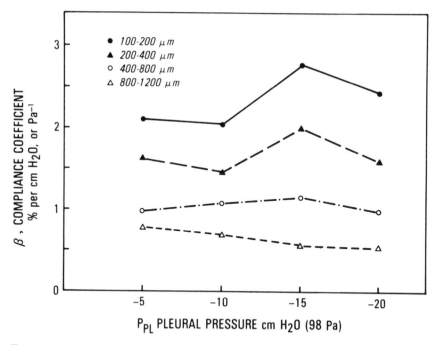

FIGURE 6.7:5. Variation of the compliance constant of the pulmonary veins of the cat with transpulmonary pressure, $p_A - p_{PL}$. In the experiment, the alveolar gas pressure, p_A, was zero. Hence the compliance coefficient β is plotted against the pleural pressure for four groups of vessels in the diameter ranges 100 to 200, 200 to 400, 400 to 800, and 800 to 1,200 μm. From Yen and Foppiano (1981). Reproduced by permission.

Elastomer Casting Method

For vessels smaller than 100 μm in diameter, reading the diameter from x-ray film becomes increasingly difficult. Hence we used the silicone elastomer method. To prepare the specimens, the lung was inflated and perfused with a liquid silicone rubber of low viscosity, freshly catalyzed with 3% stannous ethylhexoate and 1.5% ethyl silicate. Then the flow was stopped and the perfusion pressure equilibrated at a selected value. The elastomer was allowed to solidify. After solidification, histological specimens were prepared and the vessel diameters were measured. The pressure–diameter relationship was then determined for vessels of identified hierarchy (order number). In this method it is not possible to follow a single vessel and observe its change in diameter with changing blood pressure. The best one can do is to identify the hierarchy of the vascular tree, measure the mean diameter of the vessels of each order, and examine the change in the mean diameter with the changing pressure difference $p - p_A$. Our results are presented in Table 4.10:1 on p. 255. The estimated compliance constants of these small vessels are also listed in Table 6.14:3, on p. 397.

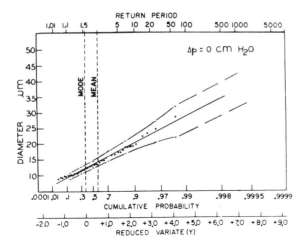

FIGURE 6.7:6. Extreme-value distribution of the smallest noncapillary pulmonary blood vessels. Solid circles are the experimentally observed blood vessel diameters. The regression line is a visual best fit. From Sobin et al. (1978). Reproduced by permission.

One special vessel in each histological photograph is easily identified, namely, the smallest noncapillary blood vessel, either the smallest arteriole or the smallest venule. To these smallest vessels, the *extreme-value* statistics can be applied, [see Fung (1993), p. 117; and Sobin et al. (1977, 1978)]. An accurate statement of the expected diameter of the smallest noncapillary vessel in a given size of the sample can be made. Figure 6.7:6 shows an example of the data plotted on Gumbel extreme-value probability distribution paper. Figure 6.7:7 shows the variation in the mean diameter of the

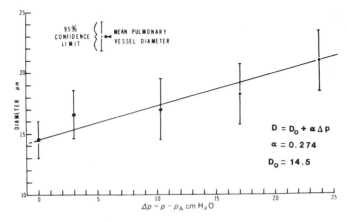

FIGURE 6.7:7. Plot of the mean diameter and 95% confidence level of the smallest vessels obtained by extreme-value statistics for the five transmural pressures, $p - p_A = 0$, 3, 10.25, 17, and 23.75. p, p_A are blood and airway pressures, respectively. From Sobin et al. (1978). Reproduced by permission.

smallest noncapillary blood vessel in each of 50 histological slides with transmural pressure $p - p_A$. From this curve we obtain the compliance constant of these vessels of the cat as $0.274\,\mu m$ per cm H_2O, or 1.61% per cm H_2O, quite consistent with the values listed in Table 6.14:3 obtained by an entirely different method.

For data on human lung, see Huang et al. (1996), Yen (1988, 1989), Yen and Sobin (1988). For the dog, see Al-Tinawi et al. (1991), Gan and Yen (1994). Krenz et al. (1994) presented additional data. Haworth et al. (1991) made estimates based on fractal considerations.

6.8 Elasticity of Pulmonary Alveolar Sheet

As is common to most topics in science, the question of elasticity can be formulated in many different forms, varying in degree of generality and depth. For the blood flow problem, the question of elasticity we are concerned with is: When the pressure in the blood vessel is changed, how much does the blood volume change? In the sheet-flow model of the pulmonary alveoli, the problem is: When the pressures of the blood and of the alveolar air change, how does the thickness of the sheet change? How does the plane area of the sheet change? Note that the blood volume in an alveolar sheet is equal to

thickness × area × VSTR,

where the VSTR is the vascular-space–tissue ratio discussed earlier. The morphometric data presented in Section 6.3 show that the VSTR remains constant when blood pressure changes. Hence the elasticity problem can be limited to the consideration of thickness and area.

It turns out that these questions have both a simple and a complex answer. In simple terms, the thickness h varies with the pressure difference Δp (equal to the static pressure of blood minus the pressure of the alveolar air) as follows (Fig. 6.8:1):

$h = 0$ if Δp is negative, and smaller than $-\varepsilon$, where ε is
a small number about $1\,cm$ H_2O.   (1a)

$h = h_0 + \alpha\Delta p$, if Δp is positive and smaller than
certain limiting value.   (1b)

h tends to a limiting value h_∞ if Δp increases
beyond the limiting value.   (1c)

In the small range $-\varepsilon < \Delta p < 0$, h increases from 0 to h_0.
A rough approximation is $h = h_0 + \left(h_0/\varepsilon\right)\Delta p$.   (1d)

Here α, h_0, h_∞, and ε are constants independent of Δp. The parameter h_0 is the sheet thickness at zero pressure difference when the pressure decreases

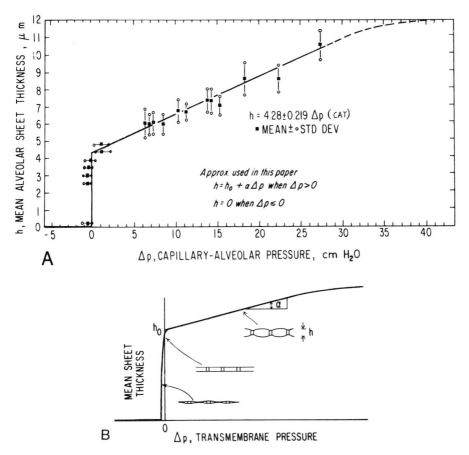

FIGURE 6.8:1. (A) Sheet thickness–pressure relationship of the cat. Equations (1a) to (1d) approximate this curve by a discontinuous curve that is composed of four line segments: a horizontal line $h = 0$ for Δp negative, which jumps to $h = h_0$ at Δp slightly greater than 0, then continues as a straight line for positive Δp until some upper limit is reached, beyond which it bends down and tends to a constant thickness. (B) The elastic deformation of the alveolar sheet is sketched for three conditions: $\Delta p < 0$, $\Delta p = 0$, and $\Delta p > 0$. The relaxed thickness, h_0, of the sheet is equal to the relaxed length of the posts. Under a positive internal (transmural) pressure, the thickness of the sheet increases; the posts are lengthened; the membranes deflect from the planes connecting the ends of the posts; and the mean thickness becomes h. From Fung and Sobin (1972a,b). Reproduced by permission of the American Heart Association, Inc.

from positive values. The parameter α is called the *compliance coefficient* of the pulmonary capillary bed. The thickness h is understood to be the mean value averaged over an area that is large compared with the posts, but small relative to the alveoli. The known values of h_0 and α are given in Table 6.8:1.

TABLE 6.8:1. The Compliance Constant, α, and the Thickness at Zero Transmural Pressure, h_0, of the Pulmonary Alveolar Vascular Sheet at Specified Values of Transpulmonary Pressure, $p_A - p_{PL}$

Animal	h_0 (μm)	α (μm/cm H$_2$O)	Reference	$p_A - p_{PL}$ (cm H$_2$O)
Cat	4.28	0.219	Figure 6.5:1	10
Dog	2.5	0.122	Fung & Sobin (1972b)*	10
	2.5	0.079	Fung & Sobin (1972b)*	25
Human	3.5	0.127	Sobin et al. (1979)	10

* Based on the data of Glazier et al. (1969) and Permutt et al. (1969).
p_A = alveolar gas pressure, p_{PL} = pleural pressure.

Also, in simple terms, the answer to the second question is that if A_0 represents the area of certain region of an alveolar septa when the static pressure of the blood is some physiologic value p_0, then the area of the same region A when blood pressure is changed to p is

$$A \doteq A_0. \tag{2}$$

In other words, the area is unaffected by the blood pressure.

The answers to our questions become more complex when we try to relate the constants h_0, α, h_∞, and A_0 to the airway pressure, the intrapleural pressure, the surfactants on the alveolar septa, the geometric parameters of the septa, the structure of the entire lung and the shape of the thoracic cage and the diaphragm, and whether a pathologic condition such as edema or emphysema exists. In order to unravel these relationships, we may take a two-pronged approach: We may study the problem theoretically and derive these relationships from the principles of mechanics, or we may lay out an extensive program of experiments in order to deduce empirical correlations among various parameters. It is easy to see that the required experimental program would be very large. For pragmatism, one should take a combined approach: Deduce as much as possible from theoretical considerations, and test as often as possible by experiments.

A theoretical analysis of the sheet elastic compliance α as a function of geometric parameters is presented by Fung and Sobin (1972a). Of major concern is the dependence of the compliance coefficient α on the tension, T (dyne/cm^{-1}), in the alveolar septa. T increases when lung volume increases. It can be shown that the larger the tension, the smaller is the compliance coefficient. Since T is the sum of the surface tension and elastic tension, an increase of surface tension would cause an increase in T and a decrease in the compliance of the alveolar sheet, thus causing a decrease in blood flow. An expansion of the lung also increases T, and, if other things remain equal, would also decrease blood flow in the alveolar sheet.

6.9 Apparent Viscosity of Blood in Pulmonary Capillaries

We need to know the flow behavior of blood before we can formulate the pulmonary blood flow problem mathematically. The general problem of blood viscosity has been discussed in *Biomechanics: Mechanical Properties of Living Tissues* (Fung, 1993). To understand the flow of blood in the pulmonary capillary sheet we must study the interaction of red blood cells with the capillary sheet. As Ernest Sechler used to say, "When you design an airplane, you must think like an airplane"; we may wish to think of blood flow by picturing ourselves as a red blood cell. As a red cell moving through the lung, what do we see? At first we see a big tunnel. Then the tunnel divides and divides again and again, and becomes smaller. At the arteriolar level the tunnel diameter is only two or three times our own diameter. Then you enter the capillary sheet and the scenery suddenly changes. You seem to have entered an underground parking garage. You are a car. The ceiling is low and there is bumper-to-bumper traffic. You have to swing right and left in order to avoid the posts.

Figure 6.9:1 shows an interpretation of a scene in a pulmonary alveolar sheet as seen by a red cell. Many red cells float in this space. One on the

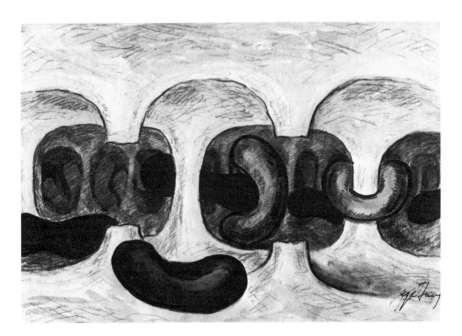

FIGURE 6.9:1. Conceptual drawing of pulmonary alveolar blood flow. The red cells move in the pulmonary alveolar vascular space, which has some resemblance to an underground parking garage. From Fung (1975c). Painted by the author.

right is caught on a post. The picture would have been more faithful if the cells had been drawn larger. To clarify the view, I have drawn them too small and too sparsely. In reality the cells should fill the height almost entirely.

To analyze the motion of the red cells in such an environment, we may begin with a dimensional analysis (see *Biomechanics: Mechanical Properties of Living Tissues*, Fung, 1993, p. 168). For our problem the variables of interest are the pressure p, the pressure gradient ∇p, the coefficient of viscosity of the plasma μ_0, the mean velocity of flow U, the angular frequency of oscillation ω, the sheet thickness, h, the width of the sheet w, the diameter of the posts ε, the distance between the posts a, the angle between the mean flow and a reference line X–X defining the postal pattern shown in Figure 6.3:3, θ, the hematocrit H, the diameter of the red cell D_c, and the characteristic elastic modulus of the red cell membrane E_c. By simple trial we see that the following parameters form a complete set of independent dimensionless parameters:

$$\frac{h^2}{\mu_0 U}\nabla p, \quad \frac{Uh\rho}{\mu_0} = N_R \left(\text{Reynolds number}\right),$$

$$\frac{D_c}{h}, \quad \frac{\mu_0 U}{E_c h} \equiv \text{cell membrane strain parameter},$$

$$\frac{h}{2}\sqrt{\frac{\omega\rho}{\mu_0}} \equiv \text{Womersley number},$$

$$H, \frac{w}{h}, \frac{h}{\varepsilon}, \frac{\varepsilon}{a}, \theta, \text{VSTR} \left[\text{see Eq.} \left(2\right) \text{of Sec. 6.3}\right].$$

Hence, according to the principle of dimensional analysis, any dimensionless relationship between the variables p, U, ... must be a relationship between these dimensionless parameters. In particular, the parameter we are most interested in, that connecting the pressure gradient and the flow velocity, may be written as

$$\frac{h^2}{\mu_0 U}\nabla p = F\left(\frac{D_c}{h}, \frac{\mu_0 U}{E_c h}, N_R, \frac{h}{2}\sqrt{\frac{\omega\rho}{\mu_0}}, H, \frac{w}{h}, \frac{h}{\varepsilon}, \frac{\varepsilon}{a}, \theta, \text{VSTR}\right), \quad (1)$$

where F is a certain function that must be determined either theoretically or experimentally.

In capillary blood flow, the Reynolds and Womersley numbers are much smaller than 1, and their effects may be neglected. The cell membrane strain parameter is the ratio of the typical shear stress in the fluid, $\mu_0 U/h$, to the elasticity modulus of the cell membrane, E_c. Its effect and the effect of D_c/h have been discussed in *Biomechanics: Mechanical Properties of Living Tissues* (Fung, 1993, pp. 180, 184, 185). In general, in narrow circular cylindrical tubes the effects of these parameters make the pressure–flow relationship nonlinear. In the pulmonary alveolar sheet, however, the red cells are far less constrained than they are in the tubes, and a model experiment

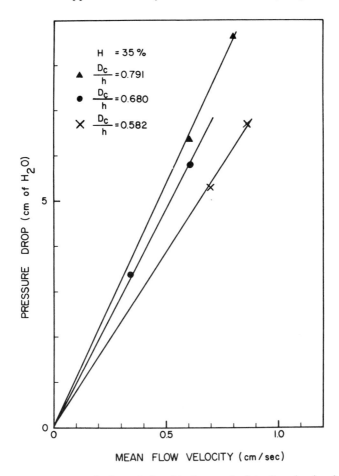

FIGURE 6.9:2. Pressure–velocity relationship for particulate flow in simulated pulmonary alveolar capillary blood vessels at a hematocrit (H) of 35%. D_c is the diameter of the red blood cell. h is the thickness of the vascular space of the interalveolar sheet. From Yen and Fung (1973). Reprinted by permission.

by Yen and Fung (1973) showed that the pressure–flow relationship is linear when D_c/h is smaller than 0.80 (Fig. 6.9:2). This is fortunate. We can then write Eq. (1) as

$$\nabla p = -\frac{\mu_0 U}{h^2} F\left(\frac{D_c}{h}, \frac{\mu_0 U}{E_c h}, H, \frac{w}{h}, \frac{h}{\varepsilon}, \frac{\varepsilon}{a}, \theta, \text{VSTR}\right). \qquad (2)$$

To investigate the function, F, Fung (1969), Lee (1969), Lee and Fung (1968), and Yen and Fung (1973) conducted theoretical and experimental studies on particulate flow in an alveolar sheet. They isolated the effects of D_c/h, $\mu_0 U/E_c h$, H, w/h, and the rest of the parameters separately, and wrote

$$\nabla p = -\frac{\mu_0 U}{h^2} F'\left(\frac{D_c}{h}, \frac{\mu_0 U}{E_c h}, H\right) k\left(\frac{w}{h}\right) f\left(\frac{h}{\varepsilon}, \frac{\varepsilon}{a}, \theta, \text{VSTR}\right). \tag{3}$$

This was further abbreviated to

$$\nabla p = -\frac{U}{h^2} \mu k f, \tag{4}$$

where μ stands for the apparent viscosity,

$$\mu = \mu_0 F'\left(\frac{D_c}{h}, \frac{\mu_0 U}{E_c h}, H\right), \tag{5}$$

and k and f are functions of the parameters indicated in Eq. (3). The ratio μ/μ_0 is called the "relative" viscosity. Yen and Fung (1973) wrote

$$\mu = \mu_0\left(1 + aH + bH^2\right), \tag{6}$$

where H is the hematocrit, a, b are functions of the ratio of the cell diameter D_c and the sheet thickness h. μ_0 is the viscosity of plasma. Their experimental results are presented in Figure 5.2:3, p. 171 of *Biomechanics: Mechanical Properties of Living Tissues*, Fung, 1993; the effect of the parameter $\mu_0 U/E_c h$ has not been investigated in detail so far.

The function k is well known (see, e.g., Purday, 1949). When $h/w < 0.2$, it is given by the equation

$$k = 12 \Big/ \left(1 - 0.63\frac{h}{w}\right). \tag{7}$$

Hence for all practical purposes k can be replaced by 12. The function f is called the *geometric friction factor*. Lee (1969) made a theoretical calculation of the function f. Yen and Fung (1973) confirmed Lee's results. f as a function of h/ε is given in Figure 6.9:3, which is obtained with the parameters ε/a, θ, and VSTR pertinent to the cat's lung, (Table 6.3:1). In the case of the cat lung, θ has little effect and f is independent of the orientation of the flow.

These experimental results were obtained in steady-state flow. Their applicability to the pulmonary microcirculation is based on the smallness of the Womersley and Reynolds numbers (on the order of 10^{-2}) in the alveolar sheet. Experimental verification that at low Reynolds number no flutter occurs when the local blood pressure is smaller than the alveolar gas pressure has been presented by Fung and Sobin (1972b).

6.10 Formulation of the Analytical Problems

"Local Mean Flow" and "Local Mean Disturbances"

To analyze the blood flow in an alveolar sheet, we first introduce the concept of local mean velocity of flow. The detailed flow field is, of course,

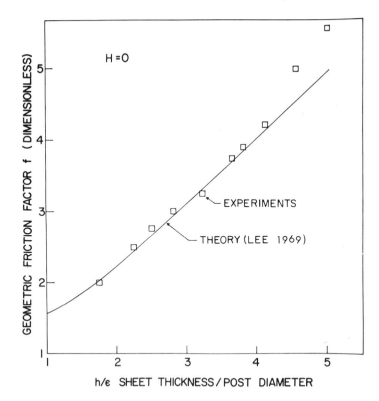

FIGURE 6.9:3. The geometric friction factor *f* as a function of the alveolar sheet structure. Theoretically and experimentally determined values of *f* are compared for an alveolar sheet for which the values of ε/a, θ, and VSTR are those of the cat and human lung, whereas the sheet thickness/post diameter ratio is varied. From Yen and Fung (1973). Reprinted by permission.

very complicated when the disturbances of individual posts, red cells, membrane deflections, and other phenomena are considered. Therefore it is mathematically expedient to separate the flow field into two parts: a local mean flow and local disturbances. It is expected that the local mean flow field will be relatively smooth and can be determined from the boundary conditions. This chapter is concerned with the local mean flow. When the mean flow field is known, the local disturbances can be computed separately in a much simplified manner.

   To be precise, we now define the local mean flow and the perturbations. Let us first define a Cartesian frame of reference x, y, z. We take the origin at a point on the midplane of the sheet; x, y axes in the sheet; and z axis perpendicular to the sheet (see Fig. 6.3:4, p. 345). At a point x, y, z, the velocity vector has three components—$u(x, y, z)$, $v(x, y, z)$, and $w(x, y, z)$—in the directions of the coordinate axes. We break u, v, w into two parts, and write

$$u(x, y, z) = U(x, y) + u'(x, y, z),$$
$$v(x, y, z) = V(x, y) + v'(x, y, z),$$
$$w(x, y, z) = W(x, y) + w'(x, y, z), \tag{1}$$

where $U(x, y)$, $V(x, y)$, and $W(x, y)$ are functions of x, y, alone. We call $U(x, y)$, $V(x, y)$ the *local mean velocity* if they represent the average value of $u(x, y, z)$, $v(x, y, z)$ over a small volume around $(x, y, 0)$; that is, if

$$U(x, y) = \frac{1}{Ah} \iint_{A(x,y)} dx' dy' \int_{-h/2}^{h/2} u(x', y', z) dz$$

$$V(x, y) = \frac{1}{Ah} \iint_{A(x,y)} dx' dy' \int_{-h/2}^{h/2} v(x', y', z) dz. \tag{2}$$

The mean velocity $W(x, y)$, defined by a similar equation, vanishes when the area A is chosen to be large enough. In these equations, h is the thickness of the sheet, $A(x, y)$ is a domain of integration containing the point (x, y), and A is the area of $A(x, y)$, as illustrated in Figure 6.3:4, p. 345. The points (x', y') lie in the area A. The domain A is chosen to be large enough that it contains a number of posts and red cells so that the velocities U and V represent the smoothed-out velocity field, and yet small enough that the variation of U, V over the entire alveolar sheet still reflects the significant features of the flow field. In practical calculations an area containing, say, 10 posts would be satisfactory for such local averaging.

The velocity components $u'(x, y, z)$, $v'(x, y, z)$, and $w'(x, y, z)$ in Eq. (1) are called the *local perturbations*. According to Eqs. (2) the mean values of the local perturbations $u'(x, y, z)$, $v'(x, y, z)$, and $w'(x, y, z)$ are zero.

The local mean pressure and the mean sheet thickness are defined in a similar manner.

Basic Equations

In the development of our theory, the first basic relation we seek is that relating the local mean pressure $p(x, y)$ to the velocities $U(x, y)$, $V(x, y)$. This can be derived from the basic equations of hydrodynamics, or alternatively, by model testing with the help of dimensional analysis. As it is shown in Section 6.9, a dimension analysis leads to Eq. (4) of Section 6.9 or

$$\frac{\partial p}{\partial x} = -\frac{\mu U}{h^2} k f_x \left(\frac{h}{\varepsilon}, \frac{\varepsilon}{a}, \theta, \text{VSTR} \right),$$

$$\frac{\partial p}{\partial y} = -\frac{\mu V}{h^2} k f_y \left(\frac{h}{\varepsilon}, \frac{\varepsilon}{a}, \theta, \text{VSTR} \right). \tag{3}$$

In this equation, μ is the apparent viscosity of blood, h is the sheet thickness, and the numerical factors k, f_x, f_y are functions of the sheet geometry

FIGURE 6.10:1. A rectangular element of the sheet showing the balance of the flow. From Fung and Sobin (1969). Reprinted by permission.

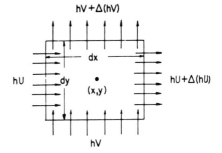

as discussed in Section 6.9. k is given by Eq. (7) of Sec. 6.9, and is equal to 12 in practical cases, and f is given in Figure 6.9:3, and lies in the range of 2 to 3.

The second basic relation we need is one that describes the conservation of mass. Consider a small rectangular element of the alveolar sheet (Fig. 6.10:1). Blood enters the left-hand side and leaves the right-hand side. In a time interval dt, the mass entering the left-hand side is equal to the product of the density, velocity, area, and dt. The velocity is U and the area is hdy, with h being the thickness of the sheet and dy the length of the edge. The density is ρ. Hence

$$\text{mass inflow at left} = \rho h U \, dy \, dt. \tag{4}$$

On the right-hand side, the mass outflow in the same time interval is

$$\left[\rho h U + \frac{\partial(\rho h U)}{\partial x} dx\right] dy \, dt. \tag{5}$$

Similarly, the mass inflow at the bottom edge is $\rho h V \, dx \, dt$, and the outflow at the top is

$$\left[\rho h V + \frac{\partial(\rho h V)}{\partial y} dy\right] dx \, dt.$$

Summing up all the inflow and outflow, we have

$$\text{net mass outflow} = \left[\frac{\partial(\rho h U)}{\partial x} + \frac{\partial(\rho h V)}{\partial y}\right] dx \, dy \, dt. \tag{6}$$

By the law of conservation of mass, this net mass outflow must be equal to the net decrease of the mass of the element. For a *steady flow* in a sheet with impermeable walls, there can be no change in the mass of the element; hence, we must have

$$\frac{\partial(\rho h U)}{\partial x} + \frac{\partial(\rho h V)}{\partial y} = 0. \tag{7}$$

If the sheet thickness h is variable, then the equation of continuity, Eq. (7), combined with the equation of motion of the fluid, Eq. (3), yields the following equation:

$$\frac{\partial}{\partial x}\left(\frac{\rho h^3}{kf_x}\frac{\partial p}{\partial x}\right) + \frac{\partial}{\partial y}\left(\frac{\rho h^3}{kf_y}\frac{\partial p}{\partial y}\right) = 0. \tag{8}$$

Now p is related to h by the constitutive Eq. (6.8:1a–d), p. 366. Using Eq. (6.8:1b) and assuming α constant, we obtain

$$\frac{\partial}{\partial x}\left(\frac{\rho h^3}{kf_x}\frac{\partial h}{\partial x}\right) + \frac{\partial}{\partial y}\left(\frac{\rho h^3}{kf_y}\frac{\partial h}{\partial y}\right) = 0. \tag{9}$$

If we ignore the spatial variation of ρ, k, and f, and assume $f_x = f_y$ for practical purposes, we obtain

$$\frac{\partial}{\partial x}\left(h^3\frac{\partial h}{\partial x}\right) + \frac{\partial}{\partial y}\left(h^3\frac{\partial h}{\partial y}\right) = 0, \tag{10}$$

or

$$\left(\frac{\partial^2}{\partial x^2} + \frac{\partial^2}{\partial y^2}\right)h^4 = 0. \tag{11}$$

Thus the fourth power of h is governed by a Laplace equation. Expressed in terms of pressure, we have, on defining

$$\Phi = h^4 = \left[h_0 + \alpha(p - p_A)\right]^4, \tag{12}$$

the result,

$$\left(\frac{\partial^2}{\partial x^2} + \frac{\partial^2}{\partial y^2}\right)\Phi = 0. \tag{13}$$

This completes the mathematical formulation for the steady-state case of sheets with impermeable wall.

If p is smaller than the alveolar gas pressure, p_A, then $h = 0$ and there will be no flow. If p is greater than the upper limiting value, then the linear relation Eq. (6.8:1b) ceases to hold, and Eqs. (11) and (13) are no longer valid. When p becomes sufficiently large, h tends to a constant: Then Eq. (8) can be used again with the factor h^3 eliminated. Thus in this limiting case the differential equation is again linear, and the pressure distribution satisfies the Laplace equation.

Generalization to Nonstationary Case and Permeable Vessel

For a nonstationary flow, and in the case in which fluid movement in or cut of the endothelium of the capillary blood vessel takes place, the equation of continuity, Eq. (7), must be modified. This modification can be obtained by considering the balance of inflow and outflow in a control volume that consists of the sides $x = $ const., $x = $ const. $+ dx$, $y = $ const., $y = $ const. $+ dy$, and the membranes $z = \pm h/2$, and the rate of change of the control volume,

$$\text{Rate of change of volume} = \left[\frac{\partial(\rho h U)}{\partial x} + \frac{\partial(\rho h V)}{\partial y}\right] dx \, dy \, dt. \quad (14)$$

The control volume is changed by (a) the transient change in thickness h and (b) the filtration across the blood–tissue barrier. The rate of increase of thickness is $\partial h/\partial t$, which induces, in a time interval dt, an increase of control volume equal to $(\partial h/\partial t)dt \, dx \, dy$. The mass transfer across the blood–tissue barrier may be assumed to obey Starling's hypothesis* $\dot{m} = K_p(\Delta p - \sigma \Delta \pi)$, where $\dot{m}$ is the rate of mass transfer per unit time per unit area, K_p is the filtration coefficient, Δp is the difference in hydrostatic pressure on the sides of the barrier, $\Delta \pi$ is the corresponding difference in osmotic pressure, and σ is the reflection coefficient. Since our attention is focused on the flow of blood, we may write

$$\dot{m} = K_p\left(p - p^*\right), \quad (15)$$

where p refers to the local mean pressure in the blood, and

$$p^* = \sigma \Delta \pi + \text{pressure in tissue space.} \quad (16)$$

Summing up both contributions we obtain, from Eq. (14),

$$\left[\frac{\partial(\rho h U)}{\partial x}\right] + \left[\frac{\partial(\rho h V)}{\partial y}\right] = -\rho\left(\frac{\partial h}{\partial t}\right) - 2K_p\left(p - p^*\right). \quad (17)$$

The factor 2 in the last term is added because the capillary wall area available for fluid exchange is about twice the area of the sheet. Eliminating U, V from Eqs. (17) and (3), we obtain, if ρ is constant,

$$\frac{\rho}{\mu}\left[\frac{\partial}{\partial x}\left(\frac{h^3}{kf_x}\frac{\partial p}{\partial x}\right) + \frac{\partial}{\partial y}\left(\frac{h^3}{kf_y}\frac{\partial p}{\partial y}\right)\right] = \rho\left(\frac{\partial h}{\partial t}\right) + 2K_p\left(p - p^*\right). \quad (18)$$

To complete the analysis we again use the equation that describes the elasticity of the alveolar sheet in the physiologic range when $p - p_{alv}$ is positive:

$$h = h_0 + \alpha\left(p - p_{alv}\right). \quad (6.8:1b)$$

Then we obtain the following basic equation:

$$\frac{\partial}{\partial x}\left(\frac{h^3}{kf_x}\frac{\partial h}{\partial x}\right) + \frac{\partial}{\partial y}\left(\frac{h^3}{kf_y}\frac{\partial h}{\partial y}\right) = \mu\alpha\left[\frac{\partial h}{\partial t} + \frac{2}{\rho}K_p\left(\frac{h - h_0}{\alpha} - p^* + p_{alv}\right)\right]. \quad (19)$$

---

*This is a linear phenomenologic law for mass transport across a membrane. See Chapter 8, Section 8.6, of the companion volume, *Biomechanics: Motion, Flow, Stress, and Growth* (Fung, 1990).

This is a nonlinear "diffusion" equation, basically different from the usual "wave" equation that describes pulsatile flow in large arteries. The impedance characteristics of the capillaries are therefore different from those of the arteries.

In general, we may use the approximation $f_x = f_y = f$. If the variation of h is not too large, we may regard k and f as constant and simplify Eq. (19) to

$$\left(\frac{\partial^2}{\partial x^2} + \frac{\partial^2}{\partial y^2}\right)h^4 = 4\mu k f \alpha \left[\frac{\partial h}{\partial t} + \frac{2K_p}{\rho\alpha}(h - h^*)\right],$$

(20)

where

$$h^* = h_0 + \alpha\left(p^* - p_{\text{alv}}\right).$$

(21)

Solutions of these equations are considered in the next sections.

6.11 An Elementary Analog of the Theory

The sheet-flow equation contains a great deal of information that can be revealed through solutions with various boundary conditions. This equation is nonlinear. But the nonlinearity is of a special kind, through the fourth power of h, which is operated on by the Laplace operator.

The nature of the solution can be brought out easily if we use a one-dimensional analog. Consider two uniform, parallel, horizontal, elastic strings as shown in Figure 6.11:1. At finite intervals these strings are tied by vertical cross members. If we pull on the horizontal strings with a tension T, the strings are stretched uniformly, and the ratio of the lengths of segments bc and ac remains a constant. This ratio is an analog of VSTR: The segments ab and cd are analogs of the posts, and bc and de are the analog of the alveolar–capillary membrane.

On the other hand, if the post segments were replaced by springs of different compliance than the strings, then the VSTR (bc/ac) would change with the tension T (Fig. 6.11:1B). The constancy of VSTR observed in the pulmonary alveolar sheets of the cat suggests that the elasticity of the postal region (segments ab, cd, etc.) is represented by the analog shown in Figure 6.11:1A.

Let the strings be loaded by an internal vertical loading of Δp per unit length, as shown in Figure 6.11:1C. The equilibrium of the string requires that

$$T \times \text{curvature of string} = \Delta p.$$

If the tension remains constant and the deflection is small, then the vertical deflection of the string is proportional to Δp.

The change of the average distance between the strings (analog of the sheet thickness) is given by the sum of the distention of the posts aa', bb', etc. and the average deflection of the strings.

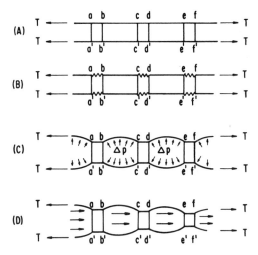

FIGURE 6.11:1. Simplified one-dimensional analog of the alveolar sheet. (A) Stretching of the membrane. (B) Stretching of the post. (C) Deflection under internal pressure. (D) Deflection due to variable internal pressure when blood flows in the channel. From Fung and Sobin (1977a). Reprinted by permission.

Now consider flow in a channel of unit width represented by Figure 6.11:1D (replace the strings with channel walls). Let the average speed of flow be U, the volume flow rate per unit width be $\dot{Q}$, the local thickness be h, and the pressure be p. Then

$$U = -\frac{h^2}{\mu k f}\frac{dp}{dx} \tag{1}$$

$$\dot{Q} = hU. \tag{2}$$

In the range of p in which $h = h_0 + \alpha p$, we have

$$\frac{dh}{dx} = \alpha\frac{dp}{dx}. \tag{3}$$

Hence,

$$\dot{Q} = -\frac{h^3}{\mu k f \alpha}\frac{dh}{dx} = -\frac{1}{4\mu k f \alpha}\frac{dh^4}{dx}. \tag{4}$$

For a steady flow, $\dot{Q}$ is a constant. Differentiating Eq. (4) with respect to x, we obtain, because $d\dot{Q}/dx = 0$, the differential equation

$$\frac{d^2 h^4}{dx^2} = 0, \tag{5}$$

which is a special case of the sheet-flow equation, Eq. (6.10:11). This differential equation is easily integrated. The general solution is

$$h^4 = c_1 x + c_2, \tag{6}$$

where c_1, c_2 are arbitrary constants. To determine c_1, c_2, we notice the boundary conditions:

a. At the "arteriole," $x = 0$, the thickness of the "sheet" is h_a,
b. At the "venule," $x = L$, the thickness of the "sheet" is h_v.

Hence,

$$h_a^4 = c_2, \quad h_v^4 = c_1 L + h_a^4. \tag{7}$$

Solving for c_1, c_2 and substituting into Eq. (6), we obtain

$$h^4 = h_a^4 - \left(h_a^4 - h_v^4\right) x / L, \tag{8}$$

or

$$h = \left[h_a^4 - \left(h_a^4 - h_v^4\right) x / L\right]^{1/4}. \tag{9}$$

This is the full solution. For various combinations of h_a and h_v the distribution of the thickness h is shown in Figure 6.11:2. It is seen that the exponent $\frac{1}{4}$ makes h rather flat near the arteriole, and constricts rather rapidly toward the venule if h_v is small.

We can obtain the flow $\dot{Q}$ from Eq. (4). If the channel length is L, then the mean flow in the whole channel is

$$\frac{1}{L} \int_0^L \dot{Q} dx = \frac{1}{48 \mu \alpha L} \left[h^4(0) - h^4(L)\right]. \tag{10}$$

For a one-dimensional channel flow of an incompressible fluid, the conservation of mass requires $\dot{Q}$ to be constant. Hence the left-hand side of Eq. (10) is exactly $\dot{Q}$. In this case, we can also integrate Eq. (4) to obtain

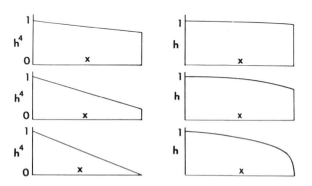

FIGURE 6.11:2. Plot of Eqs. (8) and (9) showing the variation of h^4 and h with x. With an appropriate choice of units, we assume that the thickness h_a is 1. h_v^4 at $x = L$ is assumed to be 0.75, 0.25, and 0 for the three cases shown in the figure. The corresponding values of h_v are 0.931, 0.707, and 0, respectively.

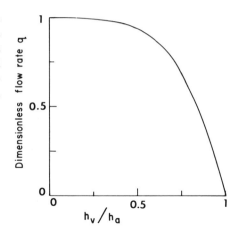

FIGURE 6.11:3. The variation of the ratio of the volume flow rate to the maximum possible with the thickness ratio h_v/h_a. $\dot{q}$ is dimensionless, defined by Eq. (12). h_v and h_a are sheet thickness at the venule and arteriole, respectively.

$$h^4 = -48\mu\alpha\dot{Q}x + \text{constant}.$$

But $h = h_a$ when $x = 0$. Hence the constant equations h_a^4, and we have the channel thickness distribution,

$$h(x) = \left(h_a^4 - 48\mu\alpha\dot{Q}x\right)^{1/4},$$

which is, of course, equivalent to Eq. (9). When $x = L$, the thickness h becomes h_v, and we obtain the exact result,

$$\dot{Q} = \frac{h_a^4 - h_v^4}{48\mu\alpha L}. \tag{11}$$

The last equation is a remarkable result. It shows that the flow depends on the difference of the fourth power of the sheet thickness at the arteriole from the fourth power of the sheet thickness at the venule. Thus if h_v is considerably smaller than h_a, its influence on the volume flow rate would be small. This is exhibited in Figure 6.11:3, in which we have plotted the dimensionless ratio of the volume flow rate divided by the maximum possible:

$$\dot{q} = \frac{\dot{Q}}{\dot{Q}_{max}}, \qquad \dot{Q}_{max} = \frac{h_a^4}{48\mu\alpha L} \tag{12}$$

against the ratio h_v/h_a. From Eqs. (11) and (12) we have, obviously,

$$\dot{q} = 1 - \left(\frac{h_v}{h_a}\right)^4. \tag{13}$$

It is seen clearly that $\dot{q}$ is significantly reduced by h_v only if h_v approaches h_a. If h_v is less than one half of h_a, $\dot{q}$ differs from 1 by less than 6%; hence, the flow is controlled essentially by h_a, the sheet thickness at the arteriole.

6.12 General Features of Sheet Flow

Rapid Variation of Sheet Thickness at the Venule, and Nearly Uniform Hydrostatic Pressure near the Arteriole

In the more general case, the nature of the solution may be illustrated by an idealized example shown in Figure 6.12:1. In the upper figure, we have sketched a hypothetical rigid alveolar sheet ($h_0 \neq 0$, $\alpha = 0$), which is opened to an arteriole and a venule at regularly spaced intervals. The pattern is periodic, and three segments are shown in the figure. At the left are shown a plan view and a vertical cross section. The thickness of the sheet is assumed to be constant. At the right are shown the streamlines, that is, the lines tangential to the velocity vectors. In the center are shown the lines of constant blood pressure. These contours of equipressure lines are drawn at intervals of constant pressure drop. The pressure at the horizontal line in the middle is halfway between the arteriole pressure and venule pressure. Other equipressure lines are orthogonal to the streamlines. The pressure gradient is inversely proportional to the distance between the equipressure lines. The velocity of flow is inversely proportional to the spacing between streamlines. These contours thus show the velocity and pressure distribution in the alveolar walls.

In the lower panel of Figure 6.12:1 we consider a different situation. We relax the assumption that the alveolar wall is rigid, and assume instead that

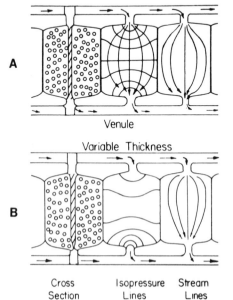

Arteriole

A

Venule

Variable Thickness

B

Cross Section Isopressure Lines Stream Lines

FIGURE 6.12:1. A type of problem that can be investigated theoretically. With given sheet dimensions and boundary conditions, the variation of the average sheet thickness, velocity distribution, and pressure distribution can be computed. Figure shows streamlines and pressure contours. From Fung and Sobin (1969). Reprinted by permission.

the sheet thickness varies linearly with the difference between the blood pressure and the air pressure in the alveolar space. In this case the pressure and velocity distributions are significantly modified. Figure 6.12:1(B) is drawn for the case in which the arteriole pressure is greater than the airway pressure, but the venule pressure is equal to the airway pressure. In the left panel, the sheet thickness is seen to contract rapidly in the neighborhood of the venule. In the middle panel, the equipressure contours are seen to crowd toward the venule. The pressure in the sheet is much more uniform than that in Figure 6.12:1(A), except in the neighborhood of the venule, where a rapid pressure drop occurs. The streamlines sketched in the figure on the right show a similar increase in the velocity near the drain into the venule.

The detailed information on velocity and pressure distribution have an important bearing on pulmonary physiology. The velocity distribution is relevant to the blood transit time in alveoli and oxygenation of red cells. The pressure distribution is important to the question of fluid transport across the alveolar wall, and hence is relevant to the question of edema and homeostasis. The thickness distribution is directly related to flow resistance. The mathematical details of the solution follow.

Computation of Flow Field

Let us consider first the problem shown in Figure 6.12:1(A), *the flow in a sheet of constant thickness*. This is an artificial problem, but its solution provides a stepping stone to the real problem of flow in an elastic sheet governed by Eq. (6.10:11).

By the periodicity of spatial geometry, it is sufficient to consider a single rectangular sheet. Let us assume that the sheet is of uniform thickness. The boundary conditions are that the borders of the sheet are streamlines, except at the openings to the arteriole and venule, where pressures are specified. To satisfy these boundary conditions, it is convenient to introduce a *stream function* $\psi(x, y)$ so that

$$hU = \frac{\partial \psi}{\partial y}, \quad hV = -\frac{\partial \psi}{\partial x}, \tag{1}$$

which satisfy the equation of continuity, Eq. (6.10:7), identically. The equation of motion is satisfied if pressure is computed from velocity through Eq. (6.10:3). On the other hand, according to Eq. (6.10:3) the pressure p may be regarded as a *velocity potential*. Furthermore, for $h = $ constant and $f_x = f_y$, Eq. (8) of Section 6.10 becomes a harmonic equation,

$$\frac{\partial^2 p}{\partial x^2} + \frac{\partial^2 p}{\partial y^2} = 0. \tag{2}$$

The potential lines $p = $ constant and the steamlines $\psi = $ constant are mutually orthogonal. Since p and ψ are conjugate to each other, ψ is governed by a harmonic equation too. The boundary condition on a boundary where a constant pressure is specified is, therefore,

$$p = \text{constant}, \quad \text{or} \quad \frac{\partial p}{\partial s} = 0, \quad \text{or} \quad \frac{\partial \psi}{\partial n} = 0, \tag{3}$$

where s and n denote tangential and normal directions, respectively. On a boundary that coincides with a streamline, the normal velocity vanishes: then the boundary condition is

$$\psi = \text{constant}, \quad \text{or} \quad \frac{\partial \psi}{\partial s} = 0, \quad \text{or} \quad \frac{\partial p}{\partial n} = 0. \tag{4}$$

Consider a square alveolar sheet of uniform thickness as shown in Figure 6.12:2, which is one quarter of a panel of Figure 6.12:1(A). The flow enters the sheet at the arteriole opening A-A′, and leaves the sheet at the venule, which is shown in Figure 6.12:1(A) but not in Figure 6.12:2. The streamlines of the draining flow are mirror images of those shown in Figure 6.12:2 reflected in the plane at the base. Let the width of the arteriole and venule opening be 1/10 of the width of the sheet. By symmetry of the boundary conditions, it is easy to see that the centerline and the solid boundaries are

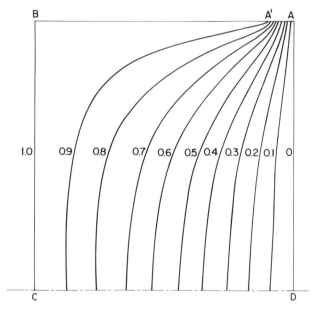

FIGURE 6.12:2. Streamlines of flow corresponding to case (A) in Figure 6.12:1. The fluid enters the sheet at AA′. No fluid penetrates the borders of A′B, BC, and the centerline AD. The boundary CD is a line of symmetry on which the pressure is constant. Numerals are the values of the stream function ψ on the streamlines $\psi = $ const. Reproduced by permission from Fung and Sobin (1969).

FIGURE 6.12:3. Geometry of an elastic alveolar sheet.

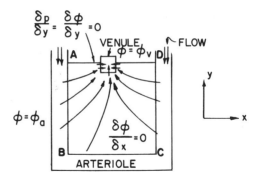

streamlines. Without loss of generality, let us choose the units of measurement in such a way that the boundary value of ψ on the centerline is 0 and that on the solid boundary is $\psi = 1$. If the flow entering and leaving the sheet is uniform, then ψ increases linearly from 0 to 1 at the arteriole opening AA' and the venule opening BB'. With these known boundary values, the values of ψ can be found easily by several standard methods. The results in Figure 6.12:2 were obtained by the relaxation method. The streamlines can then be plotted by interpolation. Since the pattern is symmetric, only one quarter of the sheet is shown in Figure 6.12:2.

From the stream functions, the velocities can be deduced according to Eq. (1). Contours of constant blood pressure lines (p = constant) are lines orthogonal to the streamlines, and thus can be deduced very easily from Figure 6.12:2.

Next, let us consider the case of real interest: *flow in an elastic sheet.* Examples in this case will demonstrate the rapid thickness variation and pressure drop in the neighborhood of the venule, especially when the so-called zone-2 condition prevails (see Sec. 6.15). Let us consider an example as shown in Figure 6.12:3. Here we assume that the precapillary forms the outer border ABCD, in which the pressure is a constant. The postcapillary drains the blood at the upper center in a rectangular opening. The upper horizontal border is assumed to be a streamline.

Let us examine a case in which the exit pressure at the venule is so much decreased that the sheet thickness at the venule becomes very small. With an appropriate choice of units, let the thickness of the sheet at the opening into the precapillary be $h_a = 1$, and that at the opening to the postcapillary be $h_v = \varepsilon$, which is a small number approaching zero. We consider the function $\Phi' = h^4$ and set the boundary conditions

$$\Phi' = h_a^4 = 1 \text{ at border adjacent to precapillary,}$$

$$\Phi' = h_v^4 = 0 \text{ at border adjacent to postcapillary,}$$

$$\frac{\partial \Phi'}{\partial n} = 0 \text{ along the top horizontal streamline.} \tag{5}$$

The last boundary condition, that the normal derivative vanishes on a streamline, can be derived from the following equations:

$$\Phi = h^4 = \left[h_0 + \alpha\left(p - p_A\right)\right]^4$$

$$p - p_A = \frac{h}{\alpha} - \frac{h_0}{\alpha}, \tag{6}$$

and

$$\operatorname{grad} p = \frac{kf}{h^2} V, \tag{7}$$

whence the vanishing of normal velocity implies

$$\left(\operatorname{grad} p\right)_n = \left(\operatorname{grad} h\right)_n = \left(\operatorname{grad} \Phi\right)_n = 0. \tag{8}$$

By symmetry, the vertical centerline is another streamline. Setting $\partial\Phi'/\partial n = 0$ on the centerline, we need to consider only half of the alveolar sheet. We then solve the harmonic equation $\nabla^2\Phi' = 0$ subjected to the boundary conditions Eqs (5) to (8). By the relaxation method, this can be done quickly. When the solution is obtained, we compute $h = (\Phi')^{1/4}$, and plot the contour lines $h(x, y) = $ constant in Figure 6.12:4 (left panel). Note that the thickness drops very sharply near the drain into the venule, but over most of the sheet the thickness is quite uniform. The line representing a thickness of 90% of h_a lies about the border of the upper right quadrant of the sheet. The thickness drops another 10% to the $0.8\,h_a$ line in about 20% of the length of the edge. Then the drop becomes faster. The 50% thickness line is seen to be very close to the opening. The final drop of 50% of thickness takes place within the last 2% of the distance.

Next let us consider a sequence of boundary thickness h_v at the venule drain,

$$h_v = 0.2, 0.4, 0.6, 0.8,$$

and a corresponding sequence of thickness h_a at the precapillary,

$$h_a = \sqrt[4]{1 + h_v^4}$$

$$h_a \cong 1.0004, 1.0063, 1.0309, 1.0896. \tag{9}$$

We define $\Phi = h^4$, and set the boundary conditions,

$$\Phi = h_a^4 = 1 + h_v^4 \text{ on precapillary borders,}$$

$$\Phi = h_v^4 \text{ on postcapillary borders,}$$

$$\frac{\partial\Phi}{\partial n} = 0 \text{ on streamlines.} \tag{10}$$

To solve the equation $\nabla^2\Phi = 0$ subjected to these boundary conditions, let us define

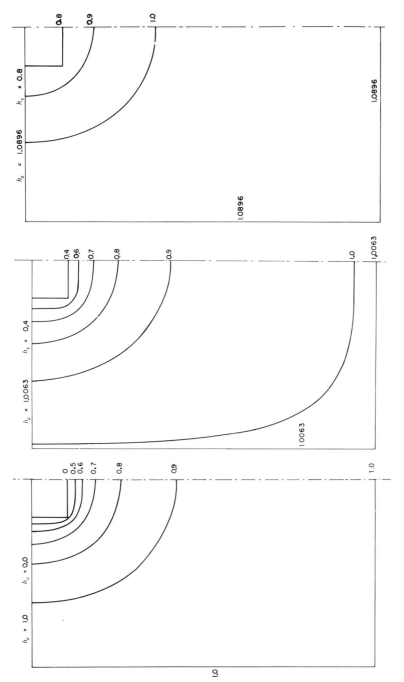

FIGURE 6.12:4. Mean sheet thickness distribution for boundary conditions specified in Figure 6.12:3. *Left:* $h_a = 1.0$, $h_v = 0$. *Middle:* $h_a = 1.0063$, $h_v = 0.4$. *Right:* $h_a = 1.0896$, $h_v = 0.8$. Numerals associated with each contour are values of the sheet thickness as a fraction of the thickness at the arteriole, h_a, of the first case. Reproduced by permission from Fung and Sobin (1969).

$$\Phi = \Phi' + h_v^4. \tag{11}$$

Then it is seen that Φ' is exactly what was solved earlier and $\Phi'^{1/4}$ has been presented in the left-hand panel of Figure 6.12:4. Hence ϕ can be computed without further analysis, and contours $h(x, y) = \Phi^{1/4} = $ constant can be plotted. Two examples are presented in the middle and right-hand panels of Figure 6.12:4.

An inspection of these figures shows that the thicker the sheet at the venule opening, the gentler is the thickness reduction. For example, the 50% reduction in thickness takes place at a distance from the drain opening at the following percentage of edge length:

h_v/h_a	0	0.199	0.397	0.582	0.734
50% line	2.1	4.1	7.9	11.1	16.8

This table shows the distance on the center line between the contour $h = h_v + \frac{1}{2}(h_a - h_v)$ and the venule drain opening, expressed as a percentage

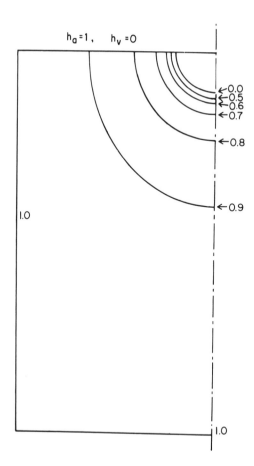

$h_a = 1, \quad h_v = 0$

←0.0
←0.5
←0.6
←0.7

←0.8

←0.9

1.0

1.0

FIGURE 6.12:5. Solution of a boundary value problem similar to that shown in Figures 6.12:3 and 6.12:4, except that the draining site is changed from a square to a circle. The mean thickness contours are shown for the case $h_a = 1, h_v = 0$. Note the similarity of the contours to those of Figure 6.12:4. Reproduced by permission from Fung and Sobin (1969). Numerals are values of the sheet thickness as fractions of h_a.

of the length of the edge AB in Figure 6.12:3. Incidentally, we see from these figures that the shape of the draining hole is not too important: The effect of the corner of the square is quickly rounded off. A circular drain will yield essentially the same result. Figure 6.12:5 illustrates the point. Compare it with Figure 6.12:4. Note the similarity of the contours in these two figures. This is a basic property of the partial differential equations of the elliptic type.

6.13 Pressure–Flow Relationship of Pulmonary Alveolar Blood Flow

We have derived a nonlinear pressure–flow relationship in the very simple case of one-dimensional flow in Section 6.11. We have shown by the examples given in Section 6.12 that the features of two-dimensional sheets are quite similar to the one-dimensional case. By integrating the equation of motion along the streamlines over the entire field, we shall derive an expression for the average flow over a sheet in the general, two-dimensional case.

We have shown in Eq. (4) of Section 6.9 that the mean flow velocity in a pulmonary alveolar sheet is

$$U = -\frac{1}{\mu kf} h^2 \text{ grad } \Delta p. \tag{1}$$

Here Δp stands for the transmural pressure, $p - p_{\text{alv}}$, μ is the apparent coefficient of viscosity of blood, k and f are numeric factors that depend on the details of the sheet structure, h is the local mean sheet thickness, and the symbol "grad" stands for the gradient operator. The flow per unit width is therefore

$$\dot{Q} = hU = -\frac{1}{\mu kf} h^3 \text{ grad } \Delta p. \tag{2}$$

Apply this to a streamtube. Let s denote the arc length along the tube, and let w be the width of the streamtube. Then the rate of flow in the tube is $\dot{Q}w$, which does not vary along the length of the streamtube, although both $\dot{Q}$ and w individually are functions of s. Now, on multiplying Eq. (2) with w, using Eq. (1b) of Section 6.8, which holds in the linear range of elasticity, and integrating from the arteriole to the venule, we obtain, with ds representing length along the streamline,

$$\int \dot{Q}w\,ds = -\int \frac{1}{\mu kf} \left[h_0 + \alpha\Delta p\right]^3 \frac{d\Delta p}{ds} w\,ds. \tag{3}$$

The integral on the left-hand side is equal to the flow in the streamtube multiplied by $\int ds = L$, the length of the streamtube. The integral on the right-hand side can be simplified according to the mean value theorem of the integral calculus. Because $w(s)$ and $\mu k f$ are finite and are functions of bounded variation, we can write the right-hand side of Eq. (3) as

$$-\frac{\overline{w}}{\overline{\mu k f}} \int \left[h_0 + \alpha \Delta p \right]^3 \frac{d\Delta p}{ds} \, ds, \tag{3a}$$

where $\overline{w}$ and $\overline{\mu k f}$ are the values of $w(s)$ and $\mu k f(s)$ evaluated at some value of s on the streamtube. $\overline{w}$ is a characteristic width of the streamtube, and can be approximated by the area of the streamtube divided by the length, L. In accordance with these remarks, and carrying out the integration in Eq. (3a), we can write Eq. (3) as

$$\text{Flow in tube} = \frac{\text{tube area}}{4 \overline{\mu k f} L^2 \alpha} \left[\left(h_0 + \alpha \Delta p_{art} \right)^4 - \left(h_0 + \alpha \Delta p_{ven} \right)^4 \right]. \tag{4}$$

Here the subscripts "art" and "ven" refer to arteriole and venule, at the entry and exit to the capillary sheet, respectively.

A comparison of Eq. (4) with the familiar Poiseuille formula given in Eq. (17) of Section 3.2 shows a great difference between the flow in an elastic sheet and the flow in a rigid rube. In a rigid tube the flow is linearly proportional to the pressure drop $(p_{art} - p_{ven})/L$. In an elastic sheet this is not the case: Eq. (4) shows that

$$\text{Flow} = \frac{\text{area}}{4 \mu k f L^2} \left(p_{art} - p_{ven} \right) \left[\left(h_0 + \alpha \Delta p_{art} \right)^3 + \left(h_0 + \alpha \Delta p_{art} \right)^2 \left(h_0 + \alpha \Delta p_{ven} \right) \right.$$

$$\left. + \left(h_0 + \alpha \Delta p_{art} \right)\left(h_0 + \alpha p_{ven} \right)^2 + \left(h_0 + \alpha \Delta p_{ven} \right)^3 \right]. \tag{5}$$

The factor in the brackets is the *conductance* of the flow and depends on the blood pressure.

Because a field of flow can be wholly covered by streamtubes, we can sum up the flow in all the streamtubes between an arteriole and a venule to obtain the flow between these two vessels. Let A be the area of alveolar sheet in question and let S be the vascular-space–tissue ratio; then the total area of the vascular space is SA, and the total flow may be written as

$$\text{Flow} = \frac{SA}{4 \mu k f \overline{L}^2 \alpha} \left[\left(h_0 + \alpha \Delta p_{art} \right)^4 - \left(h_0 + \alpha \Delta p_{ven} \right)^4 \right], \tag{6a}$$

where $\overline{L}$ is an average length of the streamtubes defined by the relation

$$\frac{SA}{\overline{L}^2} = \sum \frac{\text{area of streamtube}}{\left(\text{length of streamtube} \right)^2}, \tag{6b}$$

in which the summation covers all individual streamtubes of the sheet whose area is SA. Expressed in terms of sheet thickness, Eq. (6) is

$$\text{Flow} = \frac{1}{C} \left[h_a^4 - h_v^4 \right], \tag{7}$$

where

$$C = \frac{4\mu k f \overline{L}^2 \alpha}{SA}, \tag{8}$$

$$h_a = h_0 + \alpha \Delta p_{art}, \quad h_v = h_0 + \alpha \Delta p_{ven}, \tag{9a}$$

$$\Delta p_{art} = p_{art} - p_{alv}, \quad \Delta p_{ven} = p_{ven} - p_{alv}. \tag{9b}$$

We may compute $\overline{L}$ by associating each streamline with a specific value of the stream function, ψ, and obtain

$$\frac{1}{\overline{L}^2} = \frac{1}{\psi_2 - \psi_1} \int_{\psi_2}^{\psi_1} \frac{1}{L^2(\psi)} d\psi, \tag{10}$$

where ψ_1 and ψ_2 are the dividing streamlines that enclose the whole field of flow between the arteriole and venule in question.

Equation (6) provides an explicit formula of blood flow in the pulmonary alveoli as related to the blood rheology (through μ), alveolar area (A), alveolar structural geometry (through k, f, which depend on a, c, etc.), the vascular-space–tissue ratio (S), the arteriole and venule transmural pressures (p_{art}, p_{ven}), the average length of streamlines between an arteriole and a venule ($\overline{L}$), the compliance of the alveolar sheet with respect to blood pressure (α), and indirectly, through α, to the tension of the alveolar septa (T, which is the sum of tissue stress and surface tension).

If we write Eq. (6) in the form

$$p_{art} - p_{ven} = R \cdot (\text{flow}), \tag{11}$$

then R can be called the *resistance* of the capillary blood vessels. If both Δp_{art} and Δp_{ven} are positive, then the resistance is given by

$$R = \frac{C}{h_a^3 + h_a^2 h_v + h_a h_v^2 + h_v^3}, \tag{12}$$

which is a nonlinear function of the blood pressure.

Equation (7) is derived under the assumption of a linear thickness–pressure relationship, Eq. (1b) of Section 6.8, and is valid as long as $\Delta p_{ven} \geq 0$ or $p_{ven} \geq p_{alv}$. If the pressure in the venule is less than the alveolar gas pressure (see Sec. 6.15), then "sluicing" or "waterfall" occurs. The condition of sluicing is discussed in Section 6.17, where we conclude that a sluicing gate must be located at the junction of a capillary sheet and a draining venule. A good approximation of the maximum flow in waterfall condition is given by Eq. (7) with the h_v term omitted. Thus

$$\text{Flow} = \frac{1}{C} h_a^4. \tag{13}$$

This is called the *condition of waterfall* because it is analogous to a waterfall in nature, whose flow depends only on the head, and not on the height, of the fall.

Equation (7) exhibits the essence of pulmonary alveolar blood flow. Because of the fourth power, the flow depends much more on the pressure at the entry (p_{art}) than that at the exit (p_{ven}). If the pressures are such that h_v is one half of h_a, then h_v^4 is only 1/16 of h_a^4, and Eq. (7) shows that the flow varies almost as the fourth power of the pressure at the entry, p_{art}. Figure 6.13:1 shows the theoretical pressure–flow relationship given by Eq. (7) and the resistance given by Eq. (12). The constants used are pertinent to the dog: $h_0 = 2.5\,\mu m$, $\alpha = 0.122\,\mu m/cm$ H_2O. The pressure p_{ven} is fixed at 3 cm H_2O; the alveolar gas pressure p_A is 0, 7, 17, or 23 cm H_2O. The entry pressure p_{art} is varied. The pleural pressure is assumed to be zero in the case of positive inflation and negative in the case of negative inflation represented by the curve for $p_{alv} = 0$. In the case of positive inflation, sluicing occurs (Section 6.17), and we have exhibited the results with either $h_v = h_0$ or $h_v = 0$. The constant C is set to be 1 for the scale on the ordinate to the left.

At the time this theory was first presented, there was no direct experimental result on pulmonary alveolar blood flow available for comparison. Data on blood flow in a whole lung did exist. In Figure 6.13:1 we plotted the experimental results by Roos et al. (1961) on an isolated lung of the

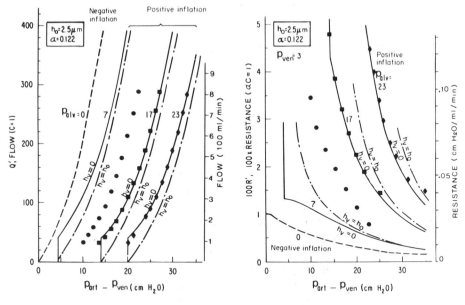

FIGURE 6.13:1. Curves referring to the ordinates to the left represent the theoretical pressure–flow relationship given by Eqs. (7) and (13), and resistance given by Eq. (12). See the text for an explanation. Q' is the flow when $C = 1$, and $R' = \alpha R$ is resistance when $\alpha C = 1$. Points associated with the ordinate to the right correspond to the experimental data of Roos et al. (1961) on the isolated lung of the dog with left atrium pressure equal to 3 cm H_2O, pleural pressure equal to 0, and alveolar pressure equal to 23 (◆), 17 (■), and 7 (●) cm H_2O. From Fung and Sobin (1972a). Reproduced by permission of American Heart Assoc. Inc.

dog on the same graph with the theoretical results except with a scale (ordinate on the right-hand side) adjusted so that one of the experimental points falls exactly on the theoretical curve for alveolar blood flow. This figure shows that the trends of the variations of flow and resistance with blood pressure are roughly the same for the whole lung as for the alveolar bed alone. In the following sections the theory of blood flow in the whole lung, and additional theories of patchy flow and sluicing are presented, and theoretical results are compared with critical experimental results.

6.14 Blood Flow in the Whole Lung

The results of the preceding section, together with the results of Chapter 3, will enable analysis of the blood flow in a whole lung. Experimental results on their elasticity can be summarized by Eq. (1b) of Section 6.8, which is valid when $p - p_A > 0$ and smaller than an upper limit of about 25 cm H_2O (for cat) or 15 cm H_2O (for human), beyond which h tends to a constant asymptotically. When $p - p_A < 0$, the capillaries are collapsed and h tends to zero. The alveolar blood flow is then given by Eqs (6.13:7) and (6.13:13).

For pulmonary arteries and veins, the results presented in Section 6.7 show that the vessel diameter D changes linearly with blood pressure p,

$$D = D_0 + \alpha p. \tag{1}$$

Here D_0 is the tube diameter when p is zero; α is the compliance constant. Then for steady flow of a viscous incompressible fluid in such a tube, the analysis of Section 3.4 yields the result Eq. (3.4:13):

$$\frac{640\mu\alpha L}{\pi}\dot{Q} = \left[D(o)\right]^5 - \left[D(L)\right]^5 = \left[D_0 + \alpha p_{entry}\right]^5 - \left[D_0 + \alpha p_{exit}\right]^5. \tag{2}$$

Equations (2) and (6.13:7) are called the *fifth* and *fourth power laws*, respectively. In calculating the static pressures, p_{entry} and p_{exit}, in Eq. (2), we must take into account the losses due to turbulence and bifurcation, which depends on the Reynolds and Womersley numbers and the effect of change of kinetic energy along the stream. For the last item we must add a pressure drop at the end of a vessel of order n by an amount equal to $\frac{1}{2}\rho v_{n+1}^2 - \frac{1}{2}\rho v_n^2$; see Eq. (18) of Section 1.11, on p. 19. Then a repeated application of Eqs. (6.13:7) and (2) will synthesize the flow in different segments into that of the whole lung. In the following, we shall apply these equations to the cat lung because morphological and elasticity data for the cat are available.

Application of Experimental Data on Vessel Elasticity

Experimental data on the elasticity of pulmonary arteries and veins of the cat are presented in Section 6.7. These data were obtained by measuring the vessel dimensions at different values of blood pressure while the pressures in the airway (p_A) and pleura (p_{PL}) were fixed. Hence the constants h_0, D_0, and α in Eqs. (1) and (3) must depend on p_A and p_{PL} because these parameters were not varied in the experiments. This dependence has not been determined fully at this time. The only data available are given in Figure 6.7:5. However, some theoretical studies of this dependence are available. Fung and Sobin (1972a) analyzed h_0 and α as functions of membrane tension in the sheet (interalveolar septa), which depends on the transpulmonary pressure, $p_A - p_{PL}$. Yen et al. (1980) argued that for pulmonary vessels whose diameter is much larger than the alveolar diameter, the vessel diameter may be written as

$$D = D_0\left[1 + \beta\left(p - p_{PL}\right)\right], \tag{3}$$

whereas for vessels whose diameter is smaller than the alveolar diameter, we should have

$$D = D_0\left[1 + \beta\left(p - p_A\right)\right], \tag{4}$$

where D_0 and β are functions of $p_A - p_{PL}$. [We changed the notation α of Yen et al. (1980, 1981) to β to avoid a conflict with Eq. (1).] The reasoning is that the pulmonary vessel diameter varies with the internal pressure p, the external pressure p_A, and the tension in the attached alveolar walls, which is approximately equal to $p_A - p_{PL}$ if averaged over a unit area. For large vessels, use of the average stress $p_A - p_{PL}$ for tethered alveolar wall is applicable, hence the "transmural" pressure becomes $p - [p_A - (p_A - p_{PL})]$ $= p - p_{PL}$, which is exhibited in Eq. (3). For small vessels, use of the average stress in the alveolar wall as a tethering force outside the vessel is inappropriate. Such a small vessel is shown in Figure 4.10:3, p. 260, as tethered by three interalveolar septa, spaced at approximately 120°. Across the vessel wall between the septa, the "transmural" pressure is $p - p_A$; hence Eq. (4). Lai-Fook (1979) has shown that the parenchyma stress is affected in the neighborhood of a large blood vessel by the elasticity of the vessel to a value that may be written as $k(p_A - p_{PL})$. Usually, $k > 1$. If this is assumed, then the "transmural" pressure across a large pulmonary vessel is $p - [p_A - k(p_A - p_{PL})]$ and Eq. (3) may be replaced by

$$D = D_0\left[1 - \beta\left(1 - k\right)p_A + \beta\left(p - p_{PL}\right)\right]. \tag{5}$$

Since the available data are expressed in terms of Eqs. (3) and (4), these equations will be used in the following analysis: Equation (3) for vessels >100 μm, Eq. (4) for vessels <100 μm. On using Eq. (2), note that $\alpha = D_0\beta$.

Application of Morphometric Data of Vascular Tree

Morphological data on the pulmonary venous and arterial tree of the cat are given in Section 6.2. To analyze blood flow, however, one must have a definite vascular circuit every time. The creation of a circuit from statistical data on branching, is, unfortunately, a nonunique process. A number of cir-

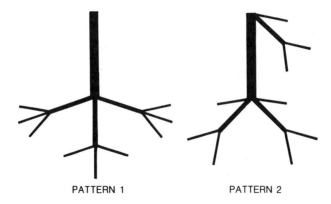

PATTERN 1   PATTERN 2

FIGURE 6.14:1. Two patterns of pulmonary vascular branching. Three orders are depicted. Vessels of the same diameter belong to the same order. From Zhuang et al. (1983), by permission.

cuits can be created that are consistent with the statistical data, but not uniquely specified by them.

For example, the two patterns shown in Figure 6.14:1 have the same branching ratio of 3 for the three orders depicted, but the distribution of flow depends on the pressures at the ends of daughter vessels. One can, however, assign a flow distribution and compute the pressure distribution, or vice versa.

We devise two circuits consistent with the statistical morphometric data. In model 1, flow in a branch of order $n + 1$ is divided equally into B_n daughter vessels of order n, with B_n being the branching ratio of order n. In model 2, we assume that each vessel of order $n + 1$ bifurcates into two vessels of order n at the exit end as shown in pattern 2 of Figure 6.14:1, whereas $B_n - 2$ remaining vessels of order n are attached to the entry end of the parent vessel (of order $n + 1$). Let the total flow be $\dot{Q}_t$. Let the total number of vessels of an order n be N_n. The flow in each vessel of order n is $\dot{Q}_t/N_n$. The largest vessels are given special attention. For the largest artery, of order 11, some flow is diverted into vessels of order 10 or smaller at the entry section, and the remaining flow in the main trunk is equal to $2\dot{Q}_t/N_{10}$. For the daughter branch of order 10, a similar division is made, and the flow is $2\dot{Q}_t/N_9$. This continues until the vessels of the smallest order are reached, in which the flow is of course, $\dot{Q}_t/N_1$. From this assumed flow pattern, we compute the pressure distribution.

Database

Tables 6.14:1 to 6.14:3 show the morphometric and elasticity data of the cat lung. D_0 is computed for vessels <100 μm according to Eq. (4) and for vessels >100 μm according to Eq. (3). Table 6.14:3 includes information given in Tables 4.10:1, 4.10:2 and Figures 4.10:2, 6.7:4, 6.7:5.

TABLE 6.14:1. Morphometric Data of Pulmonary Arteries of the Cat Measured at Transpulmonary Pressure, $p_A - p_{PL} = 10\,cm\ H_2O$

Order	Number of branches in right lung N_n	Diameter* D_{on} (cm)	Length L_n (cm)	Apparent viscosity coefficient μ_n (cp)	Compliance[†] α $(10^{-4}\,cm\ p_a^{-1})$	β $(10^{-4}\,p_a^{-1})$
1	300,358	0.0024	0.0116	2.5	0.00463	1.928
2	97,519	0.0044	0.0262	3.0	0.00848	1.928
3	31,662	0.0073	0.0433	3.5	0.01407	1.928
4	9,736	0.0122	0.0810	4.0	0.02352	1.928
5	2,925	0.0192	0.151	4.0	0.02154	1.122
6	774	0.0352	0.272	4.0	0.02802	0.796
7	202	0.0533	0.460	4.0	0.03807	0.714
8	49	0.0875	0.819	4.0	0.09818	1.122
9	12	0.1519	1.426	4.0	0.4045	2.663
10	4	0.2486	1.187	4.0	0.6620	2.663
11	1	0.5080	2.500	4.0	1.353	2.663

* Yen et al. (1984) gives the diameter data of vessels of orders 1 to 4 measured at $p - p_A = -7\,cm\ H_2O$, whereas those of orders 5 to 11 were measured at $p - p_{PL} = 3\,cm\ H_2O$. In this table, D_{on} are the diameters at zero "transmural" pressure, defined as $p - p_A = 0$ for orders 1 to 4, and $p - p_{PL} = 0$ for orders 5 to 11, and are computed from the data of Yen et al. (1980) according to the equations $D = D_o[(1 + \beta(p - p_A)]$ for orders 1 to 4 and $D = D_o[(1 + \beta(p - p_{PL})]$ for orders 5 to 11.

[†] For the method of computing compliance, see the notes in Table 6.14.3. $\alpha = \beta D_{on}$.

For pulmonary capillaries of the cat, Sobin et al. (1972) give $h_0 = 4.28\,\mu m$, $\alpha = 0.219\,\mu m$ per cm H_2O, and VSTR (vascular space-tissue ratio) = 0.916. Sobin et al. (1980) give the average path length $\bar{L} = 556 \pm 285$ (SD) μm. Based on the same topological map used by Sobin et al. (1980), we calcu-

TABLE 6.14:2. Morphometric Data of Pulmonary Veins of the Cat at $p_A - p_{PL} = 10\,cm\ H_2O$

Order	Number of branches N_n	Diameter* D_{on} (cm)	Length L_n (cm)	Apparent viscosity coefficient μ_n (cp)
1	282,733	0.0025	0.0086	2.5
2	86,241	0.0046	0.0247	3.0
3	26,306	0.0077	0.0496	3.5
4	8,024	0.0127	0.1545	4.0
5	2,348	0.0251	0.2380	4.0
6	656	0.0432	0.3810	4.0
7	171	0.0642	0.4950	4.0
8	46	0.1040	0.7610	4.0
9	13	0.1727	1.5120	4.0
10	4	0.3010	1.9240	4.0
11	1	0.4491	2.5000	4.0

* The diameter values are computed from those of Yen and Foppiano (1981) according to the formulas described in footnote (1) of Table 6.14:1.

TABLE 6.14:3. Compliance of Pulmonary Veins of the Cat Obtained in Experiments with $p_A = 0$, p_{PL} Specified, and Variable Local Blood Pressure[*,†]

| | Compliance $p_{PL} = -10$ cm H_2O | | Compliance[‡] $p_{PL} = -15$ cm H_2O | | Compliance[‡] $p_{PL} = -20$ cm H_2O | |
Order	α $(10^{-4}$ cm $p_a^{-1})$	β $(10^{-4} p_a^{-1})$	α $(10^{-4}$ cm $p_a^{-1})$	β $(10^{-4} p_a^{-1})$	α $(10^{-4}$ cm $p_a^{-1})$	β $(10^{-4} p_a^{-1})$
1	0.00482	1.928	0.00482	1.928	0.00482	1.928
2	0.00887	1.928	0.00887	1.928	0.00887	1.928
3	0.0148	1.928	0.0148	1.928	0.0148	1.928
4	0.0331	2.080	0.0446	2.806	0.0396	2.490
5	0.0430	1.469	0.0598	2.041	0.0478	1.633
6	0.0528	1.092	0.0484	1.000	0.0469	0.969
7	0.0785	1.092	0.0719	1.000	0.0697	0.969
8	0.0810	0.724	0.0628	0.561	0.0833	0.745
9	0.1346	0.724	0.1043	0.561	0.1385	0.745
10	0.2346	0.724	0.1817	0.561	0.2414	0.745
11	0.3504	0.724	0.2715	0.561	0.3606	0.745

[*] For orders 1 to 3 the data are from Sobin et al. (1978). $\alpha = \beta \cdot D_{on}$.

[†] For orders 4 to 11, Yen and Foppiano (1981) give the elastic properties by the formular $D = D_{10}[1 + \beta(p - p_{PL} - 10 \text{ cm } H_2O)]$. Hence $\alpha = D_{10}\beta$. Note that, however, β was written as α in Yen and Foppiano (1981).

[‡] The compliance at $p_{PL} = -15$ and -20 cm H_2O was measured in Yen and Foppiano (1981) and was expressed in terms of β. But the values of D_{on} are unknown at these transpulmonary pressures; hence α is computed by multiplying β with D_{10n} at $p_{PL} = -10$ cm H_2O computed according to the equation in the preceding footnote. There is a dearth of information on the dependence of D_{on} on the transpulmonary pressure.

lated the total capillary area of the right lung of the cat to be 0.42 m^2 by stereological methods.

The apparent viscosity values given in Tables 6.14:1 and 6.14:2 are estimated using the hypothesis that the hematocrit varies from about 45% in larger vessels with diameter >100 μm (orders 4 to 11) to about 30% in the capillaries. The variation of the apparent viscosity of blood in the pulmonary capillary sheet with hematocrit has been determined by Yen and Fung (1973) in model experiments, from whose data we calculate that when the hematocrit is 30%, the apparent viscosity is 1.92 cp. On the other hand, at a hematocrit of 45% in large vessels, an apparent viscosity of 4.0 cp is assumed. The apparent viscosity of blood in small vessels on the order of 1 to 3 is obtained by linear interpolation and is listed in Tables 6.14:1 and 6.14:2.

Results

Steady flow is considered. Using the data presented earlier, we compute the flow in vessels of successive orders, and then use Eqs. (6.13:7) and (2) to calculate the blood pressures, and formulas (3) to (5) to obtain vessel diameters at different pressures. Two circuits are analyzed, called models 1 and 2, analogous to patterns 1 and 2 shown in Figure 6.14:1. Analysis of steady flow in model 1 is rigorous. Pressure loss due to entry flow and changes due to kinetic energy ($\frac{1}{2}\rho v^2$) are negligible for the cat. The analysis of model 2 is approximate due to the ad hoc assumption about the flow distribution. In the following, the results refer to model 1 of the right lung of the cat, unless stated otherwise.

Figure 6.14:2 shows the relationship between flow and pulmonary arterial pressure under six different transpulmonary pressures and two left atrial pressures. In the cases labeled "positive inflations," the pleural pressure p_{PL} is 0 (atmospheric), the airway pressure p_A is indicated in the figure, and the left atrial pressure p_v is fixed at 3 cm H_2O. In "negative inflations," $p_A = 0$, p_{PL} is shown in the figure, and p_v is fixed at 2 cm H_2O. These values of p_A, p_{PL}, and p_v are chosen so that the results may be compared with an experiment to be discussed later.

The curves of Figure 6.14:2 show that the relationship between flow ($\dot{Q}$) and arterial pressure (p_a) is nonlinear. $\dot{Q}$ increases more rapidly than increasing p_a. At fixed values of p_a, flow decreases with increasing transpul-

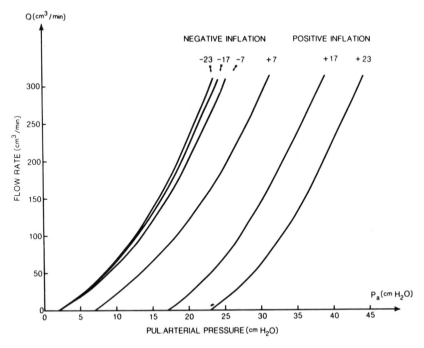

FIGURE 6.14:2. Relationship between flow and pulmonary arterial pressure under six different transpulmonary pressures. In "negative inflations," the left atrial pressure P_v is fixed at 2 cm H_2O; in "positive inflations," P_v is fixed at 3 cm H_2O. From Zhuang et al. (1983), by permission. Data refer to the cat.

FIGURE 6.14:3. The pressure–flow relationships in cat lung for two different branching models 1 and 2 explained in the text. Both curves refer to a negative inflation at $p_A = 0$, $p_{PL} = -7$ cm H_2O, $p_v = 2$ cm H_2O. From Zhuang et al. (1983), by permission.

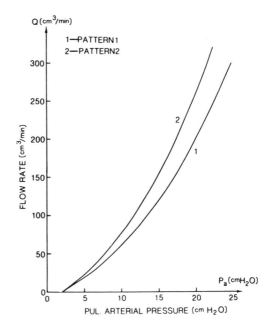

monary pressure in positive inflation, but it increases with increasing transpulmonary pressure in negative inflation.

Figure 6.14:3 shows the difference between the pressure–flow relationship for the two branching models, 1 and 2, illustrated in Figure 6.14:1. The calculation is referred to a negative inflation, $p_A = 0$, $p_{PL} = -7$ cm H_2O, $p_v = 2$ cm H_2O. It is seen that at any arterial pressure, p_a, flow is higher in a circuit of model 2 than that in model 1; in other words, the resistance to flow is lower in model 2.

Figure 6.14:4 shows flow versus a variable pulmonary venous pressure, with a fixed arterial pressure of 26 cm H_2O, $p_{PL} = 0$, and three values of p_A: 7, 17, and 23 cm H_2O. By definition, the flow is said to be in zone-2 condition when $p_v \le p_A$. In the calculation of Figure 6.14:4, the sluicing gates are assumed to lie at the exit ports of the capillary sheets. This assumption was proposed by Fung and Sobin (1972a) using the hypothesis that the pulmonary venules and veins will not collapse when $p_v < p_A$, because of the support the vessels receive from the tension in the interalveolar septa that attach to the vessel walls. This last hypothesis was fully verified experimentally, as presented in Section 4.10. The whole alveolar septal area is supposed to be patent, however. Possible patchy filling due to collapsed capillaries, and hysteresis of capillary collapse and opening to be discussed in Sections 6.15 to 6.19, are not considered here.

Figure 6.14:5 shows the pressure distribution in the lung, plotted against the order number of the vessels, for the case in which $p_a = 20$, $p_v = 2$, $p_A = $

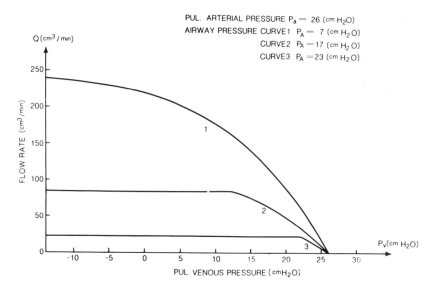

FIGURE 6.14:4. Flow versus pulmonary venous pressure in cat lung when the pulmonary arterial pressure is fixed at 26 cm H_2O, the pleural pressure p_{PL} is zero, and the alveolar gas pressure p_A is 7, 17, and 23 cm H_2O for curves 1, 2, and 3, respectively. From Zhuang et al. (1983), by permission.

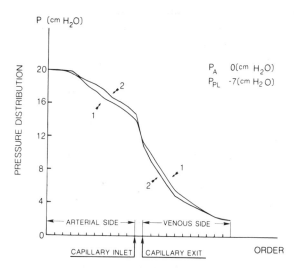

FIGURE 6.14:5. The longitudinal pressure distribution in pulmonary blood vessels of the cat for the case in which $p_a = 20$, $p_v = 2$, $p_A = 0$, and $p_{PL} = -7$ cm H_2O. Each tick mark on the horizontal axis represents the location of the exit end of each vessel in a given order. The numerals 1 and 2 refer to branching models 1 and 2 shown in Figure 6.14:1, p. 395, respectively. From Zhuang et al. (1983), by permission.

TABLE 6.14:4. Distribution of Pressure Drop in Pulmonary Blood Vessels of the Cat and Comparison with Data on the Dog in the Literature

	Arteries	Capillaries	Veins
Pattern 1	35.9%	15.4%	48.7%
Pattern 2	29.3%	21.9%	48.8%
Brody et al. (1968)	46%	34%	20%
Hakim et al. (1982)*	40.2%	15.6%	44.2%
Gaar et al. (1967)†	56%		44%

* Using the method of abrupt occlusion of inflow and outflow of blood in selected vessels in dog lung. Hakim et al. (1982) considered three compartments in the lung: the upstream or arterial compartment, (Δp_a), the downstream or venous compartment, (Δp_v), and the vascular compartment in between (Δp_m).
† Gaar et al. (1967) divided the distribution of pressure into the arterial side and the venous side from the midpoint of the capillaries.

0, and $p_{PL} = -7$ cm H_2O. Vascular circuits of both patterns 1 and 2 are considered as indicated in the figure.

If the pressure drop from the pulmonary artery of order 11 to the arteriole of order 1 is compared with the drop in the capillaries and veins, we obtain the results shown in Table 6.14:4.

Finally, we show in Figure 6.14:6 the transit time of blood in the pulmonary capillaries, calculated by the method to be presented in Section 6.20, in which $p_A = 0$, $p_{PL} = -7$, $p_v = 2$ cm H_2O, whereas the pulmonary arte-

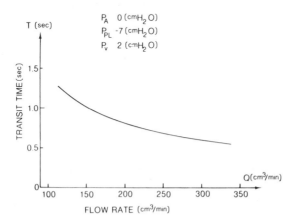

FIGURE 6.14:6. Transit time of blood in pulmonary capillaries for the case in which $p_A = 0$, $p_{PL} = -7$, and $p_v = 2$ cm H_2O. The flow rate in the half-lung and the pulmonary arterial pressure are varied. From Zhuang et al. (1983), by permission.

rial pressure or flow rate is varied. The figure shows that the transit time decreases when the flow rate increases, but does not follow an inverse proportionality relationship, owing to the distensibility of the capillary blood vessels.

Discussion

The results shown in Figure 6.14:2 may be compared with the trend shown by Roos et al. (1961) in their experiments on dog's lung (see Fig. 6.13:1). We find that the theoretical trend compares very well with the experimental result, except for the case of negative inflation with $p_A = 0, p_v = 2, p_{PL} = -23$ cm H_2O. Theoretically, the trend is continuous; at a given p_a the flow increases (i.e., the resistance decreases) as p_{PL} changes from -7 to -17 and -23. Roos et al.'s experimental results, however, show a discontinuous trend: at a given p_a, the flow increases as p_{PL} changes from -7 to -17, then decreases as p_{PL} changes from -17 to -23. The reason for this discrepancy is not clear. It could be due to an inadequate accounting of the changes in vessel diameter and elasticity as functions of p_A and p_{PL}, as discussed earlier. This suggests that the study of D_0 and α as functions of p_A and p_{PL} is an important topic for the future.

From the curves of Figure 6.14:2, we can estimate the cardiac output. Take the case in which $p_A = 0, p_{PL} = -17, p_v = 2$, and $p_a = 20$ cm H_2O. Figure 6.14:2 yields a flow of 210 ml/min for the right lung of a cat of 3.6 kg body weight. Thus, the cardiac output is 420 ml/min, or 117 ml/min/kg, a value agreeing quite well with that given by Skalak et al. (1966), and Weiner et al. (1967).

The results shown in Figure 6.14:4 may be compared with those given by Banister and Torrance (1960), Permutt and Riley (1963), and Permutt et al. (1962) for the dog lung. The flow limitation phenomenon is known as the *waterfall phenomenon*, and it is well exhibited by the theoretical curves of Figure 6.14:4. This phenomenon is similar, but differs in essential details, to the flow-limiting function of starling resistor (Starling, 1915) discussed in Chapter 4. The reduction of the maximum flow with increasing p_A is also demonstrated by the theory. But the existence of a hysteresis loop in the pressure–flow relationship when p_v first decreases and then increases, and a decrease of flow below the maximum as p_v continues to decrease below p_A are not shown by this simple analysis. To explain these discrepancies a refined theory is presented in Sections 6.15 to 6.19.

The pressure distributions shown in Figure 6.14:5 and Table 6.14:4 for the cat are compared with the experimental results by Brody et al. (1968), Gaar et al. (1967), and Hakim et al. (1982) on the dog in Table 6.14:4. See also Smith and Mitzner (1980). Further, Table 6.14:5 offers a comparison of our theoretical results with the experimental results obtained on dog lung by Bhattacharya et al. (1982), who obtained the data by two methods: micropuncture and occlusion. The theoretical data in Table 6.14:5 refer to the case in which $p_a = 16.9, p_A = 7.0$, and $p_v = 10.0$ cm H_2O for the zone-3 condition,

TABLE 6.14:5. Comparison of our Theoretical Results with Experimental Results of Bahattacharya et al. (1982)

Method	Vessels	Pressure in zone 3 cdn. cm H_2O	Pressure in zone 2 cdn. cm H_2O	Reference
Micro-puncture	Venules (20–50 μm)	11.3 ± 0.8	7.4 ± 1.2	Bhattacharya et al. (1982).
Occlusion	Venules (20–50 μm)	11.4 ± 0.5	5.2 ± 1.6	Bhattacharya et al. (1982).
Our theory	Venules of order 2 (46 μm)	11.9	6.48	This report

Arterial pressure = 16.9 cm H_2O; alveolar pressure = 7.0 cm H_2O; left atrium pressure for zone 3 condition = 10.0 cm H_2O; left atrium pressure for zone-2 condition = 1.7 cm H_2O; cdn. = condition.

and p_v = 1.7 cm H_2O for the zone-2 condition, in agreement with Bhattacharya's experimental condition. Our theoretical pressure is calculated at the exit end of a venular vessel of order 2, with a diameter of 46 μm. It is seen that our theoretical results of the cat compare quite well with the experimental results of Hakim et al. (1982) and Bhattacharya et al. (1982) on the dog, but we do not agree with Brody et al. (1968). Raj and Chen's (1986a,b) results on lamb lungs and Negrini et al.'s (1992) results on rabbit lungs are again similar in trend but different in detail. There are, of course, many possible reasons for the differences: The difference in animal species, the depth of the examined vessels below the pleura, the experimental protocol, the effect of anesthesia, and the gentleness of the experimenter's hand may be responsible.

Finally, the blood transit time in pulmonary capillaries plotted in Figure 6.14:6 may be compared with experimental data of Johnson et al. (1960) and Wagner et al. (1982). Figure 6.14:6 shows that at a physiological flow of 200 ml/min for the right lung of the cat, the transit time is about 0.80 s, which is quite close to the value 0.79 s obtained by Johnson et al. (1960), but is considerably shorter than the value given by Wagner et al. (1982) for the dog.

The preceding example illustrates the calculation of the pulmonary circulation of a whole lung of the cat on the basis of experimental data on the morphology of the vascular tree and elasticity of blood vessels of all orders. Comparison of the theoretical results with experimental results in the literature has yielded good agreement in many cases, but there are a few significant discrepancies which suggest that further work is needed. Among these are: (a) collection of data on vessel dimensions and compliance constants when airway and pleural pressures are varied, especially for small vessels, and for large inflation; (b) the hysteresis in the pressure–flow relationship and details of flow limitation in the "zone-2" condition. The last point is taken up in the following sections.

6.15 Regional Difference of Pulmonary Blood Flow

In large animals the hydrostatic pressure due to gravitation plays an important role in pulmonary blood flow. Let p_{PA} be the pressure in the pulmonary artery of the highest order immediately next to the pulmonic valve, let p_{LA} be the pressure in the left atrium, and let z be the height of a point above the level of pulmonic valve, measured along the direction of gravitational acceleration; then the pressure in the flowing blood in a blood vessel located at a level of height z is

$$p(z) = p_{PA} - \rho g z - \sum_{\text{from } PA} (\Delta p)_i \tag{1}$$

or

$$p(z) = p_{LA} - \rho g(z - z_{LA}) + \sum_{\text{from } LA} (\Delta p)_i, \tag{2}$$

where ρ is the density of the blood, g is the gravitational acceleration, $(\Delta p)_i$ is the pressure drop in a blood vessel of order i, including the effects of entry and exit at points of bifurcation and the change of dynamic pressure $(1/2)\rho v^2$, in vessels of successive orders. The summation includes all orders of vessels along the circuit starting from the pulmonary artery of the highest order in Eq. (1), or all orders of vessels along the circuit starting from the left atrium in Eq. (2). In the preceding section we have shown how to compute $(\Delta p)_i$ using the fifth-power law for arteries and veins and the fourth-power law for the capillaries. We ignored the z-terms in Section 6.14. For large animals the z-terms cannot be ignored and the analysis is more complex, although the principle remains the same.

West (1979, p. 43) divides the lung into three zones:

$$\text{zone 1:} \quad p_A > p_a > p_v, \tag{3}$$

$$\text{zone 2:} \quad p_a > p_A > p_v, \tag{4}$$

$$\text{zone 3:} \quad p_a > p_v > p_A, \tag{5}$$

where the subscripts A, a, and v stand for alveolar gas, arterial blood, and venous blood, respectively. He computes p_a, p_v as static pressures

$$p_a = p_{PA} - \rho g z, \quad p_v = p_{LA} - \rho g z. \tag{6}$$

The idea is that the "waterfall" phenomenon discussed in Section 6.13, Eq. (13), may occur in zone 2. However, since pulmonary arteries and veins remain patent when p_A exceeds p_v (see Sec. 4.10), whereas the capillaries would collapse in this condition (see Sec. 6.8), the waterfall phenomenon will occur only in the capillary sheet. Hence, it is the entry and exit condition of the capillary sheet that is significant. Using the notations p_{art} and p_{ven} for the pressures at the capillary entry from an arteriole and exit into a venule, respectively, to be computed by Eq. (2), we define

$$\text{zone 1:} \quad p_A > p_{art} > p_{ven}, \tag{7}$$

$$\text{zone 2:} \quad p_{art} > p_A > p_{ven}, \tag{8}$$

$$\text{zone 3:} \quad p_{art} > p_{ven} > p_A. \tag{9}$$

In zone 1, we expect little flow in pulmonary capillaries, because according to Eq. (7), Eq. (6.8:1d), and Figure 6.8:1, the pulmonary capillary vessels are collapsed to almost zero internal thickness. The "corner vessels" in the junctional region of the edges of neighboring alveoli where three interalveolar septa meet will remain open in zone 1 because of the tension in the interalveolar septa when the lung is inflated. Lamm et al. (1991) have found blood flow in these corner vessels in the zone-1 condition.

In zone 3, all pulmonary blood vessels are patent according to Eq. (9) and data in Sections 6.7 and 6.8. The analysis of Section 6.14 applies.

In zone 2, as defined by Eq. (8), the capillary blood vessels should be open with thickness $h > h_0$ at the arteriolar end according to the elasticity of the capillary given in Figure 6.8:1, but its thickness should be reduced to almost zero at the venular end according to Eq. (6.8:1d). The flow could be zero, or could be finite. The flow is finite if the velocity is fast and the limit of velocity × thickness is finite. Experimental evidence has shown that the flow in this condition is finite, with some degree of uncertainty, but in any case the limiting value of the flow is almost independent of the pressure in the venule, p_v. This is the waterfall phenomenon mentioned in Section 6.13, in association with Eq. (6.13:13). The limiting condition is very special for the lung, and its validity depends on the very short sluicing gate, which opens to a patent venule. The corresponding limiting case in the coronary blood vessels of the heart, where the sluicing gate is very long compared with the vessel diameter, will lead to a conclusion of zero flow. Therefore, we must analyze the sluicing gate very carefully, as we shall do in Section 6.17. The degree of uncertainty, that is, the stability or instability of the capillaries, is analyzed carefully in Section 6.18. The possible hysteresis in the pressure–flow relationship in the zone-2 condition is analyzed in Section 6.19. The resulting "patchy" flow in the alveoli is discussed in Section 6.16.

If the zones were defined by Eqs. (3) to (5) and the arterial and venous pressures were specified by Eq. (6), then the only parameter that defines the borders dividing zone 1 from zone 2, and zone 2 from zone 3, is z, the height against gravitational acceleration. By these hypotheses the stratification is horizontal. Early experiments by West and Dollery (1965), West et al. (1965) have demonstrated rough horizontal stratification. Their experiments have shown the existence of a *zone 4* at the bottom of the lung next to the diaphragm of a sitting man, in which the blood flow per unit volume is reduced, presumably due to the interference of the diaphragm. The physiological and pathological significances of regional differences are discussed fully in West (1977a,b, 1979, 1982).

In a standing human adult, the apex of the lung is normally in the zone-1 condition. The base of the lung above the diaphragm is in the zone-3 condition. A midregion sufficiently high above the left atrium and below the apex is normally in the zone-2 condition. When a person lies down in bed, the whole lung is normally in the zone-3 condition. The lung of an astronaut in space flight at zero gravity is in the zone-3 condition.

We are convinced, however, that the zonal differentiation should be based on Eqs. (7) to (9), with pressures in the arteriole and venule, p_{art} and p_{ven}, computed by Eqs. (1) and (2). The pressure drop along the pathway of blood makes a difference as to where the waterfall phenomenon sets in. By these new definitions, each zone is no longer necessarily simply connected. Zone 2 may consist of many pieces of volume scattered in the lung. In other words, zone 2 may be multiply connected. So are the other zones.

When we use Eq. (8) as the definition for the zone-2 condition, the height above ground is not the only factor that divides the lung into three zones. The pressures at arterioles and venules depend on the closeness of the capillary bed to the main artery. In fact, Hakim et al. (1987, 1988a,b, 1989) have shown that the regional difference of pulmonary blood flow per unit volume of lung tissue is not stratified by horizontal planes, but by curved surfaces, with the main pulmonary artery at the core.

An example of Hakim et al.'s (1989) result is shown in Figure 6.15:1. They injected 25 mCi of ^{99m}Tc-labeled albumin microspheres (2×10^6 spheres of 15 to 20 μm diameter) into the superior vena cava (via the proximal port of a Swan-Ganz catheter) at end expiration over five breaths. Then the animal was exanguinated rapidly, the chest was opened, the lungs were removed, the major vessels and all extra parenchymal tissue were dissected and removed, the lung was drained passively of blood and inflated to full capacity by warm air (50°C, 35 cm H_2O pressure), the lung surface was punctured at multiple sites to allow the air to pass through, and the sampled tissue was dried for 18 to 20 h. Thereafter the trachea was cut and the lung was put in the supine position at the center of rotation of a gamma camera. Tomographic sections were reconstructed.

Figure 6.15:1 shows tomographic images of a midcoronal and a midsagittal slice in an animal with a normal cardiac output. The lung was in zone 3, with the possibility of a small upper portion being in zone 2. The blood flow within each voxel (the volume of $3.9 \times 3.9 \times 11.7$ mm) is expressed as the percent of the voxel with maximum activity within the slice. In the coronary slice (see Fig. 6.15:1, top) the effect of gravity is seen, but the stratification is not entirely horizontal. Peaks exist in the central regions of the individual lobes. There is a decline in activity toward the periphery. The sagittal slice (see Fig. 6.15:1, bottom) shows that peak ac-

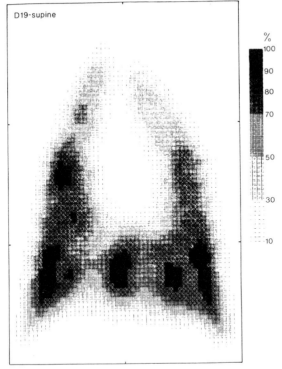

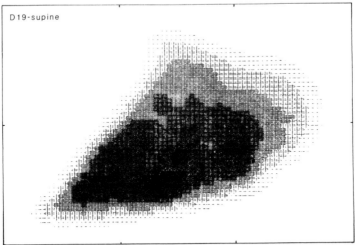

FIGURE 6.15:1. Tomographic images of a midcoronal (*top*) and a midsagittal (*bottom*) slice from one animal with normal cardiac output. Flow (or activity) is expressed as a percent of maximum flow in each slice and is divided into six shades as shown on the scale. Pixels with flow <10% of maximum were excluded. From Hakim, Lisbona, & Dean (1989), reproduced by permission.

tivity is present in the central regions but tends to be closer to the dorsal border because of the influence of gravity. These features do not contradict the role of gravity, but exhibit additional details that reflect the influence of the morphology of the vascular tree, the location of the trunk vessels in the central regions of the lobes, and the varying lengths of the resistance vessels from the core to the periphery.

6.16 Patchy Flow in the Lung

Mathematically, the basic sheet-flow equation, Eq. (6.10:9) or (6.10:11), has a simple solution

$$h = 0. \tag{1}$$

This solution is trivial in the one-dimensional case because it implies absence of flow. But in the three-dimensional structure of alveoli this is a very significant solution: $h = 0$ can be imposed in any area, provided that the remaining area can handle the flow, because $h = 0$ is an allowable boundary condition for the differential equations (6.10:9) and (6.10:11). In other words, in the three-dimensional structure of the lung, the solution $h = 0$ can be embedded in limited regions anywhere.

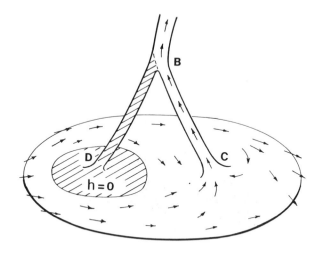

FIGURE 6.16:1. Schematic drawing showing a possible condition of flow from an alveolar sheet into a venous tree. Two terminal venules intersect the sheet. The blood pressure at the point B is smaller than the alveolar gas pressure, that is, $p - p_A < 0$. Blood flows from the sheet into the venule BC; the junction C is thus a sluicing gate. A portion of the sheet around the terminal D of the venule BD however, is closed. The vascular sheet thickness is zero in the shaded region, where there is no flow. The blood pressure at D is equal to that at B, whereas that at C is greater than that at B.

The concept of a healthy lung having collapsed patches of alveolar capillaries is illustrated in Figure 6.16:1. Here two terminal venules intersect a capillary sheet, and both remain patent. One (the venule BC) is connected to an open capillary sheet, and the other is connected to a closed patch with no flow. The hemodynamics of this system is certainly different from the one in which all capillaries are open. Hence, we must know how the closed patch got there and how to remove it.

If the condition of the lung is zone 1, then the pulmonary capillaries are expected to be collapsed. If the condition of the lung is zone 2, then the capillaries may be open, or may be closed. The determination needs a great deal of thought, and we shell devote the next three sections to its study. If the lung is in the zone-3 condition, then if a capillary sheet is open with flow, it can be expected to continue to remain open. But if a capillary sheet is collapsed, then it may remain closed unless some special efforts are made to open it. A procedure that may open a collapsed capillary sheet must supply energy to separate the adhered endothelium, and work against any external loads. Perfusing the lung with a large flow and cycling a number of times will usually accomplish this objective. However, pulmonary capillary sheets may be collapsed for a variety of pathological reasons, and reopening requires careful thought.

6.17 Analysis of Flow Through a Pulmonary Sluicing Gate

The *waterfall* phenomenon in the lung is unique because the *sluicing gates* in the lung are very short and have unique mechanical properties. Corresponding sluicing gates in the skeletal and heart muscles, due to their long cylindrical geometry, lead to an entirely different flow condition. The hemodynamic behavior of sluicing gates depends on the anatomy of the gates and the structure of the fluid. In this section, we look into the details of the pulmonary gates, and in Chapters 7 and 8 the gates in muscles are considered.

The location of the sluicing gates at the capillary–venule junctions has been clarified in the preceding section. The unique features of the pulmonary sluicing gates are as follows:

1. The pulmonary veins and venules remain patent (i.e., they do not collapse) in the normal range of venous blood pressure and alveolar gas pressure (see Sec. 4.10 and 6.7).
2. The pulmonary capillary sheets do collapse when the alveolar gas pressure exceeds the static pressure of the blood (see Sec. 6.8).
3. The pulmonary capillary sheet is "two-dimensional" at the sheet-and-venule junction, that is, along a line perpendicular to the direction of flow, so that the blood cells are confined to a "sheet," (see Section 6.6, Figure 6.6:3, p. 356), not to a narrow tube (Sec. 7.4 and 8.6).

4. The capillary sheet–venule connections are discussed in Section 6.6. Figures 6.6:1 and 6.6:2 show that there are two types of connections— edge draining and central draining—as illustrated in Figure 6.6:3.

Figure 6.17:1 illustrates the capillary–venule junction in the case of edge draining in which a capillary sheet connects to a venule. If there is no flow in the sheet, the sheet is of uniform thickness. When flow exists, however, the pressure drops and the thickness narrows along the streamline, and the walls of the capillary sheet are curved. Let each wall of the capillary sheet be regarded as an elastic membrane. Let the principal stress resultants in each membrane be T_1 and T_2, let the principal curvatures of the membrane be $1/R_1$ and $1/R_2$, and let the principal axes of these stress resultants and curvatures coincide. Then the expression

$$\frac{T_1}{R_1} + \frac{T_2}{R_2} \tag{1}$$

is equivalent to an effective lateral pressure tending to distend the sheet. Hence, Eqs. (1a to 1d of Sec. 6.8, p. 366) should be replaced by [Fung and Zhuang (1986)]:

$$h = h_0 + \alpha\left(p - p_A + \frac{T_1}{R_1} + \frac{T_2}{R_2}\right)$$

$$\text{when } p - p_A + \frac{T_1}{R_1} + \frac{T_2}{R_2} > 0, \tag{2}$$

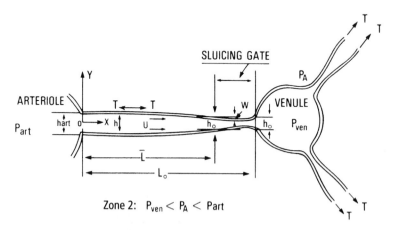

Zone 2: $P_{ven} < P_A < P_{art}$

FIGURE 6.17:1. Idealized geometry of a sluicing gate in a pulmonary alveolar septum that is drained by a venule along its edge. The figure illustrates a cross section of the interalveolar septa and the venule. Symbols are explained in the text. From Fung and Zhuang (1986), reproduced by permission.

$$h = h_0 - \alpha'\left(p_A - p - \frac{T_1}{R_1} - \frac{T_2}{R_2} \right)$$

$$\text{when } -0.5 \text{ cm H}_2\text{O} < p - p_A + \frac{T_1}{R_1} + \frac{T_2}{R_2} < 0, \tag{3}$$

$$h = 0$$

$$\text{when } p - p_A + \frac{T_1}{R_1} + \frac{T_2}{R_2} < -0.5 \text{ cm H}_2\text{O}. \tag{4}$$

Here α, α' are constants. Let p be the local blood pressure, p_A be the alveolar gas pressure, and p_{art}, p_{ven} be the arteriolar and venular pressures, respectively. A region of the lung is said to be in *zone-2 condition* if

$$p_{art} > p_A > p_{ven} \tag{6.15:8}$$

in that region. The blood pressure p is less than p_{art}. It decreases along the streamline. When $p = p_A$ the thickness of the capillary sheet is decreased to h_0 if the effect of the product of the membrane tension and curvature is negligible. Then p becomes $< p_A$, and h becomes $< h_0$. Eventually, p becomes equal to p_{ven}. Then, further downstream, it becomes less than p_{ven}, and the sheet thickness would have collapsed to $h = 0$ if the product of the membrane tension and curvature were not large enough to pull the walls apart.

What is a sluicing gate? A sluicing gate is a point at which the capillary sheet thickness is less than h_0. This gate exists at the venular end of a capillary in the zone-2 condition. If the membrane tension and curvature were sufficiently large, it is possible that the sluicing gate's opening is wide enough to allow a flow to go through. If the tension or the curvature is too small, the gate may become so narrow that the flow becomes choked. An analysis is needed to determine the outcome.

Edge Draining Equations

Let a coordinate axis x be chosen with the origin located at the arteriolar end of the capillary sheet (Fig. 6.17:1). In the sluicing gate region where $h < h_0$, we use Eq. (3) to describe the elastic behavior of the capillary sheet. In the present case R_1 is finite and R_2 is ∞. If w represents the deflection of the wall from the plane $y = h_0/2$, then

$$h = h_0 - 2w. \tag{5}$$

By reference to a book on differential geometry, it can be seen that the curvature of the membrane is

$$\frac{1}{R_1} = \frac{d^2w}{dx^2}\left[1 + \left(\frac{dw}{dx} \right)^2 \right]^{-3/2}. \tag{6}$$

If the slope $|dw/dx|$ is small compared with 1, we can use the approximation,

$$\frac{1}{R_1} = \frac{d^2w}{dx^2} = -\frac{1}{2}\frac{d^2h}{dx^2} \tag{7}$$

and write Eq. (3) as

$$\frac{T_1}{2}\frac{d^2h}{dx^2} = p - p_A + \frac{h_0 - h}{\alpha'}. \tag{8}$$

On the other hand, the momentum equation of the sheet flow is given by Eq. (3) of Section 6.10,

$$\frac{dp}{dx} = -\frac{\mu k f U}{h^2} = -\frac{\mu k f \dot{Q}}{h^3}, \tag{9}$$

where U is mean velocity of flow and $\dot{Q} = Uh$. The equation of continuity (Eq. 7 of Sec. 6.10, with the fluid density ρ a constant) is reduced to

$$\frac{\partial h U}{\partial x} = 0 \quad \text{or} \quad \dot{Q} = \text{const.} \tag{10}$$

Combining equations (8), (9), and (10), we obtain the basic equation,

$$\frac{T_1}{2}\frac{d^3h}{dx^3} + \frac{1}{\alpha'}\frac{dh}{dx} + \frac{\mu k f \dot{Q}}{h^3} = 0. \tag{11}$$

The boundary conditions are

$$h\Big|_{x=\bar{L}} = h\Big|_{x=L_0} = h_0, \tag{12}$$

$$\left(p - p_A\right)\Big|_{x=\bar{L}} = 0. \tag{13}$$

Central Draining Equations

Some interalveolar septa are orthogonal to the venules, as illustrated in Figure 6.17:2. To derive an approximate solution, we assume that the flow is axisymmetrical in the sluicing region. With polar coordinates whose axis coincides with the axis of the venule, we assume the deflection surface of the membrane to be axisymmetric and describe it by $w(r)$, so that $h = h_0 - 2w$. Assume the membrane tension per unit length, T, to be uniform everywhere initially. Assume also that when the membrane deflects the perturbation on the tension is so small that the tension remains a constant. The equation of equilibrium of the membrane is, according to Eq. (3) and the theory of elasticity,

$$T\left(\frac{1}{R_1} + \frac{1}{R_2}\right) = p_A - p - \frac{h_0 - h}{\alpha'}, \tag{14}$$

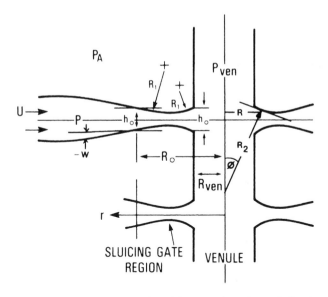

FIGURE 6.17:2. Idealized geometry of a sluicing gate in a sheet drained centrally by a venule perpendicular to it. This section cuts the venule longitudinally and the inter-alveolar septa transversely. The symbols are explained in the text. From Fung and Zhuang (1986), reproduced by permission.

where R_1 is the radius of curvature of the meridian and R_2 is the radius of the transverse curvature. It is well known that (see Fig. 6.17:2)

$$R_2 = \frac{r}{\sin \phi}, \quad \tan \phi = \frac{dw}{dr}, \tag{15}$$

where ϕ is the angle between a normal to the membrane and its axis of revolution. Further, if ϕ is small,

$$\frac{1}{R_1} = \frac{d^2 w}{dr^2}, \quad \frac{1}{R_2} = \frac{1}{r} \frac{dw}{dr}, \tag{16}$$

and the foregoing equations can be combined into

$$\frac{T}{2} \frac{1}{r} \frac{d}{dr} \left(r \frac{dh}{dr} \right) = p - p_A + \frac{h_0 - h}{\alpha'}. \tag{17}$$

The momentum equation relating the mean velocity of flow to the pressure gradient remains to be [Eq. (3) of Sec. 6.10]

$$U = -\frac{1}{\mu k f} h^2 \, \text{grad} \, p. \tag{18}$$

The equation of continuity in the axisymmetric case is

$$\frac{drUh}{dr} = 0 \quad \text{or} \quad -2\pi rUh = \dot{Q}. \tag{19}$$

Correspondingly, Eq. (18) becomes

$$\frac{dp}{dr} = \frac{2\pi\mu kf\dot{Q}}{rh^3}. \tag{20}$$

The boundary conditions are

$$h = h_0 \quad \text{when} \quad r = R_{\text{ven}} \quad \text{and} \quad r = R_0, \tag{21}$$

where R_{ven} is the radius of the venule, R_0 is the radius at which p equals p_A.

Method of Solution

The method of iteration is used to solve these equations. As a first step, we assume h equal to h_0 and integrate Eqs. (9) or (18) to obtain the pressure distribution p. On substituting and $h = h_0$ into the right-hand side of Eqs. (8) or (17), we integrate that equation to obtain the thickness distribution of capillary sheet, h. Then we iterate until convergence is obtained.

Membrane Tension, T

The tension T includes the tension due to stretching of the interalveolar septa and the surface tension due to the liquid–air interfaces. It depends on transpulmonary pressure. Lee and Flicker (1974) have shown how to obtain T by making observations on the visceral pleura. They assume that the subpleural alveoli have the same dimensions, shape, and mechanical properties as the inner alveoli. For equilibrium of a small element of the pleura, the balance of forces perpendicular to the pleura yields the relation

$$T = \frac{\left(p_A - p_{\text{PL}}\right)}{D}, \tag{22}$$

where p_A is the alveolar pressure, p_{PL} is the pleural pressure, and D is the ratio of the average length of the alveolar perimeter on the pleural surface to the average projected alveolar gas area on the pleural surface. D is a function of the degree of inflation of the lung.

According to Lee and Flicker (1974), when the volume of the inflated dog lung is 10 times that of the lung tissue, $D = 0.04\,\mu\text{m}^{-1}$. Hence, for the dog, when $p_A - p_{\text{PL}} = 10\,\text{cm H}_2\text{O}$, $T = 24.5$ dyne/cm.

Results

With the following constants pertinent to the cat,

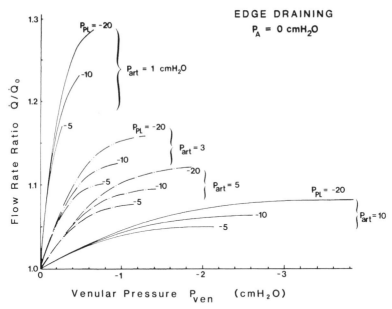

FIGURE 6.17:3. The pressure–flow relationship in the case of edge draining. The flow $\dot{Q}$ is divided by $\dot{Q}_0$, and the ratio is plotted against the venular pressure, while p_A, p_{art}, p_{PL} take on values indicated in the figure. $\dot{Q}_0$ is flow when $p_A = p_{ven} = 0$. All pressures in cm H_2O. From Fung and Zhuang (1986), reproduced by permission.

$$\mu = 1.92 \text{ cp}, \quad k = 12, \qquad\qquad h_0 = 4.28 \,\mu m,$$
$$f = 1.6, \qquad \alpha = 0.219 \,\mu m/cm \ H_2O, \quad \alpha' = 8.56 \,\mu m/cm \ H_2O$$
$$\overline{L} = 556 \,\mu m, \quad D = \overline{L}/\overline{A} = 40 \text{ cm}^{-1},$$

we obtained typical results shown in Figures 6.17:3 and 6.17:4.

Edge Draining

Figure 6.17:3 shows the pressure–flow relationship for various combinations of the arteriolar pressure, p_{art}, pleural pressure, p_{PL}, airway pressure, $p_A = 0$ (atmospheric), and venular pressure, p_{ven}. The flow ($\dot{Q}$) is calculated and its ratio to the flow at $p_{ven} = 0$, $\dot{Q}_0$, is plotted. $\dot{Q}_0$, can be calculated by the original sheet flow theory by setting $p_{ven} - p_A = 0$ in Eq. (2). It is seen that $\dot{Q}$ continues to increase as p_{ven} is lowered below p_A. Each curve terminates at an upper end, beyond which we did not obtain a convergent result. The throat of the sluicing gate narrows rapidly in this region. The gate is presumed closed at the terminal point. At higher levels of transpulmonary pressure (lower P_{PL}) one can obtain greater levels of $\dot{Q}/\dot{Q}_0$ by lowering venular pressure. This is because there is greater tension in the curved walls of the alveolar sheet; therefore the sluicing region has a smaller tendency to become narrowed as outflow pressure decreases.

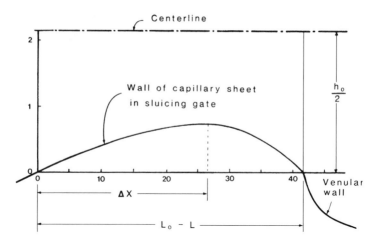

FIGURE 6.17:4. The shape of the lower wall of the sluicing gate in the edge draining case, with $p_A = 0$, $p_{PL} = -20$ cm H_2O, $p_{art} = 10$ cm H_2O, $\dot{Q} = 0.836 \times 10^{-4}$ cm^2/s, $L_0 = 556 \mu$m. $L_0 - L$ = length of sluicing gate = 41.65μm. The horizontal coordinate = distance from the place where $p - p_{alv} = 0$ and $h = h_0$, in units of micrometers. The vertical coordinate = sluicing gate wall location in units of micrometers. From Fung and Zhuang (1986), reproduced by permission.

The shape of the sluicing gate is illustrated in Figure 6.17:4 for a special case. The flow comes in from the left. The part of the channel from the arteriole to the section where the thickness is reduced to h_0 is not shown. On the right-hand end the sheet is anchored to the venule, which remains patent. In this region the blood pressure in the channel is lower than the external pressure, p_A.

The most interesting characteristic of the sluicing gate is the width of the narrowest section, that is, the throat. As the venular pressure is reduced the throat width decreases, and its location moves to the right.

The *central draining* case leads to similar results which are recorded in Fung and Zhuang (1986).

These results show that the pulmonary sluicing gates are affected by the tension in the walls. The gates are short, and the walls may have high curvature. These features tend to help the gates to remain open as the blood pressure drops below the alveolar gas pressure. Red blood cells can squeeze through these thin slits much easier than they can negotiate narrow cylindrical tubes. Comparing the pulmonary sluicing gates with other collapsible vessels considered in Chapter 4, we see the unique importance of wall tension, small size, the near zero Reynolds and Womersley numbers, and the absence of wave equation and flutter. The absence of flutter at very low Reynolds number has been verified experimentally by Fung and Sobin, 1972b.

6.18 Stability of a Collapsing Pulmonary Alveolar Sheet

We now show that when an interalveolar septum collapses at one place, the whole septum (extending from an arteriole to a venule) has to collapse. In the preceding section, sluicing at the capillary–venule junction is analyzed. If the sluicing channel collapses, the sheet may appear either as in (A) in Figure 6.18:1, or as in (B). The theme is to prove that (B) is unstable.

Stability analysis is based on classical thermodynamics. All sources of energy are identified first. When an alveolar sheet collapses ($h \to 0$) the endothelial surfaces of the capillary blood vessels may adhere. Adhesive contacts between cell membranes are mediated by a wide variety of electrostatic, electrodynamic, and intermolecular forces. These forces do work when the adhesive contact is formed. Thermodynamically this work represents a reduction of free energy, which can be denoted by W_{CB}, with the subscript CB suggesting "chemical bonds" or "crossbridges." As the area of contact increases, work is done by the external forces and strain energy is

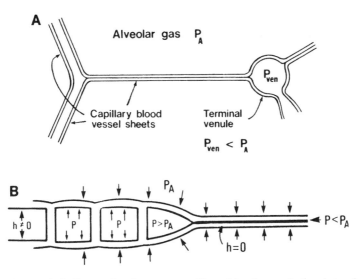

FIGURE 6.18:1. (A) Sheet of pulmonary capillary blood vessels (an interalveolar septum) is collapsed. Collapse is arrested by two open septa shown on the left. (B) A septum with one part adjacent to a venule is collapsed. Deformation of walls is illustrated. A collapsed lumen is represented by a thicker line to remind the reader that membranes are separated by posts, which do not collapse, and thus some very small areas around posts must remain open. In the text it is shown that this type of collapse is unstable and a collapse, once started, will continue to spread until arrested by something else, such as the case shown in A. P_A = alveolar pressure; P_{ven} = venule pressure; h = capillary sheet thickness. From Fung and Yen (1986), reproduced by permission.

stored in the membranes. Let these be denoted by W_P and W_D, respectively. W_{CB}, W_P, and W_D are functions of the geometry of the contact area and may depend on whether the contact is spreading or reducing because adhesion is an irreversible process. For a specific geometric pattern and a specified direction of motion (either spreading or reducing), the W's are functions of the area of adhesion, A_c. A small increment in the contact area, δA_c, will cause small changes in the energies, δW_{CB}, δW_D, and δW_P. If the sum $\delta W_{CB} + \delta W_D + \delta W_P$ is denoted by δW, and δW is negative, there will be a tendency for the contact area to increase. If the first derivatives of the W's exist, then to the first order $\delta W_{CB} = (\partial W_{CB}/\partial A_c)\delta A_c$ and the increment of energy may be written as

$$\delta W_{CB} + \delta W_D + \delta W_P = \left(\frac{\partial W_{CB}}{\partial A_c} + \frac{\partial W_D}{\partial A_c} + \frac{\partial W_P}{\partial A_c} \right)\delta A_c \tag{1}$$

If the quantity in parentheses vanishes, then the spreading or reseparation process will lose its driving force. Hence at an equilibrium condition,

$$\frac{\partial}{\partial A_c}\left(W_{CB} + W_D + W_P \right) = \frac{\partial W}{\partial A_c} = 0 \tag{2}$$

from which the area of contact, A_c, may be determined. According to thermodynamics, the equilibrium state is stable if the second derivative, $\partial^2 W/\partial A_c^2$, is positive. The equilibrium is unstable if the second derivative is negative.

Surface Energy

If we assume that the properties of the endothelium are the same everywhere, then the surface energy of adhesion is directly proportional to the area, A_c, such that

$$W_{CB} = -\gamma A_c, \tag{3}$$

where γ is a constant whose value may depend on whether the bonds are being formed or broken.

Deformation Pattern of Collapsed Sheet

Consider an idealized case of edge draining [Fig. 6.18:2(A)]. Assume that a portion of length ξ next to the venule is collapsed. In a transition region of length ΔL, the wall bending is significant. We need an estimation of ΔL. We note that the bending of the wall is due to the forces in the posts. The load–deflection curve of the posts is given by Eqs. (1a) to (1d) of Section 6.8 and Figure 6.8:1, pp. 366, 367. Upon change of the thickness of the capillary sheet from $h(x)$ to $y(x)$, the compressive force of the posts per unit area of the sheet is changed to

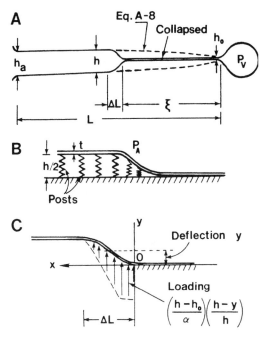

FIGURE 6.18:2. Schematic drawings for analysis of deflection pattern of collapsed interalveolar septa. (*A*) Dotted lines indicate walls of the sheet before collapsing; solid lines indicate the walls of a collapsed sheet. h = sheet thickness; ξ = collapsed region; L = length of sheet; ΔL = length of transition zone; P_V = venous pressure. (B) Equilibrium of a wall in the transition zone. Posts are drawn as springs, which balances transmural pressure, $P-P_A$ (alveolar pressure). (*C*) Out-of-balance spring forces causing deflection of wall when the right-hand side of the sheet ($x < 0$) is collapsed. From Fung and Yen (1986), reproduced by permission.

$$\left(\frac{h-h_0}{\alpha}\right)\left(\frac{h-y}{h}\right),$$

where α is the compliance constant of the sheet in Eq. (6.8:1b). Thus in the free-body diagram Figure 6.18:2(B), the load that causes bending of the sheet wall as a cantilevered beam clamped at $x = 0$ is shown in Figure 6.18:2(C). If the thickness of the wall is t, The bending rigidity is $B = Et^3/12(1 - v^2)$, where E = Young's modulus and v = Poisson's ratio. Then y is governed by the differential equation describing the bending of the wall,

$$B\frac{d^4 y}{dx^4} = \frac{h-h_0}{\alpha}f(x), \qquad \text{where } f(x) = 1 - \frac{y}{h}. \qquad (4)$$

Here h is treated as a constant in the transition zone. Let us define the transition zone by the boundary conditions that the shear and curvature are zero at $x = \Delta L$ (see Fig. 6.18:2(C)) because the membrane is undisturbed to the left, then by integration of Eq. (4) we obtain

$$\frac{d^2 y}{dx^2} = \frac{h - h_0}{\alpha B} \int_{\Delta L}^{x} \int_{\Delta L}^{u} f(\tau) \, d\tau \, du, \tag{5}$$

where τ and u are dummy variables for integration. Again, by the boundary condition that the slope and deflection are zero at $x = 0$, and introducing two more dummy variables, v and w, we obtain

$$y(x) = \frac{(h - h_0)}{\alpha B} \int_{0}^{x} dw \int_{0}^{w} dv \int_{\Delta L}^{v} du \int_{\Delta L}^{u} f(\tau) \, d\tau. \tag{6}$$

Since $f(x)$ is a function of $y(x)$, this is an integral equation that can be solved by iteration. For the estimation of ΔL, however, $f(x)$ is just some function of x with a value that varies between 0 and 1. Hence the quadruple integral in Eq. (6) is of the order of magnitude $(\Delta L)^4$. Thus, since $y = h$ when $x = \Delta L$, we obtain from Eq. (6)

$$\alpha B h = \text{const.}(h - h_0)(\Delta L)^4.$$

Solving this for ΔL, we obtain

$$\Delta L \sim \left(\frac{\alpha B h}{h - h_0} \right)^{0.25}. \tag{7}$$

This shows that ΔL is independent of h if $h \gg h_0$, but tends to be large if $h \to h_0$.

The value of h is the sheet thickness given by Eq. (9) of Section 6.11,

$$h = \left(\frac{h_a^4}{L} \xi + h_0^4 \right)^{0.25}, \tag{8}$$

in which ξ is the distance measured from the venule (see Fig. 6.18:2A), and L is the distance between the arteriole and the venule, that is, the width of the sheet. When collapse occurs at ξ, the value of h to be used in Eq. (7) is given by Eq. (8). Thus if $\xi \to 0$, then $h \to h_0$ and ΔL is large. If ξ/L is finite (e.g., between 0.5 and 1) and $h_a \gg h_0$, then $h \gg h_0$ and ΔL tend to be independent of h and ξ. This is a fact of importance to the following discussion, which focuses attention on the case of finite ξ/L.

Work Done by an External Load

When a portion of the sheet next to the venule is collapsed, the area of contact per unit length of the venule is $A_c = \xi$. In the process of collapsing the work done by the pressure is

$$W_P = \int_{0}^{\xi} d\xi \int_{0}^{h} (P_A - P) \, dh, \tag{9}$$

where P is the pressure in the vessel, equal to $P_A + (h - h_0)/\alpha$ before collapse, and P_V after collapse, except in the transition zone, where P tends to approach the arterial pressure, P_a, after collapse. Thus W_P consists of two parts, one in the occluded area and another in the transition zone, W_{P1} and W_{P2}, respectively. Since P is bounded between the limits described earlier, we can use the mean-value theorem of integral calculus to write the first integral in Eq. (9) as $(P_A - \overline{P})h$, where $\overline{P}$ is a number between P_v and $P_A + (h - h_0)/\alpha$. Thus

$$W_{P1} = \int_0^\xi \left(P_A - \overline{P}\right) h\, d\xi. \tag{10}$$

In this integral h is given by Eq. (8). Both h and $\overline{P}$ are continuous functions of ξ and of bounded variation. Hence we can use the mean-value theorem again. Further, we ignore h_0 in Eq. (8) for simplicity. Then, writing $\overline{P}'$ for a number between P_v and $P_A + (h - h_0)/\alpha$, we obtain

$$W_{P1} \doteq \left(P_A - \overline{P}'\right) \frac{h_a}{L^{0.25}} \int_0^\xi \xi^{0.25}\, d\xi = \frac{4}{5}\left(P_A - \overline{P}'\right) h_a L^{-0.25} \xi^{1.25}. \tag{11}$$

The part W_{P2} comes from the transition zone at the leading edge of the collapsed area (of width ΔL), and is of the order of

$$W_{P2} = \left(P_A - \overline{P}''\right) \int_{\Delta L} h\, d\xi \doteq \Delta L \cdot \left(P_A - \overline{P}''\right) h$$

$$\doteq \Delta L \left(P_A - \overline{P}''\right) \frac{h_a}{L^{0.25}} \xi^{0.25}, \tag{12}$$

when ξ/L is finite and h_0 can be neglected in Eq. (8). The numbers, $\overline{P}$ and $\overline{P}'$, may be different, but both are bounded between P_a and P_V. Hence

$$W_P = \frac{4}{5}\left(P_A - \overline{P}'\right) h_a L^{-0.25} \xi^{1.25} - \Delta L \left(\overline{P}'' - P_A\right) h_a L^{-0.25} \xi^{0.25}. \tag{13}$$

Strain Energy

The strain energy due to deformation of the wall is mainly concentrated at the transition region ΔL shown in Figure 6.18:1(A). It is equal to the integral of the product of the bending rigidity and the square of the curvature over the membrane. From Eqs. (5) and (8) and for finite ξ/L we have

$$W_D = B \overline{\left(\frac{d^2 y}{dx^2}\right)^2} \Delta L \sim c\, \frac{B\left(h - h_0\right)^2}{\alpha^2 B^2} \Delta L^3 \doteq c\, \frac{h_a^2 L^{-0.25}}{\alpha^2 B} \Delta L^3 \xi^{0.50}, \tag{14}$$

where the overbar means the mean value and c is a constant.

Energy Balance

The variation of W_{CB}, W_{P1}, W_{P2}, and W_D with respect to ξ is illustrated in Figure 6.18:3. The trend for W_{P1} and W_{P2} to change with increasing $P_A - P_v$ and $Pa - PA$, respectively, is indicated by arrows, on the assumption that $\overline{P}'$ is closer to P_v and $\overline{P}''$ is closer to P_a. The adhesive energy due to chemical bonds is proportional to the contact area when the contact is spreading (sheet collapsing). The energy for breaking the bonds to separate the adhered membranes is assumed to be larger than that for adhering.

The resultant, $W_{CB} + W_{P1} + W_{P2} + W_D$, is sketched in Figure 6.18:4. It is seen that the curve bulges upward. The equilibrium state at *point E*, where the slope $\partial W/\partial \xi$ is zero, has a negative curvature, $\partial^2 W/\partial \xi^2 < 0$. Hence the equilibrium is unstable, and the collapse, once started, will continue to spread. A plausible place where the collapse can be arrested is the junction where the collapsing septum meets two other septa, as illustrated in Figure 6.18:1(A). At this junction the membranes have large curvatures, which, in association with the tension in the membranes, tend to pull the membranes apart and keep the septa open. Furthermore, the posts are absent at the junction, making it possible for the membranes to form a good seal. However, it is possible that the negative pressure is transmitted through the junction into the neighboring septa, causing them to collapse also.

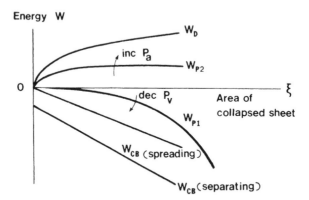

FIGURE 6.18:3. Sketches of energies given by Eqs. (3), (10), (11), and (14). Because the length of the transition zone (ΔL) was estimated under the assumption that the collapsed region (ξ)/L is finite, these equations are not accurate when ξ tends to 0. Equations (3) to (14) are derived for a sheet draining edgewise into a venule. A similar analysis of a centrally drained sheet is more complex but yields similar trends if ξ is replaced by r^2, the square of the radius of the collapsed area. W_{CB} = free energy with $_{CB}$ denoting crossbridges; W_P and W_D = functions of geometry of contact area; P_a = arterial pressure. From Fung and Yen (1986), reproduced by permission.

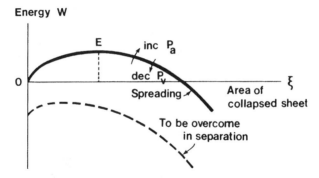

FIGURE 6.18:4. Resultant energy, W, which is the sum of W_{CB}, W_{P1}, W_{P2}, and W_D in a sheet that is drained edgewise by a venule. Equilibrium *point E* where the slope is 0 is seen to be unstable (the curve has a negative curvature at this point). Hence collapse of the sheet, once started, will spread until the whole sheet is collapsed. A similar conclusion is reached for a sheet that is drained centrally by a venule. P_a and P_V = arterial and venous pressures. From Fung and Yen (1986), reproduced by permission.

Experimental Evidence

Evidence of total collapse (and the absence of partial collapse) of some pulmonary interalveolar septa coexisting with open septa is seen in histological micrographs of lungs with blood flow in the zone-2 condition (see Warrell et al., 1972; reproduced by permission in Fung, 1990, p. 218).

6.19 Hysteresis in the Pressure–Flow Relationship of Pulmonary Blood Flow in Zone-2 Condition

With sluicing gate and patchy filling clarified in the preceding sections, we can explain the hysteresis in waterfall phenomenon first observed by Permutt et al. (1962). Figure 6.19:1 shows the results of an experiment by Fung and Yen (1986). The following discussion is intended to apply the theories presented in Sections 6.15 to 6.18 to explain the features revealed by Figure 6.19:1.

We sketch the different zones of flow in these experiments in Figure 6.19:2, with the *venular* pressure, p_{ven}, plotted on the abscissa, instead of the pressure in the large vein, p_v, as in Figure 6.19:1. Then the line $p_{ven} - p_A = 0$ divides zone 2 from zone 3. In zone 2, some capillaries are collapsed. When patchy collapse occurs, it has four effects: (a) the number of stagnant venules (in which there is no flow) increases, (b) the perfused area of the capillary sheet decreases, (c) the average length of the streamlines between arterioles and venules increases, and (d) the flow through the sluicing gates

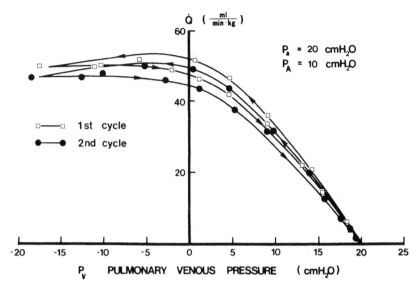

FIGURE 6.19:1. Pressure–flow relationship in right lung of cat. Dots and squares show experimental data of an isolated lung perfused with macrodox (with papaverine added to relax smooth muscle), with the arterial pressure fixed at 20 cm H_2O, pleural pressure fixed at 0 (atmospheric), airway pressure (P_A) fixed at 10 cm H_2O, and venous pressure (P_V) at left atrium gradually decreased from 20 cm H_2O to -18 cm H_2O. After reaching a minimum, venous pressure was increased again gradually until 20 cm H_2O. After a short period of rest, another cycle was done on same lung. Note that pressure in a venule (P_{ven}) can differ considerably from pressure in a large vein, P_V. The difference is equal to flow times resistance, which varies with P_V because of vascular elasticity, in veins. Experiment 5-8-85. $\dot{Q}$ = flow. From Fung and Yen (1986), reproduced by permission.

increases. These effects of (b) and (c) affect the flow according to Eq. (6a) of Section 6.13, p. 390 by changing A and $\bar{L}^2$. The effect of (d) is revealed in Section 6.17. Quantitatively, the decrease of flow along the line BC in Figure 6.19:2 is mainly due to the decrease of the perfused alveolar area. The percentage decrease in flow is approximately equal to the percentage of the area of the collapsed sheets in the total alveolar sheet area. Thus,

$$\frac{\dot{Q}}{\dot{Q}_0} \doteq \frac{A - A_c}{A},\tag{1}$$

where $\dot{Q}$ is the total flow, $\dot{Q}_0$ is the flow computed from Eq. (6a) of Section 6.13 (flow at point B in Fig. 6.19:2), A is the total alveolar sheet area, and A_c is the area of the collapsed sheet (where $h = 0$).

Now, assume that in the zone-3 condition A_c is zero when $P_{ven} - P_A = 0$. For further decrease of P_{ven} let us designate $P_{ven} - P_A$ by ΔP. A_c/A increases with increasing $|\Delta P|$. Assume that it tends to a constant F when $|\Delta P| \to \infty$.

The curves in Figure 6.19:1 suggest a bell-shaped relationship between A_c/A and ΔP. Hence we propose the relationship,

$$\frac{A_c}{A} = F\left(1 - e^{-\Delta P^2/2\sigma^2}\right) \tag{2}$$

or

$$\frac{\dot{Q}}{\dot{Q}_0} = 1 - F + Fe^{-\Delta P^2/2\sigma^2}. \tag{3}$$

This is a Gaussian curve. The parameter F means the limiting value of the fraction of alveolar area that is collapsed when P_{ven} becomes very small. The parameter σ means the value of ΔP at which the *curve BC* in Figure 6.19:2 has a point of inflection. Note that, since P_{ven} is equal to P_V plus the product of flow and resistance, and since in zone 2 the flow is essentially constant, ΔP is approximately equal to P_V minus its value at waterfall. Hence we can apply Eq. (3) to the experimental results, such as those shown in Figure 6.19:1. It is quite easy to determine the parameters F and σ from the experimental data. The results obtained by Fung and Yen (1986) from a set of experiments on cat lung are shown in Table 6.19:1. It is seen that, on average, $F = 0.104$ and $\sigma = 4.45$ cm H_2O.

When the venous pressure is decreased to a certain level (point C in Fig. 6.19:2) and then increased again (CD), how much of the collapsed capillary sheet will be reopened? The answer depends on how much additional force is required to separate the adhered endothelial cell membrane. It appears

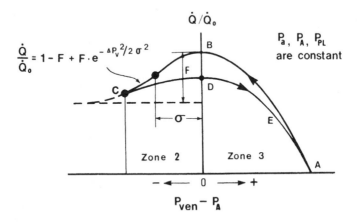

FIGURE 6.19:2. Reduction of area of perfused alveolar sheets due to collapse of capillary blood vessels and adhesion of endothelial cells in the zone-2 condition is directly related to reduction of flow through the lung. A mathematical expression for reduced flow and reduced alveolar area in region BC (zone 2) is given in Eqs. (2) and (3). From Fung and Yen (1986), reproduced by permission.

TABLE 6.19:1. Experimental Results on Pulmonary Blood Flow in Zone-2 Cat Lung that Has Been Isolated and Perfused with Macrodex

					Curvefitting Eq. (3)	
Exp no.	P_a cm H_2O	P_A cm H_2O	$\dot{Q}_0$ ml min·kg	Pv at $\dot{Q}_0$ cm H_2O	F no dim.	σ cm H_2O
4-30-85-0	20	10	91	−0.7	0.087	2.4
5-7-85-0	20	10	63	−9.2	0.071	5.3
5-8-85-1	20	10	54	−2.5	0.093	5.3
5-8-85-2	20	10	51	−2.5	0.088	4.3
5-14-85-1	20	10	38	−6.8	0.184	4.4
5-14-85-2	20	10	46	−10.0	0.098	5.0

$F = 0.104 \pm 0.016$ (SE); $\bar{\sigma} = 4.45 \pm 0.45$ (SE). P_a = pulmonary arterial pressure; P_A = alveolar pressure; $\dot{Q}_0$ = maximum flow. See the text for further explanation. From Fung and Yen (1986). Reproduced by permission.

from Figure 6.19:1 that little reseparation can be achieved in zone 2. If there were no reseparation at all, then CD must be a straight horizontal line. The stroke DE is in zone 3. The data in Figure 6.19:1 are consistent with the assumption that there was no reseparation, even in the zone-3 condition. To open collapsed capillaries, one must raise arterial and venous pressures and give the lung a large flow and perform a number of cycles of flow through them.

Comparison of theoretical and experimental results is shown in Figures 6.19:3 and 6.19:4. Figure 6.19:3 shows the data from a cat lung cycled in zone 3. The hysteresis loops are very narrow in this case. Figure 6.19:4 shows the data of Experiment 5-8-85-1 already given in Figure 6.19:1. Theoretical curves were computed with morphometric and elasticity data given in Sections 6.2 to 6.8 for the right lung of a cat, and $A = 0.84\,m^2$ for half lung. The exact value of the alveolar surface area A of the test lung was unknown. In every case shown in Figures 6.19:3 and 6.19:4 the value of the surface area A is adjusted to make one theoretically predicted point coincide with one experimental point. Then all the rest of the data can be compared. The agreement is reasonable.

Estimation of the Size of Collapsed Alveolar Septa*

According to the analysis in Section 6.18, the process of collapsing of the intervalveolar septa under the zone-2 condition consists of quantum jumps of one septum at a time, beginning with those septa directly connected to the terminal venules, but may be extended to the next adjoined septa, and then to farther septa in successive stages, because the pulmonary capillary sheet is a continuous multiply connected sheet.

---

*Fung and Yen (1986).

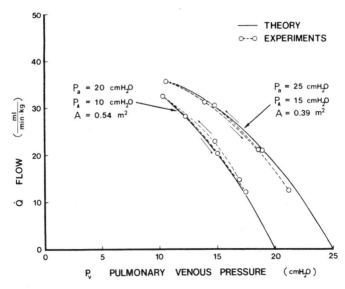

FIGURE 6.19:3. Pressure–flow relationship in two right lungs of cat in zone 3 condition. Each lung was tested first by cycling venous pressure (P_V) to zone 2 condition. The value of P_V at which the flow reached a peak was noted. This value was interpreted as corresponding to the condition at which the venule pressure (P_{ven}) = alveolar pressure (P_A). Same lung was then tested by cycling P_V in a range in which P_{ven} > P_A. For example, in the case in which $P_A = 15$ cm H_2O, it was found that $P_V = 8.0$ cm H_2O when flow ($\dot{Q}$) reached the peak value of 48 ml/min kg. Hence, when P_V was cycled between 22 and 10 cm H_2O so that the flow condition was in zone 3; i.e., P_{ven} > P_A throughout, although $P_V < P_A$ in part of the cycle. *Lower curve* was from an experiment which had a $\dot{Q}_{max}$ of 44 ml/min kg when P_V was 2.4 cm H_2O. Theoretical curves were drawn with alveolar wall surface area A (half lung) indicated in figure. Good fit is obtained by adjusting the values of A. P_a = arterial pressure. From Fung and Yen (1986). Reproduced by permission.

To estimate F, we should first determine the number of alveoli drained directly by each terminal venule and then determine the area of the intervalveolar septa per alveolus draining directly into the terminal venule. This is A_c under the assumption that only those septa draining directly into the terminal venule can collapse. Dividing A_c by the total area per alveolus, A, we obtain F.

We define the terminal venules to be those venules into which the pulmonary capillary sheets drain directly. All order-1 venules are terminal. Figure 6.6:1 on p. 355 shows that in the cat lung, the order-3 venules, with average diameter 77 μm (see Table 6.14:2), also communicate directly with capillary sheets; hence, they are also terminal. Thus we consider all cat pulmonary venules with diameter less than 100 μm to be terminal. Hence, we have the required database outlined in Sections 6.4 to 6.6. According to Zhuang et al. (1985), we have, for the cat lung:

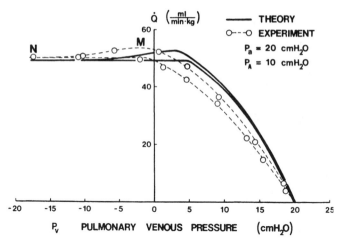

FIGURE 6.19:4. Comparison of theoretical and experimental results in pulmonary blood flow. Theoretical curves were computed with morphometric and elasticity data for right lung of cat, and $A = 0.84 \, m^2$ for half lung. In zone 2 condition, Eq. (2) was used to compute collapsed alveolar surface area (A_c). Curve fitting yields $\sigma = 5.3$ and $F = 0.093$. In return stroke, alveolar surface area is taken to be $A - A_c$; i.e., collapsed area is deducted from initial value to total alveolar wall surface area of right lung under assumption that collapsed alveolar sheets are not reopened until pulmonary arterial pressure is increased in zone 3 condition. Value of A was so selected that a reasonable agreement between theory and experiment is obtained. From Fung and Yen (1986). Reproduced by permission.

$$\frac{\text{Total no. of alveoli in a volume of lung}}{\text{No. of all arterioles of diameter} < 100 \, \mu m \text{ in the volume}} = 24.5,$$

$$\frac{\text{Total no. of alveoli in a volume of lung}}{\text{No. of all venules of diameter} < 100 \, \mu m \text{ in the volume}} = 17.8.$$

But the arterial and venous regions are separate (see Sec. 6.5). The volume of the arterial zone is $43 \pm 7.5\%$ of the total volume of the lung parenchyma (Sobin et al., 1980). Hence,

$$\frac{\text{No. of alveoli in arterial region}}{\begin{array}{c}\text{No. of all arterioles of diameter} < 100 \, \mu m \\ \text{in arterial region}\end{array}} = 24.5 \times 0.43 = 10.5,$$

$$\frac{\text{No. of alveoli in venous region}}{\begin{array}{c}\text{No. of all venules of diameter} < 100 \, \mu m \\ \text{in venous region}\end{array}} = 17.8 \times 0.57 = 10.1.$$

Hence,

$$\frac{\text{No. of alveoli in arterial region}}{\text{No. of terminal arterioles in arterial region}} = 10.5,$$

$$\frac{\text{No. of alveoli in venous region}}{\text{No. of terminal venules in venous region}} = 10.1.$$

Each alveolus (14-hedron) has one wall perforated for ventilation. The remaining walls are shared by neighboring alveoli. Hence, in 10.1 alveoli there are $10.1 \times 13/2 = 65$ interalveolar septa draining into each terminal venule in the venous region. The average terminal pulmonary venule of the cat has an average diameter of $28.0\,\mu m$ and length of $148\,\mu m$ (computed from Table 6.14:2). The average diameter of the "equivalent sphere" of a cat alveolus is $136\,\mu m$ (Zhuang et al., 1985), which is comparable with the length of the average terminal venule. Hence, depending on the placement of the terminal venule relative to the alveoli, each terminal venule can drain 1 to 12 interalveolar septa (computed by trial and error using Fig. 6.4:2). Dividing 1 to 12 by 65, we see that the number of interalveolar septa that drain directly into a terminal venule varies from 1.5% to 18.5% of the total depending on the placement of the terminal venule relative to the alveoli. Comparing with the experimental average value of $F = 10.4\%$, we see that under the experimental condition reported by Fung and Yen (1986), either many interalveolar septa not directly draining into a terminal venule must have collapsed, or the terminal venule is so strategically placed as to intersect a maximum possible number of interalveolar septa for drainage.

6.20 Distribution of Transit Time in the Lung

The length of time it takes for a red cell to go through the lung, that is, the transit time, obviously depends on the path it takes in the lung. Since different cells take different paths, their transit times are not a constant, but vary from streamline to streamline. For a given lung, we speak of the transit time distribution in a manner exactly analogous to the statistical distribution of a random variable. If $f(t)$ is the frequency function of the transit time, then by definition $f(t)dt$ is equal to the fraction of the fluid that enters the lung whose transit time lies between t and $t + dt$.

Since the velocity distribution is given by the sheet flow theory, we can derive a theoretical transit-time distribution for blood in the pulmonary alveoli. As an example, let us consider the case illustrated in Figure 6.12:1, p. 382, whose solution is given in Figure 6.12:2, p. 384. In this case, we have $h = $ constant, and the velocity along any streamline, $\psi = $ constant, is $\partial\psi/\partial n$, where n is the distance perpendicular to the streamline. Let s be the distance measured along the streamline, then the transit time, t, for a particle along that streamline, is

$$t = \int \frac{ds}{velocity} = \int \frac{ds}{\partial\psi/\partial n}. \tag{1}$$

If we consider two neighboring streamlines, ψ = constant = c and ψ = $c + \Delta\psi$, and let the distance between these two streamlines be Δn, then the transit time along these streamlines is

$$t_{\psi=c} \simeq \int_{\psi=c} \frac{\Delta n \, ds}{\Delta\psi} = \frac{1}{\Delta\psi} \int_{\psi=c} \Delta n \, ds$$

$$= \frac{\left(\text{area between the streamlines}\right)}{\Delta\psi}. \tag{2}$$

$\Delta\psi$ is by definition equal to the quantity of flow between the streamlines, $\psi = c$ and $\psi = c + \Delta\psi$. Hence if we divide up the entire flow field by a succession of streamlines separated by a constant interval $\Delta\psi$ from each other, then Eq. (2) shows that the transit time of each streamtube is proportional to the area between successive streamlines. But if $f(t)$ is the frequency function of the transit time in the alveolar sheet, then $\Delta\psi = f(t)\Delta t$, and we can compute $f(t)$ by dividing through $\Delta\psi$ with $\Delta t = t_{\psi=c+\Delta\psi} - t_{\psi=c}$.

With these relations we can determine the frequency function for the present example as follows: The area between successive streamlines in the sheet shown in Figure 6.13:1 is determined with a planimeter. Each streamtube contains 10% of the flow. The inverse of the difference of areas between successive streamlines is proportional to the frequency function. The result is shown by the histogram of Figure 6.20:1, which can be represented approximately by the empirical formula

$$f(\tau) = \frac{1}{20}\delta(\tau - 1) + 1.678 e^{-2(\tau-1)} + \frac{1}{45} e^{-0.2(\tau-1)} \tag{3}$$

for $\tau \geq 1$, while $f(\tau) = 0$ for $\tau < 1$. Here τ = (transit time)/(minimum transit time) and δ = unit impulse function. If we return to physical units with t in seconds and let t_{min} denote the minimum transit time, then

$$f(t) = \frac{1}{t_{min}} f(\tau) = \frac{1}{t_{min}} f\left(\frac{t}{t_{min}}\right). \tag{4}$$

The mean transit time, denoted by $\bar{t}$, is

$$\bar{t} = \int_0^\infty t f(t) \, dt = t_{min} \int_1^\infty \tau f(\tau) \, d\tau. \tag{5}$$

For the empirical formula given earlier, we have

$$\bar{t} = 1.475 t_{min}.$$

These formulas are derived from a sheet of uniform thickness. If the sheet thickness is variable, a stream function can be defined as in Eq. (6.12:1), p. 383, so that

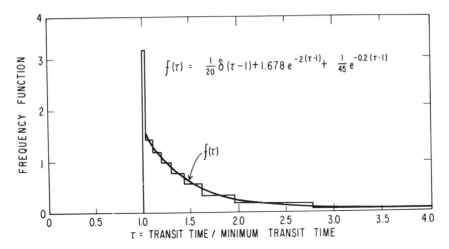

FIGURE 6.20:1. Theoretical transit time distribution for blood flow in an alveolar sheet illustrated in Fig. 6.12:2, p. 384. Computation gives the histogram that is fitted by the curve $f(\tau)$. τ is the ratio of transit time to the minimum transit time through the sheet. Reproduced by permission from Fung and Sobin (1972b). © American Heart Association, Inc.

$$hU = \frac{\partial \psi}{\partial y}, \quad hV = -\frac{\partial \psi}{\partial x}. \tag{6}$$

Then the transit time along a streamline, $\psi = c$, is given by

$$t = \int \frac{h \Delta n\, ds}{\Delta \psi}. \tag{7}$$

Thus for a field covered by streamlines spaced at constant $\Delta \psi$, the transit time in an individual streamtube is proportional to the volume of the tube. Since we have shown that the alveolar sheet remains quite uniform in thickness over a wide range of pressure, it is expected that the variation in thickness will not affect the frequency distribution function significantly.

We can relate the mean transit time to the physical parameters of the alveolar sheet. By a process entirely analogous to that employed in deriving Eq. (7) of Section 6:13, we obtain the mean velocity of flow in an alveolar sheet:

$$\overline{U} = \frac{1}{3\mu k f \alpha \overline{L}} \left[h_a^3 - h_v^3 \right], \tag{8}$$

where $\overline{L}$ is the mean path length of the streamlines between the arteriole and venule,

$$\frac{1}{\overline{L}} = \frac{1}{(\psi_2 - \psi_1)} \int_{\psi_1}^{\psi_2} \frac{1}{L(\psi)}\, dx. \tag{9}$$

Then

$$\bar{t} = \frac{\bar{L}}{\bar{U}} = \frac{3\mu k f \alpha \bar{L}^2}{h_a^3 - h_v^3}. \tag{10}$$

It is obvious that $\bar{t}$ should be proportional to the coefficient of viscosity, μ, and the friction factor, f, and increase with increasing length of the path, $\bar{L}$; but the reason for $\bar{t}$ to depend directly on the square of the mean path length between the arteriole and venule, $\bar{L}^2$, and inversely on $h_a^3 - h_v^3 = ([h_o + \alpha\Delta p_{art}]^3 - [h_0 + \alpha\Delta p_{ven}]^3)$, is more subtle, and requires some thinking for its assimilation.

For example, if we take the data of the dog, $\alpha = 0.122\,\mu$m/cm H_2O or $0.122 \times 10^{-3}\,\mu$m/dyne, $h_0 = 2.5\,\mu$m, $f = 1.8$, $k = 12$, and $\bar{L} = 556\,\mu$m (Sobin et al., 1980, for the cat), $p_{ven} = p_A = 0$ so that $h_v = h_0$, then the mean transit time in alveolar sheet is $\bar{t} = 1.89$ s when $p_{art} = 10$ cm H_2O, and $\bar{t} = 0.811$ s when $p_{art} = 20$ cm H_2O. If we use $\bar{L} = 273\,\mu$m according to Miyamoto and Moll (1971), then $\bar{t} = 0.455$ and 0.195 s in these two cases, respectively.

Transit-Time Distribution in the Whole Lung

Cumming et al. (1969) have computed the transit-time distribution from morphological data on a human lung. Their results on a particle passing from the pulmonary valve down to terminal branches of the order of $50\,\mu$m with a pulmonary blood flow of 80 ml/s are sketched in the upper left-hand corner of Figure 6.20:2. Adding to this the transit time in the alveolar sheet and the veins, we can obtain the transit time through the entire lung. Since

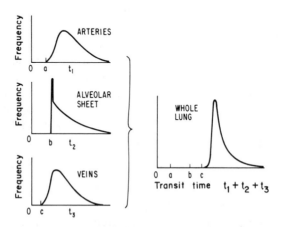

FIGURE 6.20:2. Convolution of transit times, t_1, t_2, and t_3, of blood in the pulmonary arteries, capillaries, and veins, respectively. *Upper left:* Arterial; from Cumming et al. (1969). *Middle left:* Capillary. *Lower left:* Venous. *Right-hand panel:* Result of convolution. Reproduced by permission from Fung and Sobin (1972b).

it is well known that the distribution function of the sum of several random variables is the convolution of the distribution functions of the individual variables, we see that if the transit times, t_1, t_2, and t_3, in the arteries, alveoli, and veins, respectively, are distributed as shown in Figure 6.20:2, the transit time in the whole lung from the pulmonary valve to the left atrium will be distributed as shown on the right-hand side of the figure.

As an illustration of the mathematical procedure, consider the following case. Let the frequency functions for t_1, t_2, and t_3 be approximated by

$$f_1(t) = \frac{\alpha^3}{2}(t-a)^2 e^{-\alpha(t-a)} H(t-a),$$

$$f_2(t) = \varepsilon\delta(t-b) + (1-\varepsilon)\beta e^{-\beta(t-b)} H(t-b),$$

$$f_3(t) = \frac{\gamma^3}{2}(t-c)^2 e^{-\gamma(t-a)} H(t-c), \tag{11}$$

where $H(t-a)$ is the unit-step function that equals 0 when $t < a$ and 1 when $t \geq a$. The characteristic function of $f_1(t)$ is*

$$\int_{-\infty}^{\infty} e^{ixt} f_1(t)dt = \frac{\alpha^3 e^{ixa}}{(\alpha-ix)^3}. \tag{12}$$

Similarly the characteristic function of f_2 and f_3 are, respectively,

$$f_2 = \varepsilon e^{ixb} + (1-\varepsilon)\frac{\beta e^{ixb}}{(\beta-ix)} \quad \text{and} \quad f_3 = \frac{\gamma^3 e^{ixc}}{(\gamma-ix)^3}. \tag{13}$$

Since the frequency function of $t_1 + t_2 + t_3$ is the convolution $f_1 * f_2 * f_3$, the characteristic function of $t_1 + t_2 + t_3$ is the product of the three characteristic functions,

$$\frac{\alpha^3\gamma^3 e^{ix(a+b+c)}}{(\alpha-ix)^3(\gamma-ix)^3}\left[\varepsilon + (1-\varepsilon)\frac{\beta}{(\beta-ix)}\right]. \tag{14}$$

The frequency function of $t_1 + t_2 + t_3$ is the inverse Fourier transformation of the above. The calculation becomes very simple in the case $\alpha = \beta = \gamma$; then the frequency function of $t_1 + t_2 + t_3$ is

$$f(t) = \frac{\varepsilon\alpha^6}{5!}\tau^5 e^{-\alpha\tau} + (1-\varepsilon)\frac{\alpha^7}{6!}\tau^6 e^{-\alpha\tau} \tag{15}$$

for $t \geq a+b+c$, while it is 0 for $t < a+b+c$, where

$$\tau = t - (a+b+c). \tag{16}$$

---

* See, for example, Cramer (1946), p. 235. The f_1, f_2, and f_3 are the famous χ^2 distributions of degrees of freedom 6, 2, and 6, respectively. The resultant of convolution of the functions given in Eq. (11) is a sum of two χ^2 distributions of degrees of freedom 12 and 14. These functions are tabulated.

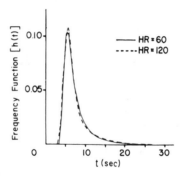

FIGURE 6.20:3. The frequency function, $h(t)$, of transit times in pulmonary circulation in open-chest dogs at constant cardiac output and left atrial pressure, showing the lack of effect of heart rate. Reproduced by permission from Maseri et al. (1970). © American Heart Association, Inc.

The frequency function Eq. (15) is more peaked than $f_1(t)$ and $f_2(t)$. The general case in which $\alpha \neq \beta \neq \gamma$ can be resolved by using partial fractions.

The frequency function of transit times through the lung can be measured by the indicator-dilution method. Tancredi and Zierler (1971) reported that at a given p_{alv} and p_{LA} a family of the density functions of transit times can be transformed to a nearly coincident function, $f(t/\bar{t})$ as described in our Eq. (4). Figure 6.20:3 shows a frequency function of transit times through the whole pulmonary circulation from the pulmonary artery to the left atrium, obtained by Maseri et al. (1970) for the dog. Its general form is in agreement with the theoretic curve. Among other things it shows that the frequency function is not affected by the heart rate or by the presence or absence of respiratory movements.

6.21 Pulmonary Blood Volume

By integrating the sheet thickness over the sheet area, we obtain the vascular volume. The features shown in Figure 6.12:1 and the discussion in Section 6.12 suggest that the alveolar blood volume would be directly related to the pulmonary arterial pressure, whereas the pulmonary venous pressure would have only a minor effect, because any decrease in sheet thickness due to a lowering of p_{ven} is localized to the immediate neighborhood of the venule.

A theoretically predicted pulmonary capillary blood volume as a function of p_{art}, p_{ven}, and p_{alv} is shown in the left-hand panel of Figure 6.21:1, with the constants and geometry pertinent to the dog lung. It is seen that the blood volume varies almost linearly with $p_{art} - p_{alv}$, whereas very little effect is shown by $p_{ven} - p_{alv}$. This result may be compared with the experimental results of Permutt et al. (1969), which are shown on the right-hand side of

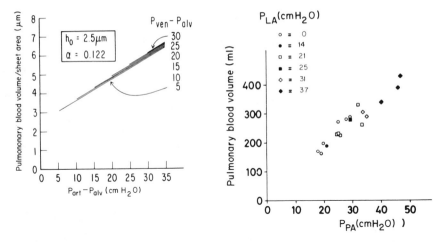

FIGURE 6.21:1. *Left:* Theoretical relationship between the pulmonary alveolar blood volume and the arterial and venous pressures. The ordinate is the pulmonary blood volume per unit area of the sheet. The abscissa is the transmembrane pressure at the arteriole end of the sheet. The transmembrane pressure at the venule end is seen to have only a minor effect on the blood volume. From Figures 4 and 5 of Fung and Sobin (1972b), by permission. *Right:* Experimental results showing the relationship between pulmonary blood volume and pulmonary artery pressure in one dog at a variety of left atrial pressures. From Permutt et al. (1969), by permission.

the figure. The pulmonary blood volume shown here, however, is the total pulmonary blood volume obtained by the indicator-dilution method, not merely the capillary blood volume. But the related work of Permutt et al. (1969) on the steady-state carbon-monoxide diffusing capacity, which is proportional to the capillary blood volume, shows the same trend.

6.22 Pulsatile Blood Flow in the Lung

A typical set of records of pulmonary pressures and flows in a conscious dog is shown in Figure 6.22:1. In the large pulmonary arteries and veins, the Reynolds and Womersley numbers are much larger than 1, and the premises assumed in Chapter 3 apply. Hence the general method of pulse wave analysis presented in Chapter 3 can be used. Extensive theoretical and experimental studies have been reported by Bergel and Milnor (1965), Milnor (1972), Milnor et al. (1966, 1969), and Patel et al. (1963), Pollack et al. (1968), Skalak (1969), Wiener et al. (1966).

For blood flow in pulmonary capillaries, the Reynolds and Womersley numbers are much smaller than 1, so that the inertial forces are much smaller than the forces from pressure and viscous stresses. Hence the equation of motion of the blood in the capillaries is the same as if the flow were

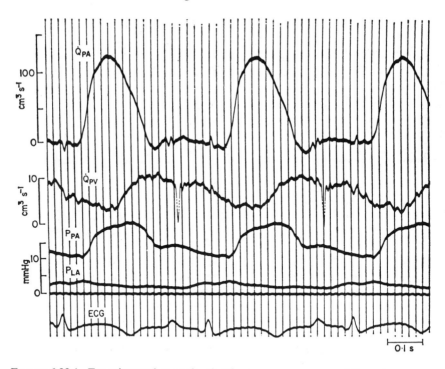

FIGURE 6.22:1. Experimental records of pulmonary pressures and flows in a conscious dog, at rest. $\dot{Q}$ = blood flow; P = pressures; PA = pulmonary artery; PV = pulmonary vein, within 2 cm of the left atrium; LA = left atrium; ECG = electrocardiogram, lead I. Sharp downward spikes in the pulmonary venous flow tracing and smaller spikes in the pulmonary arterial flow during diastole are electrocardiographic signals. Timing lines appear at intervals of 0.02 s. From Milnor (1972). Reproduced by permission.

steady. But the balance of mass is described by Eq. (17) of Section 6.10, p. 377, which depends on the rate of change of sheet thickness $\partial h/\partial t$. The basic equation of sheet flow, Eq. (20) of Section 6.10, involves the time derivative $\partial h/\partial t$ and is not quasisteady. If permeability of water across the endothelium can be neglected, the basic equation is

$$\left(\frac{\partial^2}{\partial x^2}+\frac{\partial^2}{\partial y^2}\right)h^4 = 4\mu k f\alpha\frac{\partial h}{\partial t}. \tag{1}$$

This is a nonlinear differential equation and does not have a harmonic solution with respect to time. Hence, strictly speaking, the usual concept of impedance does not apply. Only in small perturbations can the basic equations be linearized and the concept of impedance be useful. Linearization can be justified if and only if the amplitude of the thickness fluctuations is small compared with the mean pulmonary alveolar sheet thickness. This condition is met if the amplitude of the pressure oscillation is small compared with the mean pulmonary arterial pressure. Under this restriction, the solution of Eq. (1) may be set in the following form:

$$h(x, y, t) = h_{SI}(x, y) + e^{i\omega t} H(x, y). \tag{2}$$

We assume $H(x, y)$ to be much smaller than $h_{SI}(x, y)$, which is the solution of the equation for a steady flow between impervious walls:

$$\left(\frac{\partial^2}{\partial x^2} + \frac{\partial^2}{\partial y^2}\right) h_{SI}^4 = 0. \tag{3}$$

Equation (3) has been treated in Sections 6.11 to 6.13. The subscripts SI stand for steady and impervious. Substituting Eq. (2) into Eq. (1) and retaining only the first powers of H, we obtain the basic equation,

$$\left(\frac{\partial^2}{\partial x^2} + \frac{\partial^2}{\partial y^2}\right) h_{SI}^3 H = \mu k f \alpha \omega i H. \tag{4}$$

Similar to Eq. (2) the pressure and flow per unit width (with components $\dot{q}_x, \dot{q}_y$) can also be represented as the sum of the steady-impervious terms and the oscillatory terms,

$$p(x, y, t) = p_{SI}(x, y) + e^{i\omega t} P(x, y),$$
$$\dot{q}_x(x, y, t) = Uh = \dot{q}_{xSI}(x, y) + e^{i\omega t} \dot{Q}_x(x, y),$$
$$\dot{q}_y(x, y, t) = Vh = \dot{q}_{ySI}(x, y) + e^{i\omega t} \dot{Q}_y(x, y). \tag{5}$$

We introduce the dimensionless frequency parameter:

$$\Omega = \mu k f \alpha \omega L^2 / h_0^3, \tag{6}$$

where L is the length of the sheet and h_0 is the sheet thickness as the transmural pressure tends to zero from positive values. Blood enters each sheet at the arteriole and exits at the venule. Let the oscillatory pressure and flow at the arteriolar and venular edges of the sheet be denoted by P_a, $\dot{Q}_a$, P_v, and $\dot{Q}_v$, respectively. The relationship between these quantities can be expressed in the following matrix form:

$$\left\{\begin{matrix} P_a \\ P_v \end{matrix}\right\} = \left[\begin{matrix} Z_{11} & Z_{12} \\ Z_{21} & Z_{22} \end{matrix}\right] \left\{\begin{matrix} \dot{Q}_a \\ \dot{Q}_v \end{matrix}\right\}, \tag{7}$$

$$\left\{\begin{matrix} \dot{Q}_a \\ \dot{Q}_v \end{matrix}\right\} = \left[\begin{matrix} Y_{11} & Y_{12} \\ Y_{21} & Y_{22} \end{matrix}\right] \left\{\begin{matrix} P_a \\ P_v \end{matrix}\right\}, \tag{8}$$

$$\left\{\begin{matrix} P_a \\ \dot{Q}_a \end{matrix}\right\} = \left[\begin{matrix} A_{11} & A_{12} \\ A_{21} & A_{22} \end{matrix}\right] \left\{\begin{matrix} P_v \\ \dot{Q}_v \end{matrix}\right\}. \tag{9}$$

With P as the analog of voltage and $\dot{Q}$ the analog of current, we can use the terminology of the "four-terminal" network theory and call the Z_{ij} impedances, the Y_{ij} conductances, and the A_{ij} the elements of a *transfer matrix* ($i, j = 1$ and 2). For example, $Z_{11} = P_a/\dot{Q}_a$ when $\dot{Q}_v = 0$; hence it is the impedance between the arterial pressure and arterial flow when oscillation in venous flow is eliminated. $Z_{12} = P_a/\dot{Q}_v$ when $\dot{Q}_a = 0$; hence it is the transfer impedance between arterial pressure and venous flow when there is no fluctuation in arterial flow. Other coefficients can be similarly interpreted. The A matrix is the most useful, because with it the capillary network can be inserted between arteries and veins, thus completing the circuit.

Further details and extension to three-dimensional cases are presented in Fung (1972). Gan and Yen (1994) have made a thorough analysis of pulmonary vascular impedance in the dog. Olman et al. (1994) analyzed the effect of thromboembolism. The books edited by Will et al. (1987), and Wagner and Weir (1994) give a wealth of information and interesting reading on historical perspectives. Other books of importance are Crystal and West (1991), Daly and Hebb (1966), Fishman (1963), Fishman and Hecht (1968), Flokow and Neil (1971), Giuntini (1970), Nagaishi (1972), von Hayek (1960), West (1979, 1977a,b).

6.23 Fluid Movement in the Interestitial Space of the Pulmonary Alveolar Sheet

The basic equation of sheet flow, Eq. (19) of Section 6.10, p. 377, takes into account the permeability of the endothelium and the exchange of fluid between the vascular and tissue spaces. To deal with the problem of pulmonary edema in the interstitium or alveoli, it is necessary to solve this equation. The solution forms part of the general theory of fluid movement in the tissue space treated in Chapters 8 and 9 of the companion volume, *Biomechanics: Motion, Flow, Stress, and Growth*, (Fung, 1990).

The subject of lung water and solute exchange has an extensive literature; see reviews in Fishman (1972), Fishman and Hecht (1968), Giuntini (1970), Staub (1978), Staub and Schultz (1968), and Staub et al. (1967). Compared with the tissue space in other organs, the interstitium of the pulmonary alveolar sheets has the unique feature that its volume is considerably smaller than that of the vascular space. The solution of Eq. (6.10:19) is presented by Fung (1974), and the results corroborate well with the observations of Guyton and Lindsey (1959), Schultz (1959), and Staub et al. (1967).

References

Al-Tinawi, A., Madden, J.A., Dawson, C.A., Linehan, J.H., Harder, D.R., and Rickaby, D.A. (1991). Distensibility of small arteries of dog lung. *J. Appl. Physiol.* **71**: 1714–1722.

Banister, J., and Torrance, R.W. (1960). The effects of the tracheal pressure upon flow: Pressure relations in the vascular bed of isolated lungs. *Q. J. Exp. Physiol.* **45**: 352–367.

Bergel, D.H., and Milnor, W.R. (1965). Pulmonary vascular impedance in the dog. *Circ. Res.* **16**: 401–415.

Bhattacharya, J., Overholser, K., Gropper, M., and Staub, N.C. (1982). Comparison of pressures measured by micropuncture and venous occlusion in Zones III and II of the isolated dog lung. *Fed. Proc.* **41**: 1685 (abstract).

Brody, J.S., Stemmler, E.J., and duBois, A.B. (1968). Longitudinal distribution of vascular resistance in the pulmonary arteries, capillaries, and veins. *J. Clin. Invest.* **47**: 783–784.

Cramer, H. (1946). *Mathematical Method of Statistics*. Princeton University Press, Princeton, NJ.

Crystal, R.G., and West, J.B. (1991). *The Lung, Scientific Foundations*. Raven Press, New York.

Cumming, G., Henderson, R., Horsfield, K., and Singhal, S.S. (1969). The functional morphology of the pulmonary circulation. In *the Pulmonary Circulation and Interstitial Space* (Fishman, A.P., and Hecht, H.H., eds.), University of Chicago Press, Chicago, IL, pp. 327–338.

Dale, P.J., Mathews, F.L., and Schroter, R. (1980). Finite element analysis of lung alveolus. *J. Biomech.* **13**: 856–873.

Daly, I., de B., and Hebb, C. (1966). *Pulmonary and Bronchial Vascular Systems*. Williams & Wilkins, Baltimore, MD.

Debes, J.C., and Fung, Y.C. (1992). Effect of temperature on the biaxial mechanics of excised lung parenchyma of the dog. *J. Appl. Physiol.* **73**: 1171–1180.

Debes, J.C., and Fung, Y.C. (1995). Biaxial mechanics of excised canine pulmonary arteries. *Am. J. Physiol.* **269**: H433–H442.

Fishman, A.P. (1963). Dynamics of the pulmonary circulation. In *Handbook of Physiology*, Sec. 2. *Circulation*, Vol. II (W.H. Hamilton and P. Dow, eds.), American Physiological Society, Washington, D.C., pp. 1667–1743.

Fishman, A.P. (1972). Pulmonary edema: The water exchange function of the lung. *Circulation* **46**: 390–408.

Fishman, A.P., and Hecht, H.H. (eds.) (1968). *The Pulmonary Circulation and Interstitial Space*. University of Chicago Press, Chicago, IL.

Folkow, B., and Neil, E. (1971). *Circulation*. Oxford University Press, New York.

Fung, Y.C. (1969). Studies on the blood flow in the lung. In *Proceedings of the Second Canadian Congress of Applied Mechanics*, Waterloo, Canada, pp. 433–453.

Fung, Y.C. (1972). Theoretical pulmonary microvascular impedance. *Ann. Biomed. Eng.* **1**: 221–245.

Fung, Y.C. (1974). Fluid in the interstitial space of the pulmonary alveolar sheet. *Microvasc. Res.* **7**: 89–113.

Fung, Y.C. (1975a). Does the surface tension make the lung inherently unstable? *Circ. Res.* **37**: 497–502.

Fung, Y.C. (1975b). Stress, deformation, and atelectasis of the lung. *Circ. Res.* **37**: 481–496.

Fung, Y.C. (1975c). 1975 Eugene Landis Lecture: Microcirculation as seen by a red cell. *Microvasc. Res.* **10**: 246–264.

Fung, Y.C. (1988). A model of the lung structure and its validation. *J. Appl. Physiol.* **64**: 2132–2141.

Fung, Y.C. (1989). Connection of micro-and-macromechanics of the lung. In *Microvascular Mechanics: Hemodynamics of Systematic and Pulmonary Microcirculation* (J.S. Lee and T.C. Skalak, eds.), Springer-Verlag, New York. Chapter 13, pp. 191–217.

Fung, Y.C. (1990). *Biomechanics: Motion, Flow, Stress, and Growth*. Springer-Verlag, New York.

Fung, Y.C. (1993). *Biomechanics: Mechanical Properties of Living Tissues*, 2nd ed. Springer-Verlag, New York.

Fung, Y.C., and Liu, S.Q. (1991). Change in the zero-stress state of rat pulmonary arteries in hypoxic pulmonary hypertension. *J. Appl. Physiol.* **70**: 2455–2470.

Fung, Y.C., and Liu, S.Q. (1992). Strain distribution in small blood vessels with zero-stress state taken into consideration. *Am. J. Physiol.* **262**: H544–H552.

Fung, Y.C., and Sobin, S.S. (1969). Theory of sheet flow in lung alveoli. *J. Appl. Physiol.* **26**: 472–488.

Fung, Y.C., and Sobin, S.S. (1972a). Elasticity of the pulmonary alveolar sheet. *Circ. Res.* **30**: 451–469.

Fung, Y.C., and Sobin, S.S. (1972b). Pulmonary alveolar blood flow. *Circ. Res.* **30**: 470–490.

Fung, Y.C., and Sobin S.S. (1977a). Pulmonary alveolar blood flow. In *Bioengineering Aspects of Lung Biology* (J.B. West, ed.), Marcel Dekker, New York, pp. 267–358.

Fung, Y.C., and Sobin, S.S. (1977b). Mechanics of pulmonary circulation. In *Cardiovascular Flow Dynamics and Measurements* (N.H.C. Hwang and N.A. Norman, eds.), University Park Press, Baltimore, MD, pp. 665–730.

Fung, Y.C., and Yen, R.T. (1986). A new theory of pulmonary blood flow in zone 2 condition. *J. Appl. Physiol.* **60**: 1638–1650.

Fung, Y.C., and Zhuang, F.Y. (1986). An analysis of the sluicing gate in pulmonary blood flow. *J. Biomech. Eng.* **108**: 175–182.

Fung, Y.C., Fronek, K., and Patitucci, P. (1979). Pseudoelasticity of arteries and the choice of its mathematical expression. *Am. J. Physiol.* **237**: H620–H631.

Fung, Y.C., Sobin, S.S., Tremer, H., Yen, M.R.T., and Ho, H.H. (1983). Patency and compliance of pulmonary veins when airway pressure exceeds blood pressure. *J. Appl. Physiol.* **54**: 1538–1549.

Gaar, Jr., K.A., Taylor, A.E., Owens, L.-J., and Guyton, A.C. (1967). Pulmonary capillary pressure and filtration coefficient in the isolated perfused lung. *Am. J. Physiol.* **213**: 910–914.

Gan, R.Z., and Yen, R.T. (1994). Vascular impedance analysis in dog lung with detailed morphometric and elasticity data. *J. Appl. Physiol.* **77**: 706–717.

Gan, R.Z., Tian, Y., Yen, R.T., and Kassab, G.S. (1993). Morphometry of the dog pulmonary venous tree. *J. Appl. Physiol.* **75**: 432–440.

Giuntini, C. (ed.) (1970). *Central Hemodynamics and Gas Exchange*. Minerva Medica, Torino, Italy.

Glazier, J.B., Hughes, J.M.B., Maloney, J.E., and West, J.B. (1969). Measurements of capillary dimensions and blood volume in rapidly frozen lungs. *J. Appl. Physiol.* **26**: 65–76.

Guyton, A.C., and Lindsey, A.W. (1959). Effect of elevated left atrial pressure and decreased plasma protein concentration on the development of pulmonary edema. *Circ. Res.* **7**: 649–657.

Hakim, T.S., Michel, R.P., and Chang, H.K. (1982). Partition of pulmonary vascular resistance in dog by arterial and venous occlusion. *J. Appl. Physiol.* **52**: 710–715.

Hakim, T.S., Dean, G.W., and Lisbona, R. (1988a). Quantification of spatial blood flow distribution in isolated canine lung. *Invest. Radiol.* **23**: 498–504.

Hakim, T.S., Dean, G.W., and Lisbona, R. (1988b). Effect of body posture on spatial distribution of pulmonary blood flow. *J. Appl. Physiol.* **64**: 1160–1170.

Hakim, T.S., Lisbona, R., and Dean, G.W. (1987). Gravity-independent inequality in pulmonary blood flow in humans. *J. Appl. Physiol.* **63**: 1114–1121.

Hakim, T.S., Lisbona, R., and Dean, G.W. (1989). Effect of cardiac output on gravity-dependent and nondependent inequality in pulmonary blood flow. *J. Appl. Physiol.* **66**: 1570–1578.

Hansen, J.E., and Ampaya, E.P. (1975). Human air space shapes, sizes, areas, and volumes. *J. Appl. Physiol.* **38**: 990–995.

Hansen, J.E., Ampaya, E.P., Bryant, G.H., and Navin, J.J. (1975). The branching pattern of airways and air spaces in a single human terminal bronchiole. *J. Appl. Physiol.* **38**: 983–989.

Haworth, S.T., Linehan, J.H., Bronikowski, T.A., and Dawson, C.A. (1991). A hemodynamic model representation of the dog lung. *J. Appl. Physiol.* **70**: 15–26.

Horsfield, K. (1978). Morphometry of the small pulmonary arteries in man. *Circ. Res.* **42**: 593–597.

Huang, W., Yen, R.T., McLaurine, M., and Bledsoe, G. (1996). Morphometry of the human pulmonary vasculature. *J. Appl. Physiol.*

Jiang, Z.L., Kassab, G.S., and Fung, Y.C. (1994). Diameter-defined Strahler system and connectivity matrix of the pulmonary arterial tree. *J. Appl. Physiol.* **76**: 882–892.

Johnson, Jr., R.L., Spicer, W.S., Bishop, J.M., and Forster, R.E. (1960). Pulmonary capillary blood volume, flow and diffusing capacity during exercise. *J. Appl. Physiol.* **15**: 893–902.

Kassab, G.S., Rider, C.A., Tang, N.A., and Fung, Y.C. (1993). Morphometry of pig coronary arterial trees. *Am. J. Physiol.* **265**: H350–H365.

Krenz, G.S., Lin, J.M., Dawson, C.A., and Linehan, J.H. (1994). Impact of parallel heterogeneity on a continuum model of the pulmonary arterial tree. *J. Appl. Physiol.* **77**: 660–670.

Lai-Fook, S.J. (1979). A continuum mechanics analysis of pulmonary vascular interdependence in isolated dog lobes. *J. Appl. Physiol. Respirat. Environ. Exerc. Physiol.* **46**: 419–429.

Lamm, W.J.E., Kirk, K.R., Hanson, W.L., Wagner, Jr., W.W., and Albert, R.K. (1991). Flow through zone 1 lungs utilizes alveolar corner vessels. *J. Appl. Physiol.* **70**: 1518–1523.

Lee, J.S. (1969). Slow viscous flow in a lung alveoli model. *J. Biomech.* **2**: 187–198.

Lee, J.S., and Flicker, E. (1974). Equilibrium of forces acting on subpleural alveoli. *J. Appl. Physiol.* **36**: 366–374.

Lee, J.S., and Fung, Y.C. (1968). Experiments on blood flow in lung alveoli models. Paper No. 68-WA/BHF-2, American Society of Mechanical Engineers, New York pp. 1–8.

Liu, S.Q., and Fung, Y.C. (1989). Relationship between hypertension, hypertrophy, and opening angle of zero-stress state of arteries following aortic construction. *J. Biomech. Eng.* **111**: 325–335.

Maloney, J.E., and Castle, B.L. (1969). Pressure-diameter relations of capillaries and small blood vessels in frog lung. *Respir. Physiol.* **7**: 150–162.

Maseri, A., Caldini, P., Permutt, S., and Zierler, K.L. (1970). Frequency function of transit times through dog pulmonary circulation. *Circ. Res.* **26**: 527–543.

Miller, W.S. (1947). *The Lung.* Thomas, Springfield, IL.

Milnor, W.R. (1972). Pulmonary hemodynamics. In *Cardiovascular Fluid Dynamics*, Vol. 2 (D.H. Bergel, ed.), Academic Press, New York, pp. 299–340.

Milnor, W.R., Bergel, D.H., and Bargainer, J.D. (1966). Hydraulic power associated with pulmonary blood flow and its relation to heart rate. *Circ. Res.* **19**: 467–480.

Milnor, W.R., Conti, C.R., Lewis, K.B., and O'Rourke, M.F. (1969). Pulmonary arterial pulse wave velocity and impedance in man. *Circ. Res.* **25**: 637–649.

Miyamoto, Y., and Moll, W.A. (1971). Measurements of dimensions and pathway of red blood cells in rapidly frozen lungs *in situ. Respir. Physiol.* **12**: 141–156.

Nagaishi, C. (1972). *Functional Anatomy and Histology of the Lung.* University Park Press, Baltimore, MD.

Negrini, D., Gonano, C., and Miserocchi, G. (1992). Microvascular pressure profile in intact in situ lung. *J. Appl. Physiol.* **72**: 332–339.

Oldmixon, E.H., Butler, J.P., and Hoppin, Jr., F.G. (1988). Dihedral angle between alveolar septa. *J. Appl. Physiol.* **64**: 299–307.

Olman, M.A., Gan, R.Z., Yen, R.T., Villespin, I., Maxwell, R., Pedersen, C., Konopka, R., Debes, J., and Moser, K.M. (1994). Effect of chronic thromboembolism on the pulmonary artery pressure-flow relationship in dogs. *J. Appl. Physiol.* **76**: 875–881.

Orsos, F. (1936). Die Grüstsystem der Lunge und deren physiologische und pathologische Bedeutung. *Beitr. Klin. Tuberk. Spezif. Tuberk. Forsch.* **87**: 568–609.

Patel, D.J., de Freitas, F.M., and Fry, D.L. (1963). Hydraulic input impedance to aorta and pulmonary artery in dogs. *J. Appl. Physiol.* **18**: 134–140.

Permutt, S., and Riley, R.L. (1963). Hemodynamics of collapsible vessels with tone: The vascular waterfall. *J. Appl. Physiol.* **18**: 924–932.

Permutt, S., Bromberger-Barnea, B., and Bane, H.N. (1962). Alveolar pressure, pulmonary venous pressure, and the vascular waterfall. *Med. Thorac.* **19**: 239–260.

Permutt, S., Caldini, P., Maseri, A., Palmer, W.H., Sasamori. T., and Zierler, K. (1969). Recruitment versus distensibility in the pulmonary vascular bed. In *The Pulmonary Circulation and Interstitial Space* (A.P. Fishman and H.H. Hecht, eds.), University of Chicago Press, Chicago, IL, pp. 375–387.

Pollack, G.H., Reddy, R.V., and Noordergraaf, A. (1968). Input impedance, wave travel, and reflections in the human pulmonary arterial tree: Studies using an electrical analog. *IEEE Trans. Biomed. Eng.* **BME-15**: 151–164.

Purday, H.F.P. (1949). *An Introduction to the Mechanics of Viscous Flow.* Dover, New York, pp. 16–18.

Raj, J.V., and Chen, P. (1986a). Micropuncture measurements of microvascular pressure in isolated lamb lungs during hypoxia. *Circ. Res.* **59**: 398–404.

Raj, J.V., and Chen, P. (1986b). Microvascular pressures measured by micropuncture in isolated perfused lamb lungs. *J. Appl. Physiol.* **61**: 2194–2201.

Roos, A., Thomas, Jr., L.J., Nagel, E.L., and Prommas, D.C. (1961). Pulmonary vascular resistance as determined by lung inflation and vascular pressures. *J. Appl. Physiol.* **16**: 77–84.

Rosenquist, T.H., Bernick, S., Sobin, S.S., and Fung, Y.C. (1973). The structure of the pulmonary interalveolar microvascular sheet. *Microvasc. Res.* **5**: 199–212.

Schultz, H. (1959). *The Submiscroscopic Anatomy and Pathology of the Lung.* Springer-Verlag, Berlin.

Singhal, S., Henderson, R., Horsfield, K., Harding, K., and Cumming, G. (1973). Morphometry of the human pulmonary arterial tree. *Circ. Res.* **33**: 190–197.

Skalak, R. (1969). Wave propagation in the pulmonary circulation. In *The Pulmonary Circulation and Interstitial Space* (A.P. Fishman and H.H. Hecht, eds.), University of Chicago Press, Chicago, IL, pp. 361–373.

Skalak, R., Wiener, F., Morkin, E., and Fishman, A.P. (1966). The energy distribution in the pulmonary circulation. Part I. Theory. *Phys. Med. Biol.* **11**: 287–294; Part II: Experiments. **11**: 437–449.

Smith, J.C., and Mitzner, W. (1980). Analysis of pulmonary vascular interdependence in excised dog lobes. *J. Appl. Physiol.: Respirat. Envir. Exerc. Physiol.* **48**: 450–467.

Sobin, S.S., and Tremer, H.M. (1966). Functional geometry of the microcirculation. *Fed. Proc.* **15**: 1744–1752.

Sobin, S.S., Lindal, R.G., and Bernick, S. (1977). The pulmonary arteriole. *Microvasc. Res.* **14**: 227–239.

Sobin, S.S., Tremer, H.M., and Fung, Y.C. (1970). The morphometric basis of the sheet-flow concept of the pulmonary alveolar microcirculation in the cat. *Circ. Res.* **26**: 397–414.

Sobin, S.S., Fung, Y.C., and Tremer, H.M. (1982). The effect of incomplete fixation of elastin on the appearance of pulmonary alveoli. *J. Biomech. Eng.* **104**: 68–71.

Sobin, S.S., Fung, Y.C., Tremer, H.M., and Rosenquist, T.H. (1972). Elasticity of the pulmonary alveolar microvascular sheet in the cat. *Circ. Res.* **30**: 440–450.

Sobin, S.S., Lindal, R.G., Fung, Y.C., and Tremer, H.M. (1978). Elasticity of the smallest noncapillary pulmonary blood vessels in the cat. *Microvasc. Res.* **15**: 57–68.

Sobin, S.S., Tremer, H.M., Lindal, R.G., and Fung, Y.C. (1979). Distensibility of human pulmonary capillary blood vessels in the interalveolar septa (abstract). *Fed. Proc.* **38**: 990.

Sobin, S.S., Fung, Y.C., Lindal, R.G., Tremer, H.M., and Clark, L. (1980). Topology of pulmonary arterioles, capillaries, and venules in the cat. *Microvasc. Res.* **19**: 217–233.

Stamenovic, D., and Wilson, T.A. (1985). A strain energy function for lung parenchyma. *J. Biomech. Eng.* **107**: 81–86.

Starling, E.H. (1915). *The Linacre lecture on the law of the heart, given at Cambridge, 1915.* Longmans, Green & Co., London, 1918. In *Starling on The Heart* (Chapman, C.B., and Mitchell, J.H., eds.), facs. reprints. Dawson, London, 1965, pp. 119–147.

Staub, N.C. (ed.) (1978). *Lung Water and Solute Exchange.* Marcel Dekker, New York.

Staub, N.C., and Schultz, E.L. (1968). Pulmonary capillary length in dog, cat, and rabbit. *J. Appl. Physiol.* **5**: 371–378.

Staub, N.C., Nagano, H., and Pearce, M.L. (1967). Pulmonary edema in dogs, especially the sequence of fluid accumulation in lungs. *J. Appl. Physiol.* **22**: 227–240.

Tancredi, R., and Zierler, K.L. (1971). Indicator-dilution, flow-pressure and volume-pressure curves in excised dog lung. *Fed. Proc.* **30**: 380 (abstract).

Underwood, E.E. (1970). *Quantitative Stereology.* Addison-Wesley Pub. Co.

Vawter, D.L., Fung, Y.C., and West, J.B. (1979). Constitutive equation of lung tissue elasticity. *J. Biomech. Eng.* **101**: 38–45.

von Hayek, H. (1960). *The Human Lung.* Hefner, New York.

Wagner, Jr., W.W., and Weir, E.K. (1994). *The Pulmonary Circulation and Gas Exchange,* Futura Pub., Armonk, NY.

Wagner, Jr., W.W., Latham, L.P., Gillespie, M.N., and Guenther, J.P. (1982). Direct measurement of pulmonary capillary transit times. *Science* **218**: 379–381.

Warrell, D.A., Evans, J.W., Clarke, R.O., Kingaby, G.P., and West, J.B. (1972). Pattern of filling in the pulmonary capillary bed. *J. Appl. Physiol.* **32**: 346–356.

Weibel, E.R. (1963). *Morphometry of the Human Lung.* Academic Press, New York.

Weibel, E.R. (1973). Morphological basis of alveolar-capillary gas exchange. *Physiol. Res.* **53**: 419–495.

Weiner, D.E., Verrier, R.L., Miller, D.T., and Lefer, A.M. (1967). Effect of adrenalectomy on hemodynamics and regional blood flow in the cat. *Am. J. Physiol.* **213**: 473–476.

West, J.B. (1977a). *Regional Differences in the Lung.* Academic Press, New York.

West, J.B. (ed.) (1977b). *Bioengineering Aspects of the Lung,* Marcel Dekker, New York.

West, J.B. (1979). *Respiratory Physiology—the Essentials.* 2nd ed. Williams & Wilkins, Baltimore, MD.

West, J.B. (1982). *Pulmonary Pathophysiology—the Essentials.* 2nd ed. Williams & Wilkins, Baltimore, MD.

West, J.B., and Dollery, C.T. (1965). Distribution of blood flow and the pressure-flow relations of the whole lung. *J. Appl. Physiol.* **20**: 175–183.

West, J.B., Dollery, C.T., and Naimark, A. (1964). Distribution of blood in isolated lung: Relation to vascular and alveolar pressure. *J. Appl. Physiol.* **19**: 713–724.

West, J.B., Dollery, C.T., Matthews, C.M.E., and Zardini, P. (1965). Distribution of blood flow and ventilation in saline-filled lung. *J. Appl. Physiol.* **20**: 1107–1117.

Wiener, F., Morkin, E., Skalak, R., and Fishman, A.P. (1966). Wave propagation in the pulmonary circulation. *Circ. Res.* **19**: 834–850.

Will, J.A., Dawson, C.A., Weir, E.K., and Buckner, C.K. (eds.) (1987). *The Pulmonary Circulation in Health and Disease,* Academic Press, San Diego.

Wilson, T.A., and Bachofen, H.C. (1982). A model for mechanical structure of the alveolar duct. *J. Appl. Physiol.* **52**: 1064–1070.

Wright, R.R. (1961). Elastic tissue of normal and emphysematous lungs. A tridimensional histologic study. *Am. J. Pathol.* **39**: 355–366.

Yen, R.T. (1988). Elastic properties of pulmonary blood vessels. In *Respiratory Physiology* (H.K. Chang and M. Paiva, eds.), Marcel Dekker, New York, Chapter 14, pp. 533–539.

Yen, R.T. (1989). Elasticity of microvessels in postmortem human lungs. In *Microvascular Mechanics* (J.S. Lee and T.C. Shalak, eds.), Springer-Verlag, New York, Chapter 12, pp. 175–190.

Yen, M.R.T., and Foppiano, L. (1981). Elasticity of small pulmonary veins in the cat. *J. Biomech. Eng.* **103**: 38–42.

Yen, M.R.T., and Fung, Y.C. (1973). Model experiments on apparent blood viscosity and hematocrit in pulmonary alveoli. *J. Appl. Physiol.* **35**: 510–517.

Yen, R.T., and Sobin, S.S. (1988). Elasticity of arterioles and venules in postmortem human lungs. *J. Appl. Physiol.* **64**: 611–619.

Yen, M.R.T., Fung, Y.C., and Bingham, N. (1980). Elasticity of small pulmonary arteries in the cat. *J. Biomech. Eng.* **102**: 170–177.

Yen, R.T., Zhuang, F.Y., Fung, Y.C., Ho, H.H., Tremer, H., and Sobin, S.S. (1983). Morphometry of the cat's pulmonary venous tree. *J. Appl. Physiol.* **55**: 236–242.

Yen, R.T., Zhuang, F.Y., Fung, Y.C., Ho, H.H., Tremer, H., and Sobin, S.S. (1984). Morphometry of the cat's pulmonary arteries. *J. Biomech. Eng.* **106**: 131–136.

Young, J. (1930). Malpighi's "de Pulmonibus." *Proc. Roy. Soc. Med.* **23**, Part 1, 1–14.

Zhuang, F.Y., Fung, Y.C., and Yen, R.T. (1983). Analysis of blood flow in cat's lung with detailed anatomical and elasticity data. *J. Appl. Physiol.: Respirat. Environ. Exerc. Physiol.* **55**: 1341–1348.

Zhuang, F.Y., Yen, M.R.T., Fung, Y.C., and Sobin, S.S. (1985). How many pulmonary alveoli are supplied by a single arteriole and drained by a single venule? *Microvasc. Res.* **29**: 18–31.

7
Coronary Blood Flow

7.1 Introduction

While the heart pumps blood to serve the whole body, the heart muscle itself must be perfused. Yet the perfusion of the heart is unique. Most of the coronary blood vessels are embedded in the myocardium. In every cardiac cycle, the blood in these vessels are squeezed by the muscle cells. The interaction between the muscle cells and the blood vessels dominates the coronary blood flow. None of the individual coronary blood vessels functions as a free tube. The heart muscle cells and the interstitial connective tissues and fluid impose normal and shear stresses and physical constraints on the external surfaces of the coronary blood vessels. Since, however, there is no stress gauge to measure the forces of interaction directly, theoretical analysis plays a major role in assessing the boundary conditions at the external surface of the coronary blood vessel in vivo.

In this chapter we follow the principle of continuum mechanics. The anatomy, histology, and mechanical properties of the materials of the organ are described first. Then mathematical models are formulated so that questions about the functions of the organ can be expressed as boundary-value problems of mathematics. We then solve these problems, predict the outcome, and validate the conclusions against experimental results. If successful, then the model so established will be useful to medicine.

An alternative to the continuum approach is the lumped parameters, analog electric circuit approach. We refer the readers to Spaan's excellent book, *Coronary Blood Flow* (1991), for the lumped-parameters approach. Spaan's book presents a critical review of all aspects of coronary blood flow, including the continuum approach, and an extensive bibliography.

7.2 Morphometry of Coronary Arteries

The topological structure of the coronary arteries is tree-like; that of the coronary capillaries is, however, net-like. As in the case of the lung, we describe the geometry of the tree according to the way it branches, and the geometry of the net according to the way it is knit. The objective of the morphometry of the tree and the net is to use the data for hemodynamics. Each blood vessel has a diameter and a length. According to Poiseuille's formula, for a given blood pressure gradient, the flow varies with the fourth power of the diameter, and the first power of the length. Hence, we give priority to diameter. The branching pattern of the arterial and venous trees is best described by the *diameter-defined Strahler system* which has been described in Section 6.2. In this system, the smallest arteries (arterioles) are designated an order number of 1. When two order 1 vessels meet to generate a confluent vessel, the confluent is called a vessel of order 2 if its diameter is larger than those of order 1 by a certain amount. When an order 2 vessel meets another vessel of order 1, the number of the confluent vessel remains at 2 without change. When two order 2 vessels meet, the confluent vessels is designated an order number of 3, if its diameter is larger than that of the order 2 by a certain amount; and so on. In general, the rule to decide order numbers is as follows. Let the mean and standard deviation of the diameters of the arteries of an arbitrary order n be denoted by D_n and SD_n, respectively. When a vessel of order n meets another vessel of order n or smaller, the confluent vessel's order becomes $n + 1$ if, and only if, its diameter is greater than

$$\left[\left(D_n + SD_n\right) + \left(D_{n+1} - SD_{n+1}\right)\right]\Big/2 . \tag{1}$$

Otherwise its number remains as n. A process of iteration is required to determine the order number of all arteries. The first step is to ignore Eq. (1), and to apply the simpler rule that when a vessel of order n meets another vessel of order n, the confluent vessel is called vessel of order $n + 1$, and that when a vessel of order n meets a vessel of order $n - 1$, the order number of the confluent remains as n. From the result, the mean and standard deviation, D_n and SD_n, respectively, are computed. The next step is to apply Eq. (1) to every node, beginning with $n = 1$, and correct the order numbers whenever necessary. Then D_n and SD_n are calculated again, and the process iterated until convergence is obtained. Normally, one or two cycles of iteration will be sufficient.

Then we define every blood vessel between two successive points of bifurcation as a *segment*. In many instances, several vessel segments of the same order number are connected in series, and function as a longer vessel of the same diameter. Each such combination is called an *element*. In devising circuits for analysis of blood flow in a tree, all elements of a given order are considered as parallel. The relationship between the total number of elements of order n, $E(n)$, and the total number of segments of order number n, $S(n)$, is

$$S(n) = E(n) \cdot R(n) \tag{2}$$

where $R(n)$ is the ratio of the number of segments of order n divided by the number of elements of the same order n.

The way in which vessels of one order number are connected to vessels of other order numbers is represented by a *connectivity matrix* $\mathbf{C}(m, n)$. The component $C(m, n)$ in row m and column n is the ratio of the total number of elements of order m derived from elements of order n, divided by the total number of elements in order n. In hemodynamics, this kind of parental relationship must be known. For example, a steady flow in a rigid vessel element of order m that is derived from a larger vessel of order n is given by Poiseuille's formula:

$$\dot{q}_{mn} = \frac{\pi D_m^4}{128 \mu_m L_m} \left[P_n - P_m \right]. \tag{3}$$

Here, P_m donates the blood pressure at the exit end of blood flow in an element of order m. Since this particular element is connected to a larger vessel of order n, the inlet pressure to this element is the exit pressure of order n, with $n > m$. The number of vessel elements of order m that are connected to vessels of order n is $C(m, n)N_n$ in which N_n is the total number of vessel elements of order number n. The total flow in all vessels of order m is the sum:

$$\dot{Q}_m = \sum_{n>m} \dot{q}_{mn} C(m, n) N_n. \tag{4}$$

The connectivity matrix gives information on how the elements of order $n, n - 1, \ldots 1$, spring directly from the elements of order n. If the total number of elements of order n is N_n, then the total number of elements of order $n, n - 1$, etc., spring from vessels of order n is $C(n, n)N_n, C(n - 1, n)N_n$, etc., respectively. Considering all the vessels in a tree, we see that the total number of elements of order m is given by

$$N_m = \sum_{n=m}^{k} C(m, n) N_n \tag{5}$$

where the summation is from $n = m$ to the highest order of the tree, k. This formula is useful when we use pruning of the tree to reduce the labor of counting from the actual specimen, and to take care of the missing branches from the stubs of branches broken off in the specimen. The number of extrapolated branches are given by

$$N'_m = \sum_{n=m}^{k} C(m, n) \left[N'_n + N_{n,cut} \right] \tag{6}$$

where N'_m and N'_n are the extrapolated number of elements of order m and n, respectively, in the missing subtree, and $N_{n,cut}$ is the number of elements of order n pruned off or broken off.

Further generalization of connectivity matrix is possible. An element of order n may have daughter vessels of order $n - 1, n - 2, \ldots$, branching out at axial locations $x_1, x_2, \ldots$, with branching angles $\beta_1, \beta_2, \ldots$. Statistical data on the axial locations and branching angles are obviously significant to hemodynamics. We may define x as a local coordinate along a vessel element of order n, with $x = 0$ at the end coming from the root (or main trunk) of the tree, and $x = 1$ at the full-length end of the element. We may define the branching angle β as an angle between a vector tangent to the axis of the element of order n in the direction of blood flow to the axial tangent-vector of the daughter vessel. Then a *Longitudinal Fraction Matrix* $\mathbf{L}(m, n)$ may be defined as a square matrix of the mean and standard deviation of the dimensionless longitudinal coordinates x with $m < n$. A *Branching Angle Matrix* $\mathbf{B}(m, n)$ may be defined as a square matrix of the values of the branching angle between element of order n and daughters of order m with $m < n$. One may wish to know the spatial distribution of the branching angles in greater detail. For this purpose, a cylindrical-polar coordinate system may be used. Let $\mathbf{z}$ be a unit vector in the axial direction, $\mathbf{r}$ be a unit radial vector at the branching point (node). Then, as before, let $\mathbf{t}$ be a unit vector tangent to a vessel of order n, and $\mathbf{B}$ be a unit vector tangent to the branch at the node. Then

$$\mathbf{t} \cdot \mathbf{B} = \text{cosine of angle between the vessel and the branch.}$$

$$\mathbf{t} \times \mathbf{B} = \text{normal vector of the plane containing } \mathbf{t} \text{ and } \mathbf{B}.$$

$$(\mathbf{t} \times \mathbf{B}) \cdot \mathbf{r} = \text{cosine of the angle between the normal of the plane}$$
$$\mathbf{t} \times \mathbf{B} \text{ and the radial axis.}$$

$$(\mathbf{t} \times \mathbf{B}) \cdot \mathbf{z} = \text{cosine of the angle between } \mathbf{t} \times \mathbf{B} \text{ and z-axis.}$$

Matrices of these quantities are needed to present the branching angle data.

In practice, this system works because at a constant blood pressure each element of coronary artery is, to a good degree of approximation, a circular cylinder of constant diameter. Diameter changes occur only at branching. When morphometry is done on specimens prepared at the end-diastolic condition and each vessel is assigned an order number, that order number of the vessel is kept throughout the cardiac cycle.

Data obtained by Kassab et al. (1993a, 1994a) from pigs weighing 30.1 ± 0.7 kg are presented below. Data on small arteries of the orders of 1 to 4 were obtained from histological specimens at the end-diastolic condition. Data on larger arteries of order 4 and upwards were obtained from polymer casts of the vasculature at the end-diastolic condition. See original references for details of specimen preparations. Very briefly, the animal was heparinized and under proper anesthesia and midline sternotomy, after an incision was made in the pericardium, the heart was supported in a pericardial cradle and arrested with a saturated KCl solution given through a jugular vein. The coronaries were immediately perfused with a cardioplegic rinsing solution. The heart was then excised and placed in a cold (0°C) saline bath. Three major coronary arteries: the *right coronary artery (RCA), the left anterior descending artery (LAD),* and the *left circumflex artery (LCX),*

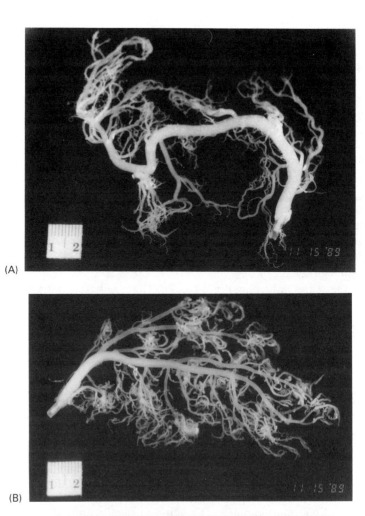

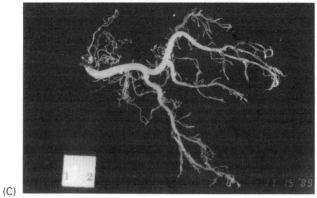

(A)

(B)

(C)

FIGURE 7.2:1. Casts of the three coronary arterial trees of the pig heart. A. The right coronary artery (RCA). B. The left anterior descending artery (LAD). C. The left circumflex artery (LCX). Scale is 1 cm. From Kassab et al. (1993a), by permission.

Figure 7.2:1, were cannulated under saline to avoid air bubbles, perfused first with a cardioplegic rinsing solution, and then perfused with a liquid polymer that was catalyzed with 6.0% stannous 2-ethylhexoate and 3.0% ethyl silicate to solidify in a specific period of time. This polymer has no measurable change of volume upon catalyzing and solidifcation, is nonexothermic in polymerization, is nontoxic to the endothelium, and has a nearly zero contact angle with blood and the vascular endothelium, so that in perfusion it will draw itself into capillary blood vessels by surface tension. The perfusion pressure was maintained at 80 mm Hg until the elastomer was hardened.

For histological studies of small vessels, myocardial tissue was removed from the left and right ventricles of four pigs, mounted on a freezing microtome, cut serially in sections and 60 to 80 micrometers thick. The sections were floated in water, then in ethyl alcohol, finally in numbered vials containing methyl salicylate to render the myocardium transparent and the silicone elastomer-filled microvasculature visible under light microscopy.

For morphometry of vessels of order 4 and above, a cast of the whole tree without capillaries and veins was obtained by mixing 0.1% of the colloidal silica Cab-O-Sil (Kodak Co.) with the catalyzed silicone elastomer to perfuse the arterial bed. The colloidal silica formed agglomerated particles that occluded the capillaries. After the elastomer was hardened, the tissue was corroded with a 30% KOH solution for several days, and washed. The capillaries and veins were washed away, and an arterial tree cast was obtained. These casts were dissected and viewed with a stereo microscope, displayed on a video monitor, measured, and analyzed. Data on larger vessels of order 4 and up were obtained from the casts.

Recognition of arterioles, venules, and capillaries in histological slides is easy because the branching patterns of the arterial and venous trees and the capillaries are very distinctive, as can be seen by comparing Figure 7.2:2 with Figures 7.3:1 and 7.4:1. The smallest vessels, whose topology is not treelike, are capillary blood vessels. Three criteria are used to distinguish capillaries from arterioles: (a) orientation: most capillaries are oriented in the direction of the muscle fiber; (b) tortuosity: arterioles are tortuous, whereas capillaries are linear; and (c) topology: the branching pattern of the capillaries is not treelike.

Branching Pattern According to the Diameter-Defined Strahler System

Figure 7.2:2 shows that the coronary arterioles of the pig bifurcate at almost right angles from the small penetrating arteries and either take an oblique course to the nearest capillary bed or wind around a muscle fiber to proceed to a more distant muscle fiber. The branching of the whole arterial tree was found to be 98% bifurcations and 2% trifurcations, with little arcading or anastomoses.

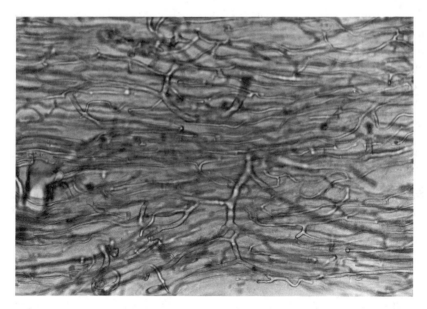

FIGURE 7.2:2. Coronary arterioles (a), venules (v), and capillaries. Photomicrograph of pig left ventricular tissue in a 70-μm-thick section taken 3.9 mm from the epicardial surface. It shows several arterioles and adjacent venules draining into a small vein. From Kassab et al. (1993a), by permission.

There are three types of small precapillary arterioles (order-1 vessels) that feed the capillaries. In one type, an approximately 10-μm arteriole, roughly parallel to a muscle bundle, gives rise to several capillaries. This arteriole gradually tapers in the direction of the muscle fiber while feeding as many as six capillaries along its length until it becomes as small as 8 μm in diameter and terminates in a bifurcation, giving rise to two capillaries. These arterioles can be distinguished from the capillaries by their tortuosity. The arterioles are tortuous, and the capillaries are smooth. The capillaries cannot be as tortuous because they are closely connected to the myocytes through collagen struts (Borg and Caulfield, 1981, see also Section 2.11). In a second type, a feeding tortuous arteriole 8 to 9 μm in diameter courses transversely to the muscle fiber direction and symmetrically gives rise to two capillaries at almost right angles that end up orienting in the direction of the muscle fiber. Finally, a third type of feeding arteriole consists of a combination of the first two types. In this type, a feeding tortuous arteriole may begin parallel to a muscle fiber orientation and eventually end up transverse to the muscle fiber (or vice versa), while giving rise to several capillaries along its length. The percentages of order-1 vessels of the former and latter types are 51%, 23%, and 26% in the right ventricle and 47%, 20%, and 33% in the left ventricle.

For the left ventricle of the pig, vessels less than 8 μm in diameter are capillaries, and arterial vessels greater than 9 μm in diameter are arterioles

TABLE 7.2:1. Diameters and Lengths (Mean ± SD) of Vessel Elements of Each Order in the LAD, LCX, and RCA Arteries of the Pig

	LAD		LCX		RCA	
Order	Diameter (μm)	Length (mm)	Diameter (μm)	Length (mm)	Diameter (μm)	Length (mm)
1	9.0 ± 0.73	0.115 ± 0.066	9.0 ± 0.073	0.115 ± 0.066	9.3 ± 0.84	0.125 ± 0.084
2	12.3 ± 1.3	0.136 ± 0.088	12.3 ± 1.3	0.136 ± 0.088	12.8 ± 1.4	0.141 ± 0.103
3	17.7 ± 2.2	0.149 ± 0.104	17.7 ± 2.2	0.149 ± 0.104	17.7 ± 2.1	0.178 ± 0.105
4	30.5 ± 6.0	0.353 ± 0.154	27.5 ± 6.1	0.405 ± 0.170	28.6 ± 5.4	0.253 ± 0.174
5	66.2 ± 13.6	0.502 ± 0.349	73.2 ± 14.2	0.908 ± 0.763	63.1 ± 11.3	0.545 ± 0.415
6	139 ± 24.1	1.31 ± 0.914	139 ± 26.2	1.83 ± 1.34	132 ± 22.2	1.64 ± 1.13
7	308 ± 56.6	3.54 ± 2.11	279 ± 38.4	4.22 ± 2.26	256 ± 30.1	3.13 ± 2.11
8	462 ± 40.9	4.99 ± 3.02	462 ± 56.1	6.98 ± 3.92	428 ± 47.5	5.99 ± 3.53
9	714 ± 81.8	9.03 ± 6.13	961 ± 193	21.0 ± 15.6	706 ± 75.2	9.06 ± 5.56
10	1,573 ± 361	20.3 ± 17.9	2,549	47.5	1,302 ± 239	16.1 ± 13.3
11	3,171	45.9			3,218	78.1

LAD = left anterior descending artery; LCX = circumflex artery; RCA = right coronary artery. Data from Kassab et al. (1993a).

of order 1 or higher. In the 8 to 9 μm diameter range, 30% are found to be capillaries and 70% are order-1 arterioles. For the right ventricle, in the 8 to 9 μm diameter range, 32% are found to be capillaries and 68% are order-1 arterioles.

Table 7.2:1 shows the mean ± SD of the diameter and length for the elements of the LAD, LCX, and RCA arteries of the pig. A total of 11 orders of vessels lie between the coronary capillaries and the aortic sinus.

Figure 7.2:3 shows the relationship between the mean vessel diameter and the order number of the elements of the RCA, LAD, and LCX arteries. Figure 7.2:4 shows the relationship between the mean vessel element length

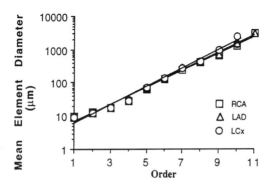

FIGURE 7.2:3. The average diameter of coronary arterial vessel elements of successive orders of the right coronary artery (RCA), left anterior descending artery (LAD), and left circumflex artery (LCX) of the pig. From Kassab et al. (1993a), by permission.

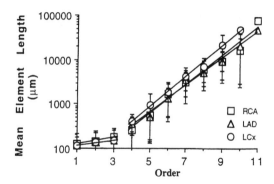

FIGURE 7.2:4. The average length of vessel elements in successive orders of coronary arteries in pig RCA, LAD, and LCX. See Figure 7.2:3 for abbreviations. From Kassab et al. (1993a), by permission.

and the order number for the RCA, LAD, and CX arteries. The curves in Figure 7.2:3 can be fitted by the equation

$$\log_{10} D_n = a + bn, \tag{7}$$

where a and b are two constants. With the method of least-squares, the empirical constants a and b are determined. The values of a, b and the correlation coefficient of the fitted curve, R^2, are listed in Table 7.2:2. Applying Eq. (7) first to D_n, and then to D_{n-1}, and subtracting, we obtain

$$\log_{10}\left(D_n/D_{n-1}\right) = b, \quad \text{or} \quad \frac{D_n}{D_{n-1}} = 10^b. \tag{8}$$

This shows that the ratio of the diameters of successive orders of arteries is a constant independent of n. The ratio, 10^b, is called the *diameter ratio*. Equations (7) and (8) are known as Horton's law. A system obeying such a law is said to be *fractal*.

TABLE 7.2:2. Empirical Constants a and b of the Eqs. (6), (9), and (10) for the Relationships Between Order Number and Mean Diameter, Length, and Number of Vessel Elements for the RCA, LAD, and LCX Arteries of the Pig

	Diameter			Length: Orders 1–3			Length: Orders 4–11			Total number		
	RCA	LAD	CX	RCA	LAD	CX	RCA	LAD	CX	RCA	LAD	CX
a	0.556	0.549	0.499	2.01	2.01	2.01	1.13	1.28	1.22	6.35	6.32	5.95
b	0.259	0.264	0.277	0.077	0.056	0.056	0.327	0.305	0.342	0.544	0.545	0.553
R^2	0.991	0.992	0.986	0.967	0.972	0.972	0.975	0.990	0.995	0.991	0.992	0.986
Ratio*	1.81	1.84	1.89	1.19	1.14	1.14	2.12	2.02	2.20	3.50	3.51	3.57

* D_n/D_{n-1}, L_n/L_{n-1}, N_{n-1}/N_n represent the diameter, length, and number ratios, respectively. See Table 7.2:1 for abbreviations. Data from Kassab et al. (1993a).

FIGURE 7.2:5. The total numbers of vessel elements in successive orders of coronary arteries in pig RCA, LAD, and LCX. See Figure 7.2:3 for abbreviations. From Kassab et al. (1993a), by permission.

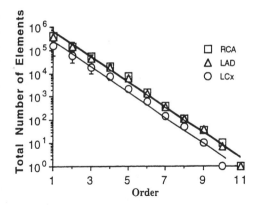

The mean element length also obeys Horton's law, but with a discontinuity in the slope at order 3 (see Fig. 7.2:4). The regression line between the element length and order number can be written as

$$\log_{10} L_n = a + bn, \tag{9}$$

and the correlation is excellent. The data on a, b and the correlation coefficient, R^2, are listed in Table 7.2:2. The ratio of the lengths of the elements of successive order numbers, L_n/L_{n-1}, again has an average value 10^b, which is called the *length ratio*. Table 7.2:2 shows that the diameter ratios of the elements of RCA, LAD, and LCX are 1.81, 1.84, and 1.89, respectively, whereas the length ratios of the elements of RCA, LAD, and LCX are 2.12, 2.02, and 2.20, respectively, for orders 4 to 11, but 1.19, 1.14, and 1.14 respectively for orders 1 to 3, respectively.

The connectivity matrix of the left anterior descending arterial tree (LAD) is given in Table 7.2:3. Connectivity matrices for LCX and RCA are given in Kassab et al. (1993a).

The total number of arterial elements of each order can be computed from information given in the preceding tables. The results, means expressed as values ± propagated errors of the mean, are presented in Table 7.2:4. When the number of elements of order n, N_n, is plotted against n on semilog paper, the result is shown in Figure 7.2:5. The regression line is again a straight line. Thus, an equation similar to Eq. (5) applies:

$$\log_{10} N_n = a - bn. \tag{10}$$

The constants a and b obtained by least-squares fitting of the data are presented in Table 7.2:2. The ratio of the numbers of elements N_{n-1}/N_n is 10^b, and is called the *numbers ratio*. It is listed in Table 7.2:2.

From these data we can compute the *total cross sectional area* (CSA), A_n, of all vessel elements of order n, which is equal to the product of the area of each element and the total number of elements. Assuming the vessel cross sections to be circular, we have

TABLE 7.2:3. Connectivity Matrix of Pig Left Anterior Descending Artery: An Element $C(m,n)$ in the mth Row and nth Column is the Ratio of the Total Number of Elements of Order m that Spring Directly from Parent Elements of Order n Divided by the Total Number of Elements of Order n

	1	2	3	4	5	6	7	8	9	10	11
0	3.18 ± 0.118	0.675 ± 0.080	0.148 ± 0.081	0	0	0	0	0	0	0	0
1	0.144 ± 0.031	2.04 ± 0.070	0.630 ± 0.116	0.071 ± 0.071	0	0	0	0	0	0	0
2		0.094 ± 0.027	2.24 ± 0.102	1.50 ± 0.274	0.063 ± 0.025	0.094 ± 0.019	0.023 ± 0.010	0	0	0	0
3			0.074 ± 0.036	2.14 ± 0.204	0.381 ± 0.056	0.098 ± 0.018	0.120 ± 0.028	0.092 ± 0.029	0.030 ± 0.030	0	0
4				0.143 ± 0.097	2.25 ± 0.097	0.425 ± 0.042	0.380 ± 0.047	0.428 ± 0.070	0.303 ± 0.102	0.167 ± 0.167	0
5					0.238 ± 0.035	2.25 ± 0.097	1.91 ± 0.111	1.56 ± 0.173	1.36 ± 0.281	0.667 ± 0.333	0
6						0.155 ± 0.022	2.50 ± 0.080	1.58 ± 0.159	1.48 ± 0.323	1.17 ± 0.654	0
7							0.116 ± 0.022	2.09 ± 0.106	1.30 ± 0.300	2.00 ± 1.09	2
8								0.061 ± 0.024	2.50 ± 0.209	2.50 ± 1.06	2
9									0.121 ± 0.058	3.33 ± 1.12	9
10										0.100 ± 0.100	5
11											0

Values are the mean ± SE. Data from Kassab et al. (1993a).

TABLE 7.2:4. Total Number of Vessel Elements in Each Order of the Right Coronary Artery (RCA), Left Anterior Descending Artery (LAD), Left Circumflex Artery (LCX) and the Whole Heart of the Pig

Order	RCA Number of vessel elements	LAD Number of vessel elements	LCX Number of vessel elements	Whole heart, RCA, and LCCA Number of vessel elements
11	1	1		2
10	10	7	1	18
9	35	37 ± 2	10	83 ± 2
8	114 ± 1	113 ± 9	51	283 ± 11
7	403 ± 5	348 ± 32	144 ± 4	909 ± 44
6	1,458 ± 44	1,385 ± 162	638 ± 51	3,524 ± 247
5	7,354 ± 649	6,386 ± 11,052	2,148 ± 312	16,093 ± 2,117
4	20,074 ± 3,739	17,985 ± 5,676	7,554 ± 2,338	46,194 ± 12,089
3	51,915 ± 13,644	44,456 ± 19,672	17,820 ± 8,001	115,638 ± 42,301
2	138,050 ± 46,070	140,293 ± 72,949	56,915 ± 29,829	339,873 ± 152,326
1	393,294 ± 158,657	368,554 ± 221,134	149,386 ± 90,276	923,339 ± 480,169

Values are the mean ± SE. Data from Kassab et al. (1993a). The *left common coronary artery* (LCCA) bifurcates into LAD and LCX.

$$A_n = \left(\pi/4\right)D_n^2 N_n \qquad (11)$$

The results are plotted in Figure 7.2:6 and fitted by fifth-order polynomials. We can compute also the *total blood volume*, V_n, in all vessel elements of order n,

$$V_n = L_n A_n \qquad (12)$$

The results are shown in Figure 7.2:7. The standard errors of A_n and V_n are computed from those of D_n, N_n, L_n, A_n according to differentials of A_n and V_n given by Eqs. (11) and (12). These data are for a pig weighing 30.1 ± 0.7 kg, heart weight 150 g, left ventricle weight 85 g.

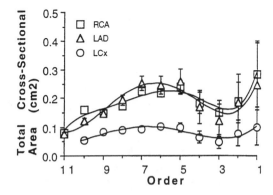

FIGURE 7.2:6. The total cross-sectional area of arterial elements in successive orders of coronary arteries in pig RCA, LAD, and LCX. See Figure 7.2:3 for abbreviations. From Kassab et al. (1993a), by permission.

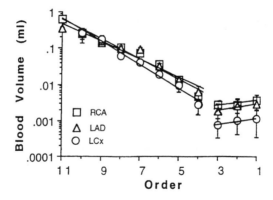

FIGURE 7.2:7. The total blood volume of arterial elements in successive orders of coronary arteries in pig RCA, LAD, and LCX. See Figure 7.2:3 for abbreviations. From Kassab et al. (1993a), by permission.

7.3 Coronary Veins

Sinusal and Thebesian Veins

There are two routes by which coronary venous flow returns to the heart. In one route, blood flows from the great cardiac vein, the posterior vein of the left ventricle, the posterior interventricular vein, the oblique vein of Marshal, and the small cardiac vein into the coronary sinus and then to the right atrium. The anterior cardiac veins on the epicardial surface empty directly into the right atrium. In another route, blood flows through the small cardiac veins of Thebesius to the endocardial surface, and drains directly into the heart chambers. Detailed anatomical studies of these coronary veins began in the 19th century (Mohl et al., 1984).

Unique Geometry of the Coronary Venules

Using the same method of animal and histological preparation described in Section 7.2, Kassab et al. (1994) obtained the photographs of the endocardial venules shown in Figure 7.3:1. The smallest venules initially lie in the direction of the capillaries that they drain, then usually break away to run obliquely toward larger veins. The characteristic branching pattern of endocardial venules has been referred to as a turnip-root pattern, ginger-root pattern, or as fingers of a hand. In Figure 7.3:1, several adjacent venules can be found draining into a nearby vein. This pattern is very different from that of the coronary arterioles, as shown in Figure 7.2:2. Hence, it is not difficult to identify the coronary arterioles, venules, and capillaries in histological specimens. Kassab et al. (1994) obtained morphometric data on venules of order 1–4 from four hearts and on veins of order 4–11 from one cast.

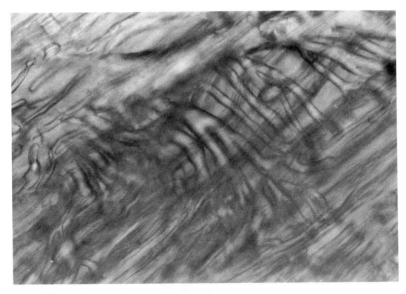

FIGURE 7.3:1. A photomicrograph of a venule in pig left ventricle in an 80-μm-thick section taken 4.2 mm from the epicardial surface, showing several adjacent venules draining into a small vein. 1 cm in photo = 42 μm in specimen. From Kassab et al. (1994a), by permission.

Shape of Coronary Veins

Figure 7.3:2 shows sketches of several unusual endocardial venous geometries. Trifurcations, quadrifications, and quintifications are more frequent in venules than arterioles. The branching of the whole arterial tree was found to be 98% bifurcations and 2% trifurcations; that of venous trees was found to be 86% bifurcations, 12.8% trifurcations, 1% quadrifications, and 0.2% quintifications. The branching pattern of the endocardial venules, however, is strictly treelike, lacking arcading.

Morphometry of the venules was performed through optical sectioning. The normal cross section of the lumen of each venule at low perfusion pressures is noncircular and roughly elliptical. An image processing system was used to measure the dimensions. We call the length of the major axis of the lumen cross section the *diameter* and the length of the minor axis of the approximating ellipse the *thickness*. Both were measured.

Data on small veins of orders 1 to 4 were obtained from histological slides. Data on larger veins of order 4 and up were obtained from hardened elastomer casts of the veins after the tissue was corroded with KOH and washed away. The arteries were carefully pruned away, leaving venous trees with clusters of capillaries. The venous casts were viewed and dissected under a stereo dissecting microscope. The images were measured in computer, the tree was sketched, and the structure was reconstructed.

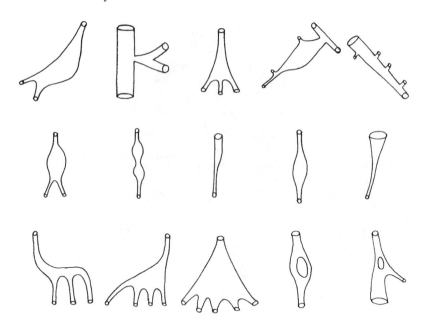

FIGURE 7.3:2. Schematic reconstruction of some coronary venous patterns. From Kassab et al. (1994a), by permission.

Two protocols were used to measure the thickness or minor axis of the lumen of the venous vessels. Either a cast vessel was cut transverse to its longitudinal axis and obtained its cross section for measurement, or rotated the vessel specimen in the petri dish on the microscope stage to visualize the thickness of the vessel. The two methods gave similar results.

Diameter-Defined Strahler System

From the data on the diameter of each segment and the sketch of the complete tree, order numbers were assigned to the venous vessels in the same way as for the arterial trees. Negative integers $(-1, -2, -3, \ldots, -n, \ldots)$ were used to designate the order numbers of the veins. The larger the absolute value of the order number, the larger is the diameter of the vein. Equations (1) to (8) of the preceding Section 7.2 are applicable to the veins, with n replaced by $-n$.

Sets of the morphometric data of the venules of the right and left ventricles, excluding the epicardial venules, are combined together since no statistical differences were found in the venular morphometric characteristics of the two ventricles. Table 7.3:1 shows the experimental results of the mean $\pm$ SD of the major axes, the major-to-minor axes ratios, and the length for the elements of sinusal and Thebesian veins. Note that the data of the smallest venules of orders $-1, -2, -3$ were obtained from histological slides in

TABLE 7.3:1. Major Axis, Length, and Major-to-Minor Axis Ratio of Vessel Elements in Each Order of Vessels in Pig Coronary Veins

	Sinusal			Thebesian		
Order	D, μm	L, mm	D/minor axis	D, μm	L, mm	D/minor axis
-1	10.6 ± 1.6	0.079 ± 0.054	1.32 ± 0.047	10.6 ± 1.6	0.079 ± 0.054	1.32 ± 0.047
-2	16.5 ± 2.7	0.092 ± 0.065	1.38 ± 0.082	16.5 ± 2.7	0.092 ± 0.065	1.38 ± 0.082
-3	29.6 ± 3.2	0.117 ± 0.71	1.48 ± 0.100	29.6 ± 3.2	0.117 ± 0.071	1.48 ± 0.100
-4	57.5 ± 11.8	0.350 ± 0.277	1.70 ± 0.060	54.8 ± 11.5	0.384 ± 0.327	1.70 ± 0.060
-5	117 ± 18.7	0.698 ± 0.649	1.83 ± 0.05	111 ± 17.2	0.684 ± 0.580	1.83 ± 0.052
-6	205 ± 25.6	1.26 ± 1.13	1.91 ± 0.069	189 ± 22.5	1.25 ± 1.11	1.91 ± 0.069
-7	317 ± 32.7	2.08 ± 1.87	1.96 ± 0.058	292 ± 30.6	2.44 ± 1.99	1.96 ± 0.058
-8	488 ± 46.5	3.62 ± 3.19	1.90 ± 0.101	549 ± 53.5	4.21 ± 3.36	1.90 ± 0.101
-9	773 ± 62.4	6.07 ± 5.11	1.82 ± 0.110	813 ± 55.6	6.23 ± 5.51	1.82 ± 0.110
-10	1,165 ± 88.2	10.2 ± 8.32	1.68 ± 0.148	1,184 ± 230	9.35 ± 6.67	1.68 ± 0.148
-11	1,804 ± 464	25.8 ± 24.2	1.45 ± 0.053			
-12	5,919	71.9	1.25 ± 0.145			

D, major axis (means ± SD); L, length of segment (means ± SD); D/minor axis, major-to-minor axis ratio (means ± SE). Data from Kassab et al. (1994a).

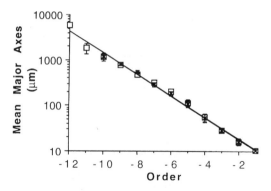

FIGURE 7.3:3. The average major axis length of the elements of coronary veins of successive orders of sinusal (□) and Thebesian (x) veins of the pig. Sinusal (□): $a = 0.803$, $b = 0.236$, $R^2 = 0.991$. Thebesian (x): $a = 0.795$, $b = 0.237$, $R^2 = 0.995$. From Kassab et al. (1994a), by permission.

which it was not possible to distinguish sinusal from Thebesian vessels. Hence, it is assumed that there is no difference between those two systems at orders of −1, −2, and −3. There are a total of 12 orders of veins in the sinusal system, whereas there are only 10 orders of veins in the Thebesian system.

The experimental results are presented in Figures 7.3:3 to 7.3:5. In Figure 7.3:3 the average major axis length of vessel elements of each order is plotted against the order number of the veins. Figure 7.3:4 presents similar

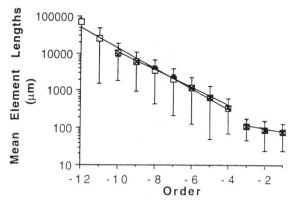

FIGURE 7.3:4. The mean and standard deviation of the length of the elements of coronary veins of successive orders of sinusal (□) and Thebesian (x) veins of the pig. Sinusal (□): $a = 1.80$, $b = 0.085$, $R^2 = 0.984$ for $-n = -1$ to -3; $a = 1.44$, $b = 0.271$, $R^2 = 0.986$ for $-n = -4$ to -12. Thebesian (x): $a = 1.80$, $b = 0.085$, $R^2 = 0.984$ for $-n = -1$ to -3; $a = 1.68$, $b = 0.236$, $R^2 = 0.992$ for $-n = -4$ to -12. From Kassab et al. (1994a), by permission.

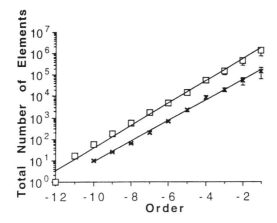

FIGURE 7.3:5. The total number of the venous elements in successive orders of sinusal (□) and Thebesian (x) veins of the pig. Sinusal (□): $a = 6.85$, $b = 0.528$, $R^2 = 0.990$. Thebesian (x): $a = 5.77$, $b = 0.485$, $R^2 = 0.997$. From Kassab et al. (1994a), by permission.

data for the length of the elements. Figure 7.3:5 shows the number of the elements. Numerical data on mean ± standard deviation are given in Kassab et al. (1994a).

Horton's law described in Eqs. (7), (9), and (10) of Section 7.2 was found to be valid for veins with a change of n to $-n$. Thus, the curves in Figures 7.3:3 to 7.3:5 can be fitted to equations

$$\log_{10} D_{-n} = a + b(-n), \quad \log_{10} L_{-n} = a + b(-n),$$

$$\log_{10} N_{-n} = a + bn, \tag{1}$$

where D is the diameter, L is the length, N is the number of vessels, and $-n$ is the order number. The empirical constants a and b for the coronary veins are listed in the legends of Figures 7.3:3 to 7.3:5. The curve fitting can be improved if a trigonomic term is added to the right-hand side of Eq. (1); for example,

$$\log_{10} D_{-n} = a + b(-n) + c \cos\left[\pi(-n)/d\right], \tag{2}$$

where c and d are two additional empirical constants. The physical meaning of c is the amplitude of oscillatory deviation from Horton's law, and d is the wave length of oscillation in terms of order numbers. For the length equation, the constants a, b for orders -1, -2, and -3 are different from a, b for orders -4, -5, -6, -7, -8, -9, and -10. The ratios D_{-n}/D_{-n+1}, L_{-n}/L_{-n+1}, and N_{-n}/N_{-n+1} are called the *major axis ratio*, the *element length ratio*, and the *numbers ratio*, respectively. Their values are given by the respective 10^b. The major axis ratios of sinusal and Thebesian veins are 1.69 and 1.65, respectively; whereas the element length ratios of sinusal and Thebesian veins are

1.77 and 1.61, respectively, for veins of orders −4, and −5 and above, but 1.18 and 1.18 for orders −1, −2, and −3.

The connectivity matrices of sinusal and Thebesian veins are given in Table 7.3:2.

These data can be used to compute the *total cross-sectional area* (CSA) and the *blood volume in the coronary veins* of each order. If the cross section of the vein is assumed to be elliptical, then A_{-n} is equal to the product of the area of each elliptic element and the total number of elements:

$$A_{-n} = \left[(\pi/4)D_{-n}^2 N_{-n}\right] \Big/ \left(\text{major axis}/\text{minor axis}\right)_{-n} \tag{3}$$

TABLE 7.3:2. Connectivity Matrix $\mathbf{C}(m, n)$ for Coronary Veins of Pig

	Sinusal veins											
Order	−1	−2	−3	−4	−5	−6	−7	−8	−9	−10	−11	−12
0	2.56	0.426	0.347	0.033	0.012	0.020	0.006	0	0	0	0	0
−1	0.105	2.47	0.773	0.315	0.104	0.103	0.089	0.064	0.066	0.028	0	0
−2		0.109	2.44	1.01	0.338	0.325	0.303	0.345	0.209	0.083	0	0
−3			0.067	2.17	0.808	0.649	0.689	0.635	0.505	0.194	0.308	0
−4				0.140	2.30	1.97	2.28	2.64	2.64	1.42	1.77	0
−5					0.097	2.01	1.81	2.08	2.15	1.64	2.38	1
−6						0.168	1.71	1.63	1.60	0.861	2.77	1
−7							0.164	1.86	1.85	1.22	2.08	0
−8								0.188	1.64	1.33	2.54	1
−9									0.231	1.33	2.38	3
−10										0.167	2.15	9
−11											0.462	8
−12												0

	Thebesian veins									
Order	−1	−2	−3	−4	−5	−6	−7	−8	−9	−10
0	2.56	0.426	0.347	0.067	0.015	0.020	0.014	0.020	0	0
−1	0.105	2.47	0.773	0.435	0.152	0.130	0.055	0.078	0.042	0
−2		0.109	2.44	0.854	0.460	0.311	0.324	0.392	0.458	0.600
−3			0.067	2.03	0.668	0.544	0.641	0.608	0.458	0.600
−4				0.300	2.01	1.76	2.15	2.49	2.25	2.80
−5					0.181	1.84	1.91	2.18	1.83	2.50
−6						0.203	1.90	1.84	1.54	2.20
−7							0.234	1.53	1.37	1.50
−8								0.216	1.37	1.40
−9									0.167	1.90
−10										0.417

Data from Kassab et al. (1994a). Standard errors are given in the original paper.

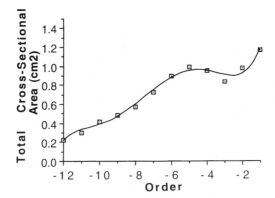

FIGURE 7.3:6. The calculated total cross-sectional area of the coronary venous elements in each order of coronary veins of a pig weighing 30.1 ± 0.7 kg (heart weight 150 g). From Kassab et al. (1994a), by permission.

which is a formula derived in Kassab et al. (1994a). The total blood volume in all elements of a given order, V_{-n}, is given by

$$V_{-n} = L_{-n}A_{-n}. \tag{4}$$

The results of the total cross section area and blood volume calculations are shown in Figures 7.3:6 and 7.3:7, respectively. The cumulative volume of the whole venous tree is the summation of V_{-n} over $-n$ from -1 to -12.

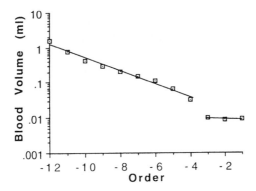

FIGURE 7.3:7. The calculated total blood volume of the venous elements in successive orders of coronary veins of a pig weighing 30.1 ± 0.7 kg (heart weight 150 g). From Kassab et al. (1994a), by permission.

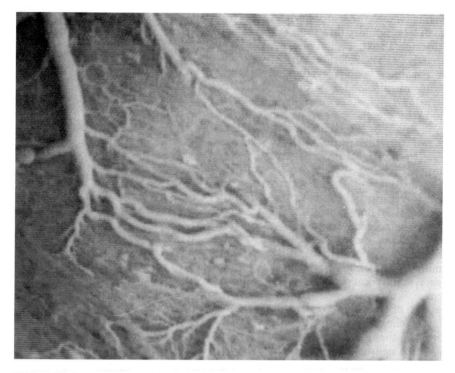

FIGURE 7.3:8. A photomicrograph of venous anastomoses at epicardial surface of the pig. An intravital view of a radiopaque Microfil perfused coronary sinusal veins. From Kassab et al. (1994a), by permission.

Arcading and Anastomoses

Many arcading venules, that is, interconnections between parts of the same vein, are found within the epicardial surface of the pig heart (see Fig. 7.3:8). Arcading veins may have multiple feeding vessels but a single draining vessel. There also exist anastomoses between different veins. Venous anastomoses, however, are not restricted to the epicardial surface. They may be endocardial, and may connect Thebesian to sinusal veins. Anastomotic veins may have multiple feeding vessels and two draining vessels. These arcades and anastomoses may serve to homogenize the local flow field. Data on their sizes and connection to the venous trees are given in Kassab et al. (1994a).

Distribution of Coronary Blood Volume and Contrast Between Arteries and Veins

In summary, the coronary venous system of the pig has a treelike branching pattern, except at the epicardial surface, where arcades and anastomoses are found connecting the sinusal veins, and at the endocardial surface, where arcades are found connecting Thebesian veins. The diameters of the first several orders of veins are larger than those of the corresponding arteries, the lengths are shorter, and the numbers are greater. These features imply that the venous system is a lower resistance system.

The total coronary blood vessel cross-sectional area (CSA) and blood volume vary throughout the cardiac cycle. For a maximally vasodilated relaxed myocardium, the total CSA and blood volume of coronary blood vessel elements of each order in which the arterial inlet pressure was 80 mm Hg and the venous outlet pressure was zero (atmospheric) are shown in Figures 7.2:6, 7.2:7, 7.3:6, and 7.3:7. It can be seen that the total CSAs of various orders of veins are larger than those of the corresponding orders of arteries. For a 150 g heart, the cumulative arterial volume is 3.0 ml, whereas the cumulative venous volume (sinusal and Thebesian trees, arcades, and anastomoses) is 4.1 ml in diastole. In diastole the Thebesian veins contain 5% of the volume of sinusal veins. These values are in agreement with our experimental measurements of the volumes of the venous casts, arterial casts, and the cast volume of the entire coronary vasculature of the pig, which yielded a capillary blood volume of 3.3 ml. The mass normalized coronary blood volume in the arterial, capillary, and venous compartments is 3.5, 3.8, and 4.9 ml/100 g of left ventricle), respectively. The total mass normalized coronary blood volume is therefore 12.2 ml/100 g LV. The mean blood volumes reported for the coronary vasculature range from 4.8 to 14 ml/100 g LV for various species (see critical review by Spaan, 1985). The lower values are underestimates since those hearts were blotted before determination of blood volume, implying that some of the blood volume from the larger arteries and veins was lost.

The pig heart has arcading veins at the epicardial and endocardial surfaces, but no arcading arteries. In other organs (Johnson, 1978) arcading veins are accompanied by arcading arteries. The dog heart, for instance, does contain arcading arteries along with arcading veins at the epicardial surface.

7.4 Coronary Capillaries

Topologically, the coronary arteries and veins are treelike, but the coronary capillaries are not. The coronary capillaries form a network with certain special types of connections among its branches, and a special three-dimensional relationship with respect to arteries and veins. Papers by Anderson et al. (1988), Anversa and Capasso (1991), Bassingthwaighte et al. (1974), Brown (1965), Ludwig (1971), Nussbaum (1912), Rakusan and Spalteholz (1907), Wearn (1928), Wicker (1990), and others are listed in the references at the end of the chapter and will be compared with some newer data.

In Chapter 5, capillary blood vessels are characterized by their small size, their structural lack of smooth muscle cells, and their function as the site of

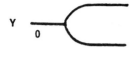

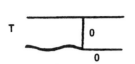

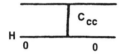

FIGURE 7.4:1. Schematic diagram of capillary segment branching showing the typical geometric patterns known as Y, T, HP (hairpin), and H types. 0, capillaries of order 0; C_{cc}, the capillary cross-connection (like the middle bar of the letter H). From Kassab and Fung (1994), by permission.

mass exchange between blood and tissue. In coronary morphometry, we recognize them simply by their non-treelike topology and their position between terminal arterioles and venules. We have seen the shapes of the terminal arterioles and venules of the pig heart in Figures 7.2:2 and 7.3:1. Hence, it is easy to recognize the coronary capillaries as being smooth and lying in the direction of muscle fibers.

Capillary Branching Pattern and Types of Nodes

The non-treelike pattern of branching of pig coronary capillaries is shown in Figure 7.4:1. The nodes can be classified into four types: Y, T, H, and hairpin (HP) based on their geometric shape. The relative frequencies of H, Y, T, and HP types in the right ventricle of the pig are 0.52, 0.21, 0.21, and 0.06, respectively, whereas those in the left ventricle of the pig are 0.53, 0.21, 0.20, and 0.06, respectively.

Capillary Vessel Segment Statistics

Because arteries are given order numbers 1, 2, 3, ... according to their diameters, and veins are given the order number −1, −2, −3, ..., we shall give capillaries an order number of 0. A capillary between two successive nodes is called a *segment*. Capillary segments attached to arterioles are designated as C_{0a} vessels. Capillary segments attached to venules are designated as C_{0v} vessels. The C_{0a} and C_{0v} capillary vessels branch further in

TABLE 7.4:1. Segment Diameters and Lengths of Pig Coronary Capillaries in Left Ventricle and Right Ventricle Free Walls

Capillary order	LV free wall		RV free wall	
	Diameter (μm)	Length (μm)	Diameter (μm)	Length (μm)
C_{0a}	6.2 ± 1.1	52.0 ± 32.3	6.5 ± 1.0	55.4 ± 40.3
C_{00}	5.7 ± 1.2	54.5 ± 43.0	6.0 ± 1.1	62.5 ± 41.2
C_{0v}	7.0 ± 1.2	45.0 ± 30.5	6.9 ± 1.2	47.5 ± 29.5
C_{cc}	5.5 ± 1.4	21.1 ± 15.5	5.7 ± 1.3	33.4 ± 28.3

Values are means ± SD. RV, right ventricle. LV, left ventricle. All capillaries are order 0: C_{0a}, those fed directly by arterioles; C_{0v}, those drained directly into venules; C_{00}, those connecting C_{0a} and C_{0v} vessels; C_{cc}, capillary cross-connection. Data from Kassab and Fung (1994). The numbers of vessels measured are given in the original paper.

patterns shown in Figure 7.4:1. We designate the large number of other capillary segments that connect to C_{0a} and C_{0v} vessels as C_0 vessels. Some C_0 capillary segments appear to be extensions of C_{0a} and C_{0v} connected in series. Other C_0 vessels that anastomose in H-type nodes are called C_{cc} vessels. Statistical data (mean ± SD) from four pigs are given in Table 7.4:1. The distances between each pair of consecutive C_{cc} vessels along the capillaries were found to be 61.2 ± 44.0 μm and 52.9 ± 39.6 μm for the right and left ventricles, respectively.

Connectivity Matrix

All C_{0a} capillaries are connected to coronary arteries but not necessarily to the smallest arterioles (of order 1). Some are connected to order-2 arteries and some to even larger arteries of order 3 or 4. The statistical data of the average number of capillaries originating from an arterial element of order n can be listed as a row matrix $\mathbf{C}(o, n)$. Similarly, a row matrix describing the connection between capillaries and venous elements is defined. Table 7.4:2 shows these connectivity matrices.

TABLE 7.4:2. Connectivity of C_{0a} and C_{0v} to Arteries and Veins in RV and LV

	Order number							
	1	2	3	4	5	6	7	8
Coronary arteries (C_{0a})								
RV	2.75	0.674	0.151	0.040				
LV	3.18	0.675	0.148	0				
RV and LV coronary veins (C_{0v})								
Sinusal	2.56	0.426	0.347	0.033	0.012	0.020	0.006	0
Thebesian	2.56	0.426	0.347	0.067	0.015	0.020	0.014	0.020

Values are the number of capillaries given out by each artery or vein of a specific order. Data from Kassab and Fung (1994). The standard errors are given in the original paper.

TABLE 7.4:3. Total Number of Capillaries Arising from Coronary Arteries and Veins in a Pig Heart Weighing 150 g

	Right coronary artery	Left anterior descending coronary artery	Left circumflex artery
Arterial capillaries	1,187,194 ± 515,552	1,273,281 ± 813,674	516,084 ± 332,019
Arterial capillaries	Whole heart	3,018,384 ± 1,697,777	
Venous capillaries	Whole heart	5,085,977 ± 2,085,250	

One of the important uses of the connectivity matrix is to compute the total number of vessels in any given order. In the case of the coronary capillaries, use of the connectivity matrices yields the total numbers of arterial capillaries listed in Table 7.4:3.

Topology of Arteriolar and Venular Zones in Myocardium

The trunks of the arterial and venous trees are separated in locations. When the branches divide repeatedly and become smaller and smaller, do the small twigs tend to be distributed uniformly in space? To answer this question, we must study the spatial distribution of the C_{0a} and C_{0v} segments. Let a random plane cross section cut the myocardium. If an arteriole is found to have two neighboring arterioles and the triangular region defined by the three arterioles as vertices contains no venules, then the three arterioles lie in an arteriolar zone. Probing arteriolar cross sections in its neighborhood, we can obtain a connected region in which all noncapillary blood vessels are arteriolar, and beyond its border there are venules. A similar treatment can be initiated with a venule. The principle of stereology then translates the features recognized in the plane cross sections to the three-dimensional space of the heart.

Figure 7.4:2 shows a typical result reconstructed from a histological section taken 1.7 mm from the epicardial surface of a right ventricle. In this figure, the muscle fibers are horizontal. If we shade the arteriolar zones while leaving the venular zones nonshaded, we obtain a two-colored map. But a two-colored map can represent only islands in an ocean. Figure 7.4:2 shows that the arteriolar zones are islands and the venular zone is an ocean. Hence, the arteriolar zones in the heart are like apples in an apple tree. This is similar to a corresponding result in the lung (see Sec. 6.13).

Mean Capillary Length

An overriding characteristic of the coronary capillaries is that most of them are parallel to the muscle fibers. Therefore, a capillary originating from a

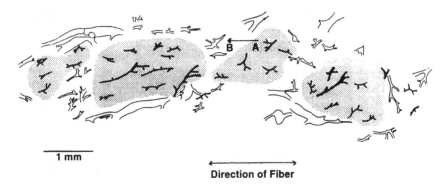

Direction of Fiber

FIGURE 7.4:2. Reconstructed spatial distribution of arterioles and venules from a histological section taken 1.7 mm from the epicardial surface of a pig right ventricle. Solid vessels, the arterioles; open vessels, the venules, shaded regions, arteriolar zones ("islands"); unshaded regions, venular zones ("ocean"). A—B indicates a vessel with an origin in the arteriolar zone and an end in the venular zone. From Kassab and Fung (1994), by permission.

point in an arteriolar zone will carry blood to a point in the venular zone in a flow that is parallel to the muscle fiber, for example, from A to B in Figure 7.4:2, which was constructed with the muscle fibers horizontal. If we assume that every point in the island has an equal chance of being point A, and that every point in the "legalized" part of the ocean in a direction parallel to the muscle fiber has an equal chance being point B, then the average length of the capillaries AB is exactly one half of the average length of the straight-line segments intercepted by the island and its immediate ocean. This length is defined as the *mean functional capillary length*, and can be obtained by measuring the distance between the centers of mass of adjacent arteriolar and venular domains. This length may be compared with the *mean capillary blood path length*, which is the mean value of the actual curved length of the capillaries between arterioles and venules. In the pig myocardium, we found the mean capillary blood path lengths to be $547 \pm 228 \mu m$ and $550 \pm 203 \mu m$ for right and left ventricles, respectively. The functional capillary lengths were found to be $501 \pm 178 \mu m$ and $512 \pm 163 \mu m$ for the right and left ventricles. These data suggest that most capillary pathways followed the shortest path from a terminal arteriole to a collecting venule. These path lengths are 8 to 10 times longer than the segmental lengths of the capillaries listed in Table 7.4:1.

Species Differences

There are similarities and differences in the morphometric features of different animal species. A comparison of the coronary capillary diameters and lengths in the left and right ventricles of various species published by many authors is given in Kassab and Fung (1994b).

7.5 Analysis of Coronary Diastolic Arterial Blood Flow with Detailed Anatomical Data

The capillary blood flow that perfuses the muscle cells of the heart comes from the coronary arterial tree. Since the tree trunk and its branches are localized, the flow in all the coronary capillary blood vessels cannot be expected to be the same. Some spatial nonuniformity in the perfusion of the myocardium of the whole heart may be expected. This nonuniformity of perfusion has obvious physiological meaning, and must depend on the anatomy of the arterial tree. Hence, let us consider the following mathematical problem: Let a mathematical model of an arterial tree be constructed that simulates all the morphometric data on the pig heart presented in Section 7.2. Let the blood flow through this arterial tree be analyzed under the following boundary conditions: The input pressure is specified at the left coronary Valsalva sinus. The exit pressure is specified at the arteriolar nodes of the coronary capillary bed. The solution will yield flow in every blood vessel. Then we can learn about the flow distribution in arteries of all orders. The flow in each vessel has a probability distribution with a mean and a standard deviation. The larger the standard deviation, the larger is the nonuniformity of blood flow.

Of course, we could consider an enlarged problem in which the coronary arterial tree is connected to a capillary bed, as described in Section 7.4, and a venous tree, as described in Section 7.3, and the boundary conditions are specified arterial input pressure at the Valsalva sinus and specified exit blood pressure in the right atrium for the sinusal veins, and exit pressures in the heart chambers where the Thebesian veins end. The solution of the enlarged problem will yield the mean value and the standard deviation of the pressure distribution at the ends of the smallest arterioles of order 1. Thus, we see that the mathematical problem of the coronary arterial blood flow formulated at the beginning of this section is one half of the enlarged problem. We can learn the methodology of solution by solving the smaller problem.

For simplicity, we assume further that the flow is steady and that the Poiseuillean formula given in Eq. (17) of Section 3.2 applies to all vessel elements. This implies certain limitations in application, as discussed in Chapter 3.

Setting up a Mathematical Model of the Arterial Tree that Simulates the Statistical Data

Figure 7.5:1(A) shows a schematic of a branching pattern of the trunk of the left common coronary artery that branches into left anterior descending and circumflex arteries. This schematic pattern is consistent with the connectivity matrix given in Table 7.2:3. In turn, each of the branches arising

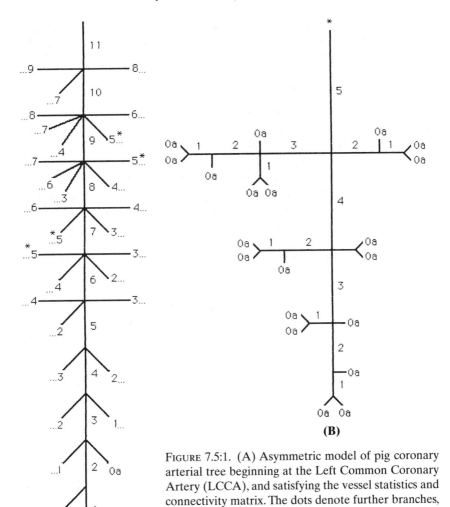

FIGURE 7.5:1. (A) Asymmetric model of pig coronary arterial tree beginning at the Left Common Coronary Artery (LCCA), and satisfying the vessel statistics and connectivity matrix. The dots denote further branches, such as the one shown in (B) for a 5..., connected to the tree trunk at an end which is marked by an asterisk (*). (B) A fifth order arteriole demonstrating all possible pathways to the capillaries as prescribed by the connectivity matrix. From Kassab et al. (1996), by permission.

from the trunk also gives rise to branches consistent with the connectivity matrix, and so on, down to the arterial capillaries (order 0a). Figure 7.5:1(B) shows a schematic of an order-5 arteriole branching down to order-0 vessels. Each element shown in Figure 7.5:1 may represent several elements in parallel, as determined by the total number of elements in each order. In Kassab and Fung (1995) it is shown that probability frequency distribution of the diameter is uniform, and that of the logarithm of the length is normal. Hence, we assume that the diameter and length of the vessels in Figure 7.5:1 are so distributed statistically.

System of Equations

Equations for a hydrodynamical analysis of flow in the circuit can be set up as follows: Let the nodes of the circuit be numbered serially. The pressure at a node number j is denoted by P_j. A vessel element joining the nodes i and j is denoted as vessel ij, whose diameter is D_{ij}, and length is L_{ij}. If there were N vessel elements connecting the nodes i and j, then these elements will be denoted by N_{ij}. Now, under the simplifying assumption that the flow in each vessel element obeys Poiseuille's formula, that is, the rate of volume flow in the vessel connecting the two nodes i and j, is given in terms of the pressure differential, ΔP_{ij}; the blood viscosity μ_{ij}, which depends on the hematocrit and vessel diameter; and the conductance coefficient, D_{ij}^4/L_{ij}, by

$$\dot{q}_{ij} = \left(\frac{\pi}{128\mu_{ij}}\right)\frac{D_{ij}^4}{L_{ij}}(P_i - P_j). \tag{1}$$

The total flow between these two nodes, $\dot{Q}_{ij}$, is the sum of the flow in the N parallel elements:

$$\dot{Q}_{ij} = N_{ij}\dot{q}_{ij}. \tag{2}$$

Hence, by Eq. (1),

$$\dot{Q}_{ij} = \left(\frac{\pi}{128}\right)Geq_{ij}(P_i - P_j), \tag{3}$$

$$Geq_{ij} = N_{ij}\frac{D_{ij}^4}{L_{ij}\mu_{ij}}. \tag{4}$$

Geq_{ij} is called *equivalent conductance coefficient* of all the vessel elements connecting the nodes i and j. The validity of Poiseuille's formula Eq. (1) is discussed throughout Chapter 3. In larger coronary arteries of orders 6 to 11, some corrections for the effect of Reynolds and Womersley numbers are needed.

If there are m_j vessels converging at the jth node, then the law of conservation of mass requires that

$$\sum_{i=1}^{m_j}\dot{Q}_{ij} = 0. \tag{5}$$

$\dot{Q}_{ij}$ is positive if the flow is into a node, and it is negative if the flow is out of a node. From Eqs. (3) and (5) we obtain a set of linear algebraic equations in pressure for all the nodes in the network, M in number, namely,

$$\sum_{i=1}^{m_j} \left[P_i - P_j \right] Geq_{ij} = 0, \qquad (j = 1, 2, \ldots M).$$ (6)

Example

Assume μ_{ij} to be four centipoise (0.004 Pa-s) in vessels of orders 11 to 5, and to decrease linearly with order number to 1.5 centipoise in the capillaries. Assume the pressure at the sinus of Valsalva = 100 mm Hg, and the pressure at the first bifurcation of the capillary network to be 26 mm Hg. Use 850 nodes. The Numerical Algorithms Group Subroutine (F04AEF-NAG Fortran Library) exploits the sparsity of the matrix in obtaining solutions of a large system of equations. It was used to solve many cases represented by Figure 7.5:1, each of which represents a possible case within the meaning of statistical sampling of a specified distribution. One hundred runs were made. Figures 7.5:2 to 7.5:4 show the results on the mean values of longitudinal pressure distribution, the pressure drop per vessel element, and the coronary blood flow per vessel. The results marked as "asymmetric model" were obtained by the method just outlined. Those marked as "symmetric model" uses an additional simplifying assumption described in the next paragraph. The uniform of pressure at the inlets to the capillary network is the assumed boundary condition. The statistical distribution deviates rapidly from the Gaussian shape as the order number increases. In the vessel of order 11, the largest coronary artery, the pressure at the entry is that of the sinus of Valsalva, and the statistical frequency function is a delta function.

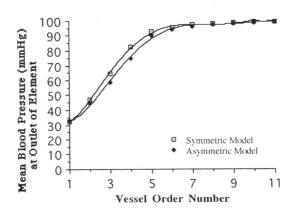

FIGURE 7.5:2. Statistical mean of the blood pressure at the outlets of arterial elements of successive orders computed by the "asymmetric" model shown in Figure 7.5:1 are marked by filled diamonds (◆). The corresponding pressures computed by a model satisfying the vessel statistics but not the connectivity matrix are marked by open squares (□). From Kassab et al. (1996), by permission.

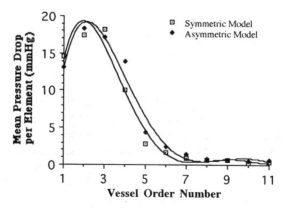

FIGURE 7.5:3. Relation between pressure drop per arterial element and the order number of the element for the arterial branches of the symmetric and asymmetric models of the left common coronary artery. From Kassab et al. (1996), by permission.

Comparison with a Simpler Model and with Experimental Results

How do the results compare with those of a simpler mathematical model of the coronary arterial tree that satisfies all the mean anatomical data, but not the connectivity matrix and standard deviations? Let us call this a *simplified symmetric model*, for which all standard deviations are zero and the connectivity matrix is one with each element in row n, column $n + 1$ equal to the branching ratio, and all other elements in the connectivity matrix $\mathbf{C}(m, n)$ replaced by zero. In this simplified symmetric model, all the vessel

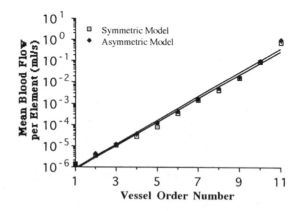

FIGURE 7.5:4. The blood flow per arterial element in successive orders of the arteries. From Kassab et al. (1996), by permission.

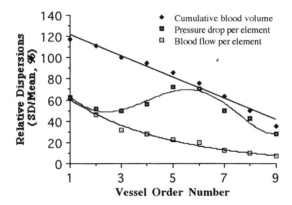

FIGURE 7.5:5. Relation between the % flow dispersion, the % pressure drop dispersion, and the % dispersion of cumulative blood volume in all elements of each order (100 × SD/mean) and the order number of the elements for the arterial branches of the asymmetric model of the left common coronary artery. From Kassab et al. (1996), by permission.

elements in every order are parallel, and the blood pressures at all the junctions between specific orders of vessels are equal. The boundary conditions are the same as before.

The results of this simplified model are plotted in Figures 7.5:2 to 7.5:4 as those of a "symmetric model." It is seen that the simplified model yields almost correct longitudinal blood pressure profile in Figure 7.5:2, and a reasonable prediction of the inaccurate pressure drop per element in small vessels of order 1 to 3, as shown in Figure 7.5:3, and a good estimation of the flow, as shown in Figure 7.5:4. (The mean flow per element is different in these two models because the boundary conditions prescribe pressures, not flow.) The pressure drop prediction by the asymmetric model is consistent with the epicardial pressure measurements reported by Chilian et al. (1986), Kanatsuke et al. (1991), and Tillmanns et al. (1981). The symmetric model has no dispersion in flow and pressure, so it cannot handle any question about nonuniformity of the flow field. The dispersion (SD/mean) of flow in the asymmetric model is shown in Figure 7.5:5. In the same figure, the dispersion of the pressure drop in each element and the cumulative blood volume of the elements are plotted. The dispersions of flow, volume, diameter, and length are related, of course, by the Poiseuillean formula, which, however, also involves the pressure differential that is not a local property. The local blood flow in the myocardium is very nonuniform as demonstrated by Bassingthwaighte et al. (1989) with the microsphere deposition technique, which shows that the level of nonuniformity depends on the volume of the tissue samples. Bassingthwaighte has explained this feature in terms of the fractal characteristics of the coronary blood vessel structure.

Experimentally, the pressure distribution can be measured only on the epicardial surface, and flow measurement in the myocardium is very difficult. See Chilian and Marcus (1982), Chilian et al. (1986), Kanatsuke et al. (1991), and Tillmanns et'al. (1981). Knowing the coronary vasculature in detail in three dimensions, one may perform theoretical analysis to extend the interpretation of the surface data to the interior of the myocardium.

7.6 Morphometry of Vascular Remodeling

In ventricular hypertrophy under hypertension, the myocardium and coronary blood vessels remodel. Since it is not possible to observe each vessel individually throughout the remodeling process, it has been difficult to record and analyze the effects of the remodeling of the morphology, materials, and mechanical properties of the coronary blood vessels on the function of the heart. The task is made easier by identifying the treelike structure of the coronary arteries and veins, and the netlike topology of the coronary capillaries. Morphometry, histology, and mechanical studies can then be done for each order of the vessel in different hearts.

Kassab et al. (1993b) have studied coronary arterial tree remodeling in right ventricular hypertrophy. They used the diameter-defined Strahler system defined in Section 7.2 to analyze the arterial tree in pig right ventricle in which hypertrophy was induced by pulmonary artery stenosis for 5 weeks. It was found that in right ventricular hypertrophy the total number of orders of vessels was one larger than that of the control; the total number

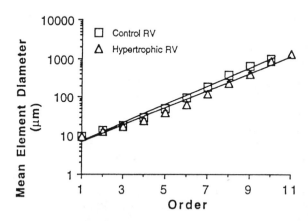

FIGURE 7.6:1. The average diameters of vessel elements in successive orders of vessels in the right ventricular branch of the right coronary artery (RCA-RVB) of control and hypertrophic pig right ventricle. From Kassab et al. (1993b), by permission.

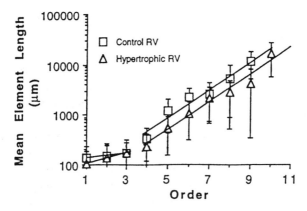

FIGURE 7.6:2. The average lengths of vessel elements in successive orders of vessels and order number of vessels in RCA-RVB of control and hypertrophic pig right ventricle. From Kassab et al. (1993b), by permission.

of elements in each order increased greatly, whereas the diameters and lengths of each order decreased somewhat. The total resistance to blood flow in the right coronary arteries decreased in hypertrophy, mainly because the total cross-sectional area of all parallel elements in each order was increased due to the great increase in the number of vessels. The decrease in the diameters of the vessels in each order is more than compensated for by the increase in the number of elements in each order in such a way that the total cross-sectional area increased and the total resistance is decreased. The quantitative features of the changes in diameter and length of the arteries of successive orders and the pressure–flow relationship of the right coronary artery are shown in Figures 7.6:1 to 7.6:3.

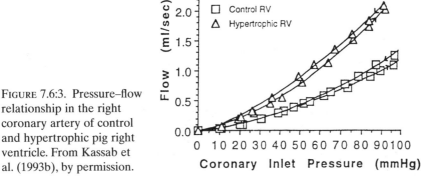

FIGURE 7.6:3. Pressure–flow relationship in the right coronary artery of control and hypertrophic pig right ventricle. From Kassab et al. (1993b), by permission.

7.7 In Vivo Measurements of the Dimensions of Coronary Blood Vessels

Kajiya and his associates (Hiramatsu et al., 1994; Yada et al., 1993, 1995) have developed a portable needle-probe video microscope with a charge-coupled device (CCD) camera to visualize the subendocardial microcirculation. The needle probe has a sheath and a doughnut-shaped balloon to lodge the probe in place against papillary muscles. A microtube at the head flushes away the intervening blood. The probe can be introduced into the left ventricle through an incision in the left atrial appendage via the mitral valve. Images of microvessels of the beating heart are magnified 200 to 400 times. A sketch of the microscope system is shown in Figure 7.7:1. The experimental configuration is shown in Figure 7.7:2.

An example of the diameter change in a subendocardial arteriole and a subendocardial venule throughout a cardiac cycle is shown in Figure 7.7:3. The phasic diameter change in the subendocardial arterioles during the cardiac cycle was from $114 \pm 46\,\mu m$ (mean $\pm$ SD) in end diastole to $84 \pm 26\,\mu m$ in end systole ($p < 0.001$, $n = 13$, ratio of change = 24%). In subendocardial venules, the diameter change was from $134 \pm 60\,\mu m$ to $109 \pm 245\,\mu m$ ($p < 0.001$, $n = 15$, ratio of change = 17%). In constrast, the diameter of subepicardial arterioles was almost unchanged (2% decrease, $n = 5$, $p < 0.01$),

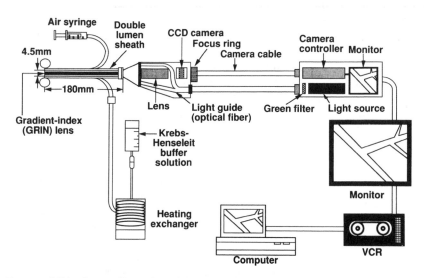

FIGURE 7.7:1. A needle-probe videomicroscope with a charge-coupled device (CCD) camera. The system consists of a needle probe, a camera, a lens and light guide, a control unit, a light source, a monitor, and videocassette recorder. A needle probe containing a gradient-index lens with a length of 180 mm was used to obtain the images of the subendocardial microcirculation of the left ventricle. From Yada et al. (1993), by permission of the authors and the American Heart Association.

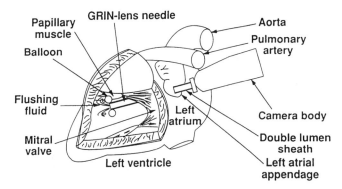

FIGURE 7.7:2. Experimental preparation. The sheathed needle probe was introduced into the left ventricle through an incision in the left atrial appendage via the mitral valve. A balloon was inflated to prevent direct compression of the endocardial microcirculation by the needle tip, and the intervening blood between the needle probe and endocardium was flushed away with a Krebs-Henseleit buffer solution. From Yada et al. (1993), by permission.

whereas the subepicardial venular diameter increased by 19% (n = 8, $p < 0.001$) from end diastole to end systole. Partial kinking and/or pinching of vessels was observed in some segments of subendocardial arterioles and venules. The percentage of change in the diameter from end diastole to end systole in the larger (>100 μm) subendocardial arterioles and venules was greater than that in the smaller (50 to 100 μm) vessels (Fig. 7.7:4).

These data clearly show the effect of heart muscle contraction on the coronary blood vessel dimensions in a beating heart. Gota et al. (1991) measured the blood flow in the myocardium and the diameters of coronary blood vessels in different layers of the myocardium during stable systolic contracture and during stable diastolic arrest. They measured blood flow with radionuclide-labeled microspheres. They concluded that cardiac contraction predominantly affects subendocardial vessels and impedes subendocardial flow more than the subepicardial flow, regardless of left ventricular pressure.

Coronary Blood Flow Throughout a Cardiac Cycle

The pulsatile blood flow in the coronary arteries and veins is unlike that in other organs: The arterial inflow is greater during diastole and the venous outflow is greater during systole, all due to the contraction of the heart muscle. Scaramucci (1695) theoretically inferred a hypothesis that blood in the myocardial vessels is squeezed by the heart muscle into the coronary veins, and are replenished or refilled from the aorta during diastole. Modern research has aimed at critically validating Scaramucci's hypothesis, clarifying the words *squeezing* and *refilling* quantitatively, and determining if this ad hoc hypothesis can be deduced from basic laws of physics from the specific geometric and mechanical properties of the materials involved.

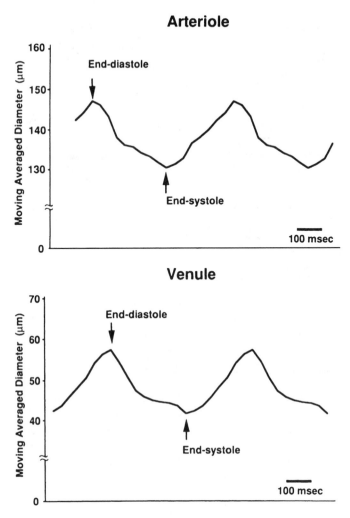

FIGURE 7.7:3. Example of the diameter change in a subendocardial arteriole and a subendocardial venule throughout a cardiac cycle. The arrows indicate end diastole and end systole. From Yada et al. (1993), by permission.

The in vivo measurements of the diameters of the coronary arterioles and venules by Kajiya et al. discussed earlier support Scaramucci's hypothesis. These authors used the same instrument to study the diameters of coronary arterioles and venules during a prolonged diastole (Hiramatsu et al., 1994). To do this, a complete atrioventricular heart block was produced by injecting formaldehyde (40%) into the atrioventricular node, and a pacing wire was sewn onto the right ventricular outflow tract. Prolonged diastole was

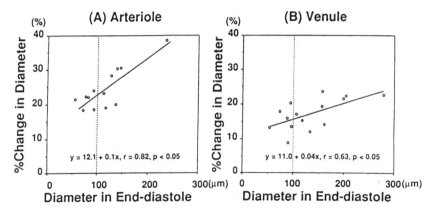

FIGURE 7.7:4. Correlation between diameter in end diastole (horizontal axis) and corresponding percent change in diameter from end diastole to end systole (vertical axis). Panel A: Subendocardial arterioles; Panel B: Subendocardial venules. From Yada et al. (1993), by permission.

produced by cessation of right ventricular pacing. During the period of prolonged diastole, the aortic pressure and coronary blood flow continued to decrease, as well as the subendocardial arteriole diameter, but the venule diameter continued to increase. After 2.2 to 3.5 s the zero-flow point was reached, but the arterioles remained patent (were not seen to have collapsed). These trends continued 8 to 9 s after zero flow. The pressure gradient and diameter at the beginning of prolonged diastole and at the zero-flow point are listed in Table 7.7:1. The trends of changes of the pressure and diameter are shown in Figure 7.7:5. The coronary blood flow and the percent changes in the diameter of coronary arterioles and venules

TABLE 7.7:1. Vascular Diameter, Aortic Pressure, and Pressure in Great Cardiac Vein at the Beginning of Prolonged Diastole and at the Zero-Flow Point

	Vascular diameter (μm)	Aortic pressure (mm Hg)	GCV pressure (mm Hg)
Arteriole			
Beginning of prolonged diastole	81 ± 25	78 ± 10	10 ± 4
Zero-flow point	$60 \pm 19*$	27 ± 6	11 ± 3
Venule			
Beginning of prolonged diastole	106 ± 44	77 ± 12	9 ± 4
Zero-flow point	$122 \pm 50*$	25 ± 6	10 ± 4

GCV indicates great cardiac vein. Values are mean $\pm$ SD ($n = 16$).
* $p < 0.01$ versus the time point at the beginning of prolonged diastole.
From Hiramatsu et al. (1994).

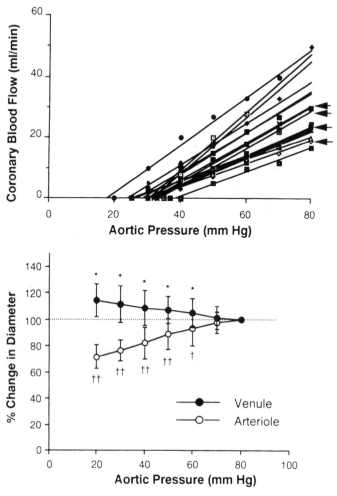

FIGURE 7.7:5. Coronary blood flow (A) and change in vascular diameter (B) during prolonged diastole plotted against aortic pressure. The arteriolar diameter was normalized by the value at aortic pressure of 80 mm Hg. Arrows indicate the diastolic values for which the identification of a zero flow in the tracing was difficult because of the resumption of beating before zero flow. *$P < 0.05$, †$P < 0.01$, and ‡$P < 0.01$ versus the values at 80 mm Hg aortic pressure. From Hiramatsu et al. (1994), by permission.

(subendocardial) are shown in the figure. The nonvanishing zero-flow arterial pressure is very clear in the upper panel. The noncollapse of the coronary arterioles and venules at the zero-flow point suggests that the sluicing mechanism of the heart is very different from that of the lung, and the flow in the capillaries must be given primary attention.

7.8 Mechanical Properties of Coronary Blood Vessels

The structure, materials, and histology of the coronary and systemic blood vessels are largely similar. Hence, one could assume that the constitutive equations of the blood vessel presented in Section 2.6, or in greater detail in *Biomechanics: Mechanical Properties of Living Tissues* (Fung, 1993), are applicable to the coronary blood vessels, and only the material constants of the intima, media, and adventitia layers of each vessel need to be determined. This is largely true, but still data on the coronary blood vessels are scarce, and very incomplete. Older data are reviewed by Cox (1982). Newer data need to be reviewed. Examples of the pressure–diameter relationship of isolated coronary arterioles of orders 4 and 5 are shown in Section 7.10, Figures 7.10:2 and 7.10:3. The pressure–diameter curves in the passive state (with nitroprusside perfusion) are seen to be very different from those in the active state. The former reflect the behavior of the connective tissues in the vessel wall, whereas the latter reflect the behavior of the smooth muscle cells. Hence, we must regard the blood vessel wall as multiphasic, and multicompartmental material, with the smooth muscle and the connective tissues occupying different compartments.

The lack of data on the material constants involved is a great handicap, and we hope it will be amended soon. Recent work by Kuo et al. on the tone of the vascular smooth muscle of coronary arterioles, however, has advanced the study of the mechanical properties by a big step. This is described in Sections 7.10 and 7.11. Jiang and He's (1990, 1991) data, however, are for connective tissues.

7.9 Mechanical Properties of Heart Muscle Cells in Directions Orthogonal to Active Tensile Force

The coronary blood vessels interact with cardiac muscle cells. The interaction takes place at the interfaces. At an interface, the boundary conditions are the continuity of surface traction (normal and shear stresses) and the compatibility of displacements. If the muscle pulls on the blood vessel at the interface and vice versa, then the traction on the muscle and vessel is tensile. If the muscle compresses the blood vessel, then the interfacial traction is compressive. In the former case we need the elasticity laws of the materials in tension. In the latter case we need their elasticity laws in compression. Tensile elasticity data of soft tissues exist. Compressive elasticity data of these tissues do not exist. Theoretical reasoning must apply.

Of all the interaction problems of the heart, we shall focus on the compression between the heart muscle and the capillary blood vessels, because this is important in coronary circulation. With regard to the problem of zero-flow pressure. Figure 7.9:1 (D) shows capillaries and muscle cells in parallel. There are two scenarios. In one, the muscle cells shorten and their diameters increase, causing an increase of the interstitial space in which capillary blood vessel lies. In the other, tension develops in muscle under isometric condition. The cell volume and diameter do not change. Question: Could the interaction between capillary and muscle be a function of active tension in the latter case?

In the old literature, capillary blood vessels were considered to be rigid tubes, because when observations were made in the mesentery, the changes in the diameter of the capillary ($\sim 10\,\mu$m) were almost unmeasurable by an optical microscope when blood pressure was changed over a range of 100 cm H_2O. Fung et al. (1966) explained this capillary rigidity by the support the capillary received from its gel-like surrounding tissue. Thus arose the "tunnel in a gel" concept of capillaries. Fung (1966) predicted that if surrounding tissue were small, the capillary rigidity will disappear. In 1972 Fung and Sobin showed that the pulmonary capillaries of the cat are distensible because they have very small surrounding tissue. (See Sec. 6.8.) In 1979, Bouskela and Wiederhielm showed that the capillaries in bat's wing are distensible for the same reason. Schmid-Schönbein et al. (1989) measured the distensibility of the capillaries in rat skeletal muscle when the fascia is removed. (Sec. 8.7, p. 532.) The distensibility of the coronary capillaries is probably similar to that of the skeletal muscle capillaries.

Now, turn to the heart muscle. The author's proposition is that owing to the cell's inhomogeneous internal structure, tension generation will create high pressure pockets which causes the cell membrane to bulge out at various places and resist external loads with an internal pressure. This is seen as follows: First, it is well known that if a taut string with tensile force T is loaded by a lateral load p (force per unit length) [Fig. 7.9:1(A)], the equation of equilibrium is

$$T \cdot \text{curvature} = p. \tag{1}$$

Think of each set of activated actin–myosin coupling as a string, and consider a muscle cell containing nucleus, mitochondria, vesicles, transverse and longitudinal tubules of sarcoplasmic reticulum, etc. [Fig. 7.9:1(B)]. When the muscle is activated, the actin–myosin strings have an average tensile stress of T (force per unit cross-sectional area), and are bent by the inclusions (mitochondria, etc.) to a curvature of $1/r$. Then, if A is the effective cross sectional area of the actin-myosin strings associated with an inclusion, the pressure in the inclusion that balances the bending of the muscle tension is given by the Laplace's formula:

$$p = \frac{TA}{r} \tag{2}$$

(A) Pressure-tension-curvature relationship

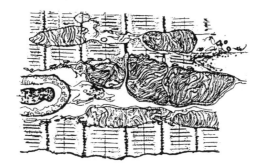

(B) Myocardium with inclusions

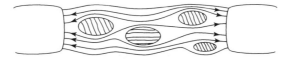

(C) Tension lines in myocardium

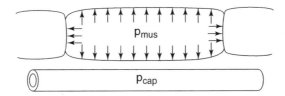

(D) Pressure in muscle and capillary

FIGURE 7.9:1. Schematic drawings for the concept that the mitochondria and other inclusions in the muscle cells cause curvature in the actin-myosin fibers. When the actin-myosin fibers generate tension, the lines of force are curved by the inclusions: the curvature causes compression in the inclusions, deforms the inclusions and the cell, and reveals as lateral forces acting on the cell membrane. The resultant appears as a pressure in the muscle cell, which can compress a neighboring capillary blood vessel. See text for details.

In the situation illustrated by Figure 7.9:1(C), most of the inclusions are in the interior of the cell, but some inclusions are next to the outer surface of the cell (cell membrane). These inclusions on the boundary will bulge out when the muscle tension is increased. If there is another body next to the surface, such as the capillary blood vessel shown in Figure 7.9:1(D), then the bulging of the inclusion will push on the capillary blood vessel with a local pressure p. Conversely, if the blood vessel dilates, it will push on the muscle cell. If the blood vessel and the muscle cell are parallel, the interaction occurs first with the bulging inclusions. If B is the fraction of the total muscle cell surface that is aposed to these boundary inclusions, then we may interpret the product $B·p$ as the *muscle pressure* that interacts with the capillary blood vessels and denote it by p_{mus}:

$$p_{mus} = Bp = \frac{TAB}{r} \tag{3}$$

A similar argument applies to the interaction between the muscle cells in a bundle. If a capillary works against a bundle of muscle cells the effective muscle pressure may increase with the size of the bundle, i.e. the product AB may increase with the bundle size. The p_{mus} is the pressure that interacts with the capillary blood vessel.

As the muscle tension T and the radius of curvature of the inclusion vary with time throughout a cardiac cycle, the viscoelastic behavior of the cell and inclusions may be important in the analysis of r as a function of T. It is seen that the muscle pressure is larger in systole, but not simply proportional to muscle tension. It is related to ventricular blood pressure, but not linearly related to it.

Whereas the muscle pressure is relevant to the interaction between parallel capillary blood vessels and myocytes, it may not be relevant in other situations. For example, one may probe the hardness of one's own biceps with fingers and conclude that the biceps are harder when the muscle is tensed up. This feeling of hardness occurs for three reasons: (1) The fingers tried to bend the muscle fiber. A tensed fiber is harder to bend than a loose fiber. Equation (1) applies. (2) If fluid is prevented from moving in or out of the biceps, then the shortening of the muscle fibers will cause bulging of the biceps' cross section, causing strain and stretch of the fascia. Tension in the fascia causes hydrodynamic pressure in the muscle. (3) With regard to the blood vessels in the biceps, we need to know much more about their geometry and mechanical properties, and the kind of interaction between myocytes and capillaries discussed in this section before we can assess their effect. Presumably in the case of biceps the effect of blood vessel is not large.

Rabbany et al. (1994) analyzed an isolated skeletal muscle cell of the giant barnacle and showed that the shortening of the muscle created bending and stretching of the cell membrane that resulted in the creation of intramuscle fluid pressure. In this case the pressure in the cell is not a direct function of the tensile force generated, but rather of shortening. Rabbany et al. (1989, 1994) applied this theory to myocytes of the heart under the assumption that the intercalated disks are rigid, indistensible, so that when the cell shortens the cell membrane bulges out, bent and stretched by the incompressible cell content. The intracellular pressure is associated with the strain energy in the cell membrane, and is a function of the shortening of the cell, not of the muscle tension.

In the human heart muscle, it is not known how rigid the intercalated disk is, and how much a role the cell membrane plays. In histological micrographs, the inclusions are much more prominent. In more complete theories of the future, all these structural components, including the collagen fibrils, must be included.

The importance of this subject will become clear in Sections 7.13 and 7.14; and has been emphasized by Spaan (1981, 1991).

7.10 Coronary Arteriolar Myogenic Response: Length–Tension Relationship of Vascular Smooth Muscle

In Section 5.14 we discussed the autoregulation of blood flow. The coronary blood flow is strongly autoregulated. Fortunately, Kuo, Chilian, and Davis have done excellent work that elucidates this phenomenon.

Kuo et al. (1990) examined isolated arterioles for the effect of pressure and flow. Their ability to cannulate the arteriole is outstanding, their instrumentation and experimental procedure are unique. Figure 7.10:1 shows a schematic diagram of their system. The test specimen was cannulated to the inflow and outflow pipettes. The pipettes were connected to two independent reservoir systems. The two reservoirs were initially set at the same hydrostatic level, so that the arteriole could be pressurized without flow by simultaneously moving the reservoirs in the same direction. Flow could be initiated by simultaneously moving the reservoirs in opposite directions. Because the resistances of both cannulation pipettes were the same, simultaneous movement of the reservoirs in equal and opposite directions did not change the midpoint luminal pressure. Hence, flow and pressure could be varied spearately.

To obtain the test specimens, they perfused a mixture of India ink and gelatin in physiological salt solution (PSS) into the pig left anterior descending artery (LAD) and the left circumflex artery (LCX) to visualize the coronary microvessels. At 4°C, arteriolar branches from the LAD or LCX (1.0 to 1.5 mm in length and 40 to 70 μm internal diameter) were selected and dissected from the surrounding tissue and transferred for further dissection to a dish (4°C) containing filtered PSS–albumin solution at pH 7.4. After removal of any remaining tissue, an arteriole was transferred to a cannulation chamber at room temperature, and was cannulated to a glass

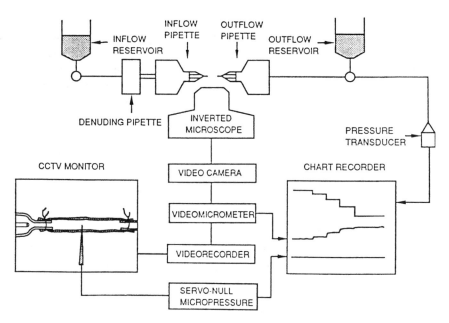

FIGURE 7.10:1. A dual-reservoir system for videomicroscopic measurement of internal diameter and pressure of isolated vessels. The chart recorder illustrates what a typical record would look like if the inflow reservoir is raised and the out-flow reservoir is lowered. From Davis, Chilian (1990), by permission.

micropipette with a $40 \mu m$ tip internal diameter and tied. The ink-gelatin was flushed out at low perfusion pressure (<20 cm H_2O). Then the other end was cannulated and secured with a suture. In the test chamber, the specimen was bathed in PSS-albumin solution at 36 to 37°C, set to its in situ length, and allowed to develop spontaneous tone at 60 cm H_2O luminal pressure without flow. A pressure of 60 cm H_2O was selected because the intraluminal pressure was found to be 40 to 50 mm Hg in arterioles of this size in vivo.

Kuo et al.'s results on flow and pressure combined are presented in Sections 7.11 and 7.12. Here we present their results on pressure alone, without flow. Without flow, they needed only one reservoir (see Fig. 7.10:1), so they connected the other end to a pressure transducer. Then they used the equipment to study the effect of endothelium denudation. They used two methods to denude: (a) intraluminal perfusions with 3-[(3-cholamido-propyl) dimethylammonio]-1-propanesulfonate (CHAPS; 0.2% or 0.5%) in physiological saline solution for 90 s, followed by washing with saline for 30 min and (b) mechanical removal with a concentric glass abrasive micropipette. A third method of intraluminal injection of an air bolus required high pressure to push the bubble through the arteriole and caused smooth muscle damage.

Figure 7.10:2 shows Kuo et al.'s results on the pressure–diameter relations of pig subepicardial arterioles before and after endothelium denudation by CHAPS perfusion. In physiological saline–albumin solution at 60 cm H_2O intraluminal pressure, the control vessels spontaneously constricted to 76.4 ± 3.3% of their diameters in nitroprusside (10^{-4} M) solution. In all control vessels, increasing intraluminal pressure decreased the diameter. The ordinate shows the lumen diameter normalized to a diameter at a pressure of 60 cm H_2O in the presence of nitroprusside. The average luminal diameter in the control state at 60 cm H_2O was 61 ± 4 μm. Tests of the smooth muscle response to acetylcholine chloride (ACh) and bradykinin (10^{-8} M) showed that the myogenic responses failed after CHAPS perfusion.

After mechanical denudation of the endothelium, the viability of vascular smooth muscle was tested using three criteria: (a) preservation of spontaneous tone, (b) constriction due to ACh, and (c) lack of dilation of preconstricted vessels to bradykinin. The pressure–diameter relationships of viable vessels are shown in Figure 7.10:3. From these data Kuo et al. (1990a) claimed that the myogenic response of small coronary arterioles to changes of blood pressure alone (without flow) does not depend on the presence of an intact, functional endothelium.

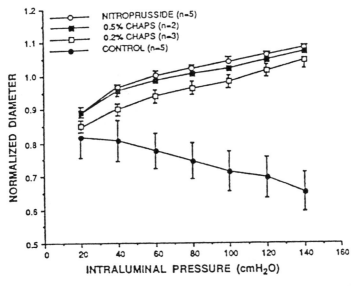

FIGURE 7.10:2. Pressure-diameter relations of subepicardial arterioles before and after CHAPS perfusion. After CHAPS (0.2% or 0.5%) perfusion, vessels behaved passively as in the nitroprusside solution (10^{-4} M). Vertical bars denote mean ± SEM. From Kuo et al. (1990a), reproduced by permission.

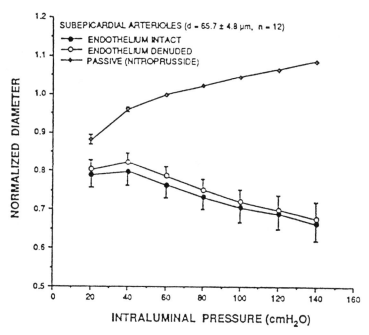

FIGURE 7.10:3. Pressure-diameter relations of subepicardial arterioles in the physiological saline-albumin solution before (closed circles) and after (open circles) mechanical denudation of endothelium. Vertical bars denote mean ± SEM. From Kuo et al. (1990a), reproduced by permission.

Smooth Muscle Length-Tension Curve Derived from Kuo et al.'s Results

I would like to interpret the results of Kuo et al. (1990a) in terms of the vascular smooth muscle length–tension relationship. If a blood vessel has lumen diameter D_i, wall thickness h, intraluminal pressure p_i, and external pressure p_e, then by static equilibrium the hoop tension T (circumferential stress resultant) in the vessel wall is*

$$T = p_i \frac{D_i}{2} - p_e \frac{D_i + h}{2}. \tag{1}$$

T is the sum of tension in the connective tissues, denoted by P, and the tension in the smooth muscle, denoted by S. P is a nonlinear function of the

---

* This equation is exact if the vessel is a circular cylindrical tube. It says, however, nothing about the stress distribution, which depends on the zero-stress state of the vessel. In general, the larger the h/D_i ratio is, the more nonuniform are the stress and strain in the wall, to which the muscle tension and change of length are related. Hence, Eqs. (1) and (2) need refinement when D_i is small. The reader should keep this need in mind when interpreting the simplified results in Figs. 7.10:4, and 7.11:3 when D_i/D_o is small.

diameter and can be extracted from the "passive" curve (nitroprusside) in Figures 7.10:2 and 7.10:3. T can be computed from the other curves given in these figures. Hence, the muscle tension, S, can be computed from experimental results according to the equation

$$S = T - P. \tag{2}$$

We assume that S is related to the length of the muscle cell, which may be assumed to be proportional to the mean diameter of the vessel $(D_i + h)/2$. Since h was not given in the paper, we shall ignore it in the following formulas, but it can be added back later. Hence, S is related to D_i.

The experimental results on the control vessels shown in Figure 7.10:2 can be fitted with a straight line and represented by the following formula:

$$\frac{D_i}{D_0} = c_0 - c_1 p_i \tag{3}$$

Where D_i is the lumen diameter, D_0 is the diameter of the passive vessel at 60 cm H_2O, and c_0, c_1 are constants. From Figure 7.10:2,

$$c_0 = 0.860$$
$$c_1 = 0.00150 \text{ per cm } H_2O = 1.53 \times 10^{-5} \text{ m}^2/\text{N} = 1.53 \times 10^{-6} \text{ cm}^2/\text{dyne}. \tag{4}$$

Similar curves shown in Figure 7.10:3 or other publications can be fitted by a polynomial, but the principle is the same.

Solving Eq. (3) for p_i and substituting the result into Eq. (1) with $p_e = 0$, we obtain

$$T = \frac{D_0}{2c_1} \left(\frac{D_i}{D_0}\right)\left[c_0 - \frac{D_i}{D_0}\right]. \tag{5}$$

Fitting the "passive (nitroprusside)" curve of Figure 7.10:3 with an exponential function well known for soft tissues (see Sec. 7.5 of Chapter 7 in Fung, *Biomechanics: Mechanical Properties of Living Tissues*, 1993),

$$p_i = c_2 e^{\alpha(D_i/D_0)}, \tag{6}$$

we find

$$\alpha = 9.902,$$
$$c_2 = 0.00300 \text{ cm } H_2O = 0.294 \text{ N/m}^2 = 2.94 \text{ dyne/cm}^2. \tag{7}$$

The tension $P = p_i D_i/2$. By Eq. (6), we have

$$P = \frac{c_2}{2} D_i e^{\alpha(D_i/D_0)}. \tag{8}$$

Hence, by Eqs. (5) and (8) we obtain the tension in the smooth muscle, S:

$$S = \frac{D_0}{2c_1}\left(\frac{D_i}{D_0}\right)\left[c_0 - \frac{D_i}{D_0}\right] - \frac{c_2 D_0}{2}\left(\frac{D_i}{D_0}\right)e^{\alpha(D_i/D_0)}. \tag{9}$$

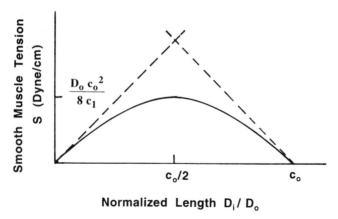

FIGURE 7.10:4. Relationship between the vascular smooth muscle tension and the muscle length described by Eq. 9.

With the numerical coefficients listed in Eqs. (4) and (7), we plot S versus D_i/D_0 in Figure 7.10:4. We see that S is practically a parabola with a peak at $(D_i/D_0) = c_0/2 \doteq 0.430$. In the practical range of interest, $D_i/D_0 > 0.430$, S decreases when D_i/D_0 increases.

Equation (9) and Figure 7.10:4 show that the length-tension relationship of the vascular smooth muscle of the coronary arterioles is similar to the well-known length–tension curves of the skeletal muscle and heart muscle (see Fig. 2.1:3). Similar to the skeletal and heart muscles, the length–tension relationship of the vascular smooth muscle can be influenced by pharmacological agents and by metabolites such as oxygen. Unlike the skeletal and heart muscles, whose normal lengths lie on the left half of the length–tension curve before the peak, the vascular smooth muscle's working range of length lies on the right half of the length–tension curve, after the peak. The slope of the curve is upward on the left half and downward on the right half. Hence, in the heart we have Starling's law, whereas in the blood vessel we have autoregulation.

The vascular smooth muscle in the blood vessel is separated from the flowing blood by only a layer of endothelial cells a few micrometers thick. Kuo et al. (1990a) have shown that when blood is not flowing, the endothelium has no effect. In other words, the length–tension relationship exists whether the endothelium is intact or not. However, blood flow imposes shear stress on the endothelium, the endothelium releases factors in response to the shear stress, and the smooth muscle reacts to the factors. Hence, we expect the length–tension curve to be influenced by the shear stress in a way similar to its being influenced by a pharmacological agent. This matter is discussed in the following section.

7.11 Vessel Dilation Due to Flow: Effect of Shear Stress on the Endothelium on Smooth Muscle Length–Tension Relationship

Using their equipment shown in Figure 7.10:1, Kuo et al. (1990a) measured the diameter changes in pig coronary arterioles due to a pressure gradient without a change of the mean pressure. Their results for 14 pig (6- to 10-week-old) subepicardial coronary arterioles perfused by physiological saline solution with albumin at 37°C with a fixed midpoint intraluminal pressure of 60 cm H_2O are shown in Figure 7.11:1. The diameters of these arterioles at 60 cm H_2O were 64.2 ± 21 (SE) μm. The diameters of the vessels maximally dilated with nitroprusside were about 89.0 μm. Their lengths were 0.8 to 1.2 mm. Figure 7.11:1(A) shows the diameter changes following the flow changes. Figure 7.11:1(B) shows the flow-induced dilation plotted against the pressure differential when the lumen pressure was fixed at 60 cm H_2O. The contrast of the results *before* (closed circles) and *after* (open circles) mechanical removal of the endothelium is shown. Clearly, the flow-induced dilation is endothelium dependent.

Figure 7.11:2(A) and 7.11:2(B) show additional results on flow-induced dilation when the midpoint luminal pressure was fixed at 100 and 20 cm H_2O, respectively. Lumen diameters were normalized to the passive (with nitroprusside) diameter at a corresponding lumen pressure (at 100 cm H_2O for A and 20 cm H_2O for B.) Again, mechanical removal of endothelium abolished the flow-induced dilation.

In the following, a formula representing the length-tension relationship of the vascular smooth muscle influenced by the blood shear acting on the endothelium is derived. This derivation follows the remark at the end of Section 7.10 that the shear stress on the endothelium acts like a pharmacological agent on the smooth muscle. Before we present the derivation, let us examine the effect of blood shear on the principal stress in the smooth muscle. First, according to Poiseuille's formula, Eq. (3.2:19), the shear stress acting on the endothelium due to blood flow, denoted by the symbol τ, is

$$\tau = -\frac{a}{2}\frac{\Delta p}{L},\qquad(1)$$

in which a is the lumen radius, and Δp is the pressure differential at the ends of a vessel of length L. In a case in Kuo et al.'s paper, an arteriole of radius $a = 24\,\mu m$, length $L = 1$ mm, $p_i = 60$ cm H_2O, $\Delta p = 60$ cm H_2O, the value of τ given by Eq. (1) is 0.72 cm H_2O, which is large compared with the known wall shear of about 16 dyne/cm^2 or 0.16 mm H_2O in normal in vivo circulation, but is much smaller than the normal stress, the lumen pressure, $p_i = 60$ cm H_2O. The hoop stress σ in the arteriole is given by

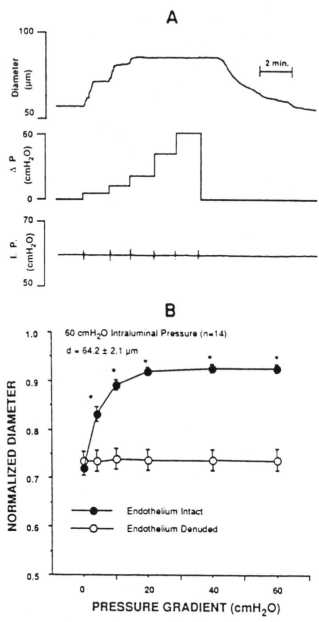

FIGURE 7.11:1. A: Experiment in which flow-induced vasodilatation was observed during production of a pressure difference (ΔP) while the mean intraluminal pressure (IP) was not altered. B: Flow-induced dilation of 14 vessels at 60 cm H_2O intraluminal pressure before (closed circles) and after (open circles) mechanical removal of endothelium. Vertical bars denote mean ± SEM. From Kuo et al. (1990a), reproduced by permission.

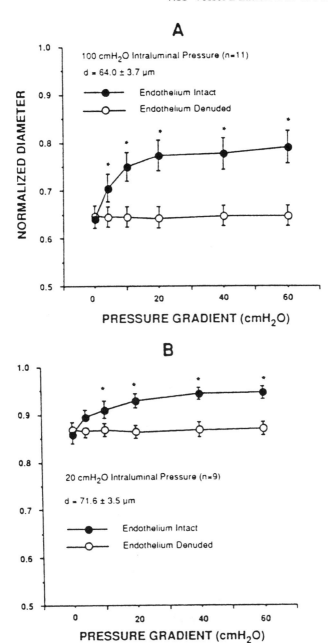

FIGURE 7.11:2. A and B: Flow-induced dilation in vessels at 100 and 20 cm H_2O intraluminal pressures, respectively. From Kuo et al. (1990a), reproduced by permission.

dividing T of Eq. (3.10:1) by the wall thickness of the vessel, h. Assuming $p_e = 0$, then

$$\sigma = ap_i/h. \tag{2}$$

If we assume $a/h = 5$, and $p_i = 60\,\text{cm H}_2\text{O}$, then $\sigma = 300\,\text{cm H}_2\text{O}$, which is much larger than the shear τ given by Eq. (1). In the vascular smooth muscle the principal stress is essentially equal to the hoop stress; hence it is hardly affected by the blood shear.

The diameter–pressure gradient curves in Figures 7.11:1 and 7.11:2 for vessels with intact endothelium can be fitted by the following formula:

$$D_i = D_{i\,\Delta p=0} + f\left(p_i\right)\left[1 - e^{-\beta\Delta p}\right], \tag{3}$$

in which D_i is the intraluminal diameter, $D_{i\,\Delta p=0}$ is the value of D_i when $\Delta p = 0$ discussed in Section 7.10, Δp is the pressure differential at the ends of an arteriole of length L, β is a constant, and $f(p_i)$ is the range of vessel lumen dilation from $\Delta p = 0$ to the maximum pressure differential $\Delta p = 60\,\text{cm}$ H_2O reported in these figures. $f(p_i)$ is a function of the mean (midpoint) pressure, p_i. From Figures 7.10:3, 7.11:1, and 7.11:2, we find that for $p_i = 20$, 60, and $100\,\text{cm H}_2\text{O}$, the passive diameters of the arterioles are 84.2, 89.0, and $100\,\mu\text{m}$, respectively, and the $f(p_i)$ are 7.58, 17.8, and $16\,\mu\text{m}$, respectively. Hence, roughly,

$$f\left(p_i\right) = a + bp_i - cp_i^2, \tag{4}$$

with $a = -2.04$, $b = 0.556$, and $c = 3.76 \times 10^{-3}$ when p_i is in units of cm H_2O.

By fitting Eq. (3) to the curves in Figures 7.11:1 and 7.11:2, we find the value of β for $p_i = 60\,\text{cm H}_2\text{O}$ to be 0.180 per cm H_2O. For $p_i = 20$, β is between 0.10 and 0.20; and for $p_i = 100$, β is between 0.118 and 0.132. More accurate determination of the constants a, b, c, and β requires greater precision in raw data. For a qualitative discussion we shall treat β as a constant independent of p_i.

Now, from Eq. (7.10:3) we have

$$D_{i\Delta p=0} = \left(c_0 - c_1 p_i\right)D_0, \tag{5}$$

with $D_0 = 89.0\,\mu\text{m}$ in Figure 7.11:1. Hence, Eq. (3) becomes

$$D_i = \left(c_0 - c_1 p_i\right)D_0 + \left(a + bp_i - cp_i^2\right)\left[1 - e^{-\beta\Delta p}\right]. \tag{6}$$

For p_i this is a quadratic equation of the form $Ax^2 + Bx + C = 0$, and we can write down the solution x in terms of A, B, and C. For simplicity in manipulation, we may limit ourselves to a narrower range of local blood pressure, say, to p_i from 40 to $90\,\text{cm H}_2\text{O}$, and treat β and $f(p_i)$ as constants:

$$D_i = D_0\left(c_0 - c_1 p_i\right) + f\left[1 - e^{-\beta\Delta p}\right]. \tag{7}$$

Then, on solving for p_i, we have

$$p_i = \frac{1}{c_1 D_0}\left[D_0 c_0 - D_i + f\left(1 - e^{-\beta \Delta p}\right)\right]. \tag{8}$$

Using Eqs. (1), (2), and (8) of Section 7.10, we obtain, finally, the stress resultant of the vascular smooth muscle, S,

$$S = \frac{D_i}{2}\frac{1}{c_1 D_0}\left[D_0 c_0 - D_i + f\left(1 - e^{-\beta \Delta p}\right)\right] - p_e \frac{(D_i + h)}{2} - \frac{c_2}{2} D_i e^{\alpha(D_i/D_0)}. \tag{9}$$

This is the length–tension curve of the vascular smooth muscle in coronary arterioles in a range of blood pressure not too far away from the normal in vivo value, say, ± 20 cm H_2O around the mean value of 60 cm H_2O. Consistent units should be used for the pressure, diameter, and the constants.

We can express the result in terms of blood shear stress, τ, acting on the endothelium by using Eq. (1) and replacing Δp in Eq. (9) by

$$\Delta p = \frac{4L}{D_i}\tau. \tag{10}$$

It is conventional to express τ in N/m^2 or dyne/cm^2. Then Δp should used the same units and β should be scaled accordingly. If the last two terms in Eq. (9) are negligible, we have

$$S = \frac{D_i}{2}\frac{1}{c_1 D_0}\left[D_0 c_0 - D_i + f\left(1 - e^{-\beta 4 L D_i^{-1}\tau}\right)\right]. \tag{11}$$

A sketch of this length–tension curve is given in Figure 7.11:3. Note that the stress resultant of the vascular smooth muscle S is equal to the product of a numerical factor $(2c_1 D_0)^{-1}$ and the product of two straight lines of equal and opposite slope; one intercept is the origin, and the other intercept is $D_0 c_0 + f(1 - e)^{-\beta \Delta p}$. From this construction the result is easily appreciated. The maximum possible muscle tension occurs when the diameter D_i is equal to the optimal value $D_{i\,\mathrm{opt}}$:

$$D_{i\,\mathrm{opt}} = \frac{1}{2}\left[D_0 c_0 + f\left(1 - e^{-\beta \Delta p}\right)\right], \tag{12}$$

and the maximum muscle tension per unit length of the vessel wall is

$$S_{\mathrm{max}} = \frac{1}{2c_1 D_0} D_{i\,\mathrm{opt}}^2 \tag{13}$$

Additional Information

Kuo et al. (1988, 1990a, 1990b, 1991a, 1991b, 1992) have published additional information about the variation of pressure and flow effects on coronary arteries with the size of the arterioles (or generation numbers), the loca-

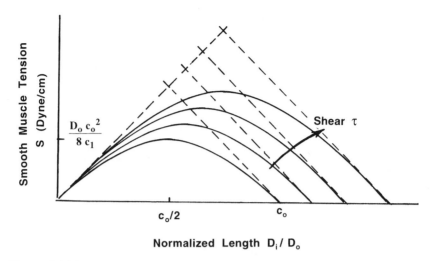

FIGURE 7.11:3. Relationship between the vascular smooth muscle tension and the muscle length influenced by the shear stress acting on the endothelium, as described by Eq. 11.

tion, and with various chemicals and disease (atherosclerosis). Davis (1993) presented data on the myogenic response gradient in an arteriolar network of hamster cheek pouch and an extensive discussion of the literature. The features of pressure and flow effects in the hamster cheek pouch are similar to those of pig coronary vessels. Davis (1993) presented data on the diameter at extremely low (toward zero) and high (to 200 cm H_2O) intraluminal pressure. At the low end of p_i, the D_i versus p_i curves became nonlinear and the vessels seemed to lose their myogenic response. Meininger and Davis (1992) reviewed the cellular mechanisms involved and the various hypotheses proposed and give the reader both a historical perspective and a summary of the current status of knowledge in the field.

With additional data, the length–tension equation derived earlier will be refined and improved. The data and formulas listed in Sections 7.10 and 7.11 are for pig coronary arterioles of order 4 to 5; with the larger end being of order 4 and the smaller end of order 5. In a new paper, Kuo et al. (1995) studied the myogenic response of pig coronary arteries of orders 4 to 6 as defined by Table 7.2:1. To summarize their results, let us limit the range of midpoint intralumen pressure p_i to the range 50 to 140 cm H_2O, so that the passive diameter D_0 and the range of maximum flow-induced dilation $f(p_i)$ are constant for each vessel. Since $f(p_i)$ and D_0 now depend on the order number N, we shall use the symbols $f(p_i|N)$, $D_0(N)$ to indicate the association of $f(p_i)$ and D_0 with the order number of the vessel. The results are as follows:

Order number N		4	5 Lower end	5 Upper end	6
$f(p_i|N)$	(%)	21 ± 3	32 ± 2	52 ± 8	22 ± 6
$D_0(N)$	(mm)	60 ± 2	96 ± 1	158 ± 5	245 ± 9

With these parameters, we can obtain the length–tension relationship for the vascular smooth muscle in these vessels from Eqs. (7.10:9) and (7.11:9). It is seen that arterioles at the larger end of order 5 are most sensitive to the shear stress of blood flow. Arteries (arterioles) of order 4 and small arteries of order 6 are less than half as sensitive.

7.12 Regulation and Autoregulation of Coronary Blood Flow

In bioengineering, when we say that the regulation of a phenomenon is understood, we mean the following: (a) A variable x, or a relationship among a set of variables $(x, y, \ldots)$ that best describes the phenomenon has been identified. (b) There exists a standard value of x or a standard relationship among $(x, y, \ldots)$ that is associated with a stable or optimal living condition. (c) There exists a sensor that can detect any deviation or error of the variable x from the standard value, or of the relationship among $(x, y, \ldots)$ from the standard one. The error is monitored all the time. (d) A mechanism to minimize the error exists. (e) The dynamics of error minimization is biologically satisfactory. In the sense just listed, the regulation of coronary blood flow is not understood. The variable x and its error sensor have not been identified, and a vigorous search for them is continuing (Johnson, 1978; Spaan, 1991).

The autoregulation of the pressure–flow relationship, as described in Section 5.14, however, is consistent with the length–tension relationship derived in Sections 7.10 and 7.11. If we plot the normal length of the coronary vascular smooth muscle cells and the tension in the muscle cells on a sheet of paper with length as abscissa and tension as ordinate, the point will lie on the right leg of the length-tension arch, as shown in Figure 7.11:3. If a step increase of local blood pressure is imposed on the vessel, then, before the smooth muscle has time to adjust itself, the blood vessel diameter will be increased because of elasticity, the resistance is reduced and the flow is increased. The vascular smooth muscle cells then shortens according to the length-tension curve, increases the resistance to flow, and returns the flow toward the normal value. This is a course of autoregulation. Further, if an artery is compressed and flow is stopped, the blood pressure is reduced, the tension in smooth muscle goes down, and the muscle cell lengthens according to the length-tension curve of Figure 7.11:3. The resistance to flow becomes small. Now if the occlusion is removed, the pressure gradient is reestablished, and the flow will be greater than normal because the resistance is low. But then the smooth muscle shortens, resistance is raised, and the flow is restored toward normal. This is a course of hyperemia.

Other factors must be considered. In the community of blood flow researchers, there is a widespread belief that oxygen concentration in the muscle must be a controlling factor of blood flow because during exercise oxygen consumption of the heart may increase by a factor of five. It is hypothesized that if oxygen is not supplied by the flow, some cells low in oxygen may call for more oxygen or threat to go into the anaerobic mode by sending out some signal or molecules to dilate the blood vessel. Such signal or molecules in muscle has not been found. People have thus been looking at other hypotheses, such as (a) the contraction itself; (b) sympathetic nerve stimulation; (c) altered cell membrane properties, leading to activation of ion channels for Ca^{2+} flux etc.; (d) modulation of biochemical cell-signaling pathways within the vascular smooth muscle, (e) a length-dependent change in contractile protein function; and (f) endothelial-dependent modulation of vascular smooth muscle tension. New literature has produced evidence for all of these; see the reviews by Cábel et al. (1994), Johnson (1991), Meininger and Davis (1992), Ping and Johnson (1992a,b, 1994).

Historically, William Bayliss (1902) was credited for the first report on the transient increases in blood volume of organs following brief periods of blood flow stoppage by local compression. He attributed the volume change to a local reduction of blood pressure, which was proposed by him as a stimulus. However, others pointed out that the vasodilatation could just as well be attributed to a buildup of vasodilator metabolites during the period of flow stoppage. The concept was largely discredited, until resurrected by Bjorn Folkow in 1949, who showed that arterioles constrict when the intraluminal pressure is elevated and dilate when the pressure falls. This sort of behavior is seen in the earthworm: You pull on it and it shortens. You push on it and it swells up. Looking for other causes of this phenomenon, Grande et al. (1977) showed that the arterial pulse provides a constrictor for arterioles. Thurau (1964) suggested that the arteriole senses a change in the circumferential wall tension because its plasma membrane is in series with the contractile element. Guharay and Sachs (1984) described the role of stretch-sensitive ion channels in muscle cells. This ninety year story shows the difficulty of interpreting the results of experiments which involve too many parameters. Thus there arose a trend to test simpler systems. The testing of isolated arterioles is an effort in this direction. Van Bavel (1989) worked on mesenteric arteries. Osol and Halpern (1985) worked on cerebral blood vessels. Kuo et al.'s work on coronary arterioles has been reviewed in sections 7.10 and 7.11. The length-tension relationship of coronary vascular smooth muscle, Eqs. (7.10:9) and (7.11:9), are derived from the data of Kuo et al. We hope that the effects of other factors can be added one by one. With these pieces of information, one can use system analysis to investigate larger systems on one hand, or seek molecular explanation on the other hand.

7.13 Pressure–Flow Relationship of Coronary Circulation

The health of a person depends on good perfusion of the heart muscle. Good perfusion of the heart must rely on the interaction between the coronary blood vessels and the myocardium, because every coronary blood vessel is either embedded in or attached to the heart muscle. The interaction has dominated coronary circulation research. Some well-known papers are: Anrep et al. (1927), Dole and Bishop (1982), Eng et al. (1982), Fibich et al. (1993). Hoffman and Spaan (1990), Judd and Levy (1991), Krams et al. (1989), Meininger et al. (1987), Nellis and Whitesell (1989), Porter (1898), Suga (1979), and Wiggers (1954). See Spaan (1991) for a well rounded account. In the following, however, we shall discuss only one of the most distinctive feature of the coronary circulation.

The most distinctive feature is the existence of a large zero-flow pressure gradient as shown in Figure 7.7:5 in Section 7.7. Hiramatsu et al. (1994) observed that at the zero-flow point the subendocardial arterioles and venules are open (are not seen to be collapsed). Earlier, Bellamy (1978) studied the zero-flow pressure phenomenon in coronary pressure–flow relationship in the circumflex coronary artery in long individual diastoles of dogs in a semibasal state, and said "the zero-flow pressure may represent the height of a vascular waterfall caused by vasomotor tone, with the resistance controlling coronary flow being $(p_c - p_{f=0}) \cdot F^{-1}$." Here p_c denotes the blood pressure in the circumflex artery, $p_{f=0}$ denotes the zero-flow pressure, and F denotes the flow (in ml/min measured in the circumflex artery), whereas the venous pressure is irrelevant. The reactive hyperemia after a short occlusion of the artery was used to vary p_c. Bellamy suggested that the cessation of flow was caused by collapse of arterioles. Hiramatsu et al. (1994) did not see the closing of arterioles on the heart surface, but intramyocardial arterioles remains to be studied. Bellamy claimed that the pressure–flow relationship is linear, so that $(p_c - p_{f=0}) F^{-1}$ is a constant resistance. Klocke et al. (1981, 1985) pointed out that the diastolic pressure–flow lines can be cuvilinear, especially at pressures close to zero-flow pressure.

Since the stress and strain in the myocardium, and pressure and flow in coronary blood vessels, are time varying and are nonuniformly distributed in a beating heart, the whole panorama cannot be represented by the measured pressure and flow in the coronary circumflex artery alone. It is difficult to pin down what occurs locally at the microcirculation level from the averages in the circumflex artery. Making unique unmistakable deductions about microcirculation from macrovascular data is difficult. For this reason,

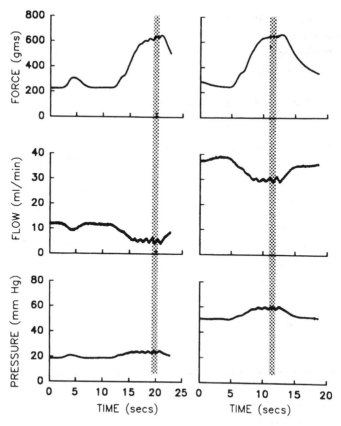

FIGURE 7.13:1. Illustration of how the pressure-flow relationship in coronary blood vessels in an interventricular septum is measured at different levels of muscle tension. Twitch and tetani from a specimen at 2 different pressure levels. Force developed during a tetanus is much greater than during a single beat. Hatched lines indicate the time at which pressure and flow were averaged (over 1s) to yield a point on pressure-flow relationship. The pressure at the venous outlet was zero (atmospheric). From Livingston et al. (1993), reproduced by permission.

Yin et al. (see Livingston et al., 1993) measured the coronary pressure–flow relationship in steady-state (tetanic) contractions in the maximally vasodilated isolated canine interventricular septum. Their use of ryanodine in the perfusate and electric stimulation allowed the production of reproducible and reversible tetani. Figure 7.13:1 provides an example of two tetani in one specimen. The figure on the left illustrates a twitch followed by a tetanus, and the figure on the right illustrates a tetanus initiated at a different level of diastolic pressure and flow. Each data point was obtained by averaging the records over 1 s. An additional protocol was used to induce two contractile states by adding 2,3-butanedione monoxime to the perfusate to obtain an intermediate contractile level.

Figure 7.13:2 shows representative diastolic and tetanized pressure–flow relationships. It is seen that tetanization significantly increases the zero-flow pressure, and the relationship is somewhat curved in the neighborhood of $p_{f=0}$. The results achieved from fitting the pooled data from 10 specimens into a quadratic expression or a cubic expression are shown in Figure 7.13:3. The effect of an intermediate level of contraction is shown in Figure 7.13:4. It is seen that the curves shift to the right when contractility is increased. In this preparation of interventricular septum, the heart muscle tissue was flat, the ventricles were cut away, and the tension of each edge of the specimen was adjusted to be apparently uniform. Hence, the spatial distribution of the interaction between muscle cells and capillary blood vessels should be quite uniform. Therefore, the results provide strong support for the concept of the existence of waterfall resulting from the muscle–microcircualtion interaction. The results show that the interaction depends on the contractility of the muscle and is independent of ventricular pressure.

The most important message given by these curves is that the pressures at the exit ends of the veins play no role at all in determining the flow in the arteries. It is as if the flows in the coronary arteries and veins are decoupled, analogous to the flow in a river with a waterfall. The venous flow is affected by contraction of heart muscle, but in its own way.

The nature of the zero-flow point on a pressure–flow curve was further clarified by Tomonaga et al. (1984). These authors continued to decrease arterial pressure during a long diastole and did not halt when the flow reached zero but continued to monitor the retrograde flow. The curves

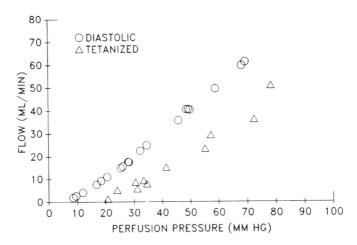

FIGURE 7.13:2. Diastolic and tetanized pressure-flow relationships while the venous outlet pressure is zero. Individual data points were obtained from a series of tetani such as those shown in Figure 7.13:1. Average weight of perfused bed was estimated to be $26 \pm 8.4 \, \text{g}$ (mean $\pm$ SD, $n = 33$), from which the flow per unit mass can be computed. From Livingston et al. (1993). Reproduced by permission.

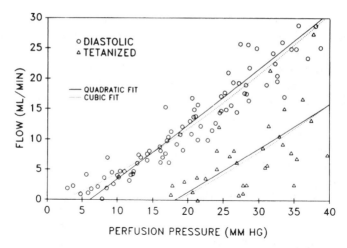

FIGURE 7.13:3. Details of how the zero-flow condition is approached from above. Quadratic (solid line) and cubic (dotted line) regression fits to pooled data from 10 specimens. From Livingston et al. (1993), reproduced by permission.

are continuous in the negative flow regime. Retrograde flow is a back flow from the blood trapped in the veins. A review of the morphometric data on the coronary veins and venules presented in Section 7.3 clearly shows that the geometry of the venules is especially favorable to endow compliance to

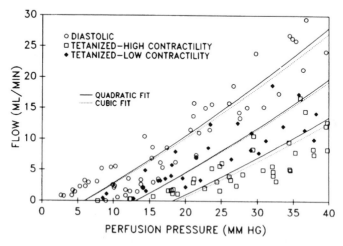

FIGURE 7.13:4. Illustration showing how the zero-flow pressure difference varies with muscle tension. Quadratic (solid line) and cubic (dotted line) regression fits to pooled data from 6 specimens. From Livingston et al. (1993). Reproduced by permission.

the venules. The features of the pressure–flow relationship in the neighborhood of the zero-flow point are discussed further in the following section.

7.14 Model of Coronary Waterfall

In the early development of a theory to explain the coronary waterfall discussed in the preceding section, there were two concepts that confused the issue. One is the belief that the capillary blood vessels are rigid tubes. The other is the belief that waterfall must be identified with the critical closing of blood vessels due to the action of vascular smooth muscle cells. We have discussed the former in Section 7.9. The capillaries are elastic, and the matter is no longer controversial. The latter has its origin in the critical closing theory of Alan Burton. Burton was a famous biophysicist whose writings in the 1940 to 1970 period brought biomechanics to physiology. His book, entitled *Physiology and Biophysics of the Circulation* (1965), was very popular. On p. 79 of that book he discussed instability of small blood vessels under vasomotor tone. His reasoning is illustrated by Figure 7.14:1. In considering the equilibrium of the vessel, Burton saw elasticity as a necessary condition for stability and active tension as destabilizing. He then went on to say, "while the cooperation of elastic tissue with the smooth muscle can stabilize the equilibrium of the wall under vasomotor tone, instability of those blood vessels which can develop large amounts of active tension (such as arterioles) will occur when the vessel is constricted enough so that the elastic fibers are unstretched. There is no elastic tension below this point of unstretched circumference. Consequently, theory predicts, and there is ample experimental verification in peripheral vascular beds, that vasomo-

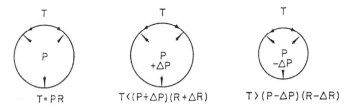

FIGURE 7.14:1. Burton drew 3 tubes to show how, without elasticity in the wall of a blood vessel, active tension (vasomotor tone) would lead to instability of equilibrium. In the first tube, the muscle tension T is equal to the product of internal pressure P and the lumen radius R, and the system is in equilibrium. In the second tube, the pressure is increased to $P + \Delta P$, the radius is increased to $R + \Delta R$, but the muscle tension is unchanged. The equilibrium is lost. The tube will continue to be enlarged, leading to blow out. In the third tube, the pressure is decreased, the muscle tension does not change, and the tube radius will continue to decrease, leading to closure.

tor tone not only tends to decrease the lumen diameter of the controlling vessels (e.g., the arterioles) but also creates a tendency to close altogether. This will happen for a given transmural pressure if the active tension rises above a critical value, or with a given degree of active tension if the pressure falls below a critical value. This has been called *critical closure* of blood vessels. . . . The critical vessels, which close in these circumstances, are undoubtedly the arterioles."

Burton's theory is based on two hypotheses: (a) Buckling is equivalent to closing. (b) The tension in the vascular smooth muscle is unaffected by the buckling of the vessel wall. Let us examine these hypotheses. The first hypothesis is an oversimplification of the real situation, which has been discussed in detail in Chapter 4, Sections 4.2 to 4.4. At a critical buckling load, a blood vessel will become unstable in the sense described in Section 4.2. The postbuckling mode may have a large deformation, but still have a finite lumen area. From the point of view of mechanics, the first hypothesis is unfounded. The second hypothesis is contradicted by the experimental findings about vascular smooth muscle. The experimental results described in Sections 7.10 and 7.11 show that if the muscle cell length is shorter then a certain optimal value, the tension decreases with further shortening; at a lower limit the tension becomes zero.

Since these two hypotheses are unfounded, the conclusions derived from them must be denied. We have to conclude that the critical closure theory is not applicable to coronary blood vessels. It is then not surprising that Burton's statement that arterioles must be closed at the zero-flow point is contradicted by direct observations in vivo, as described by Kajiya et al. (Hiramatsu et al., 1993); see Section 7.7.

Removal of these inapplicable concepts allows us to consider the entire coronary vasculature with regard to the waterfall phenomenon. Experiments described in Section 7.13 have shown that the phenomenon results from muscle–vessel interaction. At the first line of interaction are the coronary capillaries. The blood pressure in the arterial tree reaches the lowest value in the capillaries. The intramuscle cell pressure (see Sec. 7.12) acts on the outside of the blood vessels. The transmural pressure of the blood vessels is $p_{blood} - p_{mus}$. In systole, p_{mus} increases. The first place at which p_{mus} can be equal to p_{blood} is on the walls of the capillaries. When $p_{mus} > p_{blood}$ and the blood vessel is buckled, the blood vessel lumen will be distorted. In the capillaries, the red blood cells and white blood cells are squeezed, and the resistance to flow can be increased without the blood vessel having to be collapsed entirely. Quantitative studies of microcirculation in this situation have been made in skeletal muscle, see Sections 8.5, 8.10. The corresponding myocardial case should be similar. The author believes that the interaction between cardiac muscle cells and the coronary capillaries and the red and white blood cells is the most likely cause of the waterfall phenomenon in the heart.

Comparison with Waterfall in Other Organs

In Chapter 4 we discussed waterfall in systemic veins. In Chapter 6 we considered the lung in the zone-2 condition, when pulmonary alveolar gas pressure is smaller than pulmonary arteriolar blood pressure. In these cases the circuit of flow has a region with a resistance so high that it effectively decouples the downstream portion of the circuit from the upstream portion. The high resistance is caused by the collapsibility of blood vessels subjected to external pressure. In the case discussed in Chapter 4, the vein is collapsible along its entire length. In the case of the lung, the veins are patent, whereas the pulmonary capillaries are collapsible, and waterfall occurs at the junctions of capillaries and veins. In Section 6.17 we have shown that the collapsible pulmonary capillary narrows at the venular junction but flow goes through. In Section 6.18 we have shown that if some pulmonary capillaries narrowed so much as to stop flow at the venular junction, then the whole capillary will be collapsed and go out of action. These collapsed pulmonary capillaries can be "recruited" again to become patent under suitable conditions.

The external pressure on systemic vein comes from body fluid. The external pressure on pulmonary capillaries comes from the alveolar gas. The external pressure on coronary blood vessels comes from the heart muscle. Heart muscle is a solid, and its active tension generates intracellular pressure (Sec. 7.9), which acts on coronary blood vessels. Whereas sluicing in the lung occurs at a sharp gate, sluicing in the heart can take place in long narrow tunnels crowded with blood cells.

References

Anderson, W.D., Anderson, B.G., and Seguin, R.J. (1988). Microvasculature of the bear heart demonstrated by scanning electron microscopy. *Acta Anat.* **131**: 305–313.

Anrep, G.V., Cruickshank, E.W.H., Downing, A.C., and Subba, R.A. (1927). The coronary circulation in relation to the cardiac cycle. *Heart* **14**: 111–133.

Anversa, P., and Capasso, J.M. (1991). Loss of intermediate-sized coronary arteries and capillary proliferation after left ventricular failure in rats. *Am. J. Physiol.* **260** (*Heart Circ. Physiol.* 29): H1552–H1560.

Bassingthwaighte, J.B., King, R.B., and Roger, S.A. (1989). Fractal nature of regional myocardial blood flow heterogeneity. *Circ. Res.* **65**: 578–590.

Bassingthwaighte, J.B., Yipintsoi, T., and Harvey, R.B. (1974). Microvasculature of the dog left ventricular myocardium. *Microvasc. Res.* **7**: 229–249.

Bayliss, W.M. (1902). On the local reaction of the arterial wall to changes of internal pressure. *J. Appl. Physiol.* **28**: 220–231.

Bellamy, R.E. (1978). Diastolic coronary pressure-flow relations in the dog. *Circ. Res.* **43**: 92–101.

Borg, T.K., and Caulfield, J.B. (1981). The collagen matrix of the heart. *Fed. Proc.* **40**: 2037–2041.

Bouskela, E., and Wiederhielm, C. (1979). Microvascular myogenic reaction in the wing of the intact unanesthetized bat. *Am. J. Physiol.* **239**: H59–H65.

Brown, R. (1965). The pattern of the microcirculatory bed in the ventricular myocardium of domestic mammals. *Am. J. Anat.* **116**: 355–374.

Burton, A.C. (1965). *Physiology and Biophysics of the Circulation.* Year Book Medical Publishers, Chicago.

Cábel, M., Smiesko, V., and Johnson, P.C. (1994). Attenuation of blood flow-induced dilation in arterioles after muscle contraction. *Am. J. Physiol.* **266**: H2114–H2121.

Chilian, W.M., and Marcus, M.L. (1982). Phasic coronary flow velocity in intramural and epicardial coronary arteries. *Circ. Res.* **50**: 775–781.

Chilian, W.M., Eastham, C.L., and Marcus, M.L. (1986). Microvascular distribution of coronary vascular resistance in beating left ventricle. *Am. J. Physiol.* **251**: H779–H788.

Cox, R.H. (1982). Mechanical properties of the coronary vascular wall and the contractile process. In *The Coronary Artery*, (S. Kalsner, ed.), Oxford Univ. Press, New York, Ch. 2, pp. 59–90.

Davis, M.J. (1993). Myogenic response gradient in an arteriolar network. *Am. J. Physiol.* **264**: H2168–H2179.

Dole, W.P., and Bishop, V.S. (1982). Influence of autoregulation and capacitance on diastolic coronary artery pressure-flow relationships in the dog. *Circ. Res.* **51**: 261–270.

Eng, C., Jentzer, J.H., and Kirk, E.S. (1982). The effects of the coronary capacitance on the interpretation of diastolic pressure-flow relationships. *Circ. Res.* **50**: 334–341.

Fibich, G., Lanir, Y., and Liron, N. (1993). Mathematical model of blood flow in a coronary capillary. *Am. J. Physiol.* **265**: H1829–H1840.

Folkow, B. (1949). Intravascular pressure as a factor regulating the tone of the small vessels. *Acta Physiol. Scand.* **17**: 289–310.

Fung, Y.C. (1966). Theoretical considerations of the elasticity of red cells and small blood vessels. *Fed. Proc.* **25**: 1761–1772.

Fung, Y.C. (1993). *Biomechanics: Mechanical Properties of Living Tissues.* 2nd ed., Springer-Verlag, New York.

Fung, Y.C., Zweifach, B.W., and Intaglietta, M. (1966). Elastic environment of the capillary bed. *Circ. Res.* **19**: 441–461.

Goto, M., Flynn, A.E., Doucette, J.W., Jansen, C.M.A., Stork, M.M., Coggins, D.L., Muehrcke, D.D., Husseini, W.K., and Hoffman, J.I.E. (1991). Cardiac contraction affects deep myocardial vessels predominantly. *Am. J. Physiol.* **261**: H1417–H1429.

Grande, P.-O., Lundvall, J., and Mellander, S. (1977). Evidence for a rate-sensitive regularory mechanism in myogenic microvascular control. *Acta Physiol. Scand.* **99**: 432–447.

Guharay, F., and Sachs, F. (1984). Stretch-activated single ion channel currents in tissue-cultured embryonic chick skeletal muscle. *J. Physiol. Lond.* **352**: 685–701.

Hiramatsu, O., Goto, M., Yada, T., Kimura, A., Tachibane, H., Ogasawara, Y., Tsujioka, K., and Kajiya, F. (1994). Diameters of subendocardial arterioles and venules during prolonged diastole in canine left ventricles. *Circ. Res.* **75**: 393–399.

Hoffman, J.I.E., and Spaan, J.A.E. (1990). Pressure-flow relations in coronary circulation. *Physiol. Rev.* **70**: 331–390.

Jiang, Z.L., and He, G.C. (1990). Microstructural components in canine coronary arteries and veins. *Acta Anat. Sinica* **21**: 348–352.

Jiang, Z.L., and He, G.C. (1991). Biomechanical properties of dog coronary arteries. I. Uniaxial loading. *Chinese J. Biomed. Eng.* **10**: 9–18.

Johnson, P.C. (ed.) (1978). *Peripheral Circulation*. John Wiley, New York.

Johnson, P.C. (1991). The myogenic response. *News in Physiol Sci.* **6**: 41–42.

Judd, R.M., and Levy, B.L. (1991). Effects of barium-induced cardiac contraction on large- and small-vessel intramyocardial blood volume. *Circ. Res.* **68**: 217–225.

Kanatsuka, H., Lamping, K.G., Eastham, C.L., Marcus, M.L., and Dellsperger, K.C. (1991). Coronary microvascular resistance in hypertensive cats. *Circ. Res.* **68**: 726–733.

Kassab, G.S., and Fung, Y.C. (1994). Topology and dimensions of pig coronary capillary network. *Am. J. Physiol.* **267** *(Heart Circ. Physiol. 36)*: H319–H325.

Kassab, G.S., and Fung, Y.C. (1995). The pattern of coronary arteriolar bifurcations and the uniform shear hypothesis. *Anna. Biomed. Eng.* **23**: 13–20.

Kassab, G.S., Rider, C.A., Tang, N.A., and Fung, Y.C. (1993a). Morphometry of pig coronary arterial trees. *Am. J. Physiol.* **265** *(Heart Circ. Physiol. 34)*: H350–H365.

Kassab, G.S., Imoto, K., White, F.C., Rider, C.A., Fung, Y.C., and Bloor, C.M. (1993b). Coronary arterial tree remodeling in right ventricular hypertrophy. *Am. J. Physiol.* **265** *(Heart Circ. Physiol.)*: H366–H377.

Kassab, G.S., Lin, D.H., and Fung, Y.C. (1994a). Morphometry of pig coronary venous systems. *Am. J. Physiol.* **267** *(Heart Circ. Physiol. 36)*: H2100–H2113.

Kassab, G.S., Lin, D.H., and Fung, Y.C. (1994b). Consequences of pruning in morphometry of coronary vasculature. *Ann. Biomed. Eng.* **22**: 398–403.

Kassab, G.S., Berkley, J., and Fung, Y.C. (1996). Analysis of pig's coronary arterial blood flow with detailed anatomical data. *Ann. Biomed. Eng.* **25**.

Klocke, F.J., Mates, R.E., Canty, Jr., J.M., and Ellis, A.K. (1985). Coronary pressure-flow relationships. Controversial issues and probable implications. *Circ. Res.* **56**: 310–333.

Klocke, F.J., Weinstein, I.R., Klocke, J.F., Ellis, A.K., Kraus, D.R., Mates, R.E., Canty, J.M., Anbar, R.D., Romanowski, P.R., Wallmeyer, K.W., and Echt, M.P. (1981). Zero-flow pressures and pressure-flow relationships during single long diastoles in the canine coronary bed before and during maximum vasodilatation. *J. Clin. Invest.* **68**: 970–980.

Krams, R., Sipkema, P., and Westerhof, N. (1989). Can coronary systolic-diastolic flow difference be predicted by left ventricular pressure or time varying intramyocardial elastance? *Basic Res. Cardiol.* **84**: 149–159.

Kuo, L., Davis, M.J., and Chilian, W.M. (1988). Myogenic activity in isolated subepicardial and subendocardial coronary arterioles. *Am. J. Physiol.* **255**: H1558–H1562.

Kuo, L., Chilian, W.M., and Davis, M.J. (1990a). Coronary arteriolar myogenic response is independent of endothelium. *Circ. Res.* **66**: 860–866.

Kuo, L., Davis, M.J., and Chilian, W.M. (1990b). Endothelium-dependent, flow induced dilation of isolated coronary arteries. *Am. J. Physiol.* **259**: H1063–H1070.

Kuo, L., Chilian, W.M., and Davis, M.J. (1991a). Interaction of pressure-and flow-induced responses in porcine coronary resistance vessels. *Am. J. Physiol.* **261**: H1706–H1715.

Kuo, L., Davis, M.J., and Chilian, W.M. (1991b). Alteration of arteriolar responses during atherosclerosis. In *Resistance Arteries, Structure and Function*, (M.J. Mulvany, et al., eds.), Elsevier Sci. Pub.

Kuo, L., Davis, M.J., Cannon, M.S., and Chilian, W.M. (1992). Pathophysiological consequences of atherosclerosis extend into the coronary microcirculation. *Circ. Res.* **70**: 465–476.

Kuo, L., Davis, M.J., and Chilian, W.M. (1995). Longitudinal gradients for endothelium-dependent and-independent vascular responses in the coronary microcirculation. *Circulation* **92**: 518–525.

Livingston, J.Z., Resar J.R., and Yin, F.C.P. (1993). Effect of tetanic myocardial contraction on coronary pressure-flow relationships. *Am. J. Physiol.* **265**: H1215–H1226.

Ludwig, G. (1971). Capillary pattern of the myocardium. *Methods Achive. Exp. Pathol.* **5**: 238–271.

Meininger, G.A., and Davis, M.J. (1992). Cellular mechanisms involved in the vascular myogenic response. *Am. J. Physiol.* **263**: H647–H659.

Meininger, G.A., Mack, C.A., Fehr, K.L., and Bohlen, H.G. (1987). Myogenic vasoregulation overrides local metabolic control in resting rat skeletal muscle. *Circ. Res.* **60**: 861–870.

Mohl, W., Wolner, E., and Glogar, D. (eds.) (1984). *The Coronary Sinus.* Springer-Verlag, New York.

Nellis, S.H., and Whitesell, L. (1989). Phasic pressures and diameters in small epicardial veins of the unrestrained heart. *Am. J. Physiol.* **257**: H1056–H1061.

Nussbaum, A. (1912). Über das Gefass-system des Herzens. *Arch. Mikrobiol. Anat.* **80**: 450–477.

Osol, G., and Halpern, W. (1985). Myogenic properties of cerebral blood vessels from normotensive and hypertensive rats. *Am. J. Physiol.* **249**: H914–H921.

Ping, P., and Johnson, P.C. (1992). Role of myogenic response in enhancing autoregulation of flow during sympathetic nerve stimulation. *Am. J. Physiol.* **263**: H1177–H1184.

Ping, P., and Johnson, P.C. (1992). Mechanism of enhanced myogenic response in arterioles during sympathetic nerve stimulation. *Am. J. Physiol.* **263**: H1185–H1189.

Ping, P., and Johnson, P.C. (1994). Arteriolar network response to pressure reduction during sympathetic nerve stimulation in cat skeletal muscle. *Am. J. Physiol.* **266**: H1251–H1259.

Porter, W.T. (1898). The influence of the heart beat on the flow of blood through the walls of the heart. *Am. J. Physiol.* **1**: 145–163.

Rabbany, S.Y., Funai, J.T., and Noordergraaf, A. (1994). Pressure generation in a contracting myocyte. *Heart and Vessels* **9**: 169–174.

Rabbany, S.Y., Kresh, J.Y., and Noordergraaf, A. (1989). Intramyocardial pressure: interaction of myocardial fluid pressure and fiber stress. *Am. J. Physiol.* **257**: H357–H364.

Rakusan, K., and Wicker, P. (1990). Morphometry of the small arteries and arterioles in the rat heart: effects of chronic hypertension and exercise. *Cardiovasc. Res.* **24**: 278–284.

Scaramucci, J. (1695). Theoremata familiaria viros eruditos consulentia de variis physico-medicis lucubrationibus juxta leges mecanicas. *Apud. Joannem Baptistan Bustum*, pp. 70–81.

Schmid-Schönbein, G.W., Lee, S.Y., and Sutton, D. (1989). Dynamic viscous flow in distensible vessels of skeletal muscle microcirculation: Application to pressure and flow transients. *Biorheology* **26**: 215–227.

Spaan, J.A.E. (1985). Coronary diastolic pressure-flow relation and zero flow pressure explained on the basis of intramyocardial compliance. *Circ. Res.* **56**: 293–309.

Spaan, J.A.E. (1991). *Coronary Blood Flow.* Kluwer Academic Pub., Dordrecht, The Netherlands.

Spaan, J.A.E., Breuls, N.P.W., and Laird, J.D. (1981). Diastolic-systolic flow differences are caused by intramyocardial pump action in the anesthetized dog. *Circ. Res.* **49**: 584–593.

Spalteholz, W. (1907). Die Koronararterien des Herzens. *Verh. Anat. Ges.* **21**: 141–153.

Suga, H. (1979). Total mechanical energy of a ventricular model and cardiac oxygen consumption. *Am. J. Physiol.* **236**: H498–H505.

Thurau, K. (1964). Autoregulation of renal blood flow and glomerular filtration rate, including data on peritubular capillary pressures and wall tension. *Circ. Res.* **14**, Suppl. **1**: 132–141.

Tillmanns, H., Ikeda, S., Hansen, H., Sarma, J.S.M., Fauvel, J.-M., and Bing, R.J. (1974). Microcirculation in the ventricle of the dog and turtle. *Circ. Res.* **34**: 561–569.

Tillmanns, H., Steinhausen, M., Leinberger, H., Thederan, H., and Kubler, W. (1981). Pressure measurements in the terminal vascular bed of the epimyocardium of rats and cats. *Circ. Res.* **49**: 1202–1211.

Tomonaga, G., Tsujioka, K., Ogasawara, Y., Nakai, M., Mito, K., Hiramatsu, O., and Kajiya, F. (1984). Dynamic characteristics of diastolic pressure-flow relation in the canine coronary artery. In: *The Coronary Sinus*, (W. Mohl, E. Wolner, and D. Glogar, eds.), Springer-Verlag, New York, pp. 79–85.

Van Bavel, E. (1989). Metabolic and myogenic control of blood flow studied on isolated small arteries. PhD Thesis. Univ. Amsterdam, The Netherlands.

Wearn, J.T. (1928). The extent of the capillary bed of the heart. *J. Exp. Med.* **47**: 273–292.

Wiggers, C.J. (1954). The interplay of coronary vascular resistance and myocardial compression in regulating coronary flow. *Circ. Res.* **2**: 271–279.

Yada, T., Hiramatsu, O., Kimura, A., Goto, M., Ogasawaray, Y., Tsujioka, K., Yamamori, S., Ohno, K., Hosaka, H., and Kajiya, F. (1993). In vivo observation of subendocardial microvessels of the beating porcine heart using a needle-probe videomicroscope with a CCD camera. *Circ. Res.* **72**: 939–946.

8
Blood Flow in Skeletal Muscle

8.1 Introduction

Continuing the biomechanical analysis of specific organs in accordance with detailed anatomical data, we present in this chapter blood flow in skeletal muscles. A systematic approach should begin with the collection of geometric and materials data, determination of the constitutive equations of the materials, derivation of basic equations according to the laws of physics and chemistry, and formulation of meaningful boundary value problems, and then proceed to solve the equations, predict results, and validate the predictions with animal experiments. Such a program has been carried out by Schmid-Schönbein and his associates for blood flow in skeletal muscle. We present their main results here.

In this chapter we emphasize the experimental side of biomechanics. It will be seen that an essential part of biomechanics consists of anatomical, histological, and electron microscopic observations and measurements, which we have emphasized throughout the book. In this chapter, however, we bring morphometry to the center stage, and the reader will see how far morphometric results have led us.

8.2 Topology of Skeletal Muscle Vasculature

The arteries and veins of the pulmonary vasculature presented in Chapter 6 have a treelike topology, but the capillary sheets of the lung have a network topology. The arteries and veins of the coronary vasculature presented in Chapter 7 have a treelike topology, but the capillary blood vessels of the heart have a network topology. The coronary veins have some arcades and anastomoses, but they are not extensive. The distinctive feature of the anatomy of the skeletal muscle is that its arterioles and venules form arcades and its capillaries are bundlelike (Figures 8.2:1 to 8.2:4). In other words, the *treelike topology* of the vasculature of the skeletal muscle ends at the level of arterioles and venules. In muscle, the arterioles and venules

514

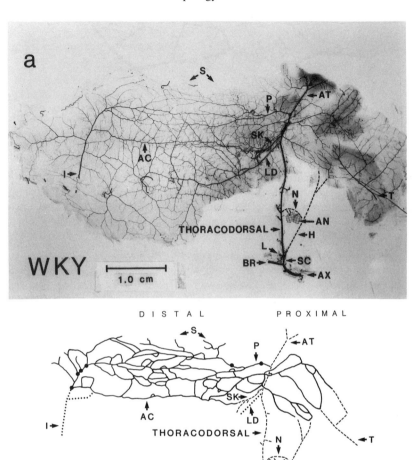

FIGURE 8.2:1. *Top*. Micrograph of artery network filled with resin in the spino-trapezius (SPI) and latissimus dorsi (LTD) muscle of WKY rat. *Bottom*. Sketch of the arcade network of the spinotrapezius and its connections with the central arteries. Black dots represent common junctions between the SPI and LTD networks. Main blood supplies are from thoracodorsal, scapular circumflex (H), transverse cervical (T), and two branches (I and S) of the intercostal arteries. Symbols for other vessels are: AC = artery in SPI, AN = axillary lymph nodes, AT = artery to adipose tissue, AX = axillary artery, BR = brachial artery, L = a branch to LTD and other muscles, LD = artery to LTD, N = connection to AN, P = large feeding artery to SPI, SC = subcostal, SK = artery to skin. From Schmid-Schönbein et al. (1986b), by permission.

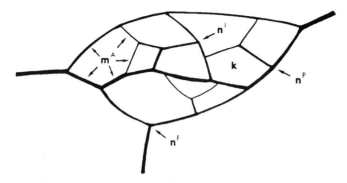

Figure 8.2:2. A schematic of the arteriolar or venular, network. n^F represents feeder vessels, n^P is a perimeter node, n^I is an interior node, k is the number of arcade loops, and m_A is the number of vessel segments comprising a single arcade loop. From Schmid-Schönbein et al. (1989a), by permission.

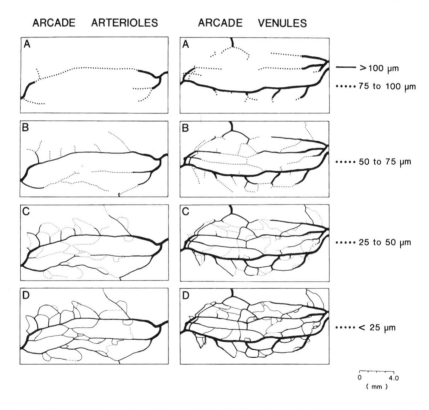

Figure 8.2:3. Tracings of arcade arterioles (*left*) and venules (*right*) from a spino-trapezius muscle in a WKY rat. For clarity no transverse arterioles or collecting venules are shown. In (A) vessels with diameter $>100\,\mu m$ are shown by continuous lines, and vessels $75–100\,\mu m$ by the dotted lines. In (B) vessels $>75\,\mu m$ are displayed by continuous line, and vessels $50–75\,\mu m$ are dotted. In (C) and (D) vessels $25–50\,\mu m$ and below $25\,\mu m$, respectively, are dotted. From Schmid-Schönbein et al. (1987a), by permission.

form networks called *arcades*. The arcades are connected by treelike *transverse arterioles* and *collecting venules* to the capillaries. The topology of the muscle capillaries are *bundlelike*. These features were recognized by Spalteholz (1888) and have been characterized quantitatively by Schmid-Schönbein and his associates in recent years.

Figure 8.2:1 shows an arteriolar network in the rat spinotrapezius muscle. It is supplied with blood by a set of feeder vessels, and is drained by a set of transverse arterioles. A schematic diagram of such an arteriolar network is shown in Figure 8.2:2 (without showing the transverse arterioles). A set of tracings of arcade arterioles and venules are shown in Figure 8.2:3.

The transverse arterioles are treelike. They take roots in the arteriolar arcade, branch dichotomously, and feed blood into capillaries. An illustration of a transverse arteriole is shown in Figure 8.2:4.

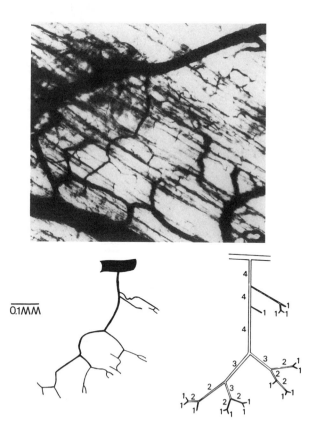

FIGURE 8.2:4. A transverse arteriole in the spinotrapezius muscle of the rat. *Top*: photographic montage. *Bottom left*, a reconstruction. *Bottom right*, a schematic branching pattern with Strahler order numbers. From Engleson et al. (1985a), by permission.

The capillary blood vessels of the skeletal muscle are interconnected over long distances (several millimeters) parallel to the muscle fibers, somewhat like the capillaries of the heart sketched in Figure 7.4:1, but longer. The network shown in Figure 8.2:5 is typical of the rat spinotrapezius muscle. It is seen that the transverse arterioles and collecting venules feed and drain the capillary bundles in alternating sequences.

Flow in the capillaries is drained by collecting venules, whose geometry is similar to that of the transverse arterioles. The collecting venules drain blood into the venular arcades, which are similar to, but considerably denser than, the arteriolar arcades (Fig. 8.2:6).

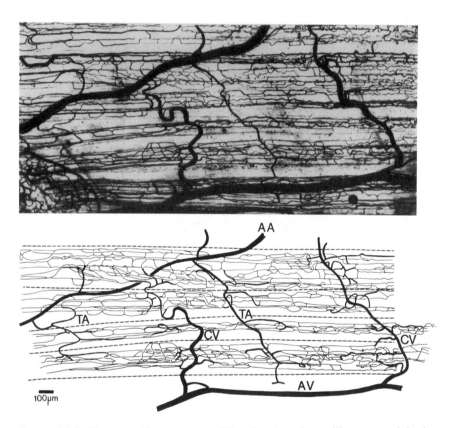

FIGURE 8.2:5. Photographic montage and line drawing of a capillary network in the rat spinotrapezius muscle. A segment of the arcade arterioles (AA), arcade venules (AV), and transverse arterioles (TA), and collecting venules (CV) are displayed. Four capillary bundles are shown; their divisions are enhanced by the dashed lines. Note the repeated arrangement of transverse arteriole and collecting venules along the capillary bundles. Diameters are not drawn to scale. From Skalak and Schmid-Schönbein (1986a), by permission.

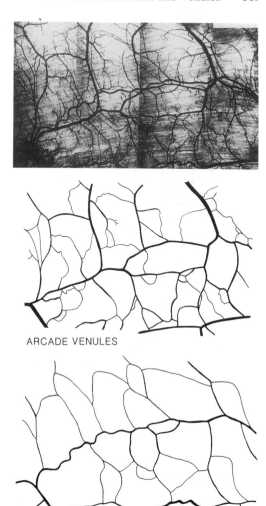

FIGURE 8.2:6. Comparison of arcade arterioles and venules in rat spinotrapezius muscle. *Top*: photographic montage of carbon–filled vasculature. *Bottom*: A reconstruction of the arcade venules and arterioles. From Engelson et al. (1985b), by permission.

ARCADE VENULES

ARCADE ARTERIOLES   1MM

8.3 Skeletal Muscle Arterioles and Venules

Statistical data about the parameters characterizing the arcades, the transverse arterioles, the collecting venules, and the capillaries have been published by Schmid-Schönbein et al. The availability of these tables makes it possible to analyze the blood flow in skeletal muscles with detailed anatomical data.

Table 8.3:1 shows the data on the feeding and arcade arterioles in the rat spinotrapezius muscle. The entries are self-explanatory. See Figure 8.2:2 for

TABLE 8.3:1. Topology of Feeding and Arcade Arterioles in Spinotrapezius Muscle

Muscle volume	$V = 115\,\text{mm}^3$
Number of feeding arterioles to the muscle	$n^F = 3$
Mean luminal diameter of feeding arterioles	$\overline{D} = 81.0\,\mu\text{m}$
Number of arterioles per arcade loop	$m^A = 5.1$
Number of internal nodes, i.e., bifurcations inside an arcade	$n^I = 30$
Number of peripheral nodes, along the arcade perimeter	$n^P = 22$
Number of nodes, total in the arcade meshwork	$n = 55$
Number of arcade arterioles excluding feeder arterioles	$n^A = 81$
Number of arcades in the meshwork	$k = 27$
Length of arcades per unit volume	$L = 1.17\,\text{mm/mm}^3$
Mean length of arcade arterioles (81 vessels)	$\overline{L} = 1{,}660\,\mu\text{m}$
Mean luminal diameter of arcade arterioles (81 vessels)	$\overline{D} = 43.3\,\mu\text{m}$
Average spacing between transverse arterioles	$z = 190\,\mu\text{m}$

From Engelson, Skalak, Schmid-Schönbein (1985a). These quantities are not all independent of each other. In the original paper, and also in Engelson, Schmid-Schönbein and Skalak (1985b), several formulas relating some of these quantities are given.

a schematic drawing of these arterioles and the use of some of the symbols. Another example is seen in Figure 8.4:1.

The transverse arterioles and collecting venules are treelike, as can be seen in Figures 8.2:4 and 8.2:5. As trees, the Strahler system described in Sections 6.2 and 7.2 can be used to characterize their branching pattern. For the relatively small trees of transverse arterioles and collecting venules of the skeletal muscle, it was found that no iteration was required to satisfy the *diameter-defined* definition of order numbers. It seems that the first try is sufficient in every case. Two examples are given in Table 8.3:2. The symbol

TABLE 8.3:2. Prediction of Transverse Arteriole Topology by Strahler Technique

	Order	No. of branches per order		Mean diameter for each order (μm)		Mean length for each order (μm)	
		Observed	Predicted	Observed	Predicted	Observed	Predicted
Transverse arteriole A		$R_B = 2.56$		$R_D = 1.64$		$R_L = 1.33$	
	1	16	(16)				
	2	6	6.3	5.4	(5.4)	123.8	(123.8)
	3	2	2.4	6.6	8.7	153.0	164.7
	4	1	1.0	13.5	14.3	219.0	219.0
Transverse arteriole B		$R_B = 2.92$		$R_D = 1.38$		$R_L = 1.32$	
	1	65	(65)				
	2	25	22.3	5.4	(5.4)	92.6	(92.6)
	3	7	7.6	6.6	7.5	168.3	126.9
	4	2	2.6	8.9	10.3	337.0	173.8
	5	1	0.9	14.0	14.2	95.0	238.1

From Engelson, Skalak, Schmid-Schönbein (1985a).

TABLE 8.3:3. Mean Diameters (± SD) of Second Through Fourth Order Collecting Venule in the WKY and SHR

Rat strain	Branching order		
	2	3	4
WKY	$8.5 \pm 2.4\,\mu$m	$12.7 \pm 3.3\,\mu$m	$19.4 \pm 2.1\,\mu$m
	$(n = 23)$	$(n = 9)$	$(n = 6)$
SHR	$7.0^* \pm 1.8\,\mu$m	$11.5 \pm 4.8\,\mu$m	$20.4 \pm 8.3\,\mu$m
	$(n = 44)$	$(n = 13)$	$(n = 3)$

*Significantly different from WKY at $p < 0.003$.
All other values differ insignificantly $(p > 0.05)$.
Data from Engelson, Schmid-Schönbein, and Zweifach (1985b).

R_B denotes the *branching number ratio*, i.e., the average ratio of the number of vessels of order n divided by the number of vessels of order $n + 1$. The average ratio of the diameter of a vessel of order $n + 1$ divided by the diameter of a vessel of order n is called the *diameter ratio* and is designated R_D. The average ratio of the length of a vessel of order $n + 1$ divided by the length of a vessel of order n is called the *length ratio* and is designated R_L. A tree whose R_B, R_D, and R_L are independent of n is said to satisfy *Horton's law*. The data in Table 8.3:2 show that the assumption that the transverse arteriole and collecting venule trees satisfy Horton's law leads to predictions that are in reasonable agreement with the experimentally observed values. Hence, it is concluded that the branching pattern of the transverse arteriole and collecting venule trees can be described by the Strahler system and obey Horton's law.

Table 8.3:3 shows the diameter (mean ± SD) data of the collecting venules of Strahler orders 2, 3, and 4. Data for two strains of rats are shown. WKY is the Wistar Kyoto rat, which is a popular strain of normal rat. SHR is the spontaneous hypertensive rat, which is genetically programmed to increase its blood pressure at a specific age. Table 8.3:4 shows other para-

TABLE 8.3:4. Rat Muscle Collecting Venules* of Orders 3, 4, and 5 (mean ± SD)

Rat strain	Order no.	R_B	N_C	R_D	R_L	L_{max}[†] (μm)	L_{total}[‡] (μm)
WKY	3.9 ± 0.6	3.15 ± 0.5	26 ± 12	1.49 ± 0.16	2.37 ± 1.54	793 ± 491	$1{,}834 \pm 1{,}169$
SHR	4.0 ± 0.8	3.11 ± 0.4	29 ± 21	2.25 ± 0.1	1.86 ± 1.13	524 ± 324	$1{,}444 \pm 1{,}203$

*Seventeen collecting venules from eight muscles.
[†] L_{max} is the longest route measured from the parent arcade venule to the most distant capillary.
[‡] L_{total} is the total length of all the vessels per tree.
R_B = branching ratio; N_C = number of capillaries; R_D = diameter ratio; R_L = length ratio.
Data from Engelson, Schmid-Schönbein, and Zweifach (1985b).

TABLE 8.3:5. Rat Muscle Venular Arcade Network Parameters (mean, $n = 3$)

Rat strain	L	K	$\overline{L}$	z	N per volume	N per arcade
WKY	2.11	1.003	815	190	12.2	5.5
SHR	2.18	1.137	684	166	13.2	5.5

L = length of arcade venules per unit volume (mm/mm³); K = number of arcade loops per unit volume (mm⁻³); $\overline{L}$ = node-to-node mean arcade venule length (μm); z = mean venous length between collecting venule (μm); N per volume = number of collecting venules per unit volume (mm⁻³); N per arcade = mean number of venules per arcade loop (m^A).
Data from Engelson, Schmid-Schönbein, and Zweifach (1985b).

meters of these collecting venules. Table 8.3:5 shows the data on venular arcade. Mean and SD are shown here. Some raw data and histograms are given in the original papers. It is shown that the diameters have log-normal distribution, and the scatter of length is large. There are consistent differences between WKY and SHR, but most differences are statistically not significant. WKY order 2 collecting venules have a diameter significantly larger ($p < 0.003$) than those of the SHR. The statistical frequency distribution of the length is highly skewed.

8.4 Capillary Blood Vessels in Skeletal Muscle

A capillary blood vessel network in rat spinotrapezius muscle is shown in Figure 8.2:5. The relationship between the capillaries and the arterioles and venules is sketched in Figure 8.4:1. We call such a capillary network a *bundle-like* network. One can recognize in Figure 8.2:5 that there exist few

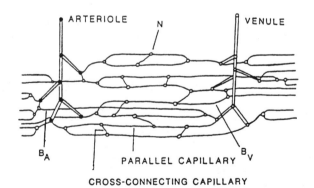

FIGURE 8.4:1. Schematic capillary bundle element. In this network, $P = 9$, $A = 8$, $V = 10$, $N = 23$, $B_A = 2$, $B_V = 2$, and $C = 0$. See text for explanations. From Skalak and Schmid-Schönbein (1986a), by permission.

or no capillary connections between neighboring capillary bundles (see the dashed lines in Fig. 8.2:5). Connections between bundles are provided by the transverse arterioles and collecting venules, which may interconnect several bundles. Figure 8.4:1 shows a segment of capillary bundle from a transverse arteriole to its neighboring collecting venule. Such a segment is called a *capillary bundle element*. An entire bundle consists of repeated bundle elements. Each element has P vessels parallel to the muscle fibers; in Figure 8.4:1, $P = 9$. In each element, the number of capillary vessels that connect to the arterioles is A; in Figure 8.4:1, $A = 8$. The total number of capillary nodes between the arteriole and venule is N; in Figure 8.4:1, $N = 23$. At the point where a transverse arteriole connects into the capillary bundle, not all capillary pathways may intersect with the arteriole, that is, the transverse arteriole may feed only a part of the bundle. The number of pathways passing the arteriole without nodal connections is designated as the number of arteriolar capillary back connections, B_A. They form bridges between neighboring capillary bundle elements. Similarly, the number of venular back connections is B_V. Finally, the number of capillary connections between neighboring bundles is designated C. In Figure 8.4:1, $B_A = 2$, $B_V = 2$, and $C = 0$.

Statistical distributions of P, A, V, N, B_A, B_V, and C are presented in dimensionless form in Skalak and Schmid-Schönbein (1986a). The statistical distribution frequency functions of capillary length and capillary diameter are also given. As can be expected, the histogram of capillary length is highly skewed. The distribution of capillary diameter, however, is almost symmetric, as is illustrated in Figure 8.4:2.

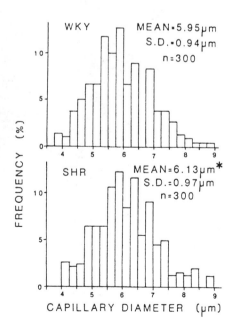

FIGURE 8.4:2. Histograms of capillary diameters which were measured at random points in the network at 62 mm Hg transmural pressure in the Wistar Kyoto rat (WKY) and the spontaneously hypertensive rat (SHR). From Skalak and Schmid-Schönbein (1986a), by permission.

The capillary blood vessels in skeletal muscle, however, are not isolated tubes. They are embedded in muscle. The relationship between the capillaries and the muscle cells is most important to the blood flow in muscle, as we have discussed in Sections 7.9, 7.13, and 7.14 in connection with the heart. It is, however, only in the skeletal muscle that more direct observations have been made. To this important topic we now turn.

Schmid-Schönbein et al. have shown that the interstitial space in skeletal muscle is very small in intact muscle, and the capillary blood vessels are in close contact with the muscle cells. See Figure 8.4:3(A) from Lee and

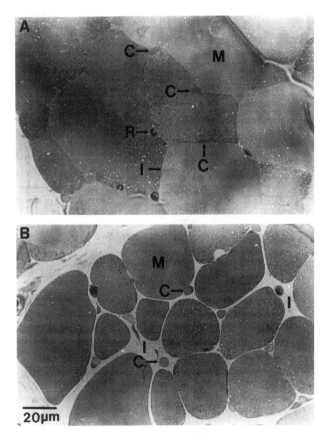

FIGURE 8.4:3. (A) Micrograph of a cross section of rat gracilis muscle fixed at 25 cm H₂O with intact fascia. The skeletal muscle fibers (M) form a compact tissue with narrow interstitial spaces (I). The capillaries (C) contain red blood cells (R). (B) Micrograph of the same muscle fixed at 25 cm H₂O after its fascia was removed. The interstitial space (I) is considerably enlarged. The capillaries do not maintain contact with the muscle fibers and appear more circular in cross-section. From Lee and Schmid-Schönbein (1995), by permission.

Schmid-Schönbein (1995). If the fascia (i.e., the connective tissue sheath enveloping the muscle) was removed, the interstitial space could increase greatly and the capillaries would appear as tubes floating in a fluid [Fig. 8.4:3(B)]. Figure 8.4:4 shows a similar result presented by Mazzoni et al. (1990). Figure 8.4:5 shows the increase of interstitial space when the fascia is removed. These pictures provide clear evidence that in an intact muscle, the capillary blood vessels are so close to the muscle cells that the distensibility of the vessels must depend on the interaction between the vessels and the muscle fibers. The concept of capillary blood vessels as isolated tubes is applicable only to injured muscles when the fascia is torn.

Lee and Schmid-Schönbein (1995) have obtained electron micrographs of capillaries in the rat gracilis muscle at various perfusion pressures. In muscles with intact fascia, the capillaries are deformed, as shown in Figure 8.4:6. Since the mass of the muscle is far bigger than that of the capillary vessel wall, it is clear that the deformation of the vessel cross section is mainly a function of the deformation of the muscle. Historically, the "tunnel

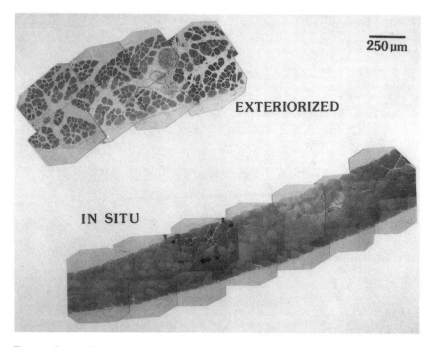

FIGURE 8.4:4. Comparison between exteriorized and in situ muscle preparations. Photomontages of micrographs of muscle cross sections. The exteriorized preparation had fascia removed and 6-0 silk sutures added for strain measurements. The in situ preparation had the fascia left intact. The large amount of white space in the exteriorized muscle reflects a large increase in the interstitial volume. From Mazzoni et al. (1990), by permission.

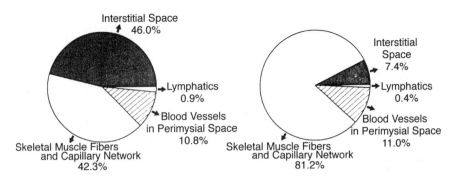

EXTERIORIZED

IN SITU
WITH INTACT FASCIA

Interstitial Space
46.0%

Lymphatics
0.9%

Blood Vessels
in Perimysial Space
10.8%

Skeletal Muscle Fibers
and Capillary Network
42.3%

Interstitial
Space
7.4%

Lymphatics
0.4%

Blood Vessels
in Perimysial Space
11.0%

Skeletal Muscle Fibers
and Capillary Network
81.2%

FIGURE 8.4:5. Fractions of tissue components in exteriorized and in situ muscle preparations. The interstitial space is increased from 7.4% in the in situ preparation to 46.0% in the exteriorized preparation. The lymphatic space is increased from 0.4% to 0.9%. From Mazzoni et al. (1990), by permission.

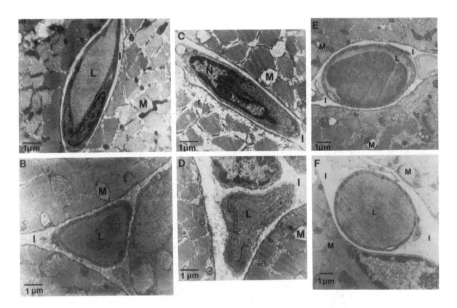

FIGURE 8.4:6. (A) Electron micrograph of a capillary cross section with a nucleus (N) located between two muscle fibers (M) fixed at 25 cm H_2O intravascular pressure with intact muscle fascia. Capillary lumen (L) is nearly elliptical and interstitial space (I) is small. (B) Capillary among three muscle fibers (M) at 25 cm H_2O with intact muscle fascia. Lumen (L) is triangular and interstitial space (I) is small. (C) Capillary fixed at zero intravascular pressure with intact fascia. Lumen (L) is collapsed and interstitial space (I) is narrow. (D) Capillary fixed at zero transmural pressure with intact muscle fascia. Lumen (L) has collapsed. (E) Capillary with transmural pressure 50 cm H_2O, and (F) capillary at 100 cm H_2O, have fascia intact, interstitial spaces small, and lumen circular. From Lee and Schmid-Schönbein (1995), by permission.

in a gel" concept of capillaries was developed by Fung (1966) and Fung et al. (1966) to explain the low compliance of capillaries in the mesentery. Obviously, in intact skeletal muscle, one may consider the corresponding "tunnel in the muscle" concept.

The analysis of resistance to flow in capillaries and the measurement of the mechanical properties of the capillaries in muscle to be presented in the following two sections are based on the observations illustrated in Figures 8.4:3 to 8.4:6.

8.5 Resistance to Flow in Capillaries

Flow in capillary blood vessels, whose geometric shape, size, and elastic environment have been described in the preceding section, can be determined theoretically or experimentally. In Section 5.6 we have examined the velocity–hematocrit relationship in the capillaries. In Figure 5.6:5 we have shown that at a bifurcation point where a capillary blood vessel branches into two daughters, the branch with slower velocity often does not get red cells. Consider these branches first. The fluid is plasma, which is a Newtonian fluid with a constant coefficient of viscosity. The flow is governed by the Stokes equation [Section 5.7, p. 291, Eq. (1)]. The Reynolds number approaches zero. The method of solution has been illustrated in Sections 5.7 to 5.9. Numerical solution of the equations yields velocity contours in the capillaries with an irregular lumen, as shown in Figure 8.5:1. From the velocity distribution, the shear strain rate and the shear stress are computed. An integration of the shear force around the periphery yields the resistance to flow. Theoretical results are shown in Figure 8.5:2, in which the hemodynamic resistance for different transmural pressures is presented relative to the average capillary resistance at 25 cm H_2O with intact fascia. A dramatic increase of the resistance at capillary pressures below normal physiological values is seen. Hence, squeezing the capillary blood vessel by the muscle cells and slowing down the flow may easily lead to flow stoppage.

8.6 Mechanical Properties of Muscle Capillaries

With fascia of a skeletal muscle removed, the capillaries appear as isolated tubes enveloped in a fluid (Fig. 8.6:1). The elasticity of the capillary blood vessel can then be measured by changing the blood pressure, as is done for the larger arteries (see Chapter 8 in *Biomechanics: Mechanical Properties of Living Tissues*, Fung, 1993). From these data Lee and Schmid-Schönbein (1995) have extracted data on the distensibility of the capillaries. When the fascia of the skeletal muscle is left intact, the capillaries are compressed by the muscle cells. When the inflating blood pressure is increased, the vessel

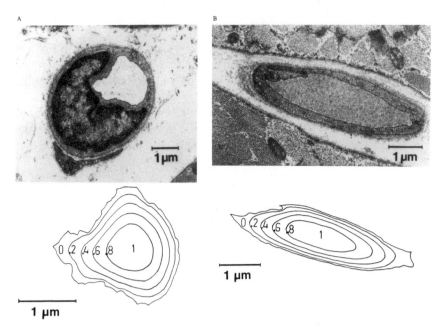

FIGURE 8.5:1. Flow in two capillary blood vessels in rat skeletal muscle. (Top). Electron micrograph. (Bottom). Velocity contours obtained by solving Stokes equation with boundaries shown in the micrographs above. The numbers indicate the velocity relative to the midstream velocity which was normalized to 1. From Lee and Schmid-Schönbein (1995), by permission.

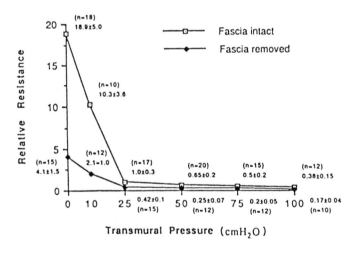

FIGURE 8.5:2. Theoretical capillary resistance (mean ± SD) for plasma flow as a function of the transmural pressure. Resistance was normalized with respect to the average resistance in the gracilis muscle capillaries at 25 cm H_2O with intact fascia. n = number of cross-sections from 3 muscles at each pressure. From Lee and Schmid-Schönbein (1995), by permission.

528

deformation is the resultant of the effects of vessel wall elasticity and the elasticity of the muscle. The experimental results can be used to evaluate the muscle elasticity in a direction perpendicular to the muscle fiber. Thus, a comparison of the distensibility of the capillary blood vessel in the fascia-removed and the fascia-intact cases is useful.

The results obtained by Lee and Schmid-Schönbein (1995) in fascia-removed muscle in resting state are given in Figure 8.6:2, in which the average capillary lumen cross-sectional area and the "aspect ratio" of the normal cross-section, that is, the ratio of the minor axis length divided by the major axis length of the normal cross section, are displayed as functions of the transmural pressure (the internal blood pressure minus the interstitial pressure, which is assumed to be zero, i.e., atmospheric). Here we see that the capillaries in skeletal muscle are distensible when the fascia of the

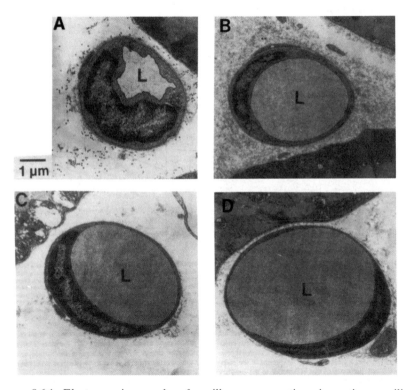

FIGURE 8.6:1. Electron micrographs of capillary cross sections in resting gracilis muscle with fascia removed. The muscle was fixed at (A), zero transmural capillary pressure; (B), 25 cm H_2O, (C), 50 cm H_2O, and (D), 100 cm H_2O. The interstitial space is enlarged. The capillaries are not in physical contact with muscle cells. From Lee and Schmid-Schönbein (1995), by permission.

muscle is removed and the capillary appears as a tube enveloped in an interstitial fluid. These results may be compared with those in other organs. We know that capillaries in the lung (see Chapter 6, Sec. 6.8), and the capillaries in the bat wings are distensible, because the surrounding tissue is small. In stretched mesentery, the capillaries appear almost rigid (see Fung, 1966, or Fung, 1993, Sec. 8.10), which was explained by the large gel-like tissue surrounding the capillaries (Fung et al., 1966) and the "tunnel in gel" concept. In contrast, Figure 8.6:2 shows that the skeletal capillaries are dis-

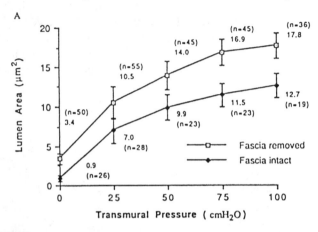

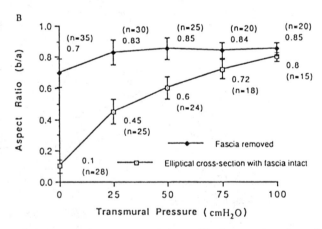

FIGURE 8.6:2. (A) Average lumen area of the capillaries as a function of the capillary transmural pressure. n = number of observations derived from three muscles at each pressure. Numbers on curves are mean lumen area. Standard deviations are shown by the vertical bars. (B) The aspect ratio (b/a) of the capillary lumen at different transmural pressures on cross sections perpendicular to the muscle fibers. The muscle was in resting state in both fascia-removed and fascia-intact cases. From Lee and Schmid-Schönbein (1995), by permission.

tensible, even in the fascia-intact case. This means that the resting, relaxed gracilis muscle surrounding the capillaries is soft in the direction normal to the muscle fiber, much softer than the stretched mesenteric connective tissue surrounding the mesenteric capillaries. One could surmise that in a contracting skeletal muscle or in a tetanized condition, the distensibility of the capillaries will be much smaller. Documentation of capillary elasticity in tetanized muscle would be valuable.

Lee and Schmid-Schönbein (1995) provided data on the inner and outer perimeters of the endothelial cells, the thickness of the cell wall, the thickness of the basement membrane, and the cross-sectional lumen area, and used the Laplace formula to compute the wall tension (membrane stress resultant), $T = Pr$, where P is the local transmural pressure and r is the approximate radius of the inner wall of the capillary blood vessel. Assuming that the circumferential stress resultant is borne by the basement membrane, they computed the mean stress $\bar{\sigma} = Pr/b$, where b is the thickness of the basement membrane. These are plotted against the membrane Green's strain $E = (\lambda^2 - 1)/2$, where λ is the stretch ratio in the circumferential direction. The results are given in Figure 8.6:3 which shows the typical nonlinearity. The thickness of the basement membrane was measured from the electron micrograph. It is nearly 600 ± 200 Å (mean $\pm$ SD) at zero transmural pressure, 400 ± 150 Å at 25 cm H_2O, 250 ± 100 Å at 50 cm H_2O, and 200 ± 80 Å at 100 cm H_2O. The authors went on to calculate the distortion

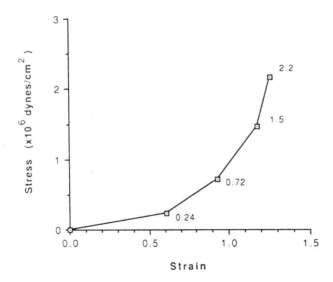

FIGURE 8.6:3. Circumferential stress-strain relationship of isolated capillary blood vessels in gracilis muscle of Sprague-Dawley rats with fascia removed. Capillary wall is not in contact with muscle fibers. It is assumed that all the circumferential membrane tension is carried by the basement membrane. From Lee and Schmid-Schönbein (1995), by permission.

of the capillary lumen as seen in the electron microscopy and showed how large the resistance to flow could be at a lower transmural pressure when the vessel was flattened, cell nuclei bulged into the lumen, and pseudopods formed. They studied the plasma lemmal vesicles (caveolae) in endothelial cells, hypothesized that the vesicles are naturally curved cell membrane, and showed that this hypothesis implies the existence of "cortical" tension in the cell wall.

8.7 Constitutive and Hemodynamic Equations of Skeletal Muscle Vasculature

The mechanical properties of blood vessels have been discussed in detail in Chapter 8 of *Biomechanics: Mechanical Properties of Living Tissues* (Fung, 1993). For the small arteries and veins and capillaries of rat spinotrapezius muscle, Skalak and Schmid-Schönbein (1986b) measured the change of vessel diameters in response to step changes of blood pressure. The muscle was superfused with papaverine (0.5 mg/ml) to abolish the smooth muscle active tension. The immediate change of diameter in response to pressure change is defined as the *elastic* response. The creep that followed is attributed to viscoelasticity. See equations in Section 2.6, pp. 56–58. They found that it is sufficient to regard the vessel wall as linearly elastic as far as the relationship between the transmural pressure P and the finite circumferential strain E is concerned. Hence,

$$P(t) = \text{blood pressure} - \text{external pressure} = \alpha E,$$

$$E(t) = \frac{1}{2}(\lambda^2 - 1), \tag{1}$$

where α is a constant and λ is the circumferential stretch ratio referred to the zero-stress state, approximately equal to the diameter of the neutral surface of the vessel wall divided by the diameter of the neutral surface when the transmural pressure is zero. To analyze the creep curves, they used the following equation of viscoelasticity:

$$P + \frac{\beta}{\alpha_1}\dot{P} = \alpha_2 E + \beta\left(1 + \frac{\alpha_2}{\alpha_1}\right)\dot{E}, \tag{2}$$

in which $\dot{P} = dP/dt$, $\dot{E} = dE/dt$; and α_1, α_2, and β are constant coefficients. The values of α_1, α_2, and β are determined and presented by Skalak and Schmid-Schönbein (1986b).

The equations of motion presented in Section 2.6 are sufficient to describe blood flow in skeletal muscle. Schmid-Schönbein (1988) considered the effect of hematocrit on the apparent coefficient of viscosity of the blood in small blood vessels, and the elastic deformation of the vessel due to blood pressure. He then applied these equations to a vascular circuit consistent with the anatomical details presented in Sections 8.2 to 8.4, and thereby obtained results that are unique for the skeletal muscle.

8.8 Pulsatile Flow in Single Vessel

Pulsation in blood flow has many important consequences. For example, in Section 8.11 we explain that pressure pulsation in arterioles and venules is probably the main reason why the terminals of our lymph system can work. In Chapter 3 we studied the features of pulsatile heart pump and wave propagation throughout the circulatory system. These waves are produced by an interplay of the inertial force of the blood, the elastic force of the blood vessel wall, and the pulsation of the heart. They persist in the skeletal muscle. Inside the muscle bundle the Reynolds and Womersley numbers of blood flow are so small that the inertial force is negligible and the pulsations reflect the viscoelasticity of the vessel wall and blood.

Schmid-Schönbein (1988) has shown, on the basis of Eqs. (1) and (2) of Section 8.7, that a step change of blood pressure will induce a creep flow that changes with time, and a harmonically oscillating pressure gradient will cause an oscillatory flow with hysteresis. The hysteresis increases with increasing frequency. Sutton and Schmid-Schönbein (1989) have shown that a step change of flow induces a transient change of pressure gradient. The change depends on the hematocrit. Sutton et al. (1989) constructed a high-precision pump to study these phenomena. One of their results is shown in Figure 8.8:1 in the form of a clockwise hysteresis loop that increases in magnitude as the frequency increases. This frequency effect has an influence on the zero-flow pressure gradient as shown in Figure 8.8:1. In Sections 7.7 and 7.13, we saw that the existence of a finite zero-flow pressure gradient is an important feature of coronary blood flow. It was not known that the zero-flow pressure gradient has something to do with the viscoelasticity of the materials and the frequency of the oscillation. The analysis of Schmid-Schönbein calls attention to the effects of stress rate and strain rate on the finite zero-flow pressure gradient phenomenon. Blood flow in the skeletal muscle also has finite zero-flow pressure gradient (Section 8.10) which, however, is much smaller than that in the coronary blood flow. Nevertheless, the mechanism is of great interest.

8.9 Blood Flow in Whole Skeletal Muscle

To analyze blood flow, anatomical data should be simulated by a circuit, and the laws of conservation of mass, momentum, and energy must be obeyed. By solving the field equations according to specified boundary conditions, one can predict the function of the system in specific circumstances. One can then validate the predictions against the results of pertinent in vivo

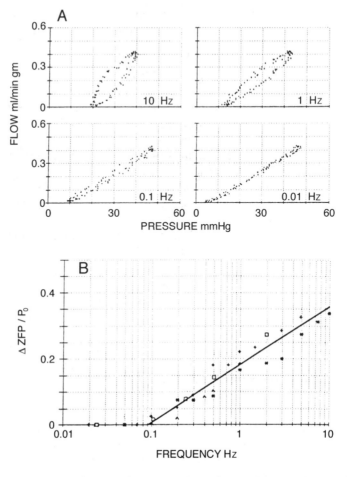

FIGURE 8.8:1. The viscoelasticity of the blood vessel–muscle system has an effect on capillary blood flow in skeletal muscle. (A) The pressure-flow relationship of capillary blood flow in spinotrapezius muscle subjected to a sinusoidally oscillating flow at four frequencies. At the same amplitude of flow, higher frequency attenuates pressure amplitude and increases hysteresis. (B) The zero-flow pressure (ZFP) increases as a function of frequency starting at 0.09 Hz. This increase (ΔZFP = ZFP − ZFP at 0.01 Hz) is normalized by the pressure amplitude at 0.01 Hz, which is denoted by P_o. The four symbols correspond to data obtained from four animals. From Sutton and Schmid-Schönbein (1989), by permission.

animal experiments to confirm or falsify the hypotheses made in formulation of the mathematical model. The task of validation is often the most difficult and most rewarding work to do. Theories and experiments have difficulties and limitations. A feasible experiment may be too difficult for theoretical analysis, or vice versa; or some needed pieces of information about the experimental animal may be unavailable. If all the difficulties

are overcome and a true validation of the theory is obtained, then the
results are often valuable. In the validation process lies the chance of
discovery.

Examples of problems of skeletal muscle circulation that can be formulated and solved theoretically and validated experimentally are the
following:

1. The pressure–flow relationship of a whole organ in steady, pulsatile, or
   transient conditions
2. The distribution of the blood pressure and blood flow in the organ
3. The relationship between the nonuniformity of blood flow and the
   demand of oxygen or other nutrients by the muscle cells
4. The possibility of local ischemia (insufficient supply of blood flow
   or oxygen to a region) and its consequences, possible recovery, or
   divergence into pathological condition.

Schmid-Schönbein and his associates used the morphological model
shown in Figure 8.4:1 for the rat spinotrapezius and gracilis muscles to solve
some of these problems. They wrote down equations of continuity and
Stokes' equation of motion, used approximations similar to Poiseuille's,
and obtained a set of equations similar to those presented in Section 7.5.
The method of solution is also similar. The circuit and the boundary conditions are different. Efficient algorithms for the inversion of sparse
matrix described by Eisenstat et al. (1982) and Vlach and Singhal (1983)
was used. The variation of blood viscosity on shear rate and vessel size, and
the effect of viscoelasticity of the blood vessel were taken into account by
a method of iteration. The purpose of such an elaborate construction is,
of coure, to reduce the number of ad hoc hypotheses, and to make the
comparison of theoretical predictions and in vivo measurements more
meaningful.

Sutton (1987) performed a series of flow experiments on rat gracilis
muscle. Figure 8.9:1 shows a comparison of model predictions and in vivo
results. Also shown for comparison purposes are data from Pappenheimer
and Maes (1942) for perfusion of an isolated dog hind limb, and from
Braakman (1988) for skeletal muscle. The agreement between theoretical
results and in vivo data is good.

Computation based on detailed morphological data yield pressures in
vessels of all sizes. Figure 8.9:2 shows a comparison between the predicted
and measured longitudinal pressure distribution. The normalized pressure
is defined by

$$\bar{p} = \frac{P_{ves} - P_{ven}}{P_{art} - P_{ven}}, \tag{1}$$

in which the subscripts "ves," "art," and "ven" indicate, respectively, vessel,
arterioles of 50 μm diameter, and veins of 50 μm diameter. Each point rep-

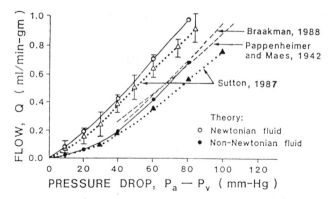

FIGURE 8.9:1. Comparison of theory and experiment on blood flow in rat skeletal muscle. *Dotted lines:* experiment by Sutton (1987). *Solid lines:* theory. *Open triangles:* experimental flow with blood plasma, bars represent the range in 10 animals. *Closed triangles:* experimental flow with a hematocrit of 15%. *Open circles:* prediction with Newtonian fluid. *Closed cricles:* prediction with non-Newtonian blood. *Dashed lines:* experimental results by other authors. From Schmid-Schönbein et al. (1989a), by permission.

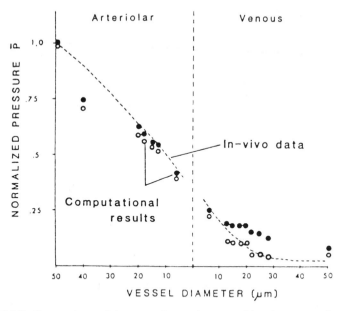

FIGURE 8.9:2. Comparison of theory and experiment on blood pressure distribution in arterioles, capillaries, and venules up to 50 μm in diameter in rat spinotrapezius muscle. The pressure, $\bar{P}$, is normalized according to Eq. (1). In vivo data are from Zweifach et al. (1981) for exteriorized spinotrapezius muscle. Computational results are shown for a Newtonian fluid (1.2 cp, *open circles*) and blood (*solid circles*). From Schmid-Schönbein et al. (1989a), by permission.

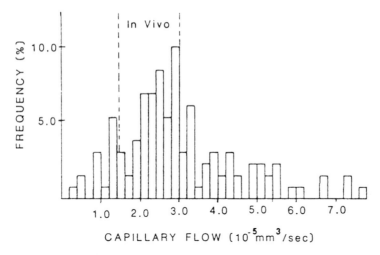

FIGURE 8.9:3. Nonuniform capillary blood flow distribution is shown by the frequency distribution of capillary flow in 100 randomly selected capillaries in the spinotrapezius muscle model with blood perfused at normal perfusion pressures. From Schmid-Schönbein et al. (1989a), by permission.

resents the mean value among at least 10 vessels selected at random within each hierarchy, both in theoretical calculations and in vivo. The agreement between the computed results and in vivo data is good in general.

Figure 8.9:3 shows the frequency distribution of capillary flow for 100 capillaries selected at random throughout the network model for the spinotrapezius muscle, for an arterio-venous pressure drop of 38 mm Hg with about 15% hematocrit. The ability to make such a prediction is a virtue of the morphological model. Without the morphological detail, it is not possible to calculate the dispersion of capillary blood flow. The dispersion, or nonhomogeneity, of capillary blood flow is important for assessment of the function of muscle cells. The magnitudes of the capillary flows are in the same range as those observed in vivo by Zweifach et al. (1981). In vivo measurements are stochastic as well, tending to measure the flow in longer, straighter capillaries.

This brief account of the blood flow in skeletal muscle is of course terribly incomplete. There is a huge literature on this subject. To begin a study, the following references are recommended: Delashaw and Duling (1988), Granger et al. (1984), Hudlická (1973), Kioller et al. (1987), Krogh (1919), Langille (1993), Lee et al. (1987), Lindbom and Arfors (1985), Popel (1987), Popel et al. (1988).

Tissue Remodeling—Arcade Arteriole Formation

It is well known that blood vessels remodel under stress (Fung, 1990, Chapter 13; and Fung, 1993, Chapter 8, Secs. 8.12 to 8.16). By analyzing the stress changes in the blood vessels of skeletal muscle when arterial pressure is increased, Price and Skalak (1994, 1995) tested the hypothesis that terminal arteriolar remodeling stimulated by elevated levels of circumferential wall stress will proceed in a network pattern that gives rise to new arcade arterioles. This hypothesis is based on Price et al.'s (1994) observation that new muscle cells are formed in terminal arterioles when the circumferential stress exceeds a certain threshold value. The new and old smooth muscle cells were identified by immunohistochemical means. Their calculations show that new smooth muscle cells will grow from the terminal arteriole to some capillaries, which cross a collecting venule and then connect to adjacent transverse arteriole trees, thus forming an arcade loop. This analysis is interesting and full of implications.

8.10 Finite Zero-Flow Arterial Pressure Gradient in Skeletal Muscle

In Section 7.13, it is seen that the finite zero-flow pressure gradient is a major feature of coronary blood flow. The same is true in skeletal muscle. Figures 8.10:1 and 8.10:2 show the detailed experimental results of Sutton and Schmid-Schönbein (1989) on rat gracilis muscle, which is a thin sheet

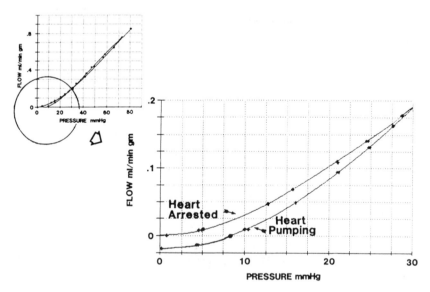

FIGURE 8.10:1. Pressure-flow curves in rat gracilis muscle obtained before and after cardiac arrest in a single animal using cell-free medium. Flow was the controlled variable. Zero-flow pressure (ZFP) was zero when heart was arrested. Retrograde flow was measured by controlling pressure. From Sutton and Schmid-Schönbein (1989), by permission.

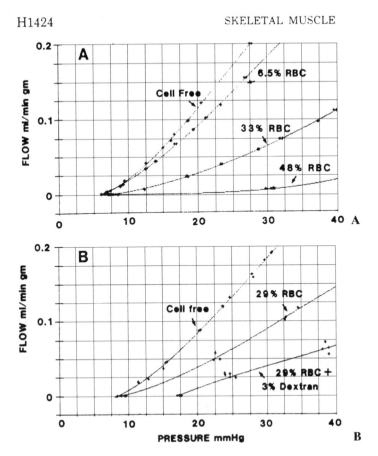

FIGURE 8.10:2. Effect of red blood cells (RBC) on finite zero-flow pressure gradient in steady flow in rat gracilis muscle. (A) The RBCs are dispersed in fluid. No significant increase in zero-flow pressure (ZFP) compared with cell-free fluid perfusion. (B) Dextran (77 kDa)–induced aggregation of RBCs increases the zero-flow pressure. From Sutton and Schmid-Schönbein (1989), by permission.

between the inner hind limb and the groin. Figure 8.10:1 shows that when the muscle is perfused with a cell-free medium that simulated the rat serum together with an anticoagulant and a vasodilator, the flow becomes zero when the arterial blood pressure is about 8 mm Hg and the venous pressure is zero (atmospheric). After arresting the heart of the animal with an anesthetic overdose, the pressure–flow curve goes through the origin and the zero-flow pressure gradient is zero. Figure 8.10:2 shows that when non-aggregated rat red blood cells are added to the perfusate, the resistance is increased but the zero-flow pressure gradient does not change. When 77 kDa dextran is added to the perfusate in a fraction of 2% to 3%, the red blood cells form aggregates and the zero-flow arterial pressure is increased by 9.4 mm Hg. Hence the zero-flow pressure is related to the apparent viscosity of blood.

Hypotheses explaining the finite zero-flow pressure gradient in coronary arteries have been reviewed in Section 7.14. Lee and Schmid-Schönbein (1995), after discussing a number of existing theories, proposed a new theory that the zero-flow pressure is caused by a communication with the central circulation via some collateral vessels. They have found by direct ink infusions of the gracilis muscle that there are some vessels ($<100 \mu m$) which connect to the central circulation while bypassing the femoral artery. These arteriolar vessels are perpendicular to the tissue beneath the gracilis muscle and are hidden from view when the muscle is left in situ. They may counteract the perfusing arteriolar pressure.

This author's opinion is that the finite zero-flow pressure gradient is related intimately to the interaction between the blood vessel and muscle, resulting in a compression on the capillaries by the muscle. A systematic study of the lateral elasticity of muscle (i.e., its elasticity with respect to normal stress acting in a direction perpendicular to the muscle tension as a function of the active tension in muscle) has not been done, but is absolutely necessary for future progress. Perhaps the results shown in Figures 8.10:1 and 8.10:2 can be explained by the muscle tension and muscle pressure. The concept of muscle pressure is introduced in Section 7.9, and is shown to be proportional to muscle tension. The muscle pressure causes compression on capillary wall, reduces the transmural pressure of the capillaries, narrows their lumens, and increases their resistance to blood flow. The capillaries are the sluicing gates. In the skeletal muscle, such a sluicing gate is a long tube, which is of course very different from the very sharp and narrow sluicing gate of the pulmonary alveoli analyzed in Section 6.17. The sharp sluicing gates in the lung permit a finite limiting flow into the vein. The long tubular sluicing gates of the skeletal muscle have a limiting flow of zero. The tubular sluicing gate does not have to be occluded completely; it is enough that it is sufficiently narrow so that the resistance is high. When an animal dies, the tension in the muscle vanishes, the muscle pressure becomes zero, and the vanishing of pressure gradient at zero flow might be expected.

8.11 Fluid Pump Mechanism in Initial Lymphatics

An anatomical detail discovered by Skalak et al. (1984), and Mazzoni et al. (1990) and shown in Figures 8.11:1 and 8.11:2 provides an explanation of how fluid in the noncontractile initial lymphatic vessels is pumped through the first lymphatic valve. Figure 8.11:1 shows the geometric relationship between the lymphatics and arcading arterioles in cat mesentery. The collecting lymphatic vessels have smooth muscle cells in their walls and valves in their lumens. The initial lymphatics, outlined by dotted curves in the

FIGURE 8.11:1. Schematic display of initial lymphatics draining into collecting lymphatics in cat mesentery. Initial lymphatics are paired with arterioles. From Schmid-Schönbein and Zweifach (1994), by permission.

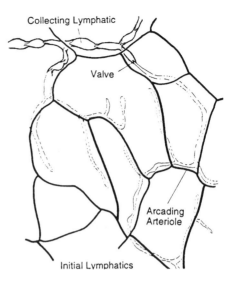

figure, are, however, microscopic condom-like tubes whose walls consist of a layer of endothelial cells and a basement membrane, similar to the walls of capillary blood vessels, without smooth muscle. The endothelium and basement membrane of the initial lymphatics have holes that are large enough for some larger molecules to pass through. The basement membrane of the initial lymphatics is attached to parenchymal cells by discrete anchoring filaments. In contrast to the initial lymphatics, the collecting lymphatics have a structure that is similar to that of the vein: There are peri-

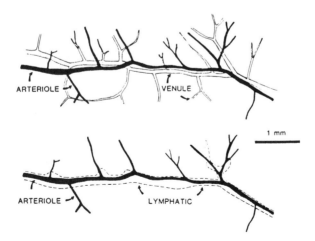

FIGURE 8.11:2. Arteriole, venule, and lymphatic in rat spinotrapezius muscle. Lymphatics were visualized by microinjection of Evans blue. Note the close association of these vessels. From Skalak et al. (1984), by permission.

odic bileaflet one-way valves and a smooth muscle medial layer, which contracts periodically to induce peristaltic motion. Whereas it is easy to understand the pumping action of peristalsis in the collecting lymphatics, it is not clear how the initial lymphatics fill themselves with interstitial fluid and how they pump fluid out into the collecting lymphatics.

The discovery of a close association between the initial lymphatic vessels and the arterioles and venules provides an answer to the pumping question. Because there are pressure and flow pulsations in the arterioles and venules, because these pressure and flow transients are influenced by organ motion and muscle contraction, and because arterioles and venules are vasoactive in such a way as to contract and dilate periodically, the arterioles and venules are vigorous organs that can serve as external pumps to help the initial lymphatics in their neighborhood. This motion pumps the fluid out of the initial lymphatics into the collecting lymphatics.

The case of a skeletal muscle is shown in Figures 8.11:3 and 8.11:4. The initial lymphatics are closely paired with arterioles, and, although they penetrate for short distances along the transverse arteriolar side branches into the skeletal muscle space, they do not extend into the capillary network. Within the skeletal muscle, all lymphatics are of the initial type. During dilation or contraction of the arterioles, the initial lymphatics are compressed or expanded accordingly, pumping fluid into the collecting lymphatics, which are located outside of the skeletal muscle parenchyma. The pumping is made possible by the lymphatic valves, which close by viscous pressure drops due to the funnel-shaped geometry and very soft leaflets (Mazzoni et al., 1987, 1990).

Other organs may rely on other mechanisms to fill the initial lymphatics with fluid. In all cases, the small size of the initial lymphatics and the low

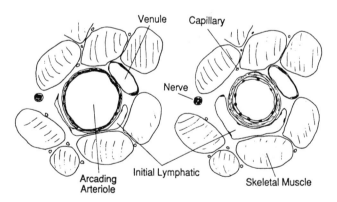

FIGURE 8.11:3. Schematic of rat skeletal muscle in area of an arcading arteriole. During dilation of arteriole, adjacent initial lymphatic is compressed (left), and it expands during constriction of arteriole (right). From Schmid-Schönbein and Zweifach (1994), by permission.

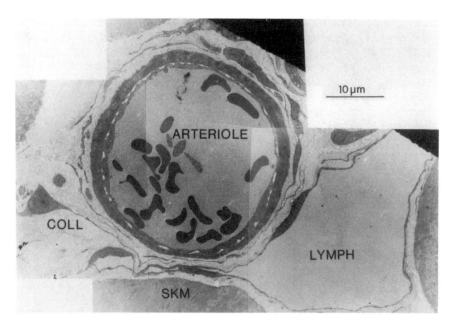

FIGURE 8.11:4. A relaxed arcading arteriole with an adjacent partially occluded lymphatic vessel. This lymphatic vessel is composed only of endothelial cells. Skeletal muscle fibers (SKM) and collagen (COLL) are visible. From Skalak et al. (1984), by permission.

speed of fluid movement imply that convective fluid movement must be governed by the pressure gradient and viscous shear stress, and not by inertial force, because both the Reynolds and Womersely numbers approach zero. Clearly, the pumping action depends on organ activity. There are many other theories on the filling of initial lymphatics. See Casley-Smith (1977), Castenholz (1984), Clough and Smaje (1978).

References

Braakmann, R. (1988). Pressure-flow relationships in skeletal muscle. Ph.D. Dissertation, University of Amsterdam.

Burton, A.C. (1951). On the physical equilibrium of small blood vessels. *Am. J. Physiol.* **164**: 319–329.

Casley-Smith, J.R. (1977). Lymph and lymphatics in: *Microcirculation* (G. Kaley and B.M. Altura, eds.), University Park Press. Baltimore, MD, pp. 423–502.

Castenholz, A. (1984). Morphological characteristics of initial lymphatics in the tongue as shown by scanning electron microscopy. *Scanning Electron Microsc.* **III**: 1343–1352.

Clough, G., and Smaje, L.H. (1978). Simultaneous measurement of pressure in the interstitium and the terminal lymphatics of the cat mesentery. *J. Physiol. Lond.* **283**: 457–468.

Delashaw, J.B., and Duling, B.R. (1988). A study of the functional elements regulating capillary perfusion in striated muscle. *Microvasc. Res.* **36**: 162–171.

Eisenstat, S.C., Gursky, M.C., Schultz, M.H., and Sherman, A.H. (1982). Yale sparse matrix package. I. The symmetric codes. *Int. J. Num. Meth. Eng.* **18**: 1145–1151.

Engelson, E.T., Schmid-Schönbein, G.W., and Zweifach, B.W. (1985b). The microvasculature in skeletal muscle. III. Venous network anatomy in normotensive and spontaneously hypertensive rats. *Int. J. Microir. Clin. Exp.* **4**: 229–248.

Engelson, E.T., Schmid-Schönbein, G.W., and Zweifach, B.W. (1986). The microvasculature in skeletal muscle. II. Arteriolar network anatomy in normotensive and hypertensive rats. *Microvasc. Res.* **31**: 356–374.

Engelson, E.T., Skalak, T.C., and Schmid-Schönbein, G.W. (1985a). The microvasculature in skeletal muscle. I. Arteriolar network in rat spinotrapezius muscle. *Microvasc. Res.* **30**: 29–44.

Fung, Y.C. (1966). Theoretical considerations of the elasticity of red cells and small blood vessels. *Fed. Proc.* **25**: 1761–1772.

Fung, Y.C. (1990). *Biomechanics: Motion, Flow, Stress, and Growth.* Springer-Verlag, New York.

Fung, Y.C. (1993). *Biomechanics: Mechanical Properties of Living Tissues.* 2nd ed. Springer-Verlag, New York.

Fung, Y.C., Zweifach, B.W., and Intaglietta, M. (1966). Elastic environment of the capillary bed. *Circ. Res.* **19**: 441–461.

Granger, H., Meininger, G.A., Borders, J.L., Morff, R.J., and Goodman, A.H. (1984). Microcirculation of skeletal muscle. *Phys. Pharm. Microcirc.* **2**: 181–265.

Hudlická, O. (1973). *Muscle Blood Flow. Its Relation to Muscle Metabolism and Function.* Swets and B.V. Zeitlinger, Amsterdam.

Kioller, A., Dawant, B., Liu, A., Popel, A.S., and Johnson, P.C. (1987). Quantitative analysis of arteriole network architecture in cat sartorius muscle. *Am. J. Physiol.* **253**: H154–H164.

Krogh, A. (1919). Number and distribution of capillaries in muscle with calculation of oxygen pressure head necessary for supplying the tissue. *Am. J. Physiol.* **52**: 409–415.

Langille, B.L. (1993). Remodeling of developing and mature arteries: endothelium, smooth muscle, and matrix. *J. Cardiovasc. Pharmacol.* **21**: 511–517.

Lee, J., and Schmid-Schönbein, G.W. (1995). Biomechanics of skeletal muscle capillaries: hemodynamic resistance, endothelial distensibility, and pseudopod formulation. *Ann. Biomed. Eng.* **23**: 226–246.

Lee, J., Salathè, E.P., and Schmid-Schönbein, G.W. (1987). Fluid exchange in skeletal muscle with viscoelastic blood vessels. *Am. J. Physiol.* **253**: H1548–H1566.

Lindbom, L., and Arfors, K.E. (1985). Mechanism and site of control for variation in the number of perfused capillaries in skeletal muscle. *Int. J. Microcirc. Clin. Exp.* **4**: 19–38.

Mazzoni, M.C., Skalak, T.C., and Schmid-Schönbein, G.W. (1987). Structure of lymphatic valves in the spinotrapezius muscle of the rat. *Blood Vessels*, 24: 304–312.

Mazzoni, M.C., Skalak, T.C., and Schmid-Schönbein, G.W. (1990). Effects of skeletal muscle fiber deformation on lymphatic volumes. *Am. J. Physiol.* **259**: H1860–H1868.

Pappenheimer, J.R., and Maes, J.P. (1942). A quantitative measure of the vasomotor tone in the hind limb muscle of the dog. *Am. J. Physiol.* **137**: 187–199.

Popel, A.S. (1987). Network Models of Peripheral Circulation. In *Handbook of Bioengineering* (R. Skalak and S. Chien, eds.), McGraw-Hill, New York, pp. 20.1–20.24.

Popel, A.S., Torres-Filho, I.P., Johnson, P.C., and Bouskela, E. (1988). A new scheme for hierarchical classification of anastomosing vessels. *Int. J. Microcirc. Clin. Exp.* **7**: 131–138.

Price, R.J., and Skalak, T.C. (1994). Circumferential wall stress as a mechanism for arteriolar rarefaction and proliferation in a network model. *Microvasc. Res.* **47**: 188–202.

Price, R.J., and Skalak, T.C. (1995). A circumferential stress-growth rule predicts arcade arteriole formation in a network model. *Microcirculation* **2**: 41–51.

Price, R.J., Owens, G.K., and Skalak, T.C. (1994). Immunohistochemical identification of arteriolar development using markers of smooth muscle differentiation: Evidence that capillary arterialization proceeds from terminal arterioles. *Circ. Res.* **75**: 520–527.

Schmid-Schönbein, G.W. (1988). A theory of blood flow in skeletal muscle. *J. Biomech. Eng.* **110**: 20–26.

Schmid-Schönbein G.W., and Murakami, H. (1985). Blood flow in contracting arterioles. *Int. J. Microcirc. Clin. Exp.* **4**: 311–328.

Schmid-Schönbein, G.W., and Zweifach, B.W. (1994). Fluid pump mechanisms in initial lymphatics. *News Physiol. Sci.* **9**: 67–71.

Schmid-Schönbein, G.W., Skalak, R., Usami, S., and Chien, S. (1980). Cell distribution in capillary networks. *Microvasc. Res.* **19**: 18–44.

Schmid-Schönbein, G.W., Skalak, T.C., Engelson, E.T., and Zweifach, B.W. (1986a). Microvascular Network Anatomy in Rat. In *Microvascular Network: Experimental and Theoretical Studies* (A.S. Popel and P.C. Johnson, eds.), Karger, Basel, pp. 38–51.

Schmid-Schönbein, G.W., Firestone, G., and Zweifach, B.W. (1986b). Network anatomy of arteries feeding the spinotrapezius muscle in normotensive and hypertensive rats. *Blood Vessels* **23**: 34–49.

Schmid-Schönbein, G.W., Skalak, T.C., and Firestone, G. (1987a). The microvasculature in skeletal muscle. V. The arteriolar and venular arcades in normotensive and hypertensive rats. *Microvasc. Res.* **34**: 385–393.

Schmid-Schönbein, G.W., Zweifach, B.W., DeLano, F.A., and Chen, P. (1987b). Microvascular tone in a skeletal muscle of spontaneously hypertensive rats. *Hypertension* **9**: H1548–H1566.

Schmid-Schönbein, G.W., Skalak, T.C., and Sutton, D.W. (1989a). Bioengineering analysis of blood flow in resting skeletal muscle. In *Microvascular Mechanics* (J.-S. Lee and T.C. Skalak, eds.), Springer Verlag, New York. pp. 65–99.

Schmid-Schönbein, G.W., Lee, S.Y., and Sutton, D. (1989b). Dynamic viscous flow in distensible vessels of skeletal muscle microcirculation: Application to pressure and flow transients. *Biorheology*. **26**: 215–227.

Skalak, T.C., Schmid-Schönbein, G.W., and Zweifach, B.W. (1984). New morphological evidence for a mechanism of lymph formation in skeletal muscle. *Microvasc. Res.* **28**: 95–112.

Skalak, T.C., and Schmid-Schönbein, G.W. (1986a). The microvasculature in skeletal muscle. IV. A model of the capillary network. *Microvasc. Res.* **32**: 333–347.

Skalak, T.C., and Schmid-Schönbein, G.W. (1986b). Viscoelastic properties of microvessels in rat spinotrapezius muscle. *J. Biomech. Eng.* **108**: 193–200.

Spalteholz, W. (1888). Die Vertheilung der Blutgefässe im Muskel. *Abh. Sächs. Ges. Wiss. Math. Phys.* **14**: 509–528.

Sutton, D.W., and Schmid-Schönbein, G.W. (1989). Hemodynamics at low flow in the resting, vasodilated rat skeletal muscle. *Am. J. Physiol.* **257**: H1419–H1427.

Sutton, D.W., and Schmid-Schönbein, G.W. (1991). The pressure-flow relation of plasma in whole organ skeletal muscle and its experimental verification. *J. Biomech. Eng.* **113**: 452–457.

Sutton, D.W., and Schmid-Schönbein, G.W. (1992). Elevation of organ resistance due to leukocyte perfusion. *Am. J. Physiol.* **262**: H1646–H1650.

Sutton, D.W., and Schmid-Schönbein, G.W. (1995). The pressure-flow relation in resting rat skeletal muscle perfused with pure erythrocyte suspensions. *Biorheology* **32**: 29–42.

Sutton, D.W., Mead, E.H., and Schmid-Schönbein, G.W. (1989). A high precision dual feedback pump for unsteady perfusion of small organs. *Ann. Biomed. Eng.* **17**: 269–278.

Vlach, J., and Singhal, K. (1983). *Computer Methods for Circuit Analysis and Design.* Van Nostrand Reinhold, New York.

Zweifach, B.W., Kovalcheck, S., DeLano, F.A., and Chen, P. (1981). Micropressure-flow relationships in a skeletal muscle of spontaneously hypertensive rats. *Hypertension* **3**: 601–614.

Author Index

Subject Index